*Independent
Component
Analysis*

Independent Component Analysis

Aapo Hyvärinen

Juha Karhunen

Erkki Oja

A Wiley-Interscience Publication

JOHN WILEY & SONS, INC.

New York / Chichester / Weinheim / Brisbane / Singapore / Toronto

Library of Congress Cataloging in Publication Data is available.

Hyvärinen, Aapo.
    Independent component analysis / Aapo Hyvärinen.
  Juha Karhunen, Erkki Oja.
      p. cm. — (Adaptive and learning systems for signal processing, communications, and control)
    ISBN 0-471-40540-X (cloth : alk. paper)

MIX
Paper from
responsible sources
FSC
www.fsc.org   FSC® C013604

Contents

Part III   EXTENSIONS AND RELATED METHODS

Preface

Independent component analysis (ICA) is a statistical and computational technique for revealing hidden factors that underlie sets of random variables, measurements, or signals. ICA defines a generative model for the observed multivariate data, which is typically given as a large database of samples. In the model, the data variables are assumed to be linear or nonlinear mixtures of some unknown latent variables, and the mixing system is also unknown. The latent variables are assumed nongaussian and mutually independent, and they are called the independent components of the observed data. These independent components, also called sources or factors, can be found by ICA.

ICA can be seen as an extension to principal component analysis and factor analysis. ICA is a much more powerful technique, however, capable of finding the underlying factors or sources when these classic methods fail completely.

The data analyzed by ICA could originate from many different kinds of application fields, including digital images and document databases, as well as economic indicators and psychometric measurements. In many cases, the measurements are given as a set of parallel signals or time series; the term blind source separation is used to characterize this problem. Typical examples are mixtures of simultaneous speech signals that have been picked up by several microphones, brain waves recorded by multiple sensors, interfering radio signals arriving at a mobile phone, or parallel time series obtained from some industrial process.

The technique of ICA is a relatively new invention. It was for the first time introduced in early 1980s in the context of neural network modeling. In mid-1990s, some highly successful new algorithms were introduced by several research groups, together with impressive demonstrations on problems like the cocktail-party effect, where the individual speech waveforms are found from their mixtures. ICA became one of the exciting new topics, both in the field of neural networks, especially unsupervised learning, and more generally in advanced statistics and signal processing. Reported real-world applications of ICA on biomedical signal processing, audio signal separation, telecommunications, fault diagnosis, feature extraction, financial time series analysis, and data mining began to appear.

Many articles on ICA were published during the past 20 years in a large number of journals and conference proceedings in the fields of signal processing, artificial neural networks, statistics, information theory, and various application fields. Several special sessions and workshops on ICA have been arranged recently [70, 348], and some edited collections of articles [315, 173, 150] as well as some monographs on ICA, blind source separation, and related subjects [105, 267, 149] have appeared. However, while highly useful for their intended readership, these existing texts typically concentrate on some selected aspects of the ICA methods only. In the brief scientific papers and book chapters, mathematical and statistical preliminaries are usually not included, which makes it very hard for a wider audience to gain full understanding of this fairly technical topic.

A comprehensive and detailed text book has been missing, which would cover both the mathematical background and principles, algorithmic solutions, and practical applications of the present state of the art of ICA. The present book is intended to fill that gap, serving as a fundamental introduction to ICA.

It is expected that the readership will be from a variety of disciplines, such as statistics, signal processing, neural networks, applied mathematics, neural and cognitive sciences, information theory, artificial intelligence, and engineering. Both researchers, students, and practitioners will be able to use the book. We have made every effort to make this book self-contained, so that a reader with a basic background in college calculus, matrix algebra, probability theory, and statistics will be able to read it. This book is also suitable for a graduate level university course on ICA, which is facilitated by the exercise problems and computer assignments given in many chapters.

Scope and contents of this book

This book provides a comprehensive introduction to ICA as a statistical and computational technique. The emphasis is on the fundamental mathematical principles and basic algorithms. Much of the material is based on the original research conducted in the authors' own research group, which is naturally reflected in the weighting of the different topics. We give a wide coverage especially to those algorithms that are scalable to large problems, that is, work even with a large number of observed variables and data points. These will be increasingly used in the near future when ICA is extensively applied in practical real-world problems instead of the toy problems or small pilot studies that have been predominant until recently. Respectively, somewhat less emphasis is given to more specialized signal processing methods involving convolutive mixtures, delays, and other blind source separation techniques than ICA.

As ICA is a fast growing research area, it is impossible to include every reported development in a textbook. We have tried to cover the central contributions by other workers in the field in their appropriate context and present an extensive bibliography for further reference. We apologize for any omissions of important contributions that we may have overlooked.

For easier reading, the book is divided into four parts.

- Part I gives the **mathematical preliminaries**. It introduces the general mathematical concepts needed in the rest of the book. We start with a crash course on probability theory in Chapter 2. The reader is assumed to be familiar with most of the basic material in this chapter, but also some concepts more specific to ICA are introduced, such as higher-order cumulants and multivariate probability theory. Next, Chapter 3 discusses essential concepts in optimization theory and gradient methods, which are needed when developing ICA algorithms. Estimation theory is reviewed in Chapter 4. A complementary theoretical framework for ICA is information theory, covered in Chapter 5. Part I is concluded by Chapter 6, which discusses methods related to principal component analysis, factor analysis, and decorrelation.

  More confident readers may prefer to skip some or all of the introductory chapters in Part I and continue directly to the principles of ICA in Part II.

- In Part II, the **basic ICA model** is covered and solved. This is the linear instantaneous noise-free mixing model that is classic in ICA, and forms the core of the ICA theory. The model is introduced and the question of identifiability of the mixing matrix is treated in Chapter 7. The following chapters treat different methods of estimating the model. A central principle is nongaussianity, whose relation to ICA is first discussed in Chapter 8. Next, the principles of maximum likelihood (Chapter 9) and minimum mutual information (Chapter 10) are reviewed, and connections between these three fundamental principles are shown. Material that is less suitable for an introductory course is covered in Chapter 11, which discusses the algebraic approach using higher-order cumulant tensors, and Chapter 12, which reviews the early work on ICA based on nonlinear decorrelations, as well as the nonlinear principal component approach. Practical algorithms for computing the independent components and the mixing matrix are discussed in connection with each principle. Next, some practical considerations, mainly related to preprocessing and dimension reduction of the data are discussed in Chapter 13, including hints to practitioners on how to really apply ICA to their own problem. An overview and comparison of the various ICA methods is presented in Chapter 14, which thus summarizes Part II.

- In Part III, different **extensions** of the basic ICA model are given. This part is by its nature more speculative than Part II, since most of the extensions have been introduced very recently, and many open problems remain. In an introductory course on ICA, only selected chapters from this part may be covered. First, in Chapter 15, we treat the problem of introducing explicit observational noise in the ICA model. Then the situation where there are more independent components than observed mixtures is treated in Chapter 16. In Chapter 17, the model is widely generalized to the case where the mixing process can be of a very general nonlinear form. Chapter 18 discusses methods that estimate a linear mixing model similar to that of ICA, but with quite different assumptions: the components are not nongaussian but have some time dependencies instead. Chapter 19 discusses the case where the mixing system includes convolutions. Further extensions, in particular models where the components are no longer required to be exactly independent, are given in Chapter 20.

- Part IV treats some **applications** of ICA methods. Feature extraction (Chapter 21) is relevant to both image processing and vision research. Brain imaging applications (Chapter 22) concentrate on measurements of the electrical and magnetic activity of the human brain. Telecommunications applications are treated in Chapter 23. Some econometric and audio signal processing applications, together with pointers to miscellaneous other applications, are treated in Chapter 24.

Throughout the book, we have marked with an asterisk some sections that are rather involved and can be skipped in an introductory course.

Several of the algorithms presented in this book are available as public domain software through the World Wide Web, both on our own Web pages and those of other ICA researchers. Also, databases of real-world data can be found there for testing the methods. We have made a special Web page for this book, which contains appropriate pointers. The address is

> www.cis.hut.fi/projects/ica/book

The reader is advised to consult this page for further information.

This book was written in cooperation between the three authors. A. Hyvärinen was responsible for the chapters 5, 7, 8, 9, 10, 11, 13, 14, 15, 16, 18, 20, 21, and 22; J. Karhunen was responsible for the chapters 2, 4, 17, 19, and 23; while E. Oja was responsible for the chapters 3, 6, and 12. The Chapters 1 and 24 were written jointly by the authors.

Acknowledgments

We are grateful to the many ICA researchers whose original contributions form the foundations of ICA and who have made this book possible. In particular, we wish to express our gratitude to the Series Editor Simon Haykin, whose articles and books on signal processing and neural networks have been an inspiration to us over the years.

Some parts of this text are based on close cooperation with other members of our research group at the Helsinki University of Technology. Chapter 21 is largely based on joint work with Patrik Hoyer, who also made all the experiments in that chapter. Chapter 22 is based on experiments and material by Ricardo Vigário. Section 13.2.2 is based on joint work with Jaakko Särelä and Ricardo Vigário. The experiments in Section 16.2.3 were provided by Razvan Cristescu. Section 20.3 is based on joint work with Ella Bingham, Section 14.4 on joint work with Xavier Giannakopoulos, and Section 20.2.3 on joint work with Patrik Hoyer and Mika Inki. Chapter 19 is partly based on material provided by Kari Torkkola. Much of Chapter 17 is based on joint work with Harri Valpola and Petteri Pajunen, and Section 24.1 is joint work with Kimmo Kiviluoto and Simona Malaroiu.

Over various phases of writing this book, several people have kindly agreed to read and comment on parts or all of the text. We are grateful for this to Ella Bingham, Jean-François Cardoso, Adrian Flanagan, Mark Girolami, Antti Honkela, Jarmo Hurri, Petteri Pajunen, Tapani Ristaniemi, and Kari Torkkola. Leila Koivisto helped in technical editing, while Antti Honkela, Mika Ilmoniemi, Merja Oja, and Tapani Raiko helped with some of the figures.

Our original research work on ICA as well as writing this book has been mainly conducted at the Neural Networks Research Centre of the Helsinki University of Technology, Finland. The research had been partly financed by the project "BLISS" (European Union) and the project "New Information Processing Principles" (Academy of Finland), which are gratefully acknowledged. Also, A. H. wishes to thank Göte Nyman and Jukka Häkkinen of the Department of Psychology of the University of Helsinki who hosted his civilian service there and made part of the writing possible.

AAPO HYVÄRINEN, JUHA KARHUNEN, ERKKI OJA

Espoo, Finland
March 2001

1

Introduction

Independent component analysis (ICA) is a method for finding underlying factors or components from multivariate (multidimensional) statistical data. What distinguishes ICA from other methods is that it looks for components that are both *statistically independent*, and *nongaussian*. Here we briefly introduce the basic concepts, applications, and estimation principles of ICA.

1.1 LINEAR REPRESENTATION OF MULTIVARIATE DATA

1.1.1 The general statistical setting

A long-standing problem in statistics and related areas is how to find a suitable representation of multivariate data. Representation here means that we somehow transform the data so that its essential structure is made more visible or accessible.

In neural computation, this fundamental problem belongs to the area of unsupervised learning, since the representation must be learned from the data itself without any external input from a supervising "teacher". A good representation is also a central goal of many techniques in data mining and exploratory data analysis. In signal processing, the same problem can be found in feature extraction, and also in the source separation problem that will be considered below.

Let us assume that the data consists of a number of variables that we have observed together. Let us denote the number of variables by m and the number of observations by T. We can then denote the data by $x_i(t)$ where the indices take the values $i = 1, ..., m$ and $t = 1, ..., T$. The dimensions m and T can be very large.

A very general formulation of the problem can be stated as follows: What could be a function from an m-dimensional space to an n-dimensional space such that the transformed variables give information on the data that is otherwise hidden in the large data set. That is, the transformed variables should be the underlying *factors* or *components* that describe the essential structure of the data. It is hoped that these components correspond to some physical causes that were involved in the process that generated the data in the first place.

In most cases, we consider linear functions only, because then the interpretation of the representation is simpler, and so is its computation. Thus, every component, say y_i, is expressed as a linear combination of the observed variables:

$$y_i(t) = \sum_j w_{ij} x_j(t), \text{ for } i = 1, ..., n, j = 1, ..., m \tag{1.1}$$

where the w_{ij} are some coefficients that define the representation. The problem can then be rephrased as the problem of determining the coefficients w_{ij}. Using linear algebra, we can express the linear transformation in Eq. (1.1) as a matrix multiplication. Collecting the coefficients w_{ij} in a matrix $\mathbf{W}$, the equation becomes

$$\begin{pmatrix} y_1(t) \\ y_2(t) \\ \vdots \\ y_n(t) \end{pmatrix} = \mathbf{W} \begin{pmatrix} x_1(t) \\ x_2(t) \\ \vdots \\ x_m(t) \end{pmatrix} \tag{1.2}$$

A basic statistical approach consists of considering the $x_i(t)$ as a set of T realizations of m random variables. Thus each set $x_i(t), t = 1, ..., T$ is a sample of one random variable; let us denote the random variable by x_i. In this framework, we could determine the matrix $\mathbf{W}$ by the statistical properties of the transformed components y_i. In the following sections, we discuss some statistical properties that could be used; one of them will lead to independent component analysis.

1.1.2 Dimension reduction methods

One statistical principle for choosing the matrix $\mathbf{W}$ is to limit the number of components y_i to be quite small, maybe only 1 or 2, and to determine $\mathbf{W}$ so that the y_i contain as much information on the data as possible. This leads to a family of techniques called principal component analysis or factor analysis.

In a classic paper, Spearman [409] considered data that consisted of school performance rankings given to schoolchildren in different branches of study, complemented by some laboratory measurements. Spearman then determined $\mathbf{W}$ by finding a single linear combination such that it explained the maximum amount of the variation in the results. He claimed to find a general factor of intelligence, thus founding factor analysis, and at the same time starting a long controversy in psychology.

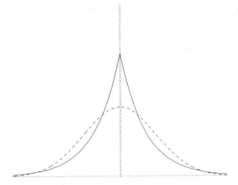

Fig. 1.1 The density function of the Laplacian distribution, which is a typical supergaussian distribution. For comparison, the gaussian density is given by a dashed line. The Laplacian density has a higher peak at zero, and heavier tails. Both densities are normalized to unit variance and have zero mean.

1.1.3 Independence as a guiding principle

Another principle that has been used for determining $\mathbf{W}$ is independence: the components y_i should be statistically independent. This means that the value of any one of the components gives no information on the values of the other components.

In fact, in factor analysis it is often claimed that the factors are independent, but this is only partly true, because factor analysis assumes that the data has a gaussian distribution. If the data is gaussian, it is simple to find components that are independent, because for gaussian data, uncorrelated components are always independent.

In reality, however, the data often does not follow a gaussian distribution, and the situation is not as simple as those methods assume. For example, many real-world data sets have *supergaussian* distributions. This means that the random variables take relatively more often values that are very close to zero or very large. In other words, the probability density of the data is peaked at zero and has heavy tails (large values far from zero), when compared to a gaussian density of the same variance. An example of such a probability density is shown in Fig. 1.1.

This is the starting point of ICA. We want to find *statistically independent* components, in the general case where the data is *nongaussian*.

1.2 BLIND SOURCE SEPARATION

Let us now look at the same problem of finding a good representation, from a different viewpoint. This is a problem in signal processing that also shows the historical background for ICA.

1.2.1 Observing mixtures of unknown signals

Consider a situation where there are a number of signals emitted by some physical objects or sources. These physical sources could be, for example, different brain areas emitting electric signals; people speaking in the same room, thus emitting speech signals; or mobile phones emitting their radio waves. Assume further that there are several sensors or receivers. These sensors are in different positions, so that each records a mixture of the original source signals with slightly different weights.

For the sake of simplicity of exposition, let us say there are three underlying source signals, and also three observed signals. Denote by $x_1(t), x_2(t)$ and $x_3(t)$ the observed signals, which are the amplitudes of the recorded signals at time point t, and by $s_1(t), s_2(t)$ and $s_3(t)$ the original signals. The $x_i(t)$ are then weighted sums of the $s_i(t)$, where the coefficients depend on the distances between the sources and the sensors:

$$x_1(t) = a_{11}s_1(t) + a_{12}s_2(t) + a_{13}s_3(t) \qquad (1.3)$$
$$x_2(t) = a_{21}s_1(t) + a_{22}s_2(t) + a_{23}s_3(t)$$
$$x_3(t) = a_{31}s_1(t) + a_{32}s_2(t) + a_{33}s_3(t)$$

The a_{ij} are constant coefficients that give the mixing weights. They are assumed *unknown*, since we cannot know the values of a_{ij} without knowing all the properties of the physical mixing system, which can be extremely difficult in general. The source signals s_i are *unknown as well*, since the very problem is that we cannot record them directly.

As an illustration, consider the waveforms in Fig. 1.2. These are three linear mixtures x_i of some original source signals. They look as if they were completely noise, but actually, there are some quite structured underlying source signals hidden in these observed signals.

What we would like to do is to find the original signals from the mixtures $x_1(t), x_2(t)$ and $x_3(t)$. This is the blind source separation (BSS) problem. *Blind* means that we know very little if anything about the original sources.

We can safely assume that the mixing coefficients a_{ij} are different enough to make the matrix that they form invertible. Thus there exists a matrix $\mathbf{W}$ with coefficients w_{ij}, such that we can separate the s_i as

$$s_1(t) = w_{11}x_1(t) + w_{12}x_2(t) + w_{13}x_3(t) \qquad (1.4)$$
$$s_2(t) = w_{21}x_1(t) + w_{22}x_2(t) + w_{23}x_3(t)$$
$$s_3(t) = w_{31}x_1(t) + w_{32}x_2(t) + w_{33}x_3(t)$$

Such a matrix $\mathbf{W}$ could be found as the inverse of the matrix that consists of the mixing coefficients a_{ij} in Eq. 1.3, if we knew those coefficients a_{ij}.

Now we see that in fact this problem is mathematically similar to the one where we wanted to find a good representation for the random data in $x_i(t)$, as in (1.2). Indeed, we could consider each signal $x_i(t), t = 1, ..., T$ as a sample of a random variable x_i, so that the value of the random variable is given by the amplitudes of that signal at the time points recorded.

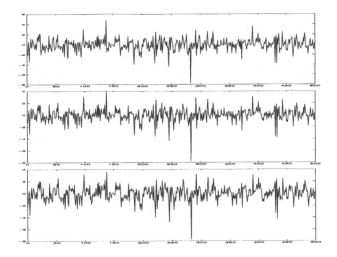

Fig. 1.2 The observed signals that are assumed to be mixtures of some underlying source signals.

1.2.2 Source separation based on independence

The question now is: How can we estimate the coefficients w_{ij} in (1.4)? We want to obtain a general method that works in many different circumstances, and in fact provides one answer to the very general problem that we started with: finding a good representation of multivariate data. Therefore, we use very general statistical properties. All we observe is the signals x_1, x_2 and x_3, and we want to find a matrix **W** so that the representation is given by the original source signals $s_1, s_2,$ and s_3.

A surprisingly simple solution to the problem can be found by considering just the statistical independence of the signals. In fact, if the signals are *not gaussian*, it is enough to determine the coefficients w_{ij}, so that the signals

$$y_1(t) = w_{11}x_1(t) + w_{12}x_2(t) + w_{13}x_3(t) \qquad (1.5)$$
$$y_2(t) = w_{21}x_1(t) + w_{22}x_2(t) + w_{23}x_3(t)$$
$$y_3(t) = w_{31}x_1(t) + w_{32}x_2(t) + w_{33}x_3(t)$$

are statistically independent. If the signals $y_1, y_2,$ and y_3 are independent, then they are equal to the original signals $s_1, s_2,$ and s_3. (They could be multiplied by some scalar constants, though, but this has little significance.)

Using just this information on the statistical independence, we can in fact estimate the coefficient matrix **W** for the signals in Fig. 1.2. What we obtain are the source signals in Fig. 1.3. (These signals were estimated by the FastICA algorithm that we shall meet in several chapters of this book.) We see that from a data set that seemed to be just noise, we were able to estimate the original source signals, using an algorithm that used the information on the independence only. These estimated signals are indeed equal to those that were used in creating the mixtures in Fig. 1.2

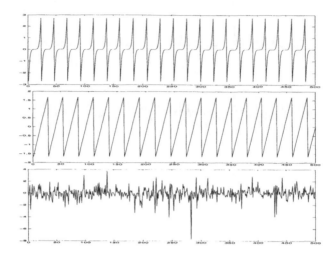

Fig. 1.3   The estimates of the original source signals, estimated using only the observed mixture signals in Fig. 1.2. The original signals were found very accurately.

(the original signals are not shown, but they really are virtually identical to what the algorithm found). Thus, in the source separation problem, the original signals were the "independent components" of the data set.

1.3   INDEPENDENT COMPONENT ANALYSIS

1.3.1   Definition

We have now seen that the problem of blind source separation boils down to finding a linear representation in which the components are statistically independent. In practical situations, we cannot in general find a representation where the components are really independent, but we can at least find components that are as independent as possible.

This leads us to the following definition of ICA, which will be considered in more detail in Chapter 7. Given a set of observations of random variables $(x_1(t), x_2(t), ..., x_n(t))$, where t is the time or sample index, assume that they are generated as a linear mixture of independent components:

$$\begin{pmatrix} x_1(t) \\ x_2(t) \\ \vdots \\ x_n(t) \end{pmatrix} = \mathbf{A} \begin{pmatrix} s_1(t) \\ s_2(t) \\ \vdots \\ s_n(t) \end{pmatrix} \tag{1.6}$$

where $\mathbf{A}$ is some unknown matrix. Independent component analysis now consists of estimating both the matrix $\mathbf{A}$ and the $s_i(t)$, when we only observe the $x_i(t)$. Note that we assumed here that the number of independent components s_i is equal to the number of observed variables; this is a simplifying assumption that is not completely necessary.

Alternatively, we could define ICA as follows: find a linear transformation given by a matrix $\mathbf{W}$ as in (1.2), so that the random variables $y_i, i = 1, ..., n$ are as independent as possible. This formulation is not really very different from the previous one, since after estimating $\mathbf{A}$, its inverse gives $\mathbf{W}$.

It can be shown (see Section 7.5) that the problem is well-defined, that is, the model in (1.6) can be estimated if and only if the components s_i are *nongaussian*. This is a fundamental requirement that also explains the main difference between ICA and factor analysis, in which the nongaussianity of the data is not taken into account. In fact, ICA could be considered as *nongaussian factor analysis*, since in factor analysis, we are also modeling the data as linear mixtures of some underlying factors.

1.3.2 Applications

Due to its generality the ICA model has applications in many different areas, some of which are treated in Part IV. Some examples are:

- In brain imaging, we often have different sources in the brain emit signals that are mixed up in the sensors outside of the head, just like in the basic blind source separation model (Chapter 22).

- In econometrics, we often have parallel time series, and ICA could decompose them into independent components that would give an insight to the structure of the data set (Section 24.1).

- A somewhat different application is in image feature extraction, where we want to find features that are as independent as possible (Chapter 21).

1.3.3 How to find the independent components

It may be very surprising that the independent components can be estimated from linear mixtures with no more assumptions than their independence. Now we will try to explain briefly why and how this is possible; of course, this is the main subject of the book (especially of Part II).

Uncorrelatedness is not enough   The first thing to note is that independence is a much stronger property than uncorrelatedness. Considering the blind source separation problem, we could actually find many different uncorrelated representations of the signals that would not be independent and would not separate the sources. Uncorrelatedness in itself is not enough to separate the components. This is also the reason why principal component analysis (PCA) or factor analysis cannot separate the signals: they give components that are uncorrelated, but little more.

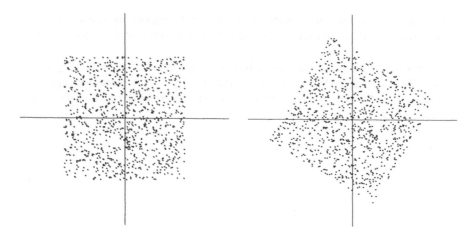

Fig. 1.4   A sample of independent components s_1 and s_2 with uniform distributions.
Horizontal axis: s_1; vertical axis: s_2.

Fig. 1.5   Uncorrelated mixtures x_1 and x_2.
Horizontal axis: x_1; vertical axis: x_2.

Let us illustrate this with a simple example using two independent components
with uniform distributions, that is, the components can have any values inside a
certain interval with equal probability. Data from two such components are plotted
in Fig. 1.4. The data is uniformly distributed inside a square due to the independence
of the components.

Now, Fig. 1.5 shows two *uncorrelated mixtures* of those independent components.
Although the mixtures are uncorrelated, one sees clearly that the distributions are not
the same. The independent components are still mixed, using an orthogonal mixing
matrix, which corresponds to a rotation of the plane. One can also see that in Fig. 1.5
the components are not independent: if the component on the horizontal axis has a
value that is near the corner of the square that is in the extreme right, this clearly
restricts the possible values that the components on the vertical axis can have.

In fact, by using the well-known decorrelation methods, we can transform any
linear mixture of the independent components into uncorrelated components, in which
case the mixing is orthogonal (this will be proven in Section 7.4.2). Thus, the trick
in ICA is to estimate the orthogonal transformation that is left after decorrelation.
This is something that classic methods cannot estimate because they are based on
essentially the same covariance information as decorrelation.

Figure 1.5 also gives a hint as to why ICA is possible. By locating the edges of
the square, we could compute the rotation that gives the original components. In the
following, we consider a couple more sophisticated methods for estimating ICA.

Nonlinear decorrelation is the basic ICA method   One way of stating how
independence is stronger than uncorrelatedness is to say that independence implies
nonlinear uncorrelatedness: If s_1 and s_2 are independent, then any nonlinear transformations $g(s_1)$ and $h(s_2)$ are uncorrelated (in the sense that their covariance is

zero). In contrast, for two random variables that are merely uncorrelated, such nonlinear transformations do not have zero covariance in general.

Thus, we could attempt to perform ICA by a stronger form of decorrelation, by finding a representation where the y_i are uncorrelated even after some nonlinear transformations. This gives a simple principle of estimating the matrix $\mathbf{W}$:

> **ICA estimation principle 1**: Nonlinear decorrelation. Find the matrix $\mathbf{W}$ so that for any $i \neq j$, the components y_i and y_j are uncorrelated, *and* the transformed components $g(y_i)$ and $h(y_j)$ are uncorrelated, where g and h are some suitable nonlinear functions.

This is a valid approach to estimating ICA: If the nonlinearities are properly chosen, the method does find the independent components. In fact, computing nonlinear correlations between the two mixtures in Fig. 1.5, one would immediately see that the mixtures are not independent.

Although this principle is very intuitive, it leaves open an important question: How should the nonlinearities g and h be chosen? Answers to this question can be found be using principles from estimation theory and information theory. Estimation theory provides the most classic method of estimating any statistical model: the *maximum likelihood* method (Chapter 9). Information theory provides exact measures of independence, such as *mutual information* (Chapter 10). Using either one of these theories, we can determine the nonlinear functions g and h in a satisfactory way.

Independent components are the maximally nongaussian components
Another very intuitive and important principle of ICA estimation is maximum nongaussianity (Chapter 8). The idea is that according to the central limit theorem, sums of nongaussian random variables are closer to gaussian that the original ones. Therefore, if we take a linear combination $y = \sum_i b_i x_i$ of the observed mixture variables (which, because of the linear mixing model, is a linear combination of the independent components as well), this will be maximally nongaussian if it equals one of the independent components. This is because if it were a real mixture of two or more components, it would be closer to a gaussian distribution, due to the central limit theorem.

Thus, the principle can be stated as follows

> **ICA estimation principle 2**: Maximum nongaussianity. Find the local maxima of nongaussianity of a linear combination $y = \sum_i b_i x_i$ under the constraint that the variance of y is constant. Each local maximum gives one independent component.

To measure nongaussianity in practice, we could use, for example, the *kurtosis*. Kurtosis is a higher-order cumulant, which are some kind of generalizations of variance using higher-order polynomials. Cumulants have interesting algebraic and statistical properties which is why they have an important part in the theory of ICA.

For example, comparing the nongaussianities of the components given by the axes in Figs. 1.4 and 1.5, we see that in Fig. 1.5 they are smaller, and thus Fig. 1.5 cannot give the independent components (see Chapter 8).

An interesting point is that this principle of maximum nongaussianity shows the very close connection between ICA and an independently developed technique called *projection pursuit*. In projection pursuit, we are actually looking for maximally nongaussian linear combinations, which are used for visualization and other purposes. Thus, the independent components can be interpreted as projection pursuit directions.

When ICA is used to extract features, this principle of maximum nongaussianity also shows an important connection to *sparse coding* that has been used in neuro-scientific theories of feature extraction (Chapter 21). The idea in sparse coding is to represent data with components so that only a small number of them are "active" at the same time. It turns out that this is equivalent, in some situations, to finding components that are maximally nongaussian.

The projection pursuit and sparse coding connections are related to a deep result that says that ICA gives a linear representation that is *as structured as possible.* This statement can be given a rigorous meaning by information-theoretic concepts (Chapter 10), and shows that the independent components are in many ways easier to process than the original random variables. In particular, independent components are easier to code (compress) than the original variables.

ICA estimation needs more than covariances There are many other methods for estimating the ICA model as well. Many of them will be treated in this book. What they all have in common is that they consider some statistics that are not contained in the covariance matrix (the matrix that contains the covariances between all pairs of the x_i).

Using the covariance matrix, we can decorrelate the components in the ordinary linear sense, but not any stronger. Thus, all the ICA methods use some form of *higher-order statistics*, which specifically means information not contained in the covariance matrix. Earlier, we encountered two kinds of higher-order information: the nonlinear correlations and kurtosis. Many other kinds can be used as well.

Numerical methods are important In addition to the estimation principle, one has to find an algorithm for implementing the computations needed. Because the estimation principles use nonquadratic functions, the computations needed usually cannot be expressed using simple linear algebra, and therefore they can be quite demanding. Numerical algorithms are thus an integral part of ICA estimation methods.

The numerical methods are typically based on optimization of some objective functions. The basic optimization method is the gradient method. Of particular interest is a fixed-point algorithm called FastICA that has been tailored to exploit the particular structure of the ICA problem. For example, we could use both of these methods to find the maxima of the nongaussianity as measured by the absolute value of kurtosis.

1.4 HISTORY OF ICA

The technique of ICA, although not yet the name, was introduced in the early 1980s by J. Hérault, C. Jutten, and B. Ans [178, 179, 16]. As recently reviewed by Jutten [227], the problem first came up in 1982 in a neurophysiological setting. In a simplified model of motion coding in muscle contraction, the outputs $x_1(t)$ and $x_2(t)$ were two types of sensory signals measuring muscle contraction, and $s_1(t)$ and $s_2(t)$ were the angular position and velocity of a moving joint. Then it is not unreasonable to assume that the ICA model holds between these signals. The nervous system must be somehow able to infer the position and velocity signals $s_1(t), s_2(t)$ from the measured responses $x_1(t), x_2(t)$. One possibility for this is to learn the inverse model using the nonlinear decorrelation principle in a simple neural network. Hérault and Jutten proposed a specific feedback circuit to solve the problem. This approach is covered in Chapter 12.

All through the 1980s, ICA was mostly known among French researchers, with limited influence internationally. The few ICA presentations in international neural network conferences in the mid-1980s were largely buried under the deluge of interest in back-propagation, Hopfield networks, and Kohonen's Self-Organizing Map (SOM), which were actively propagated in those times. Another related field was higher-order spectral analysis, on which the first international workshop was organized in 1989. In this workshop, early papers on ICA by J.-F. Cardoso [60] and P. Comon [88] were given. Cardoso used algebraic methods, especially higher-order cumulant tensors, which eventually led to the JADE algorithm [72]. The use of fourth-order cumulants has been earlier proposed by J.-L. Lacoume [254]. In signal processing literature, classic early papers by the French group are [228, 93, 408, 89]. A good source with historical accounts and a more complete list of references is [227].

In signal processing, there had been earlier approaches in the related problem of blind signal deconvolution [114, 398]. In particular, the results used in multichannel blind deconvolution are very similar to ICA techniques.

The work of the scientists in the 1980's was extended by, among others, A. Cichocki and R. Unbehauen, who were the first to propose one of the presently most popular ICA algorithms [82, 85, 84]. Some other papers on ICA and signal separation from early 1990s are [57, 314]. The "nonlinear PCA" approach was introduced by the present authors in [332, 232]. However, until the mid-1990s, ICA remained a rather small and narrow research effort. Several algorithms were proposed that worked, usually in somewhat restricted problems, but it was not until later that the rigorous connections of these to statistical optimization criteria were exposed.

ICA attained wider attention and growing interest after A.J. Bell and T.J. Sejnowski published their approach based on the infomax principle [35, 36] in the mid-90's. This algorithm was further refined by S.-I. Amari and his co-workers using the natural gradient [12], and its fundamental connections to maximum likelihood estimation, as well as to the Cichocki-Unbehauen algorithm, were established. A couple of years later, the present authors presented the fixed-point or FastICA algorithm, [210, 192,

197], which has contributed to the application of ICA to large-scale problems due to its computational efficiency.

Since the mid-1990s, there has been a growing wave of papers, workshops, and special sessions devoted to ICA. The first international workshop on ICA was held in Aussois, France, in January 1999, and the second workshop followed in June 2000 in Helsinki, Finland. Both gathered more than 100 researchers working on ICA and blind signal separation, and contributed to the transformation of ICA to an established and mature field of research.

MATHEMATICAL PRELIMINARIES

Part 1

MATHEMATICAL
PRELIMINARIES

2

Random Vectors and Independence

In this chapter, we review central concepts of probability theory, statistics, and random processes. The emphasis is on multivariate statistics and random vectors. Matters that will be needed later in this book are discussed in more detail, including, for example, statistical independence and higher-order statistics. The reader is assumed to have basic knowledge on single variable probability theory, so that fundamental definitions such as probability, elementary events, and random variables are familiar. Readers who already have a good knowledge of multivariate statistics can skip most of this chapter. For those who need a more extensive review or more information on advanced matters, many good textbooks ranging from elementary ones to advanced treatments exist. A widely used textbook covering probability, random variables, and stochastic processes is [353].

2.1 PROBABILITY DISTRIBUTIONS AND DENSITIES

2.1.1 Distribution of a random variable

In this book, we assume that random variables are continuous-valued unless stated otherwise. The *cumulative distribution function (cdf)* F_x of a random variable x at point $x = x_0$ is defined as the probability that $x \leq x_0$:

$$F_x(x_0) = P(x \leq x_0) \tag{2.1}$$

Allowing x_0 to change from $-\infty$ to ∞ defines the whole cdf for all values of x.

Clearly, for continuous random variables the cdf is a nonnegative, nondecreasing (often monotonically increasing) continuous function whose values lie in the interval

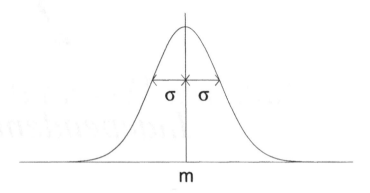

Fig. 2.1   A gaussian probability density function with mean m and standard deviation σ.

$0 \leq F_x(x) \leq 1$. From the definition, it also follows directly that $F_x(-\infty) = 0$, and $F_x(+\infty) = 1$.

Usually a probability distribution is characterized in terms of its density function rather than cdf. Formally, the *probability density function (pdf)* $p_x(x)$ of a continuous random variable x is obtained as the derivative of its cumulative distribution function:

$$p_x(x_0) = \left. \frac{dF_x(x)}{dx} \right|_{x=x_0} \tag{2.2}$$

In practice, the cdf is computed from the known pdf by using the inverse relationship

$$F_x(x_0) = \int_{-\infty}^{x_0} p_x(\xi)d\xi \tag{2.3}$$

For simplicity, $F_x(x)$ is often denoted by $F(x)$ and $p_x(x)$ by $p(x)$, respectively. The subscript referring to the random variable in question must be used when confusion is possible.

Example 2.1 The gaussian (or normal) probability distribution is used in numerous models and applications, for example to describe additive noise. Its density function is given by

$$p_x(x) = \frac{1}{\sqrt{2\pi\sigma^2}} \exp\left(-\frac{(x-m)^2}{2\sigma^2} \right) \tag{2.4}$$

Here the parameter m (mean) determines the peak point of the symmetric density function, and σ (standard deviation), its effective width (flatness or sharpness of the peak). See Figure 2.1 for an illustration.

Generally, the cdf of the gaussian density cannot be evaluated in closed form using (2.3). The term $1/\sqrt{2\pi\sigma^2}$ in front of the density (2.4) is a normalizing factor that guarantees that the cdf becomes unity when $x_0 \to \infty$. However, the values of the cdf can be computed numerically using, for example, tabulated values of the error function

$$\text{erf}(x) = \frac{1}{\sqrt{2\pi}} \int_0^x \exp\left(-\frac{\xi^2}{2}\right) d\xi \tag{2.5}$$

The error function is closely related to the cdf of a normalized gaussian density, for which the mean $m = 0$ and the variance $\sigma^2 = 1$. See [353] for details.

2.1.2 Distribution of a random vector

Assume now that $\mathbf{x}$ is an n-dimensional *random vector*

$$\mathbf{x} = (x_1, x_2, \dots, x_n)^T \tag{2.6}$$

where T denotes the transpose. (We take the transpose because all vectors in this book are column vectors. Note that vectors are denoted by boldface lowercase letters.) The components $x_1, x_2, \dots, x_n$ of the column vector $\mathbf{x}$ are continuous random variables. The concept of probability distribution generalizes easily to such a random vector. In particular, the cumulative distribution function of $\mathbf{x}$ is defined by

$$F_{\mathbf{x}}(\mathbf{x}_0) = P(\mathbf{x} \le \mathbf{x}_0) \tag{2.7}$$

where $P(.)$ again denotes the probability of the event in parentheses, and $\mathbf{x}_0$ is some constant value of the random vector $\mathbf{x}$. The notation $\mathbf{x} \le \mathbf{x}_0$ means that each component of the vector $\mathbf{x}$ is less than or equal to the respective component of the vector $\mathbf{x}_0$. The multivariate cdf in Eq. (2.7) has similar properties to that of a single random variable. It is a nondecreasing function of each component, with values lying in the interval $0 \le F_{\mathbf{x}}(\mathbf{x}) \le 1$. When all the components of $\mathbf{x}$ approach infinity, $F_{\mathbf{x}}(\mathbf{x})$ achieves its upper limit 1; when any component $x_i \to -\infty$, $F_{\mathbf{x}}(\mathbf{x}) = 0$.

The multivariate probability density function $p_{\mathbf{x}}(\mathbf{x})$ of $\mathbf{x}$ is defined as the derivative of the cumulative distribution function $F_{\mathbf{x}}(\mathbf{x})$ with respect to all components of the random vector $\mathbf{x}$:

$$p_{\mathbf{x}}(\mathbf{x}_0) = \frac{\partial}{\partial x_1} \frac{\partial}{\partial x_2} \cdots \frac{\partial}{\partial x_n} F_{\mathbf{x}}(\mathbf{x}) \bigg|_{\mathbf{x}=\mathbf{x}_0} \tag{2.8}$$

Hence

$$F_{\mathbf{x}}(\mathbf{x}_0) = \int_{-\infty}^{\mathbf{x}_0} p_{\mathbf{x}}(\mathbf{x}) d\mathbf{x} = \int_{-\infty}^{x_{0,1}} \int_{-\infty}^{x_{0,2}} \cdots \int_{-\infty}^{x_{0,n}} p_{\mathbf{x}}(\mathbf{x}) dx_n \dots dx_2 dx_1 \tag{2.9}$$

where $x_{0,i}$ is the ith component of the vector $\mathbf{x}_0$. Clearly,

$$\int_{-\infty}^{+\infty} p_{\mathbf{x}}(\mathbf{x})d\mathbf{x} = 1 \tag{2.10}$$

This provides the appropriate normalization condition that a true multivariate probability density $p_{\mathbf{x}}(\mathbf{x})$ must satisfy.

In many cases, random variables have nonzero probability density functions only on certain finite intervals. An illustrative example of such a case is presented below.

Example 2.2 Assume that the probability density function of a two-dimensional random vector $\mathbf{z} = (x, y)^T$ is

$$p_{\mathbf{z}}(\mathbf{z}) = p_{x,y}(x, y) = \begin{cases} \frac{3}{7}(2 - x)(x + y), & x \in [0, 2], \ y \in [0, 1] \\ 0, & \text{elsewhere} \end{cases}$$

Let us now compute the cumulative distribution function of $\mathbf{z}$. It is obtained by integrating over both x and y, taking into account the limits of the regions where the density is nonzero. When either $x \leq 0$ or $y \leq 0$, the density $p_{\mathbf{z}}(\mathbf{z})$ and consequently also the cdf is zero. In the region where $0 < x \leq 2$ and $0 < y \leq 1$, the cdf is given by

$$F_{\mathbf{z}}(\mathbf{z}) = F_{x,y}(x, y) = \int_0^y \int_0^x \frac{3}{7}(2 - \xi)(\xi + \eta)d\xi d\eta$$

$$= \frac{3}{7}xy \left(x + y - \frac{1}{3}x^2 - \frac{1}{4}xy \right)$$

In the region where $0 < x \leq 2$ and $y > 1$, the upper limit in integrating over y becomes equal to 1, and the cdf is obtained by inserting $y = 1$ into the preceding expression. Similarly, in the region $x > 2$ and $0 < y \leq 1$, the cdf is obtained by inserting $x = 2$ to the preceding formula. Finally, if both $x > 2$ and $y > 1$, the cdf becomes unity, showing that the probability density $p_{\mathbf{z}}(\mathbf{z})$ has been normalized correctly. Collecting these results yields

$$F_{\mathbf{z}}(\mathbf{z}) = \begin{cases} 0, & x \leq 0 \text{ or } y \leq 0 \\ \frac{3}{7}xy(x + y - \frac{1}{3}x^2 - \frac{1}{4}xy), & 0 < x \leq 2, 0 < y \leq 1 \\ \frac{3}{7}x(1 + \frac{3}{4}x - \frac{1}{3}x^2), & 0 < x \leq 2, y > 1 \\ \frac{6}{7}y(\frac{2}{3} + \frac{1}{2}y), & x > 2, 0 < y \leq 1 \\ 1, & x > 2 \text{ and } y > 1 \end{cases}$$

2.1.3   Joint and marginal distributions

The joint distribution of two different random vectors can be handled in a similar manner. In particular, let $\mathbf{y}$ be another random vector having in general a dimension m different from the dimension n of $\mathbf{x}$. The vectors $\mathbf{x}$ and $\mathbf{y}$ can be concatenated to a "supervector" $z^T = (x^T, y^T)$, and the preceding formulas used directly. The cdf that arises is called the *joint distribution function* of x and y, and is given by

$$F_{\mathbf{x},\mathbf{y}}(\mathbf{x}_0, \mathbf{y}_0) = P(\mathbf{x} \leq \mathbf{x}_0, \mathbf{y} \leq \mathbf{y}_0) \qquad (2.11)$$

Here $\mathbf{x}_0$ and $\mathbf{y}_0$ are some constant vectors having the same dimensions as x and y, respectively, and Eq. (2.11) defines the joint probability of the event $\mathbf{x} \leq \mathbf{x}_0$ and $\mathbf{y} \leq \mathbf{y}_0$.

The *joint density function* $p_{\mathbf{x},\mathbf{y}}(\mathbf{x}, \mathbf{y})$ of x and y is again defined formally by differentiating the joint distribution function $F_{\mathbf{x},\mathbf{y}}(\mathbf{x}, \mathbf{y})$ with respect to all components of the random vectors x and y. Hence, the relationship

$$F_{\mathbf{x},\mathbf{y}}(\mathbf{x}_0, \mathbf{y}_0) = \int_{-\infty}^{\mathbf{x}_0} \int_{-\infty}^{\mathbf{y}_0} p_{\mathbf{x},\mathbf{y}}(\xi, \eta) d\eta d\xi \qquad (2.12)$$

holds, and the value of this integral equals unity when both $\mathbf{x}_0 \to \infty$ and $\mathbf{y}_0 \to \infty$.

The *marginal densities* $p_{\mathbf{x}}(\mathbf{x})$ of x and $p_{\mathbf{y}}(\mathbf{y})$ of y are obtained by integrating over the other random vector in their joint density $p_{\mathbf{x},\mathbf{y}}(\mathbf{x}, \mathbf{y})$:

$$p_{\mathbf{x}}(\mathbf{x}) = \int_{-\infty}^{\infty} p_{\mathbf{x},\mathbf{y}}(\mathbf{x}, \eta) d\eta \qquad (2.13)$$

$$p_{\mathbf{y}}(\mathbf{y}) = \int_{-\infty}^{\infty} p_{\mathbf{x},\mathbf{y}}(\xi, \mathbf{y}) d\xi \qquad (2.14)$$

Example 2.3 Consider the joint density given in Example 2.2. The marginal densities of the random variables x and y are

$$p_x(x) = \int_0^1 \frac{3}{7}(2-x)(x+y)dy, \qquad x \in [0,2]$$

$$= \begin{cases} \frac{3}{7}(1 + \frac{3}{2}x - x^2) & x \in [0,2] \\ 0 & \text{elsewhere} \end{cases}$$

$$p_y(y) = \int_0^2 \frac{3}{7}(2-x)(x+y)dx, \qquad y \in [0,1]$$

$$= \begin{cases} \frac{2}{7}(2+3y), & y \in [0,1] \\ 0, & \text{elsewhere} \end{cases}$$

2.2 EXPECTATIONS AND MOMENTS

2.2.1 Definition and general properties

In practice, the exact probability density function of a vector or scalar valued random variable is usually unknown. However, one can use instead expectations of some functions of that random variable for performing useful analyses and processing. A great advantage of expectations is that they can be estimated directly from the data, even though they are formally defined in terms of the density function.

Let $g(x)$ denote any quantity derived from the random vector x. The quantity $g(x)$ may be either a scalar, vector, or even a matrix. The *expectation* of $g(x)$ is denoted by $E\{g(x)\}$, and is defined by

$$E\{g(x)\} = \int_{-\infty}^{\infty} g(x)p_x(x)dx \tag{2.15}$$

Here the integral is computed over all the components of x. The integration operation is applied separately to every component of the vector or element of the matrix, yielding as a result another vector or matrix of the same size. If $g(x) = x$, we get the expectation $E\{x\}$ of x; this is discussed in more detail in the next subsection.

Expectations have some important fundamental properties.

1. *Linearity.* Let x_i, $i = 1, \dots, m$ be a set of different random vectors, and a_i, $i = 1, \dots, m$, some nonrandom scalar coefficients. Then

$$E\{\sum_{i=1}^{m} a_i x_i\} = \sum_{i=1}^{m} a_i E\{x_i\} \tag{2.16}$$

2. *Linear transformation.* Let x be an m-dimensional random vector, and A and B some nonrandom $k \times m$ and $m \times l$ matrices, respectively. Then

$$E\{Ax\} = AE\{x\}, \qquad E\{xB\} = E\{x\}B \tag{2.17}$$

3. *Transformation invariance.* Let $y = g(x)$ be a vector-valued function of the random vector x. Then

$$\int_{-\infty}^{\infty} yp_y(y)dy = \int_{-\infty}^{\infty} g(x)p_x(x)dx \tag{2.18}$$

Thus $E\{y\} = E\{g(x)\}$, even though the integrations are carried out over different probability density functions.

These properties can be proved using the definition of the expectation operator and properties of probability density functions. They are important and very helpful in practice, allowing expressions containing expectations to be simplified without actually needing to compute any integrals (except for possibly in the last phase).

2.2.2 Mean vector and correlation matrix

Moments of a random vector x are typical expectations used to characterize it. They are obtained when $g(x)$ consists of products of components of x. In particular, the first moment of a random vector $\mathbf{x}$ is called the *mean vector* $\mathbf{m_x}$ of $\mathbf{x}$. It is defined as the expectation of $\mathbf{x}$:

$$\mathbf{m_x} = \mathrm{E}\{\mathbf{x}\} = \int_{-\infty}^{\infty} \mathbf{x} p_{\mathbf{x}}(\mathbf{x}) d\mathbf{x} \tag{2.19}$$

Each component m_{x_i} of the n-vector $\mathbf{m_x}$ is given by

$$m_{x_i} = \mathrm{E}\{x_i\} = \int_{-\infty}^{\infty} x_i p_{\mathbf{x}}(\mathbf{x}) d\mathbf{x} = \int_{-\infty}^{\infty} x_i p_{x_i}(x_i) dx_i \tag{2.20}$$

where $p_{x_i}(x_i)$ is the marginal density of the ith component x_i of $\mathbf{x}$. This is because integrals over all the other components of $\mathbf{x}$ reduce to unity due to the definition of the marginal density.

    · Another important set of moments consists of *correlations* between pairs of components of $\mathbf{x}$. The correlation r_{ij} between the ith and jth component of $\mathbf{x}$ is given by the second moment

$$r_{ij} = \mathrm{E}\{x_i x_j\} = \int_{-\infty}^{\infty} x_i x_j p_{\mathbf{x}}(\mathbf{x}) d\mathbf{x} = \int_{-\infty}^{\infty} \int_{-\infty}^{\infty} x_i x_j p_{x_i, x_j}(x_i, x_j) dx_j dx_i \tag{2.21}$$

Note that correlation can be negative or positive.

    The $n \times n$ *correlation matrix*

$$\mathbf{R_x} = \mathrm{E}\{\mathbf{x}\mathbf{x}^T\} \tag{2.22}$$

of the vector $\mathbf{x}$ represents in a convenient form all its correlations, r_{ij} being the element in row i and column j of $\mathbf{R_x}$.

The correlation matrix $\mathbf{R_x}$ has some important properties:

1. It is a *symmetric* matrix: $\mathbf{R_x} = \mathbf{R_x}^T$.

2. It is *positive semidefinite*:

$$\mathbf{a}^T \mathbf{R_x} \mathbf{a} \geq 0 \tag{2.23}$$

    for all n-vectors $\mathbf{a}$. Usually in practice $\mathbf{R_x}$ is positive definite, meaning that for any vector $\mathbf{a} \neq 0$, (2.23) holds as a strict inequality.

3. All the eigenvalues of $\mathbf{R_x}$ are real and *nonnegative* (positive if $\mathbf{R_x}$ is positive definite). Furthermore, all the eigenvectors of $\mathbf{R_x}$ are real, and can always be chosen so that they are *mutually orthonormal*.

Higher-order moments can be defined analogously, but their discussion is postponed to Section 2.7. Instead, we shall first consider the corresponding central and second-order moments for two different random vectors.

2.2.3 Covariances and joint moments

Central moments are defined in a similar fashion to usual moments, but the mean vectors of the random vectors involved are subtracted prior to computing the expectation. Clearly, central moments are only meaningful above the first order. The quantity corresponding to the correlation matrix $\mathbf{R_x}$ is called the *covariance matrix* $\mathbf{C_x}$ of $\mathbf{x}$, and is given by

$$\mathbf{C_x} = E\{(\mathbf{x} - \mathbf{m_x})(\mathbf{x} - \mathbf{m_x})^T\} \tag{2.24}$$

The elements

$$c_{ij} = E\{(x_i - m_i)(x_j - m_j)\} \tag{2.25}$$

of the $n \times n$ matrix $\mathbf{C_x}$ are called *covariances*, and they are the central moments corresponding to the correlations[1] r_{ij} defined in Eq. (2.21).

The covariance matrix $\mathbf{C_x}$ satisfies the same properties as the correlation matrix $\mathbf{R_x}$. Using the properties of the expectation operator, it is easy to see that

$$\mathbf{R_x} = \mathbf{C_x} + \mathbf{m_x}\mathbf{m}_x^T \tag{2.26}$$

If the mean vector $\mathbf{m_x} = \mathbf{0}$, *the correlation and covariance matrices become the same.* If necessary, the data can easily be made zero mean by subtracting the (estimated) mean vector from the data vectors as a preprocessing step. This is a usual practice in independent component analysis, and thus in later chapters, we simply denote by $\mathbf{C_x}$ the correlation/covariance matrix, often even dropping the subscript $\mathbf{x}$ for simplicity.

For a single random variable x, the mean vector reduces to its mean value $m_x = E\{x\}$, the correlation matrix to the second moment $E\{x^2\}$, and the covariance matrix to the *variance* of x

$$\sigma_x^2 = E\{(x - m_x)^2\} \tag{2.27}$$

The relationship (2.26) then takes the simple form $E\{x^2\} = \sigma_x^2 + m_x^2$.

The expectation operation can be extended for functions $\mathbf{g}(\mathbf{x}, \mathbf{y})$ of two different random vectors $\mathbf{x}$ and $\mathbf{y}$ in terms of their joint density:

$$E\{\mathbf{g}(\mathbf{x}, \mathbf{y})\} = \int_{-\infty}^{\infty} \int_{-\infty}^{\infty} \mathbf{g}(\mathbf{x}, \mathbf{y}) p_{\mathbf{x}, \mathbf{y}}(\mathbf{x}, \mathbf{y}) d\mathbf{y} \, d\mathbf{x} \tag{2.28}$$

The integrals are computed over all the components of $\mathbf{x}$ and $\mathbf{y}$.

Of the joint expectations, the most widely used are the *cross-correlation matrix*

$$\mathbf{R_{xy}} = E\{\mathbf{xy}^T\} \tag{2.29}$$

---

[1] In classic statistics, the correlation coefficients $\rho_j = \dfrac{c_{ij}}{(c_{ii}c_{jj})^{1/2}}$ are used, and the matrix consisting of them is called the correlation matrix. In this book, the correlation matrix is defined by the formula (2.22), which is a common practice in signal processing, neural networks, and engineering.

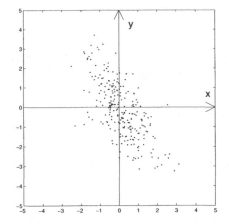

Fig. 2.2 An example of negative covariance between the random variables x and y.

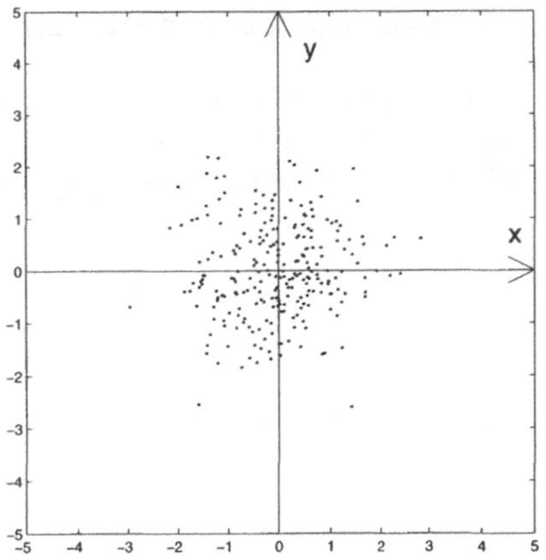

Fig. 2.3 An example of zero covariance between the random variables x and y.

and the *cross-covariance matrix*

$$\mathbf{C_{xy}} = \mathrm{E}\{(\mathbf{x} - \mathbf{m_x})(\mathbf{y} - \mathbf{m_y})^T\} \tag{2.30}$$

Note that the dimensions of the vectors $\mathbf{x}$ and $\mathbf{y}$ can be different. Hence, the cross-correlation and -covariance matrices are not necessarily square matrices, and they are not symmetric in general. However, from their definitions it follows easily that

$$\mathbf{R_{xy}} = \mathbf{R_{yx}^T}, \qquad \mathbf{C_{xy}} = \mathbf{C_{yx}^T} \tag{2.31}$$

If the mean vectors of $\mathbf{x}$ and $\mathbf{y}$ are zero, the cross-correlation and cross-covariance matrices become the same. The covariance matrix $\mathbf{C_{x+y}}$ of the sum of two random vectors $\mathbf{x}$ and $\mathbf{y}$ having the same dimension is often needed in practice. It is easy to see that

$$\mathbf{C_{x+y}} = \mathbf{C_x} + \mathbf{C_{xy}} + \mathbf{C_{yx}} + \mathbf{C_y} \tag{2.32}$$

Correlations and covariances measure the dependence between the random variables using their second-order statistics. This is illustrated by the following example.

Example 2.4 Consider the two different joint distributions $p_{x,y}(x,y)$ of the zero mean scalar random variables x and y shown in Figs. 2.2 and 2.3. In Fig. 2.2, x and y have a clear negative covariance (or correlation). A positive value of x mostly implies that y is negative, and vice versa. On the other hand, in the case of Fig. 2.3, it is not possible to infer anything about the value of y by observing x. Hence, their covariance $c_{xy} \approx 0$.

2.2.4   Estimation of expectations

Usually the probability density of a random vector $\mathbf{x}$ is not known, but there is often available a set of K samples $\mathbf{x}_1, \mathbf{x}_2, \ldots, \mathbf{x}_K$ from $\mathbf{x}$. Using them, the expectation (2.15) can be estimated by averaging over the sample using the formula [419]

$$E\{\mathbf{g}(\mathbf{x})\} \approx \frac{1}{K} \sum_{j=1}^{K} \mathbf{g}(\mathbf{x}_j) \tag{2.33}$$

For example, applying (2.33), we get for the mean vector $\mathbf{m_x}$ of $\mathbf{x}$ its standard estimator, the *sample mean*

$$\hat{\mathbf{m}}_{\mathbf{x}} = \frac{1}{K} \sum_{j=1}^{K} \mathbf{x}_j \tag{2.34}$$

where the hat over $\mathbf{m}$ is a standard notation for an estimator of a quantity.

Similarly, if instead of the joint density $p_{\mathbf{x},\mathbf{y}}(\mathbf{x}, \mathbf{y})$ of the random vectors $\mathbf{x}$ and $\mathbf{y}$, we know K sample pairs $(\mathbf{x}_1, \mathbf{y}_1), (\mathbf{x}_2, \mathbf{y}_2), \ldots, (\mathbf{x}_K, \mathbf{y}_K)$, we can estimate the expectation (2.28) by

$$E\{\mathbf{g}(\mathbf{x}, \mathbf{y})\} \approx \frac{1}{K} \sum_{j=1}^{K} \mathbf{g}(\mathbf{x}_j, \mathbf{y}_j) \tag{2.35}$$

For example, for the cross-correlation matrix, this yields the estimation formula

$$\hat{\mathbf{R}}_{\mathbf{xy}} = \frac{1}{K} \sum_{j=1}^{K} \mathbf{x}_j \mathbf{y}_j^T \tag{2.36}$$

Similar formulas are readily obtained for the other correlation type matrices $\mathbf{R_{xx}}$, $\mathbf{C_{xx}}$, and $\mathbf{C_{xy}}$.

2.3   UNCORRELATEDNESS AND INDEPENDENCE

2.3.1   Uncorrelatedness and whiteness

Two random vectors $\mathbf{x}$ and $\mathbf{y}$ are *uncorrelated* if their cross-covariance matrix $\mathbf{C_{xy}}$ is a zero matrix:

$$\mathbf{C}_{\mathbf{xy}} = E\{(\mathbf{x} - \mathbf{m_x})(\mathbf{y} - \mathbf{m_y})^T\} = 0 \tag{2.37}$$

This is equivalent to the condition

$$\mathbf{R}_{\mathbf{xy}} = E\{\mathbf{xy}^T\} = E\{\mathbf{x}\}E\{\mathbf{y}^T\} = \mathbf{m_x}\mathbf{m_y}^T \tag{2.38}$$

In the special case of two different scalar random variables x and y (for example, two components of a random vector $\mathbf{z}$), x and y are uncorrelated if their covariance c_{xy} is zero:

$$c_{xy} = \mathrm{E}\{(x - m_x)(y - m_y)\} = 0 \tag{2.39}$$

or equivalently

$$r_{xy} = \mathrm{E}\{xy\} = \mathrm{E}\{x\}\mathrm{E}\{y\} = m_x m_y \tag{2.40}$$

Again, in the case of zero-mean variables, zero covariance is equivalent to zero correlation.

Another important special case concerns the correlations between the components of a single random vector $\mathbf{x}$ given by the covariance matrix $\mathbf{C_x}$ defined in (2.24). In this case a condition equivalent to (2.37) can never be met, because each component of $\mathbf{x}$ is perfectly correlated with itself. The best that we can achieve is that different components of $\mathbf{x}$ are mutually uncorrelated, leading to the uncorrelatedness condition

$$\mathbf{C_x} = \mathrm{E}\{(\mathbf{x} - \mathbf{m_x})(\mathbf{x} - \mathbf{m_x})^T\} = \mathbf{D} \tag{2.41}$$

Here $\mathbf{D}$ is an $n \times n$ diagonal matrix

$$\mathbf{D} = \mathrm{diag}(c_{11}, c_{22}, \ldots, c_{nn}) = \mathrm{diag}(\sigma_{x_1}^2, \sigma_{x_2}^2, \ldots, \sigma_{x_n}^2) \tag{2.42}$$

whose n diagonal elements are the variances $\sigma_{x_i}^2 = \mathrm{E}\{(x_i - m_{x_i})^2\} = c_{ii}$ of the components x_i of $\mathbf{x}$.

In particular, random vectors having zero mean and unit covariance (and hence correlation) matrix, possibly multiplied by a constant variance σ^2, are said to be *white*. Thus white random vectors satisfy the conditions

$$\mathbf{m_x} = 0, \quad \mathbf{R_x} = \mathbf{C_x} = \mathbf{I} \tag{2.43}$$

where $\mathbf{I}$ is the $n \times n$ identity matrix.

Assume now that an orthogonal transformation defined by an $n \times n$ matrix $\mathbf{T}$ is applied to the random vector $\mathbf{x}$. Mathematically, this can be expressed

$$\mathbf{y} = \mathbf{Tx}, \text{ where } \mathbf{T}^T\mathbf{T} = \mathbf{TT}^T = \mathbf{I} \tag{2.44}$$

An orthogonal matrix $\mathbf{T}$ defines a rotation (change of coordinate axes) in the n-dimensional space, preserving norms and distances. Assuming that $\mathbf{x}$ is white, we get

$$\mathbf{m_y} = \mathrm{E}\{\mathbf{Tx}\} = \mathbf{T}\mathrm{E}\{\mathbf{x}\} = \mathbf{Tm}_x = 0 \tag{2.45}$$

and

$$\mathbf{C_y} = \mathbf{R_y} = \mathrm{E}\{\mathbf{Tx}(\mathbf{Tx})^T\} = \mathbf{T}\mathrm{E}\{\mathbf{xx}^T\}\mathbf{T}^T$$
$$= \mathbf{TR_x T}^T = \mathbf{TT}^T = \mathbf{I} \tag{2.46}$$

showing that **y** is white, too. Hence we can conclude that the *whiteness property is preserved under orthogonal transformations*. In fact, whitening of the original data can be made in infinitely many ways. Whitening will be discussed in more detail in Chapter 6, because it is a highly useful and widely used preprocessing step in independent component analysis.

It is clear that there also exists infinitely many ways to decorrelate the original data, because whiteness is a special case of the uncorrelatedness property.

Example 2.5 Consider the linear signal model

$$\mathbf{x} = \mathbf{As} + \mathbf{n} \tag{2.47}$$

where **x** is an n-dimensional random or data vector, **A** an $n \times m$ constant matrix, s an m-dimensional random signal vector, and **n** an n-dimensional random vector that usually describes additive noise. The correlation matrix of **x** then becomes

$$\begin{aligned}
\mathbf{R_x} &= \mathrm{E}\{\mathbf{xx}^T\} = \mathrm{E}\{(\mathbf{As} + \mathbf{n})(\mathbf{As} + \mathbf{n})^T\} \\
&= \mathrm{E}\{\mathbf{Ass}^T\mathbf{A}^T\} + \mathrm{E}\{\mathbf{Asn}^T\} + \mathrm{E}\{\mathbf{ns}^T\mathbf{A}^T\} + \mathrm{E}\{\mathbf{nn}^T\} \\
&= \mathbf{A}\mathrm{E}\{\mathbf{ss}^T\}\mathbf{A}^T + \mathbf{A}\mathrm{E}\{\mathbf{sn}^T\} + \mathrm{E}\{\mathbf{ns}^T\}\mathbf{A}^T + \mathrm{E}\{\mathbf{nn}^T\} \\
&= \mathbf{A}\mathbf{R_s}\mathbf{A}^T + \mathbf{A}\mathbf{R_{sn}} + \mathbf{R_{ns}}\mathbf{A}^T + \mathbf{R_n}
\end{aligned} \tag{2.48}$$

Usually the noise vector **n** is assumed to have zero mean, and to be uncorrelated with the signal vector **s**. Then the cross-correlation matrix between the signal and noise vectors vanishes:

$$\mathbf{R_{sn}} = \mathrm{E}\{\mathbf{sn}^T\} = \mathrm{E}\{\mathbf{s}\}\mathrm{E}\{\mathbf{n}^T\} = \mathbf{0} \tag{2.49}$$

Similarly, $\mathbf{R_{ns}} = \mathbf{0}$, and the correlation matrix of **x** simplifies to

$$\mathbf{R_x} = \mathbf{A}\mathbf{R_s}\mathbf{A}^T + \mathbf{R_n} \tag{2.50}$$

Another often made assumption is that the noise is white, which means here that the components of the noise vector **n** are all uncorrelated and have equal variance σ^2, so that in (2.50)

$$\mathbf{R_n} = \sigma^2\mathbf{I} \tag{2.51}$$

Sometimes, for example in a noisy version of the ICA model (Chapter 15), the components of the signal vector **s** are also mutually uncorrelated, so that the signal correlation matrix becomes the diagonal matrix

$$\mathbf{D_s} = \mathrm{diag}(\mathrm{E}\{s_1^2\}, \mathrm{E}\{s_2^2\}, \ldots, \mathrm{E}\{s_m^2\}) \tag{2.52}$$

where $s_1, s_2, \ldots, s_m$ are components of the signal vector **s**. Then (2.50) can be written in the form

$$\mathbf{R_x} = \mathbf{A}\mathbf{D_s}\mathbf{A}^T + \sigma^2\mathbf{I} = \sum_{i=1}^{m} \mathrm{E}\{s_i^2\}\mathbf{a}_i\mathbf{a}_i^T + \sigma^2\mathbf{I} \tag{2.53}$$

where a_i is the ith column vector of the matrix A.

The noisy linear signal or data model (2.47) is encountered frequently in signal processing and other areas, and the assumptions made on s and n vary depending on the problem at hand. It is straightforward to see that the results derived in this example hold for the respective covariance matrices as well.

2.3.2 Statistical independence

A key concept that constitutes the foundation of independent component analysis is statistical *independence*. For simplicity, consider first the case of two different scalar random variables x and y. The random variable x is independent of y, if knowing the value of y does not give any information on the value of x. For example, x and y can be outcomes of two events that have nothing to do with each other, or random signals originating from two quite different physical processes that are in no way related to each other. Examples of such independent random variables are the value of a dice thrown and of a coin tossed, or speech signal and background noise originating from a ventilation system at a certain time instant.

Mathematically, statistical independence is defined in terms of probability densities. The random variables x and y are said to be *independent* if and only if

$$p_{x,y}(x, y) = p_x(x)p_y(y) \tag{2.54}$$

In words, the joint density $p_{x,y}(x, y)$ of x and y must factorize into the product of their marginal densities $p_x(x)$ and $p_y(y)$. Equivalently, independence could be defined by replacing the probability density functions in the definition (2.54) by the respective cumulative distribution functions, which must also be factorizable.

Independent random variables satisfy the basic property

$$E\{g(x)h(y)\} = E\{g(x)\}E\{h(y)\} \tag{2.55}$$

where $g(x)$ and $h(y)$ are any absolutely integrable functions of x and y, respectively. This is because

$$E\{g(x)h(y)\} = \int_{-\infty}^{\infty} \int_{-\infty}^{\infty} g(x)h(y)p_{x,y}(x, y)dydx \tag{2.56}$$

$$= \int_{-\infty}^{\infty} g(x)p_x(x)dx \int_{-\infty}^{\infty} h(y)p_y(y)dy = E\{g(x)\}E\{h(y)\}$$

Equation (2.55) reveals that statistical independence is a much stronger property than uncorrelatedness. Equation (2.40), defining uncorrelatedness, is obtained from the independence property (2.55) as a special case where both $g(x)$ and $h(y)$ are linear functions, and takes into account second-order statistics (correlations or covariances) only. However, if the random variables have gaussian distributions, independence and uncorrelatedness become the same thing. This very special property of gaussian distributions will be discussed in more detail in Section 2.5.

Definition (2.54) of independence generalizes in a natural way for more than two random variables, and for random vectors. Let $x, y, z, \ldots$, be random vectors which may in general have different dimensions. The independence condition for $\mathbf{x}, \mathbf{y}, \mathbf{z}, \ldots$, is then

$$p_{\mathbf{x},\mathbf{y},\mathbf{z},\ldots}(\mathbf{x},\mathbf{y},\mathbf{z},\ldots) = p_{\mathbf{x}}(\mathbf{x})p_{\mathbf{y}}(\mathbf{y})p_{\mathbf{z}}(\mathbf{z})\ldots \qquad (2.57)$$

and the basic property (2.55) generalizes to

$$E\{\mathbf{g}_{\mathbf{x}}(\mathbf{x})\mathbf{g}_{\mathbf{y}}(\mathbf{y})\mathbf{g}_{\mathbf{z}}(\mathbf{z})\ldots\} = E\{\mathbf{g}_{\mathbf{x}}(\mathbf{x})\}E\{\mathbf{g}_{\mathbf{y}}(\mathbf{y})\}E\{\mathbf{g}_{\mathbf{z}}(\mathbf{z})\}\ldots$$

$$\qquad (2.58)$$

where $\mathbf{g}_{\mathbf{x}}(\mathbf{x})$, $\mathbf{g}_{\mathbf{y}}(\mathbf{y})$, and $\mathbf{g}_{\mathbf{z}}(\mathbf{z})$ are arbitrary functions of the random variables $\mathbf{x}$, $\mathbf{y}$, and $\mathbf{z}$ for which the expectations in (2.58) exist.

The general definition (2.57) gives rise to a generalization of the standard notion of statistical independence. The components of the random vector $\mathbf{x}$ are themselves scalar random variables, and the same holds for $\mathbf{y}$ and $\mathbf{z}$. Clearly, the components of $\mathbf{x}$ can be mutually dependent, while they are independent with respect to the components of the other random vectors $\mathbf{y}$ and $\mathbf{z}$, and (2.57) still holds. A similar argument applies to the random vectors $\mathbf{y}$ and $\mathbf{z}$.

Example 2.6 First consider the random variables x and y discussed in Examples 2.2 and 2.3. The joint density of x and y, reproduced here for convenience,

$$p_{x,y}(x,y) = \begin{cases} \frac{3}{7}(2-x)(x+y), & x \in [0,2], y \in [0,1] \\ 0, & \text{elsewhere} \end{cases}$$

is not equal to the product of their marginal densities $p_x(x)$ and $p_y(y)$ computed in Example 2.3. Hence, Eq. (2.54) is not satisfied, and we conclude that x and y are not independent. Actually this can be seen directly by observing that the joint density $f_{x,y}(x,y)$ given above is not factorizable, since it cannot be written as a product of two functions $g(x)$ and $h(y)$ depending only on x and y.

Consider then the joint density of a two-dimensional random vector $\mathbf{x} = (x_1, x_2)^T$ and a one-dimensional random vector $\mathbf{y} = y$ given by [419]

$$p_{\mathbf{x},\mathbf{y}}(\mathbf{x},\mathbf{y}) = \begin{cases} (x_1 + 3x_2)y, & x_1, x_2 \in [0,1], \quad y \in [0,1] \\ 0, & \text{elsewhere} \end{cases}$$

Using the above argument, we see that the random vectors $\mathbf{x}$ and $\mathbf{y}$ are statistically independent, but the components x_1 and x_2 of $\mathbf{x}$ are not independent. The exact verification of these results is left as an exercise.

2.4   CONDITIONAL DENSITIES AND BAYES' RULE

Thus far, we have dealt with the usual probability densities, joint densities, and marginal densities. Still one class of probability density functions consists of conditional densities. They are especially important in estimation theory, which will be studied in Chapter 4. Conditional densities arise when answering the following question: "What is the probability density of a random vector $\mathbf{x}$ given that another random vector $\mathbf{y}$ has the fixed value $\mathbf{y}_0$?" Here $\mathbf{y}_0$ is typically a specific realization of a measurement vector $\mathbf{y}$.

Assuming that the joint density $p_{\mathbf{x},\mathbf{y}}(\mathbf{x}, \mathbf{y})$ of $\mathbf{x}$ and $\mathbf{y}$ and their marginal densities exist, the *conditional probability density of* $\mathbf{x}$ *given* $\mathbf{y}$ is defined as

$$p_{\mathbf{x}|\mathbf{y}}(\mathbf{x}|\mathbf{y}) = \frac{p_{\mathbf{x},\mathbf{y}}(\mathbf{x}, \mathbf{y})}{p_{\mathbf{y}}(\mathbf{y})} \tag{2.59}$$

This can be interpreted as follows: assuming that the random vector $\mathbf{y}$ lies in the region $\mathbf{y}_0 < \mathbf{y} \le \mathbf{y}_0 + \Delta\mathbf{y}$, the probability that $\mathbf{x}$ lies in the region $\mathbf{x}_0 < \mathbf{x} \le \mathbf{x}_0 + \Delta\mathbf{x}$ is $p_{\mathbf{x}|\mathbf{y}}(\mathbf{x}_0|\mathbf{y}_0)\Delta\mathbf{x}$. Here $\mathbf{x}_0$ and $\mathbf{y}_0$ are some constant vectors, and both $\Delta\mathbf{x}$ and $\Delta\mathbf{y}$ are small. Similarly,

$$p_{\mathbf{y}|\mathbf{x}}(\mathbf{y}|\mathbf{x}) = \frac{p_{\mathbf{x},\mathbf{y}}(\mathbf{x}, \mathbf{y})}{p_{\mathbf{x}}(\mathbf{x})} \tag{2.60}$$

In conditional densities, the conditioning quantity, $\mathbf{y}$ in (2.59) and $\mathbf{x}$ in (2.60), is thought to be like a nonrandom parameter vector, even though it is actually a random vector itself.

Example 2.7 Consider the two-dimensional joint density $p_{x,y}(x, y)$ depicted in Fig. 2.4. For a given constant value x_0, the conditional distribution

$$p_{y|x}(y|x_0) = \frac{p_{x,y}(x_0, y)}{p_x(x_0)}$$

Hence, it is a one-dimensional distribution obtained by "slicing" the joint distribution $p(x, y)$ parallel to the y-axis at the point $x = x_0$. Note that the denominator $p_x(x_0)$ is merely a scaling constant that does not affect the shape of the conditional distribution $p_{y|x}(y|x_0)$ as a function of y.

Similarly, the conditional distribution $p_{x|y}(x|y_0)$ can be obtained geometrically by slicing the joint distribution of Fig. 2.4 parallel to the x-axis at the point $y = y_0$. The resulting conditional distributions are shown in Fig. 2.5 for the value $x_0 = 1.27$, and Fig. 2.6 for $y_0 = -0.37$.

From the definitions of the marginal densities $p_{\mathbf{x}}(\mathbf{x})$ of $\mathbf{x}$ and $p_{\mathbf{y}}(\mathbf{y})$ of $\mathbf{y}$ given in Eqs. (2.13) and (2.14), we see that the denominators in (2.59) and (2.60) are obtained by integrating the joint density $p_{\mathbf{x},\mathbf{y}}(\mathbf{x}, \mathbf{y})$ over the unconditional random vector. This also shows immediately that the conditional densities are true probability densities satisfying

$$\int_{-\infty}^{\infty} p_{\mathbf{x}|\mathbf{y}}(\xi|\mathbf{y})d\xi = 1, \quad \int_{-\infty}^{\infty} p_{\mathbf{y}|\mathbf{x}}(\eta|\mathbf{x})d\eta = 1 \tag{2.61}$$

If the random vectors $\mathbf{x}$ and $\mathbf{y}$ are statistically independent, the conditional density $p_{\mathbf{x}|\mathbf{y}}(\mathbf{x}|\mathbf{y})$ equals to the unconditional density $p_{\mathbf{x}}(\mathbf{x})$ of $\mathbf{x}$, since $\mathbf{x}$ does not depend

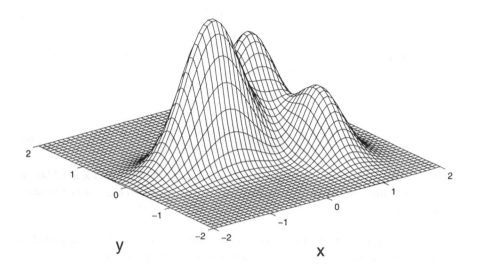

Fig. 2.4   A two-dimensional joint density of the random variables x and y.

in any way on $\mathbf{y}$, and similarly $p_{\mathbf{y}|\mathbf{x}}(\mathbf{y}|\mathbf{x}) = p_{\mathbf{y}}(\mathbf{y})$, and both Eqs. (2.59) and (2.60) can be written in the form

$$p_{\mathbf{x},\mathbf{y}}(\mathbf{x},\mathbf{y}) = p_{\mathbf{x}}(\mathbf{x})p_{\mathbf{y}}(\mathbf{y}) \qquad (2.62)$$

which is exactly the definition of independence of the random vectors $\mathbf{x}$ and $\mathbf{y}$.

In the general case, we get from Eqs. (2.59) and (2.60) two different expressions for the joint density of $\mathbf{x}$ and $\mathbf{y}$:

$$p_{\mathbf{x},\mathbf{y}}(\mathbf{x},\mathbf{y}) = p_{\mathbf{y}|\mathbf{x}}(\mathbf{y}|\mathbf{x})p_{\mathbf{x}}(\mathbf{x}) = p_{\mathbf{x}|\mathbf{y}}(\mathbf{x}|\mathbf{y})p_{\mathbf{y}}(\mathbf{y}) \qquad (2.63)$$

From this, for example, a solution can be found for the density of $\mathbf{y}$ conditioned on $\mathbf{x}$:

$$p_{\mathbf{y}|\mathbf{x}}(\mathbf{y}|\mathbf{x}) = \frac{p_{\mathbf{x}|\mathbf{y}}(\mathbf{x}|\mathbf{y})p_{\mathbf{y}}(\mathbf{y})}{p_{\mathbf{x}}(\mathbf{x})} \qquad (2.64)$$

where the denominator can be computed by integrating the numerator if necessary:

$$p_{\mathbf{x}}(\mathbf{x}) = \int_{-\infty}^{\infty} p_{\mathbf{x}|\mathbf{y}}(\mathbf{x}|\boldsymbol{\eta})p_{\mathbf{y}}(\boldsymbol{\eta})d\boldsymbol{\eta} \qquad (2.65)$$

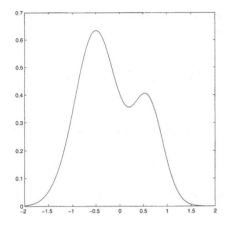

Fig. 2.5 The conditional probability density $p_{y|x}(y|x = 1.27)$.

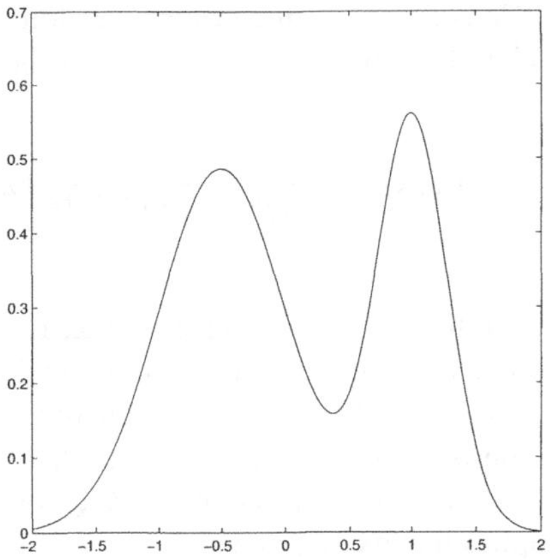

Fig. 2.6 The conditional probability density $p_{x|y}(x|y = -0.37)$.

Formula (2.64) (together with (2.65)) is called *Bayes' rule*. This rule is important especially in statistical estimation theory. There typically $p_{x|y}(x|y)$ is the conditional density of the measurement vector x, with y denoting the vector of unknown random parameters. Bayes' rule (2.64) allows the computation of the *posterior* density $p_{y|x}(y|x)$ of the parameters y, given a specific measurement (observation) vector x, and assuming or knowing the *prior* distribution $p_y(y)$ of the random parameters y. These matters will be discussed in more detail in Chapter 4.

Conditional expectations are defined similarly to the expectations defined earlier, but the pdf appearing in the integral is now the appropriate conditional density. Hence, for example,

$$E\{g(x,y)|y\} = \int_{-\infty}^{\infty} g(\xi, y) p_{x|y}(\xi|y) d\xi \qquad (2.66)$$

This is still a function of the random vector y, which is thought to be nonrandom while computing the above expectation. The complete expectation with respect to both x and y can be obtained by taking the expectation of (2.66) with respect to y:

$$E\{g(x,y)\} = E\{E\{g(x,y)|y\}\} \qquad (2.67)$$

Actually, this is just an alternative two-stage procedure for computing the expectation (2.28), following easily from Bayes' rule.

2.5 THE MULTIVARIATE GAUSSIAN DENSITY

The multivariate gaussian or normal density has several special properties that make it unique among probability density functions. Due to its importance, we shall discuss it more thoroughly in this section.

Consider an n-dimensional random vector $\mathbf{x}$. It is said to be gaussian if the probability density function of $\mathbf{x}$ has the form

$$p_{\mathbf{x}}(\mathbf{x}) = \frac{1}{(2\pi)^{n/2}(\det \mathbf{C_x})^{1/2}} \exp\left(-\frac{1}{2}(\mathbf{x} - \mathbf{m_x})^T \mathbf{C_x}^{-1}(\mathbf{x} - \mathbf{m_x})\right)$$

(2.68)

Recall that n is the dimension of $\mathbf{x}$, $\mathbf{m_x}$ its mean, and $\mathbf{C_x}$ the covariance matrix of $\mathbf{x}$. The notation $\det \mathbf{A}$ is used for the determinant of a matrix $\mathbf{A}$, in this case $\mathbf{C_x}$. It is easy to see that for a single random variable x ($n = 1$), the density (2.68) reduces to the one-dimensional gaussian pdf (2.4) discussed briefly in Example 2.1. Note also that the covariance matrix $\mathbf{C_x}$ is assumed strictly positive definite, which also implies that its inverse exists.

It can be shown that for the density (2.68)

$$E\{\mathbf{x}\} = \mathbf{m_x}, \quad E\{(\mathbf{x} - \mathbf{m_x})(\mathbf{x} - \mathbf{m_x})^T\} = \mathbf{C_x}$$

(2.69)

Hence calling $\mathbf{m_x}$ the mean vector and $\mathbf{C_x}$ the covariance matrix of the multivariate gaussian density is justified.

2.5.1 Properties of the gaussian density

In the following, we list the most important properties of the multivariate gaussian density omitting proofs. The proofs can be found in many books; see, for example, [353, 419, 407].

Only first- and second-order statistics are needed   Knowledge of the mean vector $\mathbf{m_x}$ and the covariance matrix $\mathbf{C_x}$ of $\mathbf{x}$ are sufficient for defining the multivariate gaussian density (2.68) completely. Therefore, all the higher-order moments must also depend only on $\mathbf{m_x}$ and $\mathbf{C_x}$. This implies that these moments do not carry any novel information about the gaussian distribution. An important consequence of this fact and the form of the gaussian pdf is that linear processing methods based on first- and second-order statistical information are usually optimal for gaussian data. For example, independent component analysis does not bring out anything new compared with standard principal component analysis (to be discussed later) for gaussian data. Similarly, linear time-invariant discrete-time filters used in classic statistical signal processing are optimal for filtering gaussian data.

Linear transformations are gaussian   If $\mathbf{x}$ is a gaussian random vector and $\mathbf{y} = \mathbf{Ax}$ its linear transformation, then $\mathbf{y}$ is also gaussian with mean vector $\mathbf{m_y} = \mathbf{Am_x}$ and covariance matrix $\mathbf{C_y} = \mathbf{AC_xA}^T$. A special case of this result says that any linear combination of gaussian random variables is itself gaussian. This result again has implications in standard independent component analysis: it is impossible to estimate the ICA model for gaussian data, that is, one cannot blindly separate gaussian sources from their mixtures without extra knowledge of the sources, as will be discussed in Chapter 7. [2]

Marginal and conditional densities are gaussian   Consider now two random vectors $\mathbf{x}$ and $\mathbf{y}$ having dimensions n and m, respectively. Let us collect them in a single random vector $\mathbf{z}^T = (\mathbf{x}^T, \mathbf{y}^T)$ of dimension $n + m$. Its mean vector $\mathbf{m_z}$ and covariance matrix $\mathbf{C_z}$ are

$$\mathbf{m_z} = \begin{pmatrix} \mathbf{m_x} \\ \mathbf{m_y} \end{pmatrix}, \quad \mathbf{C_z} = \begin{bmatrix} \mathbf{C_x} & \mathbf{C_{xy}} \\ \mathbf{C_{yx}} & \mathbf{C_y} \end{bmatrix} \tag{2.70}$$

Recall that the cross-covariance matrices are transposes of each other: $\mathbf{C_{xy}} = \mathbf{C_{yx}^T}$.

Assume now that $\mathbf{z}$ has a jointly gaussian distribution. It can be shown that the marginal densities $p_\mathbf{x}(\mathbf{x})$ and $p_\mathbf{y}(\mathbf{y})$ of the joint gaussian density $p_\mathbf{z}(\mathbf{z})$ are gaussian. Also the conditional densities $p_{\mathbf{x}|\mathbf{y}}$ and $p_{\mathbf{y}|\mathbf{x}}$ are n- and m-dimensional gaussian densities, respectively. The mean and covariance matrix of the conditional density $p_{\mathbf{y}|\mathbf{x}}$ are

$$\mathbf{m_{y|x}} = \mathbf{m_y} + \mathbf{C_{yx}}\mathbf{C_x^{-1}}(\mathbf{x} - \mathbf{m_x}) \tag{2.71}$$

$$\mathbf{C_{y|x}} = \mathbf{C_y} - \mathbf{C_{yx}}\mathbf{C_x^{-1}}\mathbf{C_{xy}} \tag{2.72}$$

Similar expressions are obtained for the mean $\mathbf{m_{x|y}}$ and covariance matrix $\mathbf{C_{x|y}}$ of the conditional density $p_{\mathbf{x}|\mathbf{y}}$.

Uncorrelatedness and geometrical structure.   We mentioned earlier that *uncorrelated gaussian random variables are also independent*, a property which is not shared by other distributions in general. Derivation of this important result is left to the reader as an exercise. If the covariance matrix $\mathbf{C_x}$ of the multivariate gaussian density (2.68) is not diagonal, the components of $\mathbf{x}$ are correlated. Since $\mathbf{C_x}$ is a symmetric and positive definite matrix, it can always be represented in the form

$$\mathbf{C_x} = \mathbf{E}\mathbf{D}\mathbf{E}^T = \sum_{i=1}^{n} \lambda_i \mathbf{e}_i \mathbf{e}_i^T \tag{2.73}$$

Here $\mathbf{E}$ is an orthogonal matrix (that is, a rotation) having as its columns $\mathbf{e}_1, \mathbf{e}_2, \dots, \mathbf{e}_n$ the n eigenvectors of $\mathbf{C_x}$, and $\mathbf{D} = \text{diag}(\lambda_1, \lambda_2, \dots, \lambda_n)$ is the diagonal matrix containing the respective eigenvalues λ_i of $\mathbf{C_x}$. Now it can readily be verified that applying the rotation

$$\mathbf{u} = \mathbf{E}^T(\mathbf{x} - \mathbf{m_x}) \tag{2.74}$$

---

[2]It is possible, however, to separate temporally correlated (nonwhite) gaussian sources using their second-order temporal statistics on certain conditions. Such techniques are quite different from standard independent component analysis. They will be discussed in Chapter 18.

to **x** makes the components of the gaussian distribution of **u** uncorrelated, and hence also independent.

Moreover, the eigenvalues λ_i and eigenvectors $\mathbf{e}_i$ of the covariance matrix $\mathbf{C_x}$ reveal the geometrical structure of the multivariate gaussian distribution. The contours of any pdf are defined by curves of constant values of the density, given by the equation $p_\mathbf{x}(\mathbf{x}) = \text{constant}$. For the multivariate gaussian density, this is equivalent to requiring that the exponent is a constant c:

$$(\mathbf{x} - \mathbf{m_x})^T \mathbf{C_x}^{-1} (\mathbf{x} - \mathbf{m_x}) = c \tag{2.75}$$

Using (2.73), it is easy to see [419] that the contours of the multivariate gaussian are hyperellipsoids centered at the mean vector $\mathbf{m_x}$. The principal axes of the hyperellipsoids are parallel to the eigenvectors $\mathbf{e}_i$, and the eigenvalues λ_i are the respective variances. See Fig. 2.7 for an illustration.

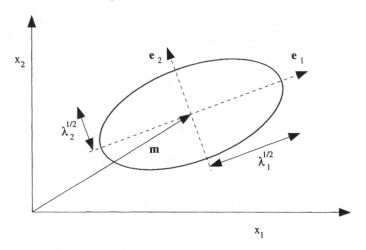

Fig. 2.7 Illustration of a multivariate gaussian probability density.

2.5.2 Central limit theorem

Still another argument underlining the significance of the gaussian distribution is provided by the central limit theorem. Let

$$x_k = \sum_{i=1}^{k} z_i \tag{2.76}$$

be a partial sum of a sequence $\{z_i\}$ of independent and identically distributed random variables z_i. Since the mean and variance of x_k can grow without bound as $k \rightarrow \infty$, consider instead of x_k the standardized variables

$$y_k = \frac{x_k - m_{x_k}}{\sigma_{x_k}} \tag{2.77}$$

where m_{x_k} and σ_{x_k} are the mean and variance of x_k.

It can be shown that the distribution of y_k converges to a gaussian distribution with zero mean and unit variance when $k \to \infty$. This result is known as the *central limit theorem*. Several different forms of the theorem exist, where assumptions on independence and identical distributions have been weakened. The central limit theorem is a primary reason that justifies modeling of many random phenomena as gaussian random variables. For example, additive noise can often be considered to arise as a sum of a large number of small elementary effects, and is therefore naturally modeled as a gaussian random variable.

The central limit theorem generalizes readily to independent and identically distributed random vectors $\mathbf{z}_i$ having a common mean $\mathbf{m_z}$ and covariance matrix $\mathbf{C_z}$. The limiting distribution of the random vector

$$\mathbf{y}_k = \frac{1}{\sqrt{k}} \sum_{i=1}^{k} (\mathbf{z}_i - \mathbf{m_z}) \tag{2.78}$$

is multivariate gaussian with zero mean and covariance matrix $\mathbf{C_z}$.

The central limit theorem has important consequences in independent component analysis and blind source separation. A typical mixture, or component of the data vector $\mathbf{x}$, is of the form

$$x_i = \sum_{j=1}^{m} a_{ij} s_j \tag{2.79}$$

where $a_{ij}, j = 1, \dots, m$, are constant mixing coefficients and $s_j, j = 1, \dots, m$, are the m unknown source signals. Even for a fairly small number of sources (say, $m = 10$) the distribution of the mixture x_k is usually close to gaussian. This seems to hold in practice even though the densities of the different sources are far from each other and far from gaussianity. Examples of this property can be found in Chapter 8, as well as in [149].

2.6 DENSITY OF A TRANSFORMATION

Assume now that both $\mathbf{x}$ and $\mathbf{y}$ are n-dimensional random vectors that are related by the vector mapping

$$\mathbf{y} = \mathbf{g}(\mathbf{x}) \tag{2.80}$$

for which the inverse mapping

$$\mathbf{x} = \mathbf{g}^{-1}(\mathbf{y}) \tag{2.81}$$

exists and is unique. It can be shown that the density $p_\mathbf{y}(\mathbf{y})$ of $\mathbf{y}$ is obtained from the density $p_\mathbf{x}(\mathbf{x})$ of $\mathbf{x}$ as follows:

$$p_\mathbf{y}(\mathbf{y}) = \frac{1}{|\det J\mathbf{g}(\mathbf{g}^{-1}(\mathbf{y}))|} p_\mathbf{x}(\mathbf{g}^{-1}(\mathbf{y})) \tag{2.82}$$

Here $J\mathbf{g}$ is the *Jacobian matrix*

$$
J\mathbf{g}(\mathbf{x}) = \begin{bmatrix} \frac{\partial g_1(\mathbf{x})}{\partial x_1} & \frac{\partial g_2(\mathbf{x})}{\partial x_1} & \cdots & \frac{\partial g_n(\mathbf{x})}{\partial x_1} \\ \frac{\partial g_1(\mathbf{x})}{\partial x_2} & \frac{\partial g_2(\mathbf{x})}{\partial x_2} & \cdots & \frac{\partial g_n(\mathbf{x})}{\partial x_2} \\ \vdots & \vdots & \ddots & \vdots \\ \frac{\partial g_1(\mathbf{x})}{\partial x_n} & \frac{\partial g_2(\mathbf{x})}{\partial x_n} & \cdots & \frac{\partial g_n(\mathbf{x})}{\partial x_n} \end{bmatrix}
\tag{2.83}
$$

and $g_j(\mathbf{x})$ is the jth component of the vector function $\mathbf{g}(\mathbf{x})$.

In the special case where the transformation (2.80) is linear and nonsingular so that $\mathbf{y} = \mathbf{A}\mathbf{x}$ and $\mathbf{x} = \mathbf{A}^{-1}\mathbf{y}$, the formula (2.82) simplifies to

$$
p_{\mathbf{y}}(\mathbf{y}) = \frac{1}{|\det \mathbf{A}|} p_{\mathbf{x}}(\mathbf{A}^{-1}\mathbf{y})
\tag{2.84}
$$

If $\mathbf{x}$ in (2.84) is multivariate gaussian, then $\mathbf{y}$ also becomes multivariate gaussian, as was mentioned in the previous section.

Other kinds of transformations are discussed in textbooks of probability theory [129, 353]. For example, the sum $z = x + y$, where x and y are statistically independent random variables, appears often in practice. Because the transformation between the random variables in this case is not one-to-one, the preceding results cannot be applied directly. But it can be shown that the pdf of z becomes the convolution integral of the densities of x and y [129, 353, 407].

A special case of (2.82) that is important in practice is the so-called probability integral transformation. If $F_x(x)$ is the cumulative distribution function of a random variable x, then the random variable

$$
z = F_x(x)
\tag{2.85}
$$

is uniformly distributed on the interval $[0, 1]$. This result allows generation of random variables having a desired distribution from uniformly distributed random numbers. First, the cdf of the desired density is computed, and then the inverse transformation of (2.85) is determined. Using this, one gets random variables x with the desired density, provided that the inverse transformation of (2.85) can be computed.

2.7   HIGHER-ORDER STATISTICS

Up to this point, we have characterized random vectors primarily using their second-order statistics. Standard methods of statistical signal processing are based on utilization of this statistical information in linear discrete-time systems. Their theory is well-developed and highly useful in many circumstances. Nevertheless, it is limited by the assumptions of gaussianity, linearity, stationarity, etc.

From the mid-1980s, interest in higher-order statistical methods began to grow in the signal processing community. At the same time, neural networks became popular with the development of several new, effective learning paradigms. A basic idea in neural networks [172, 48] is distributed nonlinear processing of the input data. A neural network consists of interconnected simple computational units called neurons. The output of each neuron typically depends nonlinearly on its inputs. These nonlinearities, for example, the hyperbolic tangent $\tanh(u)$, also implicitly introduce higher-order statistics for processing. This can be seen by expanding the nonlinearities into their Taylor series; for example,

$$\tanh(u) = u - \frac{1}{3}u^3 + \frac{2}{15}u^5 - \cdots \tag{2.86}$$

The scalar quantity u is in many neural networks the inner product $u = \mathbf{w}^T\mathbf{x}$ of the weight vector $\mathbf{w}$ of the neuron and its input vector $\mathbf{x}$. Inserting this into (2.86) shows clearly that higher-order statistics of the components of the vector $\mathbf{x}$ are involved in the computations.

Independent component analysis and blind source separation require the use of higher-order statistics either directly or indirectly via nonlinearities. Therefore, we discuss in the following basic concepts and results that will be needed later.

2.7.1 Kurtosis and classification of densities

In this subsection, we deal with the simple higher-order statistics of one scalar random variable. In spite of their simplicity, these statistics are highly useful in many situations.

Consider a scalar random variable x with the probability density function $p_x(x)$. The jth *moment* α_j of x is defined by the expectation

$$\alpha_j = \mathrm{E}\{x^j\} = \int_{-\infty}^{\infty} \xi^j p_x(\xi) d\xi, \qquad j = 1, 2, \ldots \tag{2.87}$$

and the jth *central moment* μ_j of x respectively by

$$\mu_j = \mathrm{E}\{(x - \alpha_1)^j\} = \int_{-\infty}^{\infty} (\xi - m_x)^j p_x(\xi) d\xi. \qquad j = 1, 2, \ldots \tag{2.88}$$

The central moments are thus computed around the mean m_x of x, which equals its first moment α_1. The second moment $\alpha_2 = \mathrm{E}\{x^2\}$ is the average power of x. The central moments $\mu_0 = 1$ and $\mu_1 = 0$ are insignificant, while the second central moment $\mu_2 = \sigma_x^2$ is the variance of x.

Before proceeding, we note that there exist distributions for which all the moments are not finite. Another drawback of moments is that knowing them does not necessarily specify the probability density function uniquely. Fortunately, for most of the distributions arising commonly all the moments are finite, and their knowledge is in practice equivalent to the knowledge of their probability density [315].

The third central moment

$$\mu_3 = \mathrm{E}\{(x - m_x)^3\} \tag{2.89}$$

is called the *skewness*. It is a useful measure of the asymmetricity of the pdf. It is easy to see that the skewness is zero for probability densities that are symmetric around their mean.

Consider now more specifically fourth-order moments. Higher than fourth order moments and statistics are used seldom in practice, so we shall not discuss them. The fourth moment $\alpha_4 = E\{x^4\}$ is applied in some ICA algorithms because of its simplicity. Instead of the fourth central moment $\mu_4 = E\{(x - m_x)^4\}$, the fourth-order statistics called the *kurtosis* is usually employed, because it has some useful properties not shared by the fourth central moment. Kurtosis will be derived in the next subsection in the context of the general theory of cumulants, but it is discussed here because of its simplicity and importance in independent component analysis and blind source separation.

Kurtosis is defined in the zero-mean case by the equation

$$\text{kurt}(x) = E\{x^4\} - 3[E\{x^2\}]^2 \tag{2.90}$$

Alternatively, the normalized kurtosis

$$\tilde{\kappa}(x) = \frac{E\{x^4\}}{[E\{x^2\}]^2} - 3 \tag{2.91}$$

can be used. For whitened data $E\{x^2\} = 1$, and both the versions of the kurtosis reduce to

$$\text{kurt}(x) = \tilde{\kappa}(x) = E\{x^4\} - 3 \tag{2.92}$$

This implies that for white data, the fourth moment $E\{x^4\}$ can be used instead of the kurtosis for characterizing the distribution of x. Kurtosis is basically a normalized version of the fourth moment.

A useful property of kurtosis is its additivity. If x and y are two statistically independent random variables, then it holds that

$$\text{kurt}(x + y) = \text{kurt}(x) + \text{kurt}(y) \tag{2.93}$$

Note that this additivity property does not hold for the fourth moment, which shows an important benefit of using cumulants instead of moments. Also, for any scalar parameter β,

$$\text{kurt}(\beta x) = \beta^4 \text{kurt}(x) \tag{2.94}$$

Hence kurtosis is not linear with respect to its argument.

Another very important feature of kurtosis is that it is the simplest statistical quantity for indicating the nongaussianity of a random variable. It can be shown that if x has a gaussian distribution, its kurtosis $\text{kurt}(x)$ is zero. This is the sense in which kurtosis is "normalized" when compared to the fourth moment, which is *not* zero for gaussian variables.

A distribution having zero kurtosis is called mesokurtic in statistical literature. Generally, distributions having a negative kurtosis are said to be *subgaussian* (or

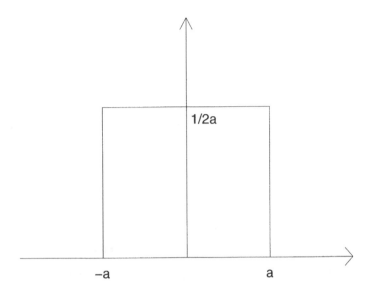

Fig. 2.8 Example of a zero-mean uniform density.

platykurtic in statistics). If the kurtosis is positive, the respective distribution is called *supergaussian* (or leptokurtic). Subgaussian probability densities tend to be flatter than the gaussian one, or multimodal. A typical supergaussian probability density has a sharper peak and longer tails than the gaussian pdf.

Kurtosis is often used as a quantitative measure of the nongaussianity of a random variable or signal, but some caution must then be taken. The reason is that the kurtosis of a supergaussian signal can have a large positive value (the maximum is infinity in principle), but the negative value of the kurtosis of a subgaussian signal is bounded below so that the minimum possible value is -2 (when variance is normalized to unity). Thus comparing the nongaussianity of supergaussian and subgaussian signals with each other using plain kurtosis is not appropriate. However, kurtosis can be used as a simple measure of nongaussianity if the signals to be compared are of the same type, either subgaussian or supergaussian.

In computer simulations, an often used subgaussian distribution is the *uniform distribution*. Its pdf for a zero-mean random variable x is

$$p_x(x) = \begin{cases} \frac{1}{2a}, & x \in [-a, a] \\ 0, & \text{elsewhere} \end{cases} \tag{2.95}$$

where the parameter a determines the width (and height) of the pdf; see Fig. 2.8. A widely used supergaussian distribution is the *Laplacian* or doubly exponential distribution. Its probability density (again assuming zero mean) is

$$p_x(x) = \frac{\lambda}{2} \exp(-\lambda|x|) \tag{2.96}$$

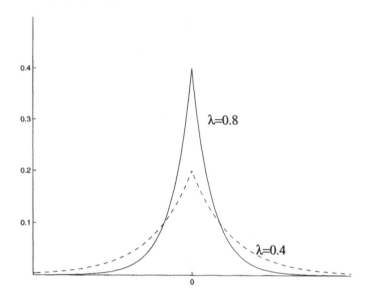

Fig. 2.9   Examples of a Laplacian density.

The only parameter $\lambda > 0$ determines both the variance and the height of the peak of the Laplacian density. It is easy to see that increasing the parameter λ decreases the variance of the Laplacian distribution and makes its peak value $\lambda/2$ at $x = 0$ higher; see Fig. 2.9.

Both the uniform and Laplacian density can be obtained as special cases of the generalized gaussian or exponential power family of pdf's [53, 256]. The general expression of the densities belonging to this family is (for zero mean)

$$p_x(x) = C \exp\left(-\frac{|x|^\nu}{\nu \mathrm{E}\{|x|^\nu\}}\right) \tag{2.97}$$

The positive real-valued power ν determines the type of distribution, and C is a scaling constant which normalizes the distribution to unit area (see [53]). (The expectation in the denominator is a normalizing constant as well.) If the parameter $\nu = 2$, the usual gaussian density is obtained. The choice $\nu = 1$ yields the Laplacian density, and $\nu \to \infty$ the uniform density. The parameter values $\nu < 2$ in (2.97) give rise to supergaussian densities, and $\nu > 2$ to subgaussian ones. Impulsive-type distributions are obtained from (2.97) when $0 < \nu < 1$.

2.7.2   Cumulants, moments, and their properties

Now we proceed to the general definition of cumulants. Assume that x is a real-valued, zero-mean, continuous scalar random variable with probability density function $p_x(x)$.

The first *characteristic function* $\varphi(\omega)$ of x is defined as the continuous Fourier transform of the pdf $p_x(x)$:

$$\varphi(\omega) = \mathrm{E}\{\exp(\jmath\omega x)\} = \int_{-\infty}^{\infty} \exp(\jmath\omega x)p_x(x)dx \tag{2.98}$$

where $\jmath = \sqrt{-1}$ and ω is the transformed variable corresponding to x. Every probability distribution is uniquely specified by its characteristic function, and vice versa [353]. Expanding the characteristic function $\varphi(\omega)$ into its Taylor series yields [353, 149]

$$\varphi(\omega) = \int_{-\infty}^{\infty} \left(\sum_{k=0}^{\infty} \frac{x^k(\jmath\omega)^k}{k!} \right) p_x(x)dx = \sum_{k=0}^{\infty} \mathrm{E}\{x^k\}\frac{(\jmath\omega)^k}{k!} \tag{2.99}$$

Thus the coefficient terms of this expansion are moments $\mathrm{E}\{x^k\}$ of x (assuming that they exist). For this reason, the characteristic function $\varphi(\omega)$ is also called the *moment generating function*.

It is often desirable to use the *second characteristic function* $\phi(\omega)$ of x, or *cumulant generating function* for reasons to be discussed later in this section. This function is given by the natural logarithm of the first characteristic function (2.98):

$$\phi(\omega) = \ln(\varphi(\omega)) = \ln(\mathrm{E}\{\exp(\jmath\omega x)\}) \tag{2.100}$$

The *cumulants* κ_k of x are defined in a similar way to the respective moments as the coefficients of the Taylor series expansion of the second characteristic function (2.100):

$$\phi(\omega) = \sum_{k=0}^{n} \kappa_k \frac{(\jmath\omega)^k}{k!} \tag{2.101}$$

where the kth cumulant is obtained as the derivative

$$\kappa_k = (-\jmath)^k \left. \frac{d^k\phi(\omega)}{d\omega^k} \right|_{\omega=0} \tag{2.102}$$

For a zero mean random variable x, the first four cumulants are

$$\kappa_1 = 0, \quad \kappa_2 = \mathrm{E}\{x^2\}, \quad \kappa_3 = \mathrm{E}\{x^3\}, \text{ and} \tag{2.103}$$
$$\kappa_4 = \mathrm{E}\{x^4\} - 3[\mathrm{E}\{x^2\}]^2$$

Hence the first three cumulants are equal to the respective moments, and the fourth cumulant κ_4 is recognized to be the kurtosis defined earlier in (2.90).

We list below the respective expressions for the cumulants when the mean $\mathrm{E}\{x\}$ of x is nonzero [319, 386, 149].

$$\kappa_1 = \mathrm{E}\{x\}$$
$$\kappa_2 = \mathrm{E}\{x^2\} - [\mathrm{E}\{x\}]^2$$
$$\kappa_3 = \mathrm{E}\{x^3\} - 3\mathrm{E}\{x^2\}\mathrm{E}\{x\} + 2[\mathrm{E}\{x\}]^3 \tag{2.104}$$
$$\kappa_4 = \mathrm{E}\{x^4\} - 3[\mathrm{E}\{x^2\}]^2 - 4\mathrm{E}\{x^3\}\mathrm{E}\{x\} + 12\mathrm{E}\{x^2\}[\mathrm{E}\{x\}]^2 - 6[\mathrm{E}\{x\}]^4$$

These formulas are obtained after tedious manipulations of the second characteristic function $\phi(\omega)$. Expressions for higher-order cumulants become increasingly complex [319, 386] and are omitted because they are applied seldom in practice.

Consider now briefly the multivariate case. Let $\mathbf{x}$ be a random vector and $p_{\mathbf{x}}(\mathbf{x})$ its probability density function. The characteristic function of $\mathbf{x}$ is again the Fourier transform of the pdf

$$\varphi(\omega) = \mathrm{E}\{\exp(\jmath\omega\mathbf{x})\} = \int_{-\infty}^{\infty} \exp(\jmath\omega\mathbf{x})p_{\mathbf{x}}(\mathbf{x})d\mathbf{x} \qquad (2.105)$$

where ω is now a row vector having the same dimension as $\mathbf{x}$, and the integral is computed over all components of $\mathbf{x}$. The moments and cumulants of $\mathbf{x}$ are obtained in a similar manner to the scalar case. Hence, moments of $\mathbf{x}$ are coefficients of the Taylor series expansion of the first characteristic function $\varphi(\omega)$, and the cumulants are the coefficients of the expansion of the second characteristic function $\phi(\omega) = \ln(\varphi(\omega))$. In the multivariate case, the cumulants are often called *cross-cumulants* in analogy to cross-covariances.

It can be shown that the second, third, and fourth order cumulants for a zero mean random vector $\mathbf{x}$ are [319, 386, 149]

$$\begin{aligned}
\mathrm{cum}(x_i, x_j) &= \mathrm{E}\{x_i x_j\} \\
\mathrm{cum}(x_i, x_j, x_k) &= \mathrm{E}\{x_i x_j x_k\} \\
\mathrm{cum}(x_i, x_j, x_k, x_l) &= \mathrm{E}\{x_i x_j x_k x_l\} - \mathrm{E}\{x_i x_j\}\mathrm{E}\{x_k x_l\} \\
&\quad - \mathrm{E}\{x_i x_k\}\mathrm{E}\{x_j x_l\} - \mathrm{E}\{x_i x_l\}\mathrm{E}\{x_j x_k\}
\end{aligned} \qquad (2.106)$$

Hence the second cumulant is equal to the second moment $\mathrm{E}\{x_i x_j\}$, which in turn is the correlation r_{ij} or covariance c_{ij} between the variables x_i and x_j. Similarly, the third cumulant $\mathrm{cum}(x_i, x_j, x_k)$ is equal to the third moment $\mathrm{E}\{x_i x_j x_k\}$. However, the fourth cumulant differs from the fourth moment $\mathrm{E}\{x_i x_j x_k x_l\}$ of the random variables x_i, x_j, x_k, and x_l.

Generally, higher-order moments correspond to correlations used in second-order statistics, and cumulants are the higher-order counterparts of covariances. Both moments and cumulants contain the same statistical information, because cumulants can be expressed in terms of sums of products of moments. It is usually preferable to work with cumulants because they present in a clearer way the additional information provided by higher-order statistics. In particular, it can be shown that cumulants have the following properties not shared by moments [319, 386].

1. Let $\mathbf{x}$ and $\mathbf{y}$ be statistically independent random vectors having the same dimension, then the cumulant of their sum $\mathbf{z} = \mathbf{x} + \mathbf{y}$ is equal to the sum of the cumulants of $\mathbf{x}$ and $\mathbf{y}$. This property also holds for the sum of more than two independent random vectors.

2. If the distribution of the random vector or process $\mathbf{x}$ is multivariate gaussian, all its cumulants of order three and higher are identically zero.

Thus higher-order cumulants measure the departure of a random vector from a gaussian random vector with an identical mean vector and covariance matrix. This property is highly useful, making it possible to use cumulants for extracting the nongaussian part of a signal. For example, they make it possible to ignore additive gaussian noise corrupting a nongaussian signal using cumulants.

Moments, cumulants, and characteristic functions have several other properties which are not discussed here. See, for example, the books [149, 319, 386] for more information. However, it is worth mentioning that both moments and cumulants have symmetry properties that can be exploited to reduce the computational load in estimating them [319].

For estimating moments and cumulants, one can apply the procedure introduced in Section 2.2.4. However, the fourth-order cumulants cannot be estimated directly, but one must first estimate the necessary moments as is obvious from (2.106). Practical estimation formulas can be found in [319, 315].

A drawback in utilizing higher-order statistics is that reliable estimation of higher-order moments and cumulants requires much more samples than for second-order statistics [318]. Another drawback is that higher-order statistics can be very sensitive to outliers in the data (see Section 8.3.1). For example, a few data samples having the highest absolute values may largely determine the value of kurtosis. Higher-order statistics can be taken into account in a more robust way by using the nonlinear hyperbolic tangent function $\tanh(u)$, whose values always lie in the interval $(-1, 1)$, or some other nonlinearity that grows slower than linearly with its argument value.

2.8 STOCHASTIC PROCESSES *

2.8.1 Introduction and definition

In this section,[3] we briefly discuss stochastic or random processes, defining what they are, and introducing some basic concepts. This material is not needed in basic independent component analysis. However, it forms a theoretical basis for blind source separation methods utilizing time correlations and temporal information in the data, discussed in Chapters 18 and 19. Stochastic processes are dealt with in more detail in many books devoted either entirely or partly to the topic; see for example [141, 157, 353, 419].

In short, stochastic or random processes are random functions of time. Stochastic processes have two basic characteristics. First, they are functions of time, defined on some observation interval. Second, stochastic processes are random in the sense that before making an experiment, it is not possible to describe exactly the waveform that is observed. Due to their nature, stochastic processes are well suited to the characterization of many random signals encountered in practical applications, such as voice, radar, seismic, and medical signals.

---

[3]An asterisk after the section title means that the section is more advanced material that may be skipped.

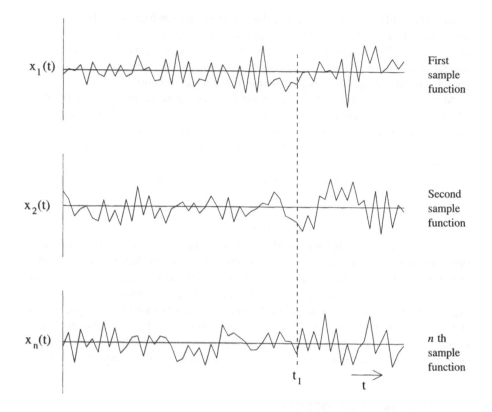

Fig. 2.10   Sample functions of a stochastic process.

Figure 2.10 shows an example of a scalar stochastic process represented by the set of sample functions $\{x_j(t)\}$, $j = 1, 2, \ldots, n$. Assume that the probability of occurrence of the ith sample function $x_i(t)$ is P_i, and similarly for the other sample functions. Suppose then we observe the set of waveforms $\{x_j(t)\}$, $j = 1, 2, \ldots, n$, simultaneously at some time instant $t = t_1$, as shown in Figure 2.10. Clearly, the values $\{x_j(t_1)\}$, $j = 1, 2, \ldots, n$ of the n waveforms at time t_1 form a discrete random variable with n possible values, each having the respective probability of occurrence P_j. Consider then another time instant $t = t_2$. We obtain again a random variable $\{x_j(t_2)\}$, which may have a different distribution than $\{x_j(t_1)\}$.

Usually the number of possible waveforms arising from an experiment is infinitely large due to additive noise. At each time instant a continuous random variable having some distribution arises instead of the discrete one discussed above. However, the time instants $t_1, t_2, \ldots$, on which the stochastic process is observed are discrete due to sampling. Usually the observation intervals are equispaced, and the resulting samples are represented using integer indices $x_j(1) = x_j(t_1)$, $x_j(2) = x_j(t_2), \ldots$ for notational simplicity. As a result, a typical representation for a stochastic process consists of continuous random variables at discrete (integer) time instants.

2.8.2 Stationarity, mean, and autocorrelation function

Consider a stochastic process $\{x_j(t)\}$ defined at discrete times $t_1, t_2, \ldots, t_k$. For characterizing the process $\{x_j(t)\}$ completely, we should know the joint probability density of all the random variables $\{x_j(t_1)\}, \{x_j(t_2)\}, \ldots, \{x_j(t_k)\}$. The stochastic process is said to be *stationary in the strict sense* if its joint density is invariant under time shifts of origin. That is, the joint pdf of the process depends only on the differences $t_i - t_j$ between the time instants $t_1, t_2, \ldots, t_k$ but not directly on them.

In practice, the joint probability density is not known, and its estimation from samples would be too tedious and require an excessive number of samples even if they were available. Therefore, stochastic processes are usually characterized in terms of their first two moments, namely the mean and autocorrelation or autocovariance functions. They give a coarse but useful description of the distribution. Using these statistics is sufficient for linear processing (for example filtering) of stochastic processes, and the number of samples needed for estimating them remains reasonable.

The *mean function* of the stochastic process $\{x(t)\}$ is defined

$$m_x(t) = E\{x(t)\} = \int_{-\infty}^{\infty} x(t)p_{x(t)}(x(t))dx(t) \tag{2.107}$$

Generally, this is a function of time t. However, when the process $\{x(t)\}$ is stationary, the probability density functions of all the random variables corresponding to different time instants become the same. This common pdf is denoted by $p_x(x)$. In such a case, the mean function $m_x(t)$ reduces to a constant mean m_x independent of time.

Similarly, the *variance function* of the stochastic process $\{x(t)\}$

$$\sigma_x^2(t) = E\{[x(t) - m_x(t)]^2\} = \int_{-\infty}^{\infty} [x(t) - m_x(t)]^2 p_{x(t)}(x(t))dx(t) \tag{2.108}$$

becomes a time-invariant constant σ_x^2 for a stationary process.

Other second-order statistics of a random process $\{x(t)\}$ are defined in a similar manner. In particular, the *autocovariance function* of the process $\{x(t)\}$ is given by

$$c_x(t, \tau) = \text{cov}[x(t), x(t - \tau)] = E\{[x(t) - m_x(t)][x(t - \tau) - m_x(t - \tau)]\} \tag{2.109}$$

The expectation here is computed over the joint probability density of the random variables $x(t)$ and $x(t - \tau)$, where τ is the constant time lag between the observation times t and $t - \tau$. For the zero lag $\tau = 0$, the autocovariance reduces to the variance function (2.108). For stationary processes, the autocovariance function (2.109) is independent of the time t, but depends on the lag τ: $c_x(t, \tau) = c_x(\tau)$.

Analogously, the *autocorrelation function* of the process $\{x(t)\}$ is defined by

$$r_x(t, \tau) = E\{x(t)x(t - \tau)\} \tag{2.110}$$

If $\{x(t)\}$ is stationary, this again depends on the time lag τ only: $r_x(t, \tau) = r_x(\tau)$. Generally, if the mean function $m_x(t)$ of the process is zero, the autocovariance and autocorrelation functions become the same. If the lag $\tau = 0$, the autocorrelation function reduces to the mean-square function $r_x(t, 0) = \mathrm{E}\{x^2(t)\}$ of the process, which becomes a constant $r_x(0)$ for a stationary process $\{x(t)\}$.

These concepts can be extended for two different stochastic processes $\{x(t)\}$ and $\{y(t)\}$ in an obvious manner (cf. Section 2.2.3). More specifically, the *cross-correlation function* $r_{xy}(t, \tau)$ and the *cross-covariance function* $c_{xy}(t, \tau)$ of the processes $\{x(t)\}$ and $\{y(t)\}$ are, respectively, defined by

$$r_{xy}(t, \tau) = \mathrm{E}\{x(t)y(t - \tau)\} \tag{2.111}$$

$$c_{xy}(t, \tau) = \mathrm{E}\{[x(t) - m_x(t)][y(t - \tau) - m_y(t - \tau)]\} \tag{2.112}$$

Several blind source separation methods are based on the use of cross-covariance functions (second-order temporal statistics). These methods will be discussed in Chapter 18.

2.8.3   Wide-sense stationary processes

A very important subclass of stochastic processes consists of *wide-sense stationary* (WSS) processes, which are required to satisfy the following properties:

1. The mean function $m_x(t)$ of the process is a constant m_x for all t.

2. The autocorrelation function is independent of a time shift: $\mathrm{E}\{x(t)x(t - \tau)\}$ $= r_x(\tau)$ for all t.

3. The variance, or the mean-square value $r_x(0) = \mathrm{E}\{x^2(t)\}$ of the process is finite.

The importance of wide-sense stationary stochastic processes stems from two facts. First, they can often adequately describe the physical situation. Many practical stochastic processes are actually at least mildly nonstationary, meaning that their statistical properties vary slowly with time. However, such processes are usually on short time intervals roughly WSS. Second, it is relatively easy to develop useful mathematical algorithms for WSS processes. This in turn follows from limiting their characterization by first- and second-order statistics.

Example 2.8 Consider the stochastic process

$$x(t) = a\cos(\omega t) + b\sin(\omega t) \tag{2.113}$$

where a and b are scalar random variables and ω a constant parameter (angular frequency). The mean of the process $x(t)$ is

$$m_x(t) = \mathrm{E}\{x(t)\} = \mathrm{E}\{a\}\cos(\omega t) + \mathrm{E}\{b\}\sin(\omega t) \tag{2.114}$$

and its autocorrelation function can be written

$$
\begin{aligned}
r_x(t, \tau) &= \mathrm{E}\{x(t)x(t - \tau)\} \\
&= \frac{1}{2}\mathrm{E}\{a^2\}[\cos(\omega(2t - \tau)) + \cos(-\omega\tau)] \\
&\quad + \frac{1}{2}\mathrm{E}\{b^2\}[-\cos(\omega(2t - \tau)) + \cos(-\omega\tau)] \\
&\quad + \mathrm{E}\{ab\}[\sin(\omega(2t - \tau))]
\end{aligned}
\tag{2.115}
$$

where we have used well-known trigonometric identities. Clearly, the process $x(t)$ is generally nonstationary, since both its mean and autocorrelation functions depend on the time t.

However, if the random variables a and b are zero mean and uncorrelated with equal variances, so that

$$
\mathrm{E}\{a\} = \mathrm{E}\{b\} = \mathrm{E}\{ab\} = 0 \qquad \mathrm{E}\{a^2\} = \mathrm{E}\{b^2\}
$$

the mean (2.114) of the process becomes zero, and its autocorrelation function (2.115) simplifies to

$$
r_x(\tau) = \mathrm{E}\{a^2\} \cos(\omega\tau)
$$

which depends only on the time lag τ. Hence, the process is WSS in this special case (assuming that $\mathrm{E}\{a^2\}$ is finite).

Assume now that $\{x(t)\}$ is a zero-mean WSS process. If necessary, the process can easily be made zero mean by first subtracting its mean m_x. It is sufficient to consider the autocorrelation function $r_x(\tau)$ of $\{x(t)\}$ only, since the autocovariance function $c_x(\tau)$ coincides with it. The autocorrelation function has certain properties that are worth noting. First, it is an even function of the time lag τ:

$$
r_x(-\tau) = r_x(\tau)
\tag{2.116}
$$

Another property is that the autocorrelation function achieves its maximum absolute value for zero lag:

$$
-r_x(0) \leq r_x(\tau) \leq r_x(0)
\tag{2.117}
$$

The autocorrelation function $r_x(\tau)$ measures the correlation of random variables $x(t)$ and $x(t - \tau)$ that are τ units apart in time, and thus provides a simple measure for the dependence of these variables which is independent of the time t due to the WSS property. Roughly speaking, the faster the stochastic process fluctuates with time around its mean, the more rapidly the values of the autocorrelation function $r_x(\tau)$ decrease from their maximum $r_x(0)$ as τ increases.

Using the integer notation for the samples $x(i)$ of the stochastic process, we can represent the last $m + 1$ samples of the stochastic process at time n using the random vector

$$
\mathbf{x}(n) = [x(n), x(n - 1), \ldots, x(n - m)]^T
\tag{2.118}
$$

Assuming that the values of the autocorrelation function $r_x(0), r_x(1), \ldots, r_x(m)$ are known up to a lag of m samples, the $(m+1) \times (m+1)$ correlation (or covariance) matrix of the process $\{x(n)\}$ is defined by

$$
\mathbf{R_x} = \begin{bmatrix} r_x(0) & r_x(1) & r_x(2) & \cdots & r_x(m) \\ r_x(1) & r_x(0) & r_x(1) & \cdots & r_x(m-1) \\ \vdots & \vdots & \vdots & \ddots & \vdots \\ r_x(m) & r_x(m-1) & r_x(m-2) & \cdots & r_x(0) \end{bmatrix} \tag{2.119}
$$

The matrix $\mathbf{R_x}$ satisfies all the properties of correlation matrices listed in Section 2.2.2. Furthermore, it is a Toeplitz matrix. This is generally defined so that on each subdiagonal and on the diagonal, all the elements of Toeplitz matrix are the same. The Toeplitz property is helpful, for example, in solving linear equations, enabling use of faster algorithms than for more general matrices.

Higher-order statistics of a stationary stochastic process $x(n)$ can be defined in an analogous manner. In particular, the cumulants of $x(n)$ have the form [315]

$$
\begin{aligned}
\mathrm{cum}_{xx}(j) &= \mathrm{E}\{x(i)x(i+j)\} \\
\mathrm{cum}_{xxx}(j,k) &= \mathrm{E}\{x(i)x(i+j)x(i+k)\} \\
\mathrm{cum}_{xxx}(j,k,l) &= \mathrm{E}\{x(i)x(i+j)x(i+k)x(i+l)\} \\
&\quad - \mathrm{E}\{x(i)x(j)\}\mathrm{E}\{x(k)x(l)\} - \mathrm{E}\{x(i)x(k)\}\mathrm{E}\{x(j)x(l)\} \\
&\quad - \mathrm{E}\{x(i)x(l)\}\mathrm{E}\{x(j)x(k)\}
\end{aligned} \tag{2.120}
$$

These definitions correspond to the formulas (2.106) given earlier for a general random vector $\mathbf{x}$. Again, the second and third cumulant are the same as the respective moments, but the fourth cumulant differs from the fourth moment $\mathrm{E}\{x(i)x(i+j)x(i+k)x(i+l)\}$. The second cumulant $\mathrm{cum}_{xx}(j)$ is equal to the autocorrelation $r_x(j)$ and autocovariance $c_x(j)$.

2.8.4   Time averages and ergodicity

In defining the concept of a stochastic process, we noted that at each fixed time instant $t = t_0$ the possible values $x(t_0)$ of the process constitute a random variable having some probability distribution. An important practical problem is that these distributions (which are different at different times if the process is nonstationary) are not known, at least not exactly. In fact, often all that we have is just one sample of the process corresponding to each discrete time index (since time cannot be stopped to acquire more samples). Such a sample sequence is called a *realization* of the stochastic process. In handling WSS processes, we need to know in most cases only the mean and autocorrelation values of the process, but even they are often unknown.

A practical way to circumvent this difficulty is to replace the usual expectations of the random variables, called *ensemble averages*, by long-term sample averages or *time averages* computed from the available single realization. Assume that this realization contains K samples $x(1), x(2), \ldots, x(K)$. Applying the preceding principle, the mean of the process can be estimated using its time average

$$\hat{m}_x(K) = \frac{1}{K} \sum_{k=1}^{K} x(k) \qquad (2.121)$$

and the autocorrelation function for the lag value l using

$$\hat{r}_x(l, K) = \frac{1}{K-l} \sum_{k=1}^{K-l} x(k+l)x(k) \qquad (2.122)$$

The accuracy of these estimates depends on the number K of samples. Note also that the latter estimate is computed over the $K - l$ possible sample pairs having the lag l that can be found from the sample set. The estimates (2.122) are unbiased, but if the number of pairs $K - l$ available for estimation is small, their variance can be high. Therefore, the scaling factor $K - l$ of the sum in (2.122) is often replaced by K in order to reduce the variance of the estimated autocorrelation values $\hat{r}_x(l, K)$, even though the estimates then become biased [169]. As $K \to \infty$, both estimates tend toward the same value.

The stochastic process is called *ergodic* if the ensemble averages can be equated to the respective time averages. Roughly speaking, a random process is ergodic with respect to its mean and autocorrelation function if it is stationary. A more rigorous treatment of the topic can be found for example in [169, 353, 141].

For mildly nonstationary processes, one can apply the estimation formulas (2.121) and (2.122) by computing the time averages over a shorter time interval during which the process can be regarded to be roughly WSS. It is important to keep this in mind. Sometimes formula (2.122) is applied in estimating the autocorrelation values without taking into account the stationarity of the process. The consequences can be drastic, for example, rendering eigenvectors of the correlation matrix (2.119) useless for practical purposes if ergodicity of the process is in reality a grossly invalid assumption.

2.8.5 Power spectrum

A lot of insight into a WSS stochastic process is often gained by representing it in the frequency domain. The *power spectrum* or *spectral density* of the process $x(n)$ provides such a representation. It is defined as the discrete Fourier transform of the autocorrelation sequence $r_x(0), r_x(1), \ldots$:

$$S_x(\omega) = \sum_{k=-\infty}^{\infty} r_x(k) \exp(-\jmath k\omega) \qquad (2.123)$$

where $\jmath = \sqrt{-1}$ is the imaginary unit and ω the angular frequency. The time domain representation given by the autocorrelation sequence of the process can be obtained from the power spectrum $S_x(\omega)$ by applying the inverse discrete-time

Fourier transform

$$r_x(k) = \frac{1}{2\pi} \int_{-\pi}^{\pi} S_x(\omega) \exp(\jmath k\omega) d\omega, \qquad k = 1, 2, \dots \tag{2.124}$$

It is easy to see that the power spectrum (2.123) is always real-valued, even, and a periodic function of the angular frequency ω. Note also that the power spectrum is a continuous function of ω, while the autocorrelation sequence is discrete. In practice, the power spectrum must be estimated from a finite number of autocorrelation values. If the autocorrelation values $r_x(k) \to 0$ sufficiently quickly as the lag k grows large, this provides an adequate approximation.

The power spectrum describes the frequency contents of the stochastic process, showing which frequencies are present in the process and how much power they possess. For a sinusoidal signal, the power spectrum shows a sharp peak at its oscillating frequency. Various methods for estimating power spectra are discussed thoroughly in the books [294, 241, 411].

Higher-order spectra can be defined in a similar manner to the power spectrum as Fourier transforms of higher-order statistics [319, 318]. Contrary to the power spectra, they retain information about the phase of signals, and have found many applications in describing nongaussian, nonlinear, and nonminimum-phase signals [318, 319, 315].

2.8.6 Stochastic signal models

A stochastic process whose power spectrum is constant for all frequencies ω is called *white noise*. Alternatively, white noise $v(n)$ can be defined as a process for which any two different samples are uncorrelated:

$$r_v(k) = \mathrm{E}\{v(n)v(n-k)\} = \begin{cases} \sigma_v^2, & k = 0 \\ 0, & k = \pm 1, \pm 2, \dots \end{cases} \tag{2.125}$$

Here σ_v^2 is the variance of the white noise. It is easy to see that the power spectrum of the white noise is $S_v(\omega) = \sigma_v^2$ for all ω, and that the formula (2.125) follows from the inverse transform (2.124). The distribution of the random variable $v(n)$ forming the white noise can be any reasonable one, provided that the samples are uncorrelated at different time indices. Usually this distribution is assumed to be gaussian. The reason is that white gaussian noise is maximally random because any two uncorrelated samples are also independent. Furthermore, such a noise process cannot be modeled to yield an even simpler random process.

Stochastic processes or time series are frequently modeled in terms of *autoregressive (AR) processes*. They are defined by the difference equation

$$x(n) = -\sum_{i=1}^{M} a_i x(n-i) + v(n) \tag{2.126}$$

where $v(n)$ is a white noise process, and $a_1, \ldots, a_M$ are constant coefficients (parameters) of the AR model. The model order M gives the number of previous samples on which the current value $x(n)$ of the AR process depends. The noise term $v(n)$ introduces randomness into the model; without it the AR model would be completely deterministic. The coefficients $a_1, \ldots, a_M$ of the AR model can be computed using linear techniques from autocorrelation values estimated from the available data [419, 241, 169]. Since the AR models describe fairly well many natural stochastic processes, for example, speech signals, they are used in many applications. In ICA and BSS, they can be used to model the time correlations in each source process $s_i(t)$. This sometimes improves greatly the performance of the algorithms.

Autoregressive processes are a special case of *autoregressive moving average (ARMA)* processes described by the difference equation

$$x(n) + \sum_{i=1}^{M} a_i x(n-i) = v(n) + \sum_{i=1}^{N} b_i v(n-i) \qquad (2.127)$$

Clearly, the AR model (2.126) is obtained from the ARMA model (2.127) when the *moving average (MA)* coefficients $b_1, \ldots, b_N$ are all zero. On the other hand, if the AR coefficients a_i are all zero, the ARMA process (2.127) reduces to a MA process of order N. The ARMA and MA models can also be used to describe stochastic processes. However, they are applied less frequently, because estimation of their parameters requires nonlinear techniques [241, 419, 411]. See the Appendix of Chapter 19 for a discussion of the stability of the ARMA model and its utilization in digital filtering.

2.9 CONCLUDING REMARKS AND REFERENCES

In this chapter, we have covered the necessary background on the theory of random vectors, independence, higher-order statistics, and stochastic processes. Topics that are needed in studying independent component analysis and blind source separation have received more attention. Several books that deal more thoroughly with the theory of random vectors exist; for example, [293, 308, 353]. Stochastic processes are discussed in [141, 157, 353], and higher-order statistics in [386].

Many useful, well-established techniques of signal processing, statistics, and other areas are based on analyzing random vectors and signals by means of their first- and second-order statistics. These techniques have the virtue that they are usually fairly easy to apply. Typically, second-order error criteria (for example, the mean-square error) are used in context with them. In many cases, this leads to linear solutions that are simple to compute using standard numerical techniques. On the other hand, one can claim that techniques based on second-order statistics are optimal for gaussian signals only. This is because they neglect the extra information contained in the higher-order statistics, which is needed in describing nongaussian data. Independent component analysis uses this higher-order statistical information, and is the reason for which it is such a powerful tool.

Problems

2.1   Derive a rule for computing the values of the cdf of the single variable gaussian (2.4) from the known tabulated values of the error function (2.5).

2.2   Let $x_1, x_2, \ldots, x_K$ be independent, identically distributed samples from a distribution having a cumulative density function $F_x(x)$. Denote by $y_1, x_2, \ldots, y_K$ the sample set $x_1, x_2, \ldots, x_K$ ordered in increasing order.

2.2.1. Show that the cdf and pdf of $y_K = \max\{x_1, \ldots, x_K\}$ are

$$F_{y_K}(y_K) = [F_x(y_K)]^K$$

$$p_{y_K}(y_K) = K[F_x(y_K)]^{K-1} p_x(y_K)$$

2.2.2. Derive the respective expressions for the cdf and pdf of the random variable $y_1 = \min\{x_1, \ldots, x_K\}$.

2.3   A two-dimensional random vector $\mathbf{x} = (x_1, x_2)^T$ has the probability density function

$$p_{\mathbf{x}}(\mathbf{x}) = \begin{cases} \frac{1}{2}(x_1 + 3x_2) & x_1, x_2 \in [0, 1] \\ 0 & \text{elsewhere} \end{cases}$$

2.3.1. Show that this probability density is appropriately normalized.
2.3.2. Compute the cdf of the random vector $\mathbf{x}$.
2.3.3. Compute the marginal distributions $p_{x_1}(x_1)$ and $p_{x_2}(x_2)$.

2.4   Computer the mean, second moment, and variance of a random variable distributed uniformly in the interval $[a, b]$ $(b > a)$.

2.5   Prove that expectations satisfy the linearity property (2.16).

2.6   Consider n scalar random variables x_i, $i = 1, 2, \ldots, n$, having, respectively, the variances $\sigma_{x_i}^2$. Show that if the random variables x_i are mutually uncorrelated, the variance σ_y^2 of their sum $y = \sum_{i=1}^n x_i$ equals the sum of the variances of the x_i:

$$\sigma_y^2 = \sum_{i=1}^n \sigma_{x_i}^2$$

2.7   Assume that x_1 and x_2 are zero-mean, correlated random variables. Any orthogonal transformation of x_1 and x_2 can be represented in the form

$$y_1 = \cos(\alpha)x_1 + \sin(\alpha)x_2$$
$$y_2 = -\sin(\alpha)x_1 + \cos(\alpha)x_2$$

where the parameter α defines the rotation angle of coordinate axes. Let $\mathrm{E}\{x_1^2\} = \sigma_1^2$, $\mathrm{E}\{x_2^2\} = \sigma_2^2$, and $\mathrm{E}\{x_1 x_2\} = \rho \sigma_1 \sigma_2$. Find the angle α for which y_1 and y_2 become uncorrelated.

2.8   Consider the joint probability density of the random vectors $\mathbf{x} = (x_1, x_2)^T$ and $\mathbf{y} = y$ discussed in Example 2.6:

$$p_{\mathbf{x},\mathbf{y}}(\mathbf{x}, \mathbf{y}) = \begin{cases} (x_1 + 3x_2)y & x_1, x_2 \in [0, 1], \quad y \in [0, 1] \\ 0 & \text{elsewhere} \end{cases}$$

2.8.1. Compute the marginal distributions $p_{\mathbf{x}}(\mathbf{x})$, $p_{\mathbf{y}}(\mathbf{y})$, $p_{x_1}(x_1)$, and $p_{x_2}(x_2)$.

2.8.2. Verify that the claims made on the independence of x_1, x_2, and y in Example 2.6 hold.

2.9   Which conditions should the elements of the matrix

$$\mathbf{R} = \begin{bmatrix} a & b \\ c & d \end{bmatrix}$$

satisfy so that $\mathbf{R}$ could be a valid autocorrelation matrix of
  2.9.1. A two-dimensional random vector?
  2.9.2. A stationary scalar-valued stochastic process?

2.10   Show that correlation and covariance matrices satisfy the relationships (2.26) and (2.32).

2.11   Work out Example 2.5 for the covariance matrix $\mathbf{C}_{\mathbf{x}}$ of $\mathbf{x}$, showing that similar results are obtained. Are the assumptions required the same?

2.12   Assume that the inverse $\mathbf{R}_{\mathbf{x}}^{-1}$ of the correlation matrix of the n-dimensional column random vector $\mathbf{x}$ exists. Show that

$$E\{\mathbf{x}^T \mathbf{R}_{\mathbf{x}}^{-1} \mathbf{x}\} = n$$

2.13   Consider a two-dimensional gaussian random vector $\mathbf{x}$ with mean vector $\mathbf{m}_{\mathbf{x}} = (2, 1)^T$ and covariance matrix

$$\mathbf{C}_{\mathbf{x}} = \begin{bmatrix} 2 & -1 \\ -1 & 2 \end{bmatrix}$$

2.13.1. Find the eigenvalues and eigenvectors of $\mathbf{C}_{\mathbf{x}}$.
2.13.2. Draw a contour plot of the gaussian density similar to Figure 2.7.

2.14   Repeat the previous problem for a gaussian random vector $\mathbf{x}$ that has the mean vector $\mathbf{m}_{\mathbf{x}} = (-2, 3)^T$ and covariance matrix

$$\mathbf{C}_{\mathbf{x}} = \begin{bmatrix} 2 & -2 \\ -2 & 5 \end{bmatrix}$$

2.15   Assume that random variables x and y are linear combinations of two uncorrelated gaussian random variables u and v, defined by

$$x = 3u - 4v$$
$$y = 2u + v$$

Assume that the mean values and variances of both u and v equal 1.

2.15.1. Determine the mean values of x and y.

2.15.2. Find the variances of x and y.

2.15.3. Form the joint density function of x and y.

2.15.4. Find the conditional density of y given x.

2.16   Show that the skewness of a random variable having a symmetric pdf is zero.

2.17   Show that the kurtosis of a gaussian random variable is zero.

2.18   Show that random variables having

2.18.1. A uniform distribution in the interval $[-a, a]$, $a > 0$, are subgaussian.

2.18.2. A Laplacian distribution are supergaussian.

2.19   The exponential density has the pdf

$$p_x(x) = \begin{cases} \beta \exp(-\beta x) & x \geq 0 \\ 0 & x < 0 \end{cases}$$

where β is a positive constant.

2.19.1. Compute the first characteristic function of the exponential distribution.

2.19.2. Using the characteristic function, determine the moments of the exponential density.

2.20   A scalar random variable x has a gamma distribution if its pdf is given by

$$p_x(x) = \begin{cases} \gamma x^{b-1} \exp(-cx) & x \geq 0 \\ 0 & x < 0 \end{cases}$$

where b and c are positive numbers and the parameter

$$\gamma = \frac{c^b}{\Gamma(b)}$$

is defined by the gamma function

$$\Gamma(b+1) = \int_0^\infty y^b \exp(-y)dy, \quad b > -1$$

The gamma function satisfies the generalized factorial condition $\Gamma(b+1) = b\Gamma(b)$. For integer values, this becomes $\Gamma(n+1) = n!$.

2.20.1. Show that if $b = 1$, the gamma distribution reduces to the standard exponential density.

2.20.2. Show that the first characteristic function of a gamma distributed random variable is

$$\varphi(\omega) = \frac{c^b}{(c - \jmath\omega)^b}$$

2.20.3. Using the previous result, determine the mean, second moment, and variance of the gamma distribution.

2.21   Let $\kappa_k(x)$ and $\kappa_k(y)$ be the kth-order cumulants of the scalar random variables x and y, respectively.

2.21.1. Show that if x and y are independent, then

$$\kappa_k(x+y) = \kappa_k(x) + \kappa_k(y)$$

2.21.2. Show that $\kappa_k(\beta x) = \beta^k \kappa_k(x)$, where β is a constant.

2.22   * Show that the power spectrum $S_x(\omega)$ is a real-valued, even, and periodic function of the angular frequency ω.

2.23   * Consider the stochastic process

$$y(n) = x(n+k) - x(n-k)$$

where k is a constant integer and $x(n)$ is a zero mean, wide-sense stationary stochastic process. Let the power spectrum of $x(n)$ be $S_x(\omega)$ and its autocorrelation sequence $r_x(0), r_x(1), \ldots$.

2.23.1. Determine the autocorrelation sequence $r_y(m)$ of the process $y(n)$.

2.23.2. Show that the power spectrum of $y(n)$ is

$$S_y(\omega) = 4S_x(\omega)\sin^2(k\omega)$$

2.24   * Consider the autoregressive process (2.126).

2.24.1. Show that the autocorrelation function of the AR process satisfies the difference equation

$$r_x(l) = -\sum_{i=1}^{M} a_i r_x(l-i), \quad l > 0$$

2.24.2. Using this result, show that the AR coefficients a_i can be determined from the Yule-Walker equations

$$\mathbf{R_x a} = -\mathbf{r_x}$$

Here the autocorrelation matrix $\mathbf{R_x}$ defined in (2.119) has the value $m = M - 1$, the vector

$$\mathbf{r_x} = [r_x(1), r_x(2), \ldots, r_x(M)]^T$$

and the coefficient vector

$$\mathbf{a} = [a_1, a_2, \ldots, a_M]^T$$

2.24.3. Show that the variance of the white noise process $v(n)$ in (2.126) is related to the autocorrelation values by the formula

$$\sigma_v^2 = r_x(0) + \sum_{i=1}^{M} a_i r_x(i)$$

Computer assignments

2.1   Generate samples of a two-dimensional gaussian random vector **x** having zero mean vector and the covariance matrix

$$\mathbf{C_x} = \begin{bmatrix} 4 & -1 \\ -1 & 2 \end{bmatrix}$$

Estimate the covariance matrix and compare it with the theoretical one for the following numbers of samples, plotting the sample vectors in each case.

2.1.1. $K = 20$.

2.1.2. $K = 200$.

2.1.3. $K = 2000$.

2.2   Consider generation of desired Laplacian random variables for simulation purposes.

2.2.1. Using the probability integral transformation, give a formula for generating samples of a scalar random variable with a desired Laplacian distribution from uniformly distributed samples.

2.2.2. Extend the preceding procedure so that you get samples of two Laplacian random variables with a desired mean vector and joint covariance matrix. (*Hint:* Use the eigenvector decomposition of the covariance matrix for generating the desired covariance matrix.)

2.2.3. Use your procedure for generating 200 samples of a two-dimensional Laplacian random variable **x** with a mean vector $\mathbf{m_x} = (2, -1)^T$ and covariance matrix

$$\mathbf{C_x} = \begin{bmatrix} 4 & -1 \\ -1 & 2 \end{bmatrix}$$

Plot the generated samples.

2.3   * Consider the second-order autoregressive model described by the difference equation

$$x(n) + a_1 x(n-1) + a_2 x(n-2) = v(n)$$

Here $x(n)$ is the value of the process at time n, and $v(n)$ is zero mean white gaussian noise with variance σ_v^2 that "drives" the AR process. Generate 200 samples of the process using the initial values $x(0) = x(-1) = 0$ and the following coefficient values. Plot the resulting AR process in each case.

2.3.1. $a_1 = -0.1$ and $a_2 = -0.8$.

2.3.2. $a_1 = 0.1$ and $a_2 = -0.8$.

2.3.3. $a_1 = -0.975$ and $a_2 = 0.95$.

2.3.4. $a_1 = 0.1$ and $a_2 = -1.0$.

3

Gradients and Optimization Methods

The main task in the independent component analysis (ICA) problem, formulated in Chapter 1, is to estimate a separating matrix $\mathbf{W}$ that will give us the independent components. It also became clear that $\mathbf{W}$ cannot generally be solved in closed form, that is, we cannot write it as some function of the sample or training set, whose value could be directly evaluated. Instead, the solution method is based on *cost functions*, also called objective functions or contrast functions. Solutions $\mathbf{W}$ to ICA are found at the minima or maxima of these functions. Several possible ICA cost functions will be given and discussed in detail in Parts II and III of this book. In general, statistical estimation is largely based on optimization of cost or objective functions, as will be seen in Chapter 4.

Minimization of multivariate functions, possibly under some constraints on the solutions, is the subject of *optimization theory*. In this chapter, we discuss some typical iterative optimization algorithms and their properties. Mostly, the algorithms are based on the gradients of the cost functions. Therefore, vector and matrix gradients are reviewed first, followed by the most typical ways to solve unconstrained and constrained optimization problems with gradient-type learning algorithms.

3.1 VECTOR AND MATRIX GRADIENTS

3.1.1 Vector gradient

Consider a *scalar valued function g* of m variables

$$g = g(w_1, ..., w_m) = g(\mathbf{w})$$

where we have used the notation $\mathbf{w} = (w_1, ..., w_m)^T$. By convention, we define $\mathbf{w}$ as a column vector. Assuming the function g is differentiable, its vector gradient with respect to $\mathbf{w}$ is the m-dimensional column vector of partial derivatives

$$\frac{\partial g}{\partial \mathbf{w}} = \begin{pmatrix} \frac{\partial g}{\partial w_1} \\ \vdots \\ \frac{\partial g}{\partial w_m} \end{pmatrix} \tag{3.1}$$

The notation $\frac{\partial g}{\partial \mathbf{w}}$ is just shorthand for the gradient; it should be understood that it does not imply any kind of division by a vector, which is not a well-defined concept. Another commonly used notation would be ∇g or $\nabla_\mathbf{w} g$.

In some iteration methods, we have also reason to use second-order gradients. We define the second-order gradient of a function g with respect to $\mathbf{w}$ as

$$\frac{\partial^2 g}{\partial \mathbf{w}^2} = \begin{pmatrix} \frac{\partial^2 g}{\partial w_1^2} & \cdots & \frac{\partial^2 g}{\partial w_1 w_m} \\ \vdots & & \vdots \\ \frac{\partial^2 g}{\partial w_m w_1} & \cdots & \frac{\partial^2 g}{\partial w_m^2} \end{pmatrix} \tag{3.2}$$

This is an $m \times m$ matrix whose elements are second order partial derivatives. It is called the *Hessian matrix* of the function $g(\mathbf{w})$. It is easy to see that it is always symmetric.

These concepts generalize to *vector-valued* functions; this means an n-element vector

$$\mathbf{g}(\mathbf{w}) = \begin{pmatrix} g_1(\mathbf{w}) \\ \vdots \\ g_n(\mathbf{w}) \end{pmatrix} \tag{3.3}$$

whose elements $g_i(\mathbf{w})$ are themselves functions of $\mathbf{w}$. The *Jacobian matrix* of $\mathbf{g}$ with respect to $\mathbf{w}$ is

$$\frac{\partial \mathbf{g}}{\partial \mathbf{w}} = \begin{pmatrix} \frac{\partial g_1}{\partial w_1} & \cdots & \frac{\partial g_n}{\partial w_1} \\ \vdots & & \vdots \\ \frac{\partial g_1}{\partial w_m} & \cdots & \frac{\partial g_n}{\partial w_m} \end{pmatrix} \tag{3.4}$$

Thus the ith column of the Jacobian matrix is the gradient vector of $g_i(\mathbf{w})$ with respect to $\mathbf{w}$. The Jacobian matrix is sometimes denoted by $J\mathbf{g}$.

For computing the gradients of *products and quotients* of functions, as well as of *composite functions*, the same rules apply as for ordinary functions of one variable.

Thus

$$\frac{\partial f(\mathbf{w})g(\mathbf{w})}{\partial \mathbf{w}} = \frac{\partial f(\mathbf{w})}{\partial \mathbf{w}}g(\mathbf{w}) + f(\mathbf{w})\frac{\partial g(\mathbf{w})}{\partial \mathbf{w}} \qquad (3.5)$$

$$\frac{\partial f(\mathbf{w})/g(\mathbf{w})}{\partial \mathbf{w}} = [\frac{\partial f(\mathbf{w})}{\partial \mathbf{w}}g(\mathbf{w}) - f(\mathbf{w})\frac{\partial g(\mathbf{w})}{\partial \mathbf{w}}]/g^2(\mathbf{w}) \qquad (3.6)$$

$$\frac{\partial f(g(\mathbf{w}))}{\partial \mathbf{w}} = f'(g(\mathbf{w}))\frac{\partial g(\mathbf{w})}{\partial \mathbf{w}} \qquad (3.7)$$

The gradient of the composite function $f(g(\mathbf{w}))$ can be generalized to any number of nested functions, giving the same chain rule of differentiation that is valid for functions of one variable.

3.1.2 Matrix gradient

In many of the algorithms encountered in this book, we have to consider scalar-valued functions g of the elements of an $m \times n$ matrix $\mathbf{W} = (w_{ij})$:

$$g = g(\mathbf{W}) = g(w_{11}, ..., w_{ij}, ..., w_{mn}) \qquad (3.8)$$

A typical function of this kind is the determinant of $\mathbf{W}$.

Of course, any matrix can be trivially represented as a vector by scanning the elements row by row into a vector and reindexing. Thus, when considering the gradient of g with respect to the matrix elements, it would suffice to use the notion of vector gradient reviewed earlier. However, using the separate concept of matrix gradient gives some advantages in terms of a simplified notation and sometimes intuitively appealing results.

In analogy with the vector gradient, the matrix gradient means a matrix of the same size $m \times n$ as matrix $\mathbf{W}$, whose ijth element is the partial derivative of g with respect to w_{ij}. Formally we can write

$$\frac{\partial g}{\partial \mathbf{W}} = \begin{pmatrix} \frac{\partial g}{\partial w_{11}} & \cdots & \frac{\partial g}{\partial w_{1n}} \\ \vdots & & \vdots \\ \frac{\partial g}{\partial w_{m1}} & \cdots & \frac{\partial g}{\partial w_{mn}} \end{pmatrix} \qquad (3.9)$$

Again, the notation $\frac{\partial g}{\partial \mathbf{W}}$ is just shorthand for the matrix gradient.

Let us look next at some examples on vector and matrix gradients. The formulas presented in these examples will be frequently needed later in this book.

3.1.3 Examples of gradients

Example 3.1 Consider the simple linear functional of $\mathbf{w}$, or inner product

$$g(\mathbf{w}) = \sum_{i=1}^{m} a_i w_i = \mathbf{a}^T \mathbf{w}$$

where $\mathbf{a} = (a_1 \dots a_m)^T$ is a constant vector. The gradient is, according to (3.1),

$$\frac{\partial g}{\partial \mathbf{w}} = \begin{pmatrix} a_1 \\ \vdots \\ a_m \end{pmatrix} \tag{3.10}$$

which is the vector $\mathbf{a}$. We can write

$$\frac{\partial \mathbf{a}^T \mathbf{w}}{\partial \mathbf{w}} = \mathbf{a}$$

Because the gradient is constant (independent of $\mathbf{w}$), the Hessian matrix of $g(\mathbf{w}) = \mathbf{a}^T \mathbf{w}$ is zero.

Example 3.2 Next consider the quadratic form

$$g(\mathbf{w}) = \mathbf{w}^T \mathbf{A} \mathbf{w} = \sum_{i=1}^{m} \sum_{j=1}^{m} w_i w_j a_{ij} \tag{3.11}$$

where $\mathbf{A} = (a_{ij})$ is a square $m \times m$ matrix. We have

$$\frac{\partial g}{\partial \mathbf{w}} = \begin{pmatrix} \sum_{j=1}^{m} w_j a_{1j} + \sum_{i=1}^{m} w_i a_{i1} \\ \vdots \\ \sum_{j=1}^{m} w_j a_{mj} + \sum_{i=1}^{m} w_i a_{im} \end{pmatrix} \tag{3.12}$$

which is equal to the vector $\mathbf{A}\mathbf{w} + \mathbf{A}^T \mathbf{w}$. So,

$$\frac{\partial \mathbf{w}^T \mathbf{A} \mathbf{w}}{\partial \mathbf{w}} = \mathbf{A}\mathbf{w} + \mathbf{A}^T \mathbf{w}$$

For symmetric $\mathbf{A}$, this becomes $2\mathbf{A}\mathbf{w}$.

The second-order gradient or Hessian becomes

$$\frac{\partial^2 \mathbf{w}^T \mathbf{A} \mathbf{w}}{\partial \mathbf{w}^2} = \begin{pmatrix} 2a_{11} & \cdots & a_{1m} + a_{m1} \\ \vdots & & \vdots \\ a_{m1} + a_{1m} & \cdots & 2a_{mm} \end{pmatrix} \tag{3.13}$$

which is equal to the matrix $\mathbf{A} + \mathbf{A}^T$. If $\mathbf{A}$ is symmetric, then the Hessian of $\mathbf{w}^T \mathbf{A} \mathbf{w}$ is equal to $2\mathbf{A}$.

Example 3.3 For the quadratic form (3.11), we might quite as well take the gradient with respect to $\mathbf{A}$, assuming now that $\mathbf{w}$ is a constant vector. Then $\frac{\partial \mathbf{w}^T \mathbf{A} \mathbf{w}}{\partial a_{ij}} = w_i w_j$. Compiling this into matrix form, we notice that the matrix gradient is the $m \times m$ matrix $\mathbf{w}\mathbf{w}^T$.

Example 3.4 In some ICA models, we must compute the matrix gradient of the *determinant* of a matrix. The determinant is a scalar function of the matrix elements consisting of multiplications and summations, and therefore its partial derivatives are relatively simple to compute. Let us prove the following: If $\mathbf{W}$ is an invertible square $m \times m$ matrix whose determinant is denoted $\det \mathbf{W}$, then

$$\frac{\partial}{\partial \mathbf{W}} \det \mathbf{W} = (\mathbf{W}^T)^{-1} \det \mathbf{W}. \tag{3.14}$$

This is a good example for showing that a compact formula is obtained using the matrix gradient; if $\mathbf{W}$ were stacked into a long vector, and only the vector gradient were used, this result could not be expressed so simply.

Instead of starting from scratch, we employ a well-known result from matrix algebra (see, e.g., [159]), stating that the inverse of a matrix $\mathbf{W}$ is obtained as

$$\mathbf{W}^{-1} = \frac{1}{\det \mathbf{W}} \operatorname{adj}(\mathbf{W}) \tag{3.15}$$

with $\operatorname{adj}(\mathbf{W})$ the so-called *adjoint* of $\mathbf{W}$. The adjoint is the matrix

$$\operatorname{adj}(\mathbf{W}) = \begin{pmatrix} W_{11} & \cdots & W_{n1} \\ W_{1n} & \cdots & W_{nn} \end{pmatrix} \tag{3.16}$$

where the scalar numbers W_{ij} are the so-called *cofactors*. The cofactor W_{ij} is obtained by first taking the $(n-1) \times (n-1)$ submatrix of $\mathbf{W}$ that remains when the ith row and jth column are removed, then computing the determinant of this submatrix, and finally multiplying by $(-1)^{i+j}$.

The determinant $\det \mathbf{W}$ can also be expressed in terms of the cofactors:

$$\det \mathbf{W} = \sum_{k=1}^{n} w_{ik} W_{ik} \tag{3.17}$$

Row i can be any row, and the result is always the same. In the cofactors W_{ik}, none of the matrix elements of the ith row appear, so the determinant is a linear function of these elements. Taking now a partial derivative of (3.17) with respect to one of the elements, say, w_{ij}, gives

$$\frac{\partial \det \mathbf{W}}{\partial w_{ij}} = W_{ij}$$

By definitions (3.9) and (3.16), this implies directly that

$$\frac{\partial \det \mathbf{W}}{\partial \mathbf{W}} = \operatorname{adj}(\mathbf{W})^T$$

But $\operatorname{adj}(\mathbf{W})^T$ is equal to $(\det \mathbf{W})(\mathbf{W}^T)^{-1}$ by (3.15), so we have shown our result (3.14).

This also implies that

$$\frac{\partial \log|\det \mathbf{W}|}{\partial \mathbf{W}} = \frac{1}{|\det \mathbf{W}|} \frac{\partial |\det \mathbf{W}|}{\partial \mathbf{W}} = (\mathbf{W}^T)^{-1} \tag{3.18}$$

see (3.15). This is an example of the matrix gradient of a composite function consisting of the log, absolute value, and det functions. This result will be needed when the ICA problem is solved by maximum likelihood estimation in Chapter 9.

3.1.4 Taylor series expansions of multivariate functions

In deriving some of the gradient type learning algorithms, we have to resort to Taylor series expansions of multivariate functions. In analogy with the well-known Taylor series expansion of a function $g(w)$ of a scalar variable w,

$$g(w') = g(w) + \frac{dg}{dw}(w' - w) + 1/2\frac{d^2g}{dw^2}(w' - w)^2 + ... \tag{3.19}$$

we can do a similar expansion for a function $g(\mathbf{w}) = g(w_1, ..., w_m)$ of m variables. We have

$$g(\mathbf{w'}) = g(\mathbf{w}) + (\frac{\partial g}{\partial \mathbf{w}})^T(\mathbf{w'} - \mathbf{w}) + 1/2(\mathbf{w'} - \mathbf{w})^T\frac{\partial^2g}{\partial \mathbf{w}^2}(\mathbf{w'} - \mathbf{w}) + ... \tag{3.20}$$

where the derivatives are evaluated at the point $\mathbf{w}$. The second term is the inner product of the gradient vector with the vector $\mathbf{w'} - \mathbf{w}$, and the third term is a quadratic form with the symmetric Hessian matrix $\frac{\partial^2g}{\partial \mathbf{w}^2}$. The truncation error depends on the distance $\|\mathbf{w'} - \mathbf{w}\|$; the distance has to be small, if $g(\mathbf{w'})$ is approximated using only the first- and second-order terms.

The same expansion can be made for a scalar function of a matrix variable. The second order term already becomes complicated because the second order gradient is a four-dimensional tensor. But we can easily extend the first order term in (3.20), the inner product of the gradient with the vector $\mathbf{w'} - \mathbf{w}$, to the matrix case. Remember that the vector inner product is defined as

$$(\frac{\partial g}{\partial \mathbf{w}})^T(\mathbf{w'} - \mathbf{w}) = \sum_{i=1}^{m}(\frac{\partial g}{\partial \mathbf{w}})_i(w'_i - w_i)$$

For the matrix case, this must become the sum $\sum_{i=1}^{m}\sum_{j=1}^{m}(\frac{\partial g}{\partial \mathbf{W}})_{ij}(w'_{ij} - w_{ij})$. This is the sum of the products of corresponding elements, just like in the vectorial inner product. This can be nicely presented in matrix form when we remember that for any two matrices, say, $\mathbf{A}$ and $\mathbf{B}$,

$$\text{trace}(\mathbf{A}^T\mathbf{B}) = \sum_{i=1}^{m}(\mathbf{A}^T\mathbf{B})_{ii} = \sum_{i=1}^{m}\sum_{j=1}^{m}(\mathbf{A})_{ij}(\mathbf{B})_{ij}$$

with obvious notation. So, we have

$$g(\mathbf{W'}) = g(\mathbf{W}) + \text{trace}[(\frac{\partial g}{\partial \mathbf{W}})^T(\mathbf{W'} - \mathbf{W})] + ... \tag{3.21}$$

for the first two terms in the Taylor series of a function g of a matrix variable.

3.2 LEARNING RULES FOR UNCONSTRAINED OPTIMIZATION

3.2.1 Gradient descent

Many of the ICA criteria have the basic form of minimizing a cost function $\mathcal{J}(\mathbf{W})$ with respect to a parameter matrix $\mathbf{W}$, or possibly with respect to one of its columns $\mathbf{w}$. In many cases, there are also constraints that restrict the set of possible solutions. A typical constraint is to require that the solution vector must have a bounded norm, or the solution matrix has orthonormal columns.

For the unconstrained problem of minimizing a multivariate function, the most classic approach is steepest descent or gradient descent. Let us consider in more detail the case when the solution is a vector $\mathbf{w}$; the matrix case goes through in a completely analogous fashion.

In gradient descent, we minimize a function $\mathcal{J}(\mathbf{w})$ iteratively by starting from some initial point $\mathbf{w}(0)$, computing the gradient of $\mathcal{J}(\mathbf{w})$ at this point, and then moving in the direction of the negative gradient or the steepest descent by a suitable distance. Once there, we repeat the same procedure at the new point, and so on. For $t = 1, 2, ...,$we have the update rule

$$\mathbf{w}(t) = \mathbf{w}(t-1) - \alpha(t)\frac{\partial \mathcal{J}(\mathbf{w})}{\partial \mathbf{w}}|_{\mathbf{w}=\mathbf{w}(t-1)} \qquad (3.22)$$

with the gradient taken at the point $\mathbf{w}(t-1)$. The parameter $\alpha(t)$ gives the length of the step in the negative gradient direction. It is often called the *step size* or *learning rate*. Iteration (3.22) is continued until it converges, which in practice happens when the Euclidean distance between two consequent solutions $\|\mathbf{w}(t) - \mathbf{w}(t-1)\|$ goes below some small tolerance level.

If there is no reason to emphasize the time or iteration step, a convenient *shorthand notation* will be used throughout this book in presenting update rules of the preceding type. Denote the difference between the new and old value by

$$\mathbf{w}(t) - \mathbf{w}(t-1) = \Delta\mathbf{w} \qquad (3.23)$$

We can then write the rule (3.22) either as

$$\Delta\mathbf{w} = -\alpha\frac{\partial \mathcal{J}(\mathbf{w})}{\partial \mathbf{w}}$$

or even shorter as

$$\Delta\mathbf{w} \propto -\frac{\partial \mathcal{J}(\mathbf{w})}{\partial \mathbf{w}}$$

The symbol $\propto$ is read "is proportional to"; it is then understood that the vector on the left-hand side, $\Delta\mathbf{w}$, has the same *direction* as the gradient vector on the right-hand side, but there is a positive scalar coefficient by which the length can be adjusted. In the upper version of the update rule, this coefficient is denoted by α. In many cases, this learning rate can and should in fact be time dependent. Yet a third very convenient way to write such update rules, in conformity with programming languages, is

$$\mathbf{w} \leftarrow \mathbf{w} - \alpha\frac{\partial \mathcal{J}(\mathbf{w})}{\partial \mathbf{w}}$$

where the symbol ← means substitution, i.e., the value of the right-hand side is computed and substituted in **w**.

Geometrically, a gradient descent step as in (3.22) means going downhill. The graph of $\mathcal{J}(\mathbf{w})$ is the multidimensional equivalent of mountain terrain, and we are always moving downwards in the steepest direction. This also immediately shows the disadvantage of steepest descent: unless the function $\mathcal{J}(\mathbf{w})$ is very simple and smooth, steepest descent will lead to the closest local minimum instead of a global minimum. As such, the method offers no way to escape from a local minimum. Nonquadratic cost functions may have many local maxima and minima. Therefore, good initial values are important in initializing the algorithm.

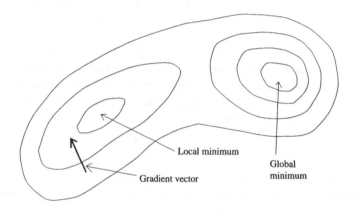

Fig. 3.1   Contour plot of a cost function with a local minimum.

As an example, consider the case of Fig. 3.1. A function $\mathcal{J}(\mathbf{w})$ is shown there as a contour plot. In the region shown in the figure, there is one local minimum and one global minimum. From the initial point chosen there, where the gradient vector has been plotted, it is very likely that the algorithm will converge to the local minimum.

Generally, the speed of convergence can be quite low close to the minimum point, because the gradient approaches zero there. The speed can be analyzed as follows. Let us denote by $\mathbf{w}^*$ the local or global minimum point where the algorithm will eventually converge. From (3.22) we have

$$\mathbf{w}(t) - \mathbf{w}^* = \mathbf{w}(t-1) - \mathbf{w}^* - \alpha(t)\frac{\partial \mathcal{J}(\mathbf{w})}{\partial \mathbf{w}}\Big|_{\mathbf{w}=\mathbf{w}(t-1)} \qquad (3.24)$$

Let us expand the gradient vector $\frac{\partial \mathcal{J}(\mathbf{w})}{\partial \mathbf{w}}$ element by element as a Taylor series around the point $\mathbf{w}^*$, as explained in Section 3.1.4. Using only the zeroth- and first-order terms, we have for the ith element

$$\frac{\partial \mathcal{J}(\mathbf{w})}{\partial w_i}\Big|_{\mathbf{w}=\mathbf{w}(t-1)} = \frac{\partial \mathcal{J}(\mathbf{w})}{\partial w_i}\Big|_{\mathbf{w}=\mathbf{w}^*} + \sum_{j=1}^{m}\frac{\partial^2 \mathcal{J}(\mathbf{w})}{\partial w_i w_j}\Big|_{\mathbf{w}=\mathbf{w}^*}[w_j(t-1) - w_j^*] + \dots$$

Now, because $\mathbf{w}^*$ is the point of convergence, the partial derivatives of the cost function must be zero at $\mathbf{w}^*$. Using this result, and compiling the above expansion into vector form, yields

$$\frac{\partial \mathcal{J}(\mathbf{w})}{\partial \mathbf{w}}\bigg|_{\mathbf{w}=\mathbf{w}(t-1)} = \mathbf{H}(\mathbf{w}^*)[\mathbf{w}(t-1) - \mathbf{w}^*] + \ldots$$

where $\mathbf{H}(\mathbf{w}^*)$ is the Hessian matrix computed at the point $\mathbf{w} = \mathbf{w}^*$. Substituting this in (3.24) gives

$$\mathbf{w}(t) - \mathbf{w}^* \approx [\mathbf{I} - \alpha(t)\mathbf{H}(\mathbf{w}^*)][\mathbf{w}(t-1) - \mathbf{w}^*]$$

This kind of convergence, which is essentially equivalent to multiplying a matrix many times with itself, is called *linear*. The speed of convergence depends on the learning rate and the size of the Hessian matrix. If the cost function $\mathcal{J}(\mathbf{w})$ is very flat at the minimum, with second partial derivatives also small, then the Hessian is small and the convergence is slow (for fixed $\alpha(t)$). Usually, we cannot influence the shape of the cost function, and we have to choose $\alpha(t)$, given a fixed cost function.

The choice of an appropriate step length or learning rate $\alpha(t)$ is thus essential: too small a value will lead to slow convergence. The value cannot be too large either: too large a value will lead to overshooting and instability, which prevents convergence altogether. In Fig. 3.1, too large a learning rate will cause the solution point to zigzag around the local minimum. The problem is that we do not know the Hessian matrix and therefore determining a good value for the learning rate is difficult.

A simple extension to the basic gradient descent, popular in neural network learning rules like the back-propagation algorithm, is to use a two-step iteration instead of just one step like in (3.22), leading to the so-called momentum method. Neural network literature has produced a large number of tricks for boosting steepest descent learning by adjustable learning rates, clever choice of the initial value, etc. However, in ICA, many of the most popular algorithms are still straightforward gradient descent methods, in which the gradient of an appropriate contrast function is computed and used as such in the algorithm.

3.2.2 Second-order learning

In numerical analysis, a large number of methods that are more efficient than plain gradient descent have been introduced for minimizing or maximizing a multivariate scalar function. They could be immediately used for the ICA problem. Their advantage is faster convergence in terms of the number of iterations required, but the disadvantage quite often is increased computational complexity per iteration. Here we consider second-order methods, which means that we also use the information contained in the second-order derivatives of the cost function. Obviously, this information relates to the curvature of the optimization terrain and should help in finding a better direction for the next step in the iteration than just plain gradient descent.

A good starting point is the multivariate Taylor series; see Section 3.1.4. Let us develop the function $J(\mathbf{w})$ in Taylor series around a point $\mathbf{w}$ as

$$J(\mathbf{w}') = J(\mathbf{w}) + [\frac{\partial J(\mathbf{w})}{\partial \mathbf{w}}]^T (\mathbf{w}' - \mathbf{w}) + \frac{1}{2}(\mathbf{w}' - \mathbf{w})^T \frac{\partial^2 J(\mathbf{w})}{\partial \mathbf{w}^2}(\mathbf{w}' - \mathbf{w}) + \dots$$
(3.25)

In trying to minimize the function $J(\mathbf{w})$, we ask what choice of the new point $\mathbf{w}'$ gives us the largest decrease in the value of $J(\mathbf{w})$. We can write $\mathbf{w}' - \mathbf{w} = \Delta \mathbf{w}$ and minimize the function $J(\mathbf{w}') - J(\mathbf{w}) = [\frac{\partial J(\mathbf{w})}{\partial \mathbf{w}}]^T \Delta \mathbf{w} + 1/2 \Delta \mathbf{w}^T \frac{\partial^2 J(\mathbf{w})}{\partial \mathbf{w}^2} \Delta \mathbf{w}$ with respect to $\Delta \mathbf{w}$. The gradient of this function with respect to $\Delta \mathbf{w}$ is (see Example 3.2) equal to $\frac{\partial J(\mathbf{w})}{\partial \mathbf{w}} + \frac{\partial^2 J(\mathbf{w})}{\partial \mathbf{w}^2} \Delta \mathbf{w}$; note that the Hessian matrix is symmetric. If the Hessian is also positive definite, then the function will have a parabolic shape and the minimum is given by the zero of the gradient. Setting the gradient to zero gives

$$\Delta \mathbf{w} = -[\frac{\partial^2 J(\mathbf{w})}{\partial \mathbf{w}^2}]^{-1} \frac{\partial J(\mathbf{w})}{\partial \mathbf{w}}$$

From this, the following second-order iteration rule emerges:

$$\mathbf{w}' = \mathbf{w} - [\frac{\partial^2 J(\mathbf{w})}{\partial \mathbf{w}^2}]^{-1} \frac{\partial J(\mathbf{w})}{\partial \mathbf{w}}$$
(3.26)

where we have to compute the gradient and Hessian on the right-hand side at the point $\mathbf{w}$.

Algorithm (3.26) is called *Newton's method*, and it is one of the most efficient ways for function minimization. It is, in fact, a special case of the well-known Newton's method for solving an equation; here it solves the equation that says that the gradient is zero.

Newton's method provides a fast convergence in the vicinity of the minimum, if the Hessian matrix is positive definite there, but the method may perform poorly farther away. A complete convergence analysis is given in [284]. It is also shown there that the convergence of Newton's method is *quadratic*; if $\mathbf{w}^*$ is the limit of convergence, then

$$\|\mathbf{w}' - \mathbf{w}^*\| \leq \gamma \|\mathbf{w} - \mathbf{w}^*\|^2$$

where γ is a constant. This is a very strong mode of convergence. When the error on the right-hand side is relatively small, its square can be orders of magnitude smaller. (If the exponent is 3, the convergence is called cubic, which is somewhat better than quadratic, although the difference is not as large as the difference between linear and quadratic convergence.)

On the other hand, Newton's method is computationally much more demanding per one iteration than the steepest descent method. The inverse of the Hessian has to be computed at each step, which is prohibitively heavy for many practical cost functions in high dimensions. It may also happen that the Hessian matrix becomes ill-conditioned or close to a singular matrix at some step of the algorithm, which induces numerical errors into the iteration. One possible remedy for this is to regularize the algorithm by adding a diagonal matrix $\delta \mathbf{I}$, with δ small, to the Hessian before inversion. This makes it better conditioned. This is the basis of the *Marquardt-Levenberg algorithm* (see, e.g., [83]).

For error functions that can be expressed as the sum of error squares, one can apply the so-called *Gauss-Newton method* instead of Newton's method. It is intermediate between the steepest descent and Newton's method with respect to both the computational load and convergence speed. Also the *conjugate gradient method* provides a similar compromise [46, 135, 284, 172, 407].

In ICA, these second-order methods in themselves are not often used, but the FastICA algorithm uses an approximation of the Newton method that is tailored to the ICA problem, and provides fast convergence with little computation per iteration.

3.2.3 The natural gradient and relative gradient

The gradient of a function $\mathcal{J}$ points in the steepest direction in the Euclidean orthogonal coordinate system. However, the parameter space is not always Euclidean but has a Riemannian metric structure, as pointed out by Amari [4]. In such a case, the steepest direction is given by the so-called *natural gradient* instead. Let us here only consider the case of nonsingular $m \times m$ matrices, that are important in ICA learning rules. It turns out that their space has a Riemannian structure with a conveniently computable natural gradient.

Assume that we are at the point $\mathbf{W}$ and wish to find a direction for a small matrix increment $\delta \mathbf{W}$ such that the value $\mathcal{J}(\mathbf{W} + \delta \mathbf{W})$ is minimized, under the constraint that the squared norm $\|\delta \mathbf{W}\|^2$ is constant. This is a very natural requirement, as any step in a gradient algorithm for minimization of function g must consist of the direction of the step and the length. Keeping the length constant, we search for the optimal direction.

The squared norm is defined as a weighted matrix inner product

$$\|\delta \mathbf{W}\|^2 = < \delta \mathbf{W}, \delta \mathbf{W} >_{\mathbf{W}}$$

such that

$$< \delta \mathbf{W}, \delta \mathbf{W} >_{\mathbf{I}} = \sum_{i,j=1}^{m} (\delta w_{ij})^2 = \text{trace}(\delta \mathbf{W}^T \delta \mathbf{W})$$

Amari argues that, due to the Riemannian structure of the matrix space, this inner product should have the following invariance:

$$< \delta \mathbf{W}, \delta \mathbf{W} >_{\mathbf{W}} = < \delta \mathbf{W} \mathbf{M}, \delta \mathbf{W} \mathbf{M} >_{\mathbf{W} \mathbf{M}}$$

for any matrix $\mathbf{M}$. Putting $\mathbf{M} = \mathbf{W}^{-1}$, this gives

$$< \delta \mathbf{W}, \delta \mathbf{W} >_{\mathbf{W}} = < \delta \mathbf{W} \mathbf{W}^{-1}, \delta \mathbf{W} \mathbf{W}^{-1} >_{\mathbf{I}} = \text{trace}((\mathbf{W}^T)^{-1} \delta \mathbf{W}^T \delta \mathbf{W} \mathbf{W}^{-1}).$$

Keeping this inner product constant, it was shown in [4] that the largest increment for $\mathcal{J}(\mathbf{W} + \delta \mathbf{W})$ is obtained in the direction of the natural gradient

$$\frac{\partial \mathcal{J}}{\partial \mathbf{W}_{nat}} = \frac{\partial \mathcal{J}}{\partial \mathbf{W}} \mathbf{W}^T \mathbf{W}$$

So, the usual gradient at point $\mathbf{W}$ must be multiplied from the right by the matrix $\mathbf{W}^T\mathbf{W}$. This results in the following natural gradient descent rule for the cost function $\mathcal{J}(\mathbf{W})$:

$$\Delta\mathbf{W} \propto -\frac{\partial\mathcal{J}}{\partial\mathbf{W}}\mathbf{W}^T\mathbf{W} \tag{3.27}$$

This kind of ICA learning rules will be discussed later in this book.

A related result was derived by Cardoso [71] from a slightly different starting-point. Let us write the Taylor series of $\mathcal{J}(\mathbf{W} + \delta\mathbf{W})$:

$$\mathcal{J}(\mathbf{W} + \delta\mathbf{W}) = \mathcal{J}(\mathbf{W}) + \text{trace}[(\frac{\partial\mathcal{J}}{\partial\mathbf{W}})^T\delta\mathbf{W}] + \ldots$$

Let us require that the displacement $\delta\mathbf{W}$ is always proportional to $\mathbf{W}$ itself, $\delta\mathbf{W} = \mathbf{D}\mathbf{W}$. We have

$$
\begin{aligned}
\mathcal{J}(\mathbf{W} + \mathbf{D}\mathbf{W}) &= \mathcal{J}(\mathbf{W}) + \text{trace}[(\frac{\partial\mathcal{J}}{\partial\mathbf{W}})^T\mathbf{D}\mathbf{W}] + \ldots \\
&= \mathcal{J}(\mathbf{W}) + \text{trace}[\mathbf{D}\mathbf{W}(\frac{\partial\mathcal{J}}{\partial\mathbf{W}})^T] + \ldots
\end{aligned}
$$

because by definition of the trace, $\text{trace}(\mathbf{M}_1\mathbf{M}_2) = \text{trace}(\mathbf{M}_2\mathbf{M}_1)$ for any matrices $\mathbf{M}_1, \mathbf{M}_2$ of a suitable size. This becomes

$$\text{trace}[\mathbf{D}(\frac{\partial\mathcal{J}}{\partial\mathbf{W}}\mathbf{W}^T)^T] = \text{trace}[(\frac{\partial\mathcal{J}}{\partial\mathbf{W}}\mathbf{W}^T)^T\mathbf{D}]$$

The multiplier for $\mathbf{D}$, or matrix $\frac{\partial\mathcal{J}}{\partial\mathbf{W}}\mathbf{W}^T$, is called the *relative gradient* by Cardoso. It is the usual matrix gradient multiplied by $\mathbf{W}^T$.

The largest decrement in the value of $\mathcal{J}(\mathbf{W} + \mathbf{D}\mathbf{W}) - \mathcal{J}(\mathbf{W})$ is now obviously obtained when the term $\text{trace}[(\frac{\partial\mathcal{J}}{\partial\mathbf{W}}\mathbf{W}^T)^T\mathbf{D}]$ is minimized, which happens when D is proportional to $-(\frac{\partial\mathcal{J}}{\partial\mathbf{W}}\mathbf{W}^T)$. Because $\delta\mathbf{W} = \mathbf{D}\mathbf{W}$ we have a gradient descent learning rule

$$\Delta\mathbf{W} \propto -(\frac{\partial\mathcal{J}}{\partial\mathbf{W}}\mathbf{W}^T)\mathbf{W}$$

which is exactly the same as the natural gradient learning rule (3.27).

3.2.4 Stochastic gradient descent

Up to now, gradient methods were considered from a general point of view, without assuming any specific form for the cost function $\mathcal{J}(\mathbf{W})$ or $\mathcal{J}(\mathbf{w})$. In this section, cost functions of a specific kind will be considered.

ICA, like many other statistical and neural network techniques, is a totally data-dependent or data-driven technique. If we do not have any observation data, then we cannot solve the problem at all. Therefore the ICA cost functions will depend on the observation data. Typically, the cost functions have the form

$$\mathcal{J}(\mathbf{w}) = \text{E}\{g(\mathbf{w}, \mathbf{x})\} \tag{3.28}$$

There $\mathbf{x}$ is the observation vector, that is modeled as a random vector with some unknown density $f(\mathbf{x})$. The expectation in (3.28) is with respect to this density. In practice, there must be a sample $\mathbf{x}(1), \mathbf{x}(2), ...$, of these vectors available. Usually this is the only data we have.

Especially in constrained optimization problems, the cost functions will be somewhat more complicated, but the core factor has the form of (3.28) and therefore we consider this simplest case first. The steepest descent learning rule becomes

$$\mathbf{w}(t) = \mathbf{w}(t-1) - \alpha(t) \frac{\partial}{\partial \mathbf{w}} \mathrm{E}\{g(\mathbf{w}, \mathbf{x}(t))\}|_{\mathbf{w}=\mathbf{w}(t-1)} \qquad (3.29)$$

Again, the gradient vector is computed at the point $\mathbf{w}(t-1)$. In practice, the expectation is approximated by the sample mean of this function over the sample $\mathbf{x}(1), ..., \mathbf{x}(T)$. This kind of algorithm, where the entire training set is used at every step of the iteration to form the expectation, is called *batch learning*.

In principle, it is easy to form the gradient and Hessian of (3.28) because first and second derivatives with respect to the elements of vector $\mathbf{w}$ can be taken inside the expectation with respect to $\mathbf{x}$; for instance,

$$\frac{\partial}{\partial \mathbf{w}} \mathrm{E}\{g(\mathbf{w}, \mathbf{x}\} \quad = \quad \frac{\partial}{\partial \mathbf{w}} \int g(\mathbf{w}, \boldsymbol{\xi}) f(\boldsymbol{\xi}) d\boldsymbol{\xi} \qquad (3.30)$$

$$= \quad \int [\frac{\partial}{\partial \mathbf{w}} g(\mathbf{w}, \boldsymbol{\xi})] f(\boldsymbol{\xi}) d\boldsymbol{\xi} \qquad (3.31)$$

It suffices that function $g(\mathbf{w}, \mathbf{x})$ is twice differentiable with respect to the elements of $\mathbf{w}$ for this operation to be allowed.

However, it may sometimes be tedious to compute the mean values or sample averages of the appropriate functions at each iteration step. This is especially problematic if the sample $\mathbf{x}(1), \mathbf{x}(2), ...$, is not fixed but new observations keep on coming in the course of the iteration. The statistics of the sample vectors may also be slowly varying, and the algorithm should be able to track this. In a learning paradigm called *on-line learning*, the whole sample is not used in batch at each step of the algorithm, but only the latest observation vector $\mathbf{x}(t)$ is used. Effectively, this means that the expectation in the learning rule (3.29) is dropped and the on-line learning rule becomes

$$\mathbf{w}(t) = \mathbf{w}(t-1) - \alpha(t) \frac{\partial}{\partial \mathbf{w}} g(\mathbf{w}, \mathbf{x})|_{\mathbf{w}=\mathbf{w}(t-1)}. \qquad (3.32)$$

This leads to highly fluctuating directions of instantaneous gradients on subsequent iteration steps, but the average direction in which the algorithm proceeds is still roughly the direction of the steepest descent of the batch cost function. Generally, stochastic gradient algorithms converge much slower than the respective steepest descent algorithms. This is compensated by their often very low computational cost. The computational load at one step of the iteration is considerably reduced; the value of the function $\frac{\partial}{\partial \mathbf{w}} g(\mathbf{w}, \mathbf{x})$ has to be computed only once, for the vector $\mathbf{x}(t)$. In the batch algorithm, when evaluating the function $\frac{\partial}{\partial \mathbf{w}} \mathrm{E}\{g(\mathbf{w}, \mathbf{x})\}$, this value must be computed T times, once for each sample vector $\mathbf{x}(t)$, then summed up and divided by T.

The trade-off is that the on-line algorithm typically needs many more steps for convergence. If the training sample is fixed, it must be used several times in the algorithm. Typically the sample vectors $\mathbf{x}(t)$ are either picked one by one in a cyclical order, or by random choice. The random ordering or shuffling is usually a better choice. By running the algorithm over sufficiently many iterations over the training set, a reasonable final accuracy can be achieved.

Examples of such stochastic gradient algorithms will be encountered later in this book. Several learning algorithms of principal component analysis networks and the well-known least-mean-squares algorithm, for example, are instantaneous stochastic gradient algorithms.

Example 3.5 We assume that $\mathbf{x}$ satisfies the ICA model $\mathbf{x} = \mathbf{A}\mathbf{s}$, with the elements of the source vector s statistically independent and $\mathbf{A}$ the mixing matrix. The problem is to solve s and $\mathbf{A}$, knowing $\mathbf{x}$ (or in practice, a sample from $\mathbf{x}$). Due to the linearity of this model, it is reasonable to look for the solution s as a linear function of $\mathbf{x}$. One way of doing this is to take a scalar variable $y = \mathbf{w}^T \mathbf{x}$ and try to solve the parameter or weight vector $\mathbf{w}$ so that y will become equal to one of the elements of s. One of the possible criteria for ICA, as we will see in Chapter 8, is to find $y = \mathbf{w}^T \mathbf{x}$ such that it has maximal fourth order moment: $\mathrm{E}\{y^4\} = \mathrm{E}\{(\mathbf{w}^T \mathbf{x})^4\} = \max$. The computational setting is such that we have a sample $\mathbf{x}(1), \mathbf{x}(2), ..., \mathbf{x}(T)$ of vectors $\mathbf{x}$, and we should solve the vector $\mathbf{w}$ that maximizes the fourth order moment of $\mathbf{w}^T \mathbf{x}$.

Now computing the gradient of the scalar function $\mathcal{J}(\mathbf{w}) = \mathrm{E}\{(\mathbf{w}^T \mathbf{x})^4\}$ gives $\frac{\partial \mathcal{J}(\mathbf{w})}{\partial \mathbf{w}} = 4\mathrm{E}\{(\mathbf{w}^T \mathbf{x})^3 \mathbf{x}\}$. Thus, a simple batch learning rule for the weight vector $\mathbf{w}$ maximizing the cost function would be

$$\mathbf{w}(t) = \mathbf{w}(t-1) + \alpha(t)\mathrm{E}\{[\mathbf{w}(t-1)^T \mathbf{x}(t)]^3 \mathbf{x}(t)\}$$

or, in our shorthand notation introduced earlier,

$$\Delta \mathbf{w} \propto \mathrm{E}\{(\mathbf{w}^T \mathbf{x})^3 \mathbf{x}\} \tag{3.33}$$

Note that the number 4 appearing in the gradient has been absorbed in the learning rate $\alpha(t)$, whose magnitude has to be determined anyway. In practice, the expectation at each step would be computed as the sample average over the sample $\mathbf{x}(1), \mathbf{x}(2), ..., \mathbf{x}(T)$ as $\mathrm{E}\{(\mathbf{w}^T \mathbf{x})^3 \mathbf{x}\} \approx 1/T \sum_{t=1}^{T} [\mathbf{w}^T \mathbf{x}(t)]^3 \mathbf{x}(t)$, where $\mathbf{w}$ is the value of the solution vector at that iteration step.

There are some comments relating to this algorithm: first, we have a positive sign in front of the gradient term because we actually wish to maximize, not minimize, the cost function. So we are moving in the direction of the gradient, which shows the direction in which the function grows fastest. Second, this is not a very good algorithm for solving the maximization problem, because the norm of $\mathbf{w}$ has not been constrained in any way; what would happen is that the norm or magnitude of $\mathbf{w}$ would grow without bounds because it has the effect of increasing the value of the cost function. A simple normalization will solve this problem, as we will see in Section 3.3.

3.2.5 Convergence of stochastic on-line algorithms *

A valid question is what exactly is the relation of the on-line learning algorithm (3.32) to the corresponding batch algorithm (3.29): Does the on-line algorithm converge to the same solution in theory? Mathematically, the two algorithms are quite different. The batch algorithm is a deterministic iteration because the random vector $\mathbf{x}$ is averaged out on the right-hand side. It can thus be analyzed with all the techniques available for one-step iteration rules, like fixed points and contractive mappings. In contrast, the on-line algorithm is a stochastic difference equation because the right-hand side is a random vector, due to $\mathbf{x}(t)$. Even the question of the convergence of the algorithm is not straightforward, because the randomness causes fluctuations that never die out unless they are deliberately frozen by letting the learning rate go to zero.

The analysis of stochastic algorithms like (3.32) is the subject of *stochastic approximation*; see, e.g., [253]. In brief, the analysis is based on the *averaged differential equation* that is obtained from (3.32) by taking averages over $\mathbf{x}$ on the right-hand side: the differential equation corresponding to (3.32) is

$$\frac{d\mathbf{w}}{dt} = -\frac{\partial}{\partial \mathbf{w}} \mathrm{E}\{g(\mathbf{w}, \mathbf{x})\} \tag{3.34}$$

This is in effect the continuous-time counterpart of the batch algorithm (3.29). Note that the right-hand side is a function of $\mathbf{w}$ only; there is no dependence on $\mathbf{x}$ or t (although, through expectation, it of course depends on the probability density of $\mathbf{x}$). Such differential equations are called *autonomous*, and they are the simplest ones to analyze.

The theory of autonomous differential equations is very well understood. The only possible points where the solution can converge are the fixed or stationary points, i.e., roots of the right-hand side, because these are the points where the change in $\mathbf{w}$ over time becomes zero. It is also well-known how by linearizing the right-hand side with respect to $\mathbf{w}$ a stability analysis of these fixed points can be accomplished. Especially important are the so-called *asymptotically stable fixed points* that are local points of attraction.

Now, if the learning rate $\alpha(t)$ is a suitably decreasing sequence, typically satisfying

$$\sum_{t=1}^{\infty} \alpha(t) = \infty \tag{3.35}$$

$$\sum_{t=1}^{\infty} \alpha^2(t) < \infty \tag{3.36}$$

and the nonlinearity $g(\mathbf{w}, \mathbf{x})$ satisfies some technical assumptions [253], then it can be shown that the on-line algorithm (3.32) must converge to one of the asymptotically stable fixed points of the differential equation (3.34). These are also the convergence points of the batch algorithm, so usually the two algorithms indeed result in the same final solution. In practice, it is often possible to analyze the fixed points even if the full convergence proof is intractable.

The conditions (3.35) are theoretical and do not necessarily work very well in practice. Sometimes the learning rate is decreased at first, but then kept at a small constant value. A good choice satisfying the conditions may be

$$\alpha(t) = \frac{\beta}{\beta + t}$$

with β an appropriate constant (e.g., $\beta = 100$). This prevents the learning rate from decreasing too quickly in the early phase of the iteration.

In many cases, however, on-line learning is used to provide a fast adaptation to a changing environment. The learning rate is then kept constant. If the input data is nonstationary, that is, its statistical structure changes as a function of time, this allows the algorithm to track these changes, and adapt quickly to the changing environment.

Example 3.6 In Chapter 6, on-line PCA is discussed. One of the learning rules is as follows:

$$\Delta \mathbf{w} \propto \mathbf{x} y - y^2 \mathbf{w} \tag{3.37}$$

where $y = \mathbf{w}^T \mathbf{x}$ and $\mathbf{x}$ is a random vector. The question is where this on-line rule might converge. We can now analyze it by forming the averaged ordinary differential equation (ODE):

$$
\begin{aligned}
\frac{d\mathbf{w}}{dt} &= \mathrm{E}\{\mathbf{x}(\mathbf{x}^T \mathbf{w}) - (\mathbf{w}^T \mathbf{x})(\mathbf{x}^T \mathbf{w})\mathbf{w}\} \\
&= \mathrm{E}\{\mathbf{x}\mathbf{x}^T\}\mathbf{w} - (\mathbf{w}^T \mathrm{E}\{\mathbf{x}\mathbf{x}^T\}\mathbf{w})\mathbf{w} \\
&= \mathbf{C_x}\mathbf{w} - (\mathbf{w}^T \mathbf{C_x}\mathbf{w})\mathbf{w}
\end{aligned}
$$

where $\mathbf{C_x} = \mathrm{E}\{\mathbf{x}\mathbf{x}^T\}$ is the covariance matrix of $\mathbf{x}$ (assumed zero-mean here). Note that the average is taken over $\mathbf{x}$, assuming $\mathbf{w}$ constant. The fixed points of the ODE are given by solutions of $\mathbf{C_x}\mathbf{w} - (\mathbf{w}^T \mathbf{C_x}\mathbf{w})\mathbf{w} = 0$. As the term $\mathbf{w}^T \mathbf{C_x}\mathbf{w}$ is a scalar, we know from elementary matrix algebra that all solutions $\mathbf{w}$ must be *eigenvectors* of the matrix $\mathbf{C_x}$. The numbers $\mathbf{w}^T \mathbf{C_x}\mathbf{w}$ are the corresponding eigenvalues. The principal components of a random vector $\mathbf{x}$ are defined in terms of the eigenvectors, as discussed in Chapter 6. With a somewhat deeper analysis, it can be shown [324] that the only asymptotically stable fixed point is the eigenvector corresponding to the largest eigenvalue, which gives the first principal component.

The example shows how an intractable stochastic on-line rule can be nicely analyzed by the powerful analysis tools existing for ODEs.

3.3 LEARNING RULES FOR CONSTRAINED OPTIMIZATION

In many cases we have to minimize or maximize a function $J(\mathbf{w})$ under some additional conditions on the solution $\mathbf{w}$. Generally, the constrained optimization problem is formulated as

$$\min \ J(\mathbf{w}), \text{ subject to } H_i(\mathbf{w}) = 0, \quad i = 1, ..., k \tag{3.38}$$

where $J(\mathbf{w})$, as before, is the cost function to be minimized and $H_i(\mathbf{w}) = 0$, $i = 1, ..., k$ give a set of k constraint equations on $\mathbf{w}$.

3.3.1 The Lagrange method

The most prominent and widely used way to take the constraints into account is the method of *Lagrange multipliers*. We form the *Lagrangian function*

$$\mathcal{L}(\mathbf{w}, \lambda_1, ..., \lambda_k) = J(\mathbf{w}) + \sum_{i=1}^{k} \lambda_i H_i(\mathbf{w}) \tag{3.39}$$

where $\lambda_1, ..., \lambda_k$ are called Lagrange multipliers. Their number k is the same as the number of separate scalar constraint equations.

The general result is that the minimum point of the Lagrangian (3.39), where its gradient is zero with respect to *both* $\mathbf{w}$ *and all the* λ_i gives the solution to the original constrained optimization problem (3.38). The gradient of $\mathcal{L}(\mathbf{w}, \lambda_1, ..., \lambda_k)$ with respect to λ_i is simply the ith constraint function $H_i(\mathbf{w})$, so putting all these to zero again gives the original constraint equations. The important point is that when we form the gradient of $\mathcal{L}(\mathbf{w}, \lambda_1, ..., \lambda_k)$ with respect to $\mathbf{w}$ and put it to zero,

$$\frac{\partial J(\mathbf{w})}{\partial \mathbf{w}} + \sum_{i=1}^{k} \lambda_i \frac{\partial H_i(\mathbf{w})}{\partial \mathbf{w}} = 0 \tag{3.40}$$

we have changed the minimization problem into a set of equations that is much easier to solve.

A possible way to solve the two sets of equations, one set given by the constraints, the other by (3.40), is, e.g., Newton iteration or some other appropriate iteration method. These methods give learning rules that resemble the ones in the previous section, but now instead of the gradient of $J(\mathbf{w})$ only, the gradient of $\mathcal{L}(\mathbf{w}, \lambda_1, ..., \lambda_k)$ will be used.

3.3.2 Projection methods

In most of the constrained optimization problems of this book, the constraints are of the equality type and relatively simple: typically, we require that the norm of $\mathbf{w}$ is constant, or some quadratic form of $\mathbf{w}$ is constant. We can then use another constrained optimization scheme: *projections on the constraint set*. This means that we solve the minimization problem with an unconstrained learning rule, which might be a simple steepest descent, Newton's iteration, or whatever is most suitable, but after each iteration step, the solution $\mathbf{w}$ at that time is projected orthogonally onto the constraint set so that it satisfies the constraints.

Example 3.7 We continue Example 3.5 here. Let us consider the following constrained problem: Assuming $\mathbf{x}$ is a random vector, maximize the fourth-order moment $E\{(\mathbf{w}^T\mathbf{x})^4\}$ under the constraint $\|\mathbf{w}\|^2 = 1$. In the terminology of (3.38), the cost function is $\mathcal{J}(\mathbf{w}) = -E\{(\mathbf{w}^T\mathbf{x})^4\}$ (note that instead of minimization, we maximize, hence the minus sign), and the only constraint equation is $H(\mathbf{w}) = \|\mathbf{w}\|^2 - 1 = 0$.

Solving this by the Lagrange method, we formulate the Lagrangian as

$$L(\mathbf{w}, \lambda) = -E\{(\mathbf{w}^T\mathbf{x})^4\} + \lambda(\|\mathbf{w}\|^2 - 1)$$

The gradient with respect to λ again gives the constraint $\|\mathbf{w}\|^2 - 1 = 0$. The gradient with respect to $\mathbf{w}$ gives $-4E\{(\mathbf{w}^T\mathbf{x})^3\mathbf{x}\} + \lambda(2\mathbf{w})$. We might try to solve the roots of this, e.g., by Newton iteration, or by a simpler iteration of the form

$$\Delta\mathbf{w} \propto E\{(\mathbf{w}^T\mathbf{x})^3\mathbf{x}\} - \frac{\lambda}{2}\mathbf{w}$$

with an appropriate learning rate and λ. When comparing this to the learning rule in Example 3.5, Eq. (3.33), we notice the additive linear term $\frac{\lambda}{2}\mathbf{w}$; by choosing λ suitably, the growth in the norm of $\mathbf{w}$ can be controlled.

However, for this simple constraint, a much simpler way is the projection on the constraint set. Let us consider simple steepest descent as in eq. (3.33). We only have to normalize $\mathbf{w}$ after each step, which is equivalent to orthogonal projection of $\mathbf{w}$ onto the unit sphere in the m-dimensional space. This unit sphere is the constraint set. The learning rule becomes

$$\mathbf{w} \leftarrow \mathbf{w} + \alpha E\{(\mathbf{w}^T\mathbf{x})^3\mathbf{x}\} \tag{3.41}$$

$$\mathbf{w} \leftarrow \mathbf{w}/\|\mathbf{w}\| \tag{3.42}$$

Exactly the same idea as in the preceding example applies to any cost function. This will be utilized heavily in the ICA learning rules in the following chapters.

Sometimes a computationally easier learning rule can be obtained from an approximation of the normalization. Consider steepest descent with the norm constraint, and for simplicity let us write the update rule as

$$\mathbf{w} \leftarrow \mathbf{w} - \alpha\mathbf{g}(\mathbf{w}) \tag{3.43}$$

$$\mathbf{w} \leftarrow \mathbf{w}/\|\mathbf{w}\| \tag{3.44}$$

where we have denoted the gradient of the cost function by $\mathbf{g}(\mathbf{w})$. Another way to write this is

$$\mathbf{w} \leftarrow \frac{\mathbf{w} - \alpha\mathbf{g}(\mathbf{w})}{\|\mathbf{w} - \alpha\mathbf{g}(\mathbf{w})\|}$$

Now, assuming the learning rate α is small as is usually the case, at least in the later iteration steps, we can expand this into a Taylor series with respect to α and get a simplified constrained learning rule [323]. Omitting some intermediate steps, the denominator becomes

$$\|\mathbf{w} - \alpha\mathbf{g}(\mathbf{w})\| \approx 1 - \alpha\mathbf{w}^T\mathbf{g}(\mathbf{w})$$

where all terms proportional to α^2 or higher powers are omitted. The final result is

$$\frac{\mathbf{w} - \alpha\mathbf{g}(\mathbf{w})}{\|\mathbf{w} - \alpha\mathbf{g}(\mathbf{w})\|} \approx \mathbf{w} - \alpha\mathbf{g}(\mathbf{w}) + \alpha(\mathbf{g}(\mathbf{w})^T\mathbf{w})\mathbf{w}$$

The resulting learning rule has one extra term compared to the unconstrained rule (3.43), and yet the norm of $\mathbf{w}$ will stay approximately equal to one.

3.4   CONCLUDING REMARKS AND REFERENCES

More information on minimization algorithms in general can be found in books dealing with nonlinear optimization, for example, [46, 135, 284], and their applications [172, 407]. The speed of convergence of the algorithms is discussed in [284, 407]. A good source for matrix gradients in general is [109]. The natural gradient is considered in detail in [118]. The momentum method and other extensions are covered in [172]. Constrained optimization has been extensively discussed in [284]. Projection on the unit sphere and the short-cut approximation for normalization has been discussed in [323, 324]. A rigorous analysis of the convergence of the stochastic on-line algorithms is discussed in [253].

Problems

3.1   Show that the Jacobian matrix of the gradient vector $\frac{\partial g}{\partial \mathbf{w}}$ with respect to $\mathbf{w}$ is equal to the Hessian of g.

3.2   The trace of an $m \times m$ square matrix $\mathbf{W}$ is defined as the sum of its diagonal elements $\sum_{i=1}^{m} w_{ii}$. Compute its matrix gradient.

3.3   Show that the gradient of trace($\mathbf{W}^T\mathbf{M}\mathbf{W}$) with respect to $\mathbf{W}$, where $\mathbf{W}$ is an $m \times n$ matrix and $\mathbf{M}$ is an $m \times m$ matrix, is equal to $\mathbf{M}\mathbf{W} + \mathbf{M}^T\mathbf{W}$.

3.4   Show that $\frac{\partial}{\partial \mathbf{W}} \log|\det \mathbf{W}| = (\mathbf{W}^T)^{-1}$.

3.5   Consider the 2×2 matrix

$$\mathbf{W} = \begin{pmatrix} a & b \\ c & d \end{pmatrix}$$

3.5.1.   Compute the cofactors with respect to the first column, compute the determinant, the adjoint matrix, and the inverse of $\mathbf{W}$ as functions of a, b, c, d.

3.5.2.   Verify in this special case that $\frac{\partial}{\partial \mathbf{W}} \log|\det \mathbf{W}| = (\mathbf{W}^T)^{-1}$.

3.6   Consider a cost function $\mathcal{J}(\mathbf{w}) = G(\mathbf{w}^T\mathbf{x})$ where we can assume that $\mathbf{x}$ is a constant vector. Assume that the scalar function G is twice differentiable.

3.6.1. Compute the gradient and Hessian of $\mathcal{J}(\mathbf{w})$ in the general case and in the cases that $G(t) = t^4$ and $G(t) = \log\cosh(t)$.

3.6.2. Consider maximizing this function under the constraint that $\|\mathbf{w}\| = 1$. Formulate the Lagrangian, its gradient (with respect to $\mathbf{w}$), its Hessian, and the Newton method for maximizing the Lagrangian.

3.7   Let $p(.)$ be a differentiable scalar function, $\mathbf{x}$ a constant vector, and $\mathbf{W} = (\mathbf{w}_1...\mathbf{w}_n)$ an $m \times n$ matrix with columns $\mathbf{w}_i$. Consider the cost function

$$\mathcal{J}(\mathbf{W}) = \sum_{i=1}^{n} \log p(u_i)$$

where $u_i = \mathbf{x}^T\mathbf{w}_i$. Show that $\frac{\partial}{\partial\mathbf{W}}\mathcal{J}(\mathbf{W}) = -\varphi(\mathbf{u})\mathbf{x}^T$ where $\mathbf{u}$ is the vector with elements u_i and $\varphi(\mathbf{u})$ is a certain function, defined element by element. Give the form of this function. (*Note:* This matrix gradient is used in the maximum likelihood approach to ICA, discussed in Chapter 9.)

3.8   Consider a general stochastic on-line ICA learning rule

$$\Delta\mathbf{W} \propto [I - g(\mathbf{y})\mathbf{y}^T]\mathbf{W}$$

where $\mathbf{y} = \mathbf{W}\mathbf{x}$ and g is a nonlinear function. Formulate
   (a) the corresponding batch learning rule,
   (b) the averaged differential equation.
   Consider a stationary point of (a) and (b). Show that if $\mathbf{W}$ is such that the elements of $\mathbf{y}$ are zero-mean and independent, then $\mathbf{W}$ is a stationary point.

3.9   Assume that we want to maximize a function $F(\mathbf{w})$ on the unit sphere, i.e., under the constraint $\|\mathbf{w}\| = 1$. Prove that at the maximum, the gradient of F must point in the same direction as $\mathbf{w}$. In other words, the gradient must be equal to $\mathbf{w}$ multiplied by a scalar constant. Use the Lagrangian method.

Computer assignments

Create a sample of two-dimensional gaussian data $\mathbf{x}$ with zero mean and covariance matrix

$$\begin{pmatrix} 3 & 1 \\ 1 & 2 \end{pmatrix}.$$

Apply the stochastic on-line learning rule (3.37), choosing a random initial point $\mathbf{w}$ and an appropriate learning rate. Try different choices for the learning rate and see how it effects the convergence speed. Then, try to solve the same problem using a batch learning rule by taking the averages on the right-hand side. Compare the computational efforts of the on-line vs. batch learning rules at one step of the iteration, and the number of steps needed for convergence in both algorithms. (*Note:* The algorithm converges to the dominant eigenvector of the covariance matrix, which can be solved in closed form.)

4

Estimation Theory

An important issue encountered in various branches of science is how to estimate the quantities of interest from a given finite set of uncertain (noisy) measurements. This is studied in estimation theory, which we shall discuss in this chapter.

There exist many estimation techniques developed for various situations; the quantities to be estimated may be nonrandom or have some probability distributions themselves, and they may be constant or time-varying. Certain estimation methods are computationally less demanding but they are statistically suboptimal in many situations, while statistically optimal estimation methods can have a very high computational load, or they cannot be realized in many practical situations. The choice of a suitable estimation method also depends on the assumed data model, which may be either linear or nonlinear, dynamic or static, random or deterministic.

In this chapter, we concentrate mainly on linear data models, studying the estimation of their parameters. The two cases of deterministic and random parameters are covered, but the parameters are always assumed to be time-invariant. The methods that are widely used in context with independent component analysis (ICA) are emphasized in this chapter. More information on estimation theory can be found in books devoted entirely or partly to the topic, for example [299, 242, 407, 353, 419].

Prior to applying any estimation method, one must select a suitable model that well describes the data, as well as measurements containing relevant information on the quantities of interest. These important, but problem-specific issues will not be discussed in this chapter. Of course, ICA is one of the models that can be used. Some topics related to the selection and preprocessing of measurements are treated later in Chapter 13.

4.1 BASIC CONCEPTS

Assume there are T scalar measurements $x(1), x(2), \dots, x(T)$ containing information about the m quantities $\theta_1, \theta_2, \dots, \theta_m$ that we wish to estimate. The quantities θ_i are called *parameters* hereafter. They can be compactly represented as the *parameter vector*

$$\boldsymbol{\theta} = (\theta_1, \theta_2, \dots, \theta_m)^T \tag{4.1}$$

Hence, the parameter vector $\boldsymbol{\theta}$ is an m-dimensional column vector having as its elements the individual parameters. Similarly, the measurements can be represented as the T-dimensional *measurement* or *data vector*[1]

$$\mathbf{x}_T = [x(1), x(2), \dots, x(T)]^T \tag{4.2}$$

Quite generally, an *estimator* $\hat{\boldsymbol{\theta}}$ of the parameter vector $\boldsymbol{\theta}$ is the mathematical expression or function by which the parameters can be estimated from the measurements:

$$\hat{\boldsymbol{\theta}} = \mathbf{h}(\mathbf{x}_T) = \mathbf{h}(x(1), x(2), \dots, x(T)) \tag{4.3}$$

For individual parameters, this becomes

$$\hat{\theta}_i = h_i(\mathbf{x}_T), \qquad i = 1, \dots, m \tag{4.4}$$

If the parameters θ_i are of a different type, the estimation formula (4.4) can be quite different for different i. In other words, the components h_i of the vector-valued function $\mathbf{h}$ can have different functional forms. The numerical value of an estimator $\hat{\theta}_i$, obtained by inserting some specific given measurements into formula (4.4), is called the *estimate* of the parameter θ_i.

Example 4.1 Two parameters that are often needed are the mean μ and variance σ^2 of a random variable x. Given the measurement vector (4.2), they can be estimated from the well-known formulas, which will be derived later in this chapter:

$$\hat{\mu} = \frac{1}{T} \sum_{j=1}^{T} x(j) \tag{4.5}$$

$$\hat{\sigma}^2 = \frac{1}{T-1} \sum_{j=1}^{T} [x(j) - \hat{\mu}]^2 \tag{4.6}$$

---

[1]The data vector consisting of T subsequent scalar samples is denoted in this chapter by $\mathbf{x}_T$ for distinguishing it from the ICA mixture vector $\mathbf{x}$, whose components consist of different mixtures.

Example 4.2 Another example of an estimation problem is a sinusoidal signal in noise. Assume that the measurements obey the measurement (data) model

$$x(j) = A\sin(\omega t(j) + \phi) + v(j), \qquad j = 1,\dots,T \qquad (4.7)$$

Here A is the amplitude, ω the angular frequency, and ϕ the phase of the sinusoid, respectively. The measurements are made at different time instants $t(j)$, which are often equispaced. They are corrupted by additive noise $v(j)$, which is often assumed to be zero mean white gaussian noise. Depending on the situation, we may wish to estimate some of the parameters A, ω, and ϕ, or all of them. In the latter case, the parameter vector becomes $\boldsymbol{\theta} = (A, \omega, \phi)^T$. Clearly, different formulas must be used for estimating A, ω, and ϕ. The amplitude A depends linearly on the measurements $x(j)$, while the angular frequency ω and the phase ϕ depend nonlinearly on the $x(j)$. Various estimation methods for this problem are discussed, for example, in [242].

Estimation methods can be divided into two broad classes depending on whether the parameters $\boldsymbol{\theta}$ are assumed to be deterministic *constants*, or *random*. In the latter case, it is usually assumed that the parameter vector $\boldsymbol{\theta}$ has an associated probability density function (pdf) $p_{\boldsymbol{\theta}}(\boldsymbol{\theta})$. This pdf, called *a priori* density, is in principle assumed to be completely known. In practice, such exact information is seldom available. Rather, the probabilistic formalism allows incorporation of useful but often somewhat vague prior information on the parameters into the estimation procedure for improving the accuracy. This is done by assuming a suitable prior distribution reflecting knowledge about the parameters. Estimation methods using the a priori distribution $p_{\boldsymbol{\theta}}(\boldsymbol{\theta})$ are often called Bayesian ones, because they utilize the Bayes' rule discussed in Section 4.6.

Another distinction between estimators can be made depending on whether they are of *batch* type or *on-line*. In batch type estimation (also called off-line estimation), all the measurements must first be available, and the estimates are then computed directly from formula (4.3). In on-line estimation methods (also called adaptive or recursive estimation), the estimates are updated using new incoming samples. Thus the estimates are computed from the recursive formula

$$\hat{\boldsymbol{\theta}}(j+1) = \mathbf{h}_1(\hat{\boldsymbol{\theta}}(j)) + \mathbf{h}_2(x(j+1), \hat{\boldsymbol{\theta}}(j)) \qquad (4.8)$$

where $\hat{\boldsymbol{\theta}}(j)$ denotes the estimate based on j first measurements $x(1), x(2),\dots, x(j)$. The correction or update term $\mathbf{h}_2(x(j+1), \hat{\boldsymbol{\theta}}(j))$ depends only on the new incoming $(j+1)$-th sample $x(j+1)$ and the current estimate $\hat{\boldsymbol{\theta}}(j)$. For example, the estimate $\hat{\mu}$ of the mean in (4.5) can be computed on-line as follows:

$$\hat{\mu}(j) = \frac{j-1}{j}\hat{\mu}(j-1) + \frac{1}{j}x(j) \qquad (4.9)$$

4.2 PROPERTIES OF ESTIMATORS

Now briefly consider properties that a good estimator should satisfy.

Generally, assessing the quality of an estimate is based on the *estimation error*, which is defined by

$$\tilde{\theta} = \theta - \hat{\theta} = \theta - \mathbf{h}(\mathbf{x}_T) \tag{4.10}$$

Ideally, the estimation error $\tilde{\theta}$ should be zero, or at least zero with probability one. But it is impossible to meet these extremely stringent requirements for a finite data set. Therefore, one must consider less demanding criteria for the estimation error.

Unbiasedness and consistency The first requirement is that the mean value of the error $E\{\tilde{\theta}\}$ should be zero. Taking expectations of the both sides of Eq. (4.10) leads to the condition

$$E\{\hat{\theta}\} = E\{\theta\} \tag{4.11}$$

Estimators that satisfy the requirement (4.11) are called *unbiased*. The preceding definition is applicable to random parameters. For nonrandom parameters, the respective definition is

$$E\{\hat{\theta} \mid \theta\} = \theta \tag{4.12}$$

Generally, conditional probability densities and expectations, conditioned by the parameter vector θ, are used throughout in dealing with nonrandom parameters to indicate that the parameters θ are assumed to be deterministic constants. In this case, the expectations are computed over the random data only.

If an estimator does not meet the unbiasedness conditions (4.11) or (4.12). it is said to be *biased*. In particular, the *bias* $\mathbf{b}$ is defined as the mean value of the estimation error:

$$\mathbf{b} = E\{\tilde{\theta}\}, \text{ or } \mathbf{b} = E\{\tilde{\theta} \mid \theta\} \tag{4.13}$$

If the bias approaches zero as the number of measurements grows infinitely large, the estimator is called *asymptotically unbiased*.

Another reasonable requirement for a good estimator $\hat{\theta}$ is that it should converge to the true value of the parameter vector θ, at least in probability,[2] when the number of measurements grows infinitely large. Estimators satisfying this asymptotic property are called *consistent*. Consistent estimators need not be unbiased; see [407].

Example 4.3 Assume that the observations $x(1), x(2), \ldots, x(T)$ are independent. The expected value of the sample mean (4.5) is

$$E\{\hat{\mu}\} = \frac{1}{T} \sum_{j=1}^{T} E\{x(j)\} = \frac{1}{T} T \mu = \mu \tag{4.14}$$

---

[2]See for example [299, 407] for various definitions of stochastic convergence.

Thus the sample mean is an unbiased estimator of the true mean μ. It is also consistent, which can be seen by computing its variance

$$\mathrm{E}\{(\hat{\mu} - \mu)^2\} = \frac{1}{T^2} \sum_{j=1}^{T} \mathrm{E}\{[x(j) - \mu]^2\} = \frac{1}{T^2} T\sigma^2 = \frac{\sigma^2}{T} \qquad (4.15)$$

The variance approaches zero when the number of samples $T \to \infty$, implying together with unbiasedness that the sample mean (4.5) converges in probability to the true mean μ.

Mean-square error It is useful to introduce a scalar-valued *loss function* $L(\tilde{\boldsymbol{\theta}})$ for describing the relative importance of specific estimation errors $\tilde{\boldsymbol{\theta}}$. A popular loss function is the squared estimation error $L(\tilde{\boldsymbol{\theta}}) = \| \tilde{\boldsymbol{\theta}} \|^2 = \| \boldsymbol{\theta} - \hat{\boldsymbol{\theta}} \|^2$ because of its mathematical tractability. More generally, typical properties required from a valid loss function are that it is symmetric: $L(\tilde{\boldsymbol{\theta}}) = L(-\tilde{\boldsymbol{\theta}})$; convex or alternatively at least nondecreasing; and (for convenience) that the loss corresponding to zero error is zero: $L(\mathbf{0}) = 0$. The convexity property guarantees that the loss function decreases as the estimation error decreases. See [407] for details.

The estimation error $\tilde{\boldsymbol{\theta}}$ is a random vector depending on the (random) measurement vector $\mathbf{x}_T$. Hence, the value of the loss function $L(\tilde{\boldsymbol{\theta}})$ is also a random variable. To obtain a nonrandom error measure, is is useful to define the *performance index* or *error criterion* $\mathcal{E}$ as the expectation of the respective loss function. Hence,

$$\mathcal{E} = \mathrm{E}\{L(\tilde{\boldsymbol{\theta}})\} \text{ or } \mathcal{E} = \mathrm{E}\{L(\tilde{\boldsymbol{\theta}}) \mid \boldsymbol{\theta}\} \qquad (4.16)$$

where the first definition is used for random parameters $\boldsymbol{\theta}$ and the second one for deterministic ones.

A widely used error criterion is the *mean-square error* (MSE)

$$\mathcal{E}_{MSE} = \mathrm{E}\{\| \boldsymbol{\theta} - \hat{\boldsymbol{\theta}} \|^2\} \qquad (4.17)$$

If the mean-square error tends asymptotically to zero with increasing number of measurements, the respective estimator is consistent. Another important property of the mean-square error criterion is that it can be decomposed as (see (4.13))

$$\mathcal{E}_{MSE} = \mathrm{E}\{\| \tilde{\boldsymbol{\theta}} - \mathbf{b} \|^2\} + \| \mathbf{b} \|^2 \qquad (4.18)$$

The first term $\mathrm{E}\{\| \tilde{\boldsymbol{\theta}} - \mathbf{b} \|^2\}$ on the right-hand side is clearly the variance of the estimation error $\tilde{\boldsymbol{\theta}}$. Thus the mean-square error $\mathcal{E}_{MSE}$ measures both the variance and the bias of an estimator $\hat{\boldsymbol{\theta}}$. If the estimator is unbiased, the mean-square error coincides with the variance of the estimator. Similar definitions hold for deterministic parameters when the expectations in (4.17) and (4.18) are replaced by conditional ones.

Figure 4.1 illustrates the bias b and standard deviation σ (square root of the variance σ^2) for an estimator $\hat{\theta}$ of a single scalar parameter θ. In a Bayesian interpretation (see Section 4.6), the bias and variance of the estimator $\hat{\theta}$ are, respectively, the mean

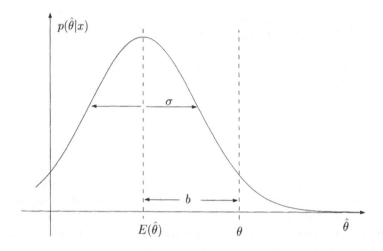

Fig. 4.1   Bias b and standard deviation σ of an estimator $\hat{\theta}$.

and variance of the *posterior* distribution $p_{\hat{\theta}|\mathbf{x}_T}(\hat{\theta} \mid \mathbf{x})$ of the estimator $\hat{\theta}$ given the observed data $\mathbf{x}_T$.

Still another useful measure of the quality of an estimator is given by the covariance matrix of the estimation error

$$\mathbf{C}_{\tilde{\theta}} = \mathrm{E}\{\tilde{\boldsymbol{\theta}}\tilde{\boldsymbol{\theta}}^T\} = \mathrm{E}\{(\boldsymbol{\theta} - \hat{\boldsymbol{\theta}})(\boldsymbol{\theta} - \hat{\boldsymbol{\theta}})^T\} \qquad (4.19)$$

It measures the errors of individual parameter estimates, while the mean-square error is an overall scalar error measure for all the parameter estimates. In fact, the mean-square error (4.17) can be obtained by summing up the diagonal elements of the error covariance matrix (4.19), or the mean-square errors of individual parameters.

Efficiency   An estimator that provides the smallest error covariance matrix among all unbiased estimators is the best one with respect to this quality criterion. Such an estimator is called an *efficient* one, because it optimally uses the information contained in the measurements. A symmetric matrix $\mathbf{A}$ is said to be smaller than another symmetric matrix $\mathbf{B}$, or $\mathbf{A} < \mathbf{B}$, if the matrix $\mathbf{B} - \mathbf{A}$ is positive definite.

A very important theoretical result in estimation theory is that there exists a lower bound for the error covariance matrix (4.19) of any estimator based on available measurements. This is provided by the Cramer-Rao lower bound. In the following theorem, we formulate the Cramer-Rao lower bound for unknown deterministic parameters.

Theorem 4.1 *[407] If $\hat{\theta}$ is any unbiased estimator of θ based on the measurement data* **x**, *then the covariance matrix of error in the estimator is bounded below by the inverse of the Fisher information matrix* **J**:

$$E\{(\theta - \hat{\theta})(\theta - \hat{\theta})^T \mid \theta\} \geq \mathbf{J}^{-1} \tag{4.20}$$

where

$$\mathbf{J} = E\left\{ \left[\frac{\partial}{\partial \theta} \ln p(\mathbf{x}_T \mid \theta)\right] \left[\frac{\partial}{\partial \theta} \ln p(\mathbf{x}_T \mid \theta)\right]^T \mid \theta \right\} \tag{4.21}$$

Here it is assumed that the inverse $\mathbf{J}^{-1}$ exists. The term $\frac{\partial}{\partial \theta} \ln p(\mathbf{x}_T \mid \theta)$ is recognized to be the gradient vector of the natural logarithm of the joint distribution[3] $p(\mathbf{x}_T \mid \theta)$ of the measurements $\mathbf{x}_T$ for nonrandom parameters θ. The partial derivatives must exist and be absolutely integrable.

It should be noted that the estimator $\hat{\theta}$ must be unbiased, otherwise the preceding theorem does not hold. The theorem cannot be applied to all distributions (for example, to the uniform one) because of the requirement of absolute integrability of the derivatives. It may also happen that there does not exist any estimator achieving the lower bound. Anyway, the Cramer-Rao lower bound can be computed for many problems, providing a useful measure for testing the efficiency of specific estimation methods designed for those problems. A more thorough discussion of the Cramer-Rao lower bound with proofs and results for various types of parameters can be found, for example, in [299, 242, 407, 419]. An example of computing the Cramer-Rao lower bound will be given in Section 4.5.

Robustness In practice, an important characteristic of an estimator is its *robustness* [163, 188]. Roughly speaking, robustness means insensitivity to gross measurement errors, and errors in the specification of parametric models. A typical problem with many estimators is that they may be quite sensitive to outliers, that is, observations that are very far from the main bulk of data. For example, consider the estimation of the mean from 100 measurements. Assume that all the measurements (but one) are distributed between -1 and 1, while one of the measurements has the value 1000. Using the simple estimator of the mean given by the sample average in (4.5), the estimator gives a value that is not far from the value 10. Thus, the single, probably erroneous, measurement of 1000 had a very strong influence on the estimator. The problem here is that the average corresponds to minimization of the squared distance of measurements from the estimate [163, 188]. The square function implies that measurements far away dominate.

Robust estimators can be obtained, for example, by considering instead of the square error other optimization criteria that grow slower than quadratically with the error. Examples of such criteria are the absolute value criterion and criteria

---

[3]We have here omitted the subscript $\mathbf{x} \mid \theta$ of the density function $p(\mathbf{x} \mid \theta)$ for notational simplicity. This practice is followed in this chapter unless confusion is possible.

that saturate as the error grows large enough [83, 163, 188]. Optimization criteria growing faster than quadratically generally have poor robustness, because a few large individual errors corresponding to the outliers in the data may almost solely determine the value of the error criterion. In the case of estimating the mean, for example, one can use the median of measurements instead of the average. This corresponds to using the absolute value in the optimization function, and gives a very robust estimator: the single outlier has no influence at all.

4.3 METHOD OF MOMENTS

One of the simplest and oldest estimation methods is the *method of moments*. It is intuitively satisfying and often leads to computationally simple estimators, but on the other hand, it has some theoretical weaknesses. We shall briefly discuss the moment method because of its close relationship to higher-order statistics.

Assume now that there are T statistically independent scalar measurements or data samples $x(1), x(2), \ldots, x(T)$ that have a common probability distribution $p(x \mid \boldsymbol{\theta})$ characterized by the parameter vector $\boldsymbol{\theta} = (\theta_1, \theta_2, \ldots, \theta_m)^T$ in (4.1). Recall from Section 2.7 that the jth moment α_j of x is defined by

$$\alpha_j = \mathrm{E}\{x^j \mid \boldsymbol{\theta}\} = \int_{-\infty}^{\infty} x^j p(x \mid \boldsymbol{\theta})dx, \qquad j = 1, 2, \ldots \tag{4.22}$$

Here the conditional expectations are used to indicate that the parameters $\boldsymbol{\theta}$ are (unknown) constants. Clearly, the moments α_j are functions of the parameters $\boldsymbol{\theta}$.

On the other hand, we can estimate the respective moments directly from the measurements. Let us denote by d_j the jth estimated moment, called the jth *sample moment*. It is obtained from the formula (see Section 2.2)

$$d_j = \frac{1}{T} \sum_{i=1}^{T} [x(i)]^j \tag{4.23}$$

The simple basic idea behind the method of moments is to equate the theoretical moments α_j with the estimated ones d_j:

$$\alpha_j(\boldsymbol{\theta}) = \alpha_j(\theta_1, \theta_2, \ldots, \theta_m) = d_j \tag{4.24}$$

Usually, m equations for the m first moments $j = 1, \ldots, m$ are sufficient for solving the m unknown parameters $\theta_1, \theta_2, \ldots, \theta_m$. If Eqs. (4.24) have an acceptable solution, the respective estimator is called the *moment estimator*, and it is denoted in the following by $\hat{\boldsymbol{\theta}}_{MM}$.

Alternatively, one can use the theoretical central moments

$$\mu_j = \mathrm{E}\{(x - \alpha_1)^j \mid \boldsymbol{\theta}\} \tag{4.25}$$

and the respective estimated *sample central moments*

$$s_j = \frac{1}{T-1} \sum_{i=1}^{T} [x(i) - d_1]^j \tag{4.26}$$

to form the m equations

$$\mu_j(\theta_1, \theta_2, \ldots, \theta_m) = s_j, \qquad j = 1, 2, \ldots, m \tag{4.27}$$

for solving the unknown parameters $\boldsymbol{\theta} = (\theta_1, \theta_2, \ldots, \theta_m)^T$.

Example 4.4 Assume now that $x(1), x(2), \ldots, x(T)$ are independent and identically distributed samples from a random variable x having the pdf

$$p(x \mid \boldsymbol{\theta}) = \frac{1}{\theta_2} \exp\left[-\frac{(x - \theta_1)}{\theta_2}\right] \tag{4.28}$$

where $\theta_1 < x < \infty$ and $\theta_2 > 0$. We wish to estimate the parameter vector $\boldsymbol{\theta} = (\theta_1, \theta_2)^T$ using the method of moments. For doing this, let us first compute the theoretical moments α_1 and α_2:

$$\alpha_1 = \mathrm{E}\{x \mid \boldsymbol{\theta}\} = \int_{\theta_1}^{\infty} \frac{x}{\theta_2} \exp\left[-\frac{(x - \theta_1)}{\theta_2}\right] dx = \theta_1 + \theta_2 \tag{4.29}$$

$$\alpha_2 = \mathrm{E}\{x^2 \mid \boldsymbol{\theta}\} = \int_{\theta_1}^{\infty} \frac{x^2}{\theta_2} \exp\left[-\frac{(x - \theta_1)}{\theta_2}\right] dx = (\theta_1 + \theta_2)^2 + \theta_2^2 \tag{4.30}$$

The moment estimators are obtained by equating these expressions with the first two sample moments d_1 and d_2, respectively, which yields

$$\theta_1 + \theta_2 = d_1 \tag{4.31}$$
$$(\theta_1 + \theta_2)^2 + \theta_2^2 = d_2 \tag{4.32}$$

Solving these two equations leads to the moment estimates

$$\hat{\theta}_{1,MM} = d_1 - (d_2 - d_1^2)^{1/2} \tag{4.33}$$
$$\hat{\theta}_{2,MM} = (d_2 - d_1^2)^{1/2} \tag{4.34}$$

The other possible solution $\hat{\theta}_{2,MM} = -(d_2 - d_1^2)^{1/2}$ must be rejected because the parameter θ_2 must be positive. In fact, it can be observed that $\hat{\theta}_{2,MM}$ equals the sample estimate of the standard deviation, and $\hat{\theta}_{1,MM}$ can be interpreted as the mean minus the standard deviation of the distribution, both estimated from the available samples.

The theoretical justification for the method of moments is that the sample moments d_j are consistent estimators of the respective theoretical moments α_j [407]. Similarly, the sample central moments s_j are consistent estimators of the true central moments μ_j. A drawback of the moment method is that it is often inefficient. Therefore, it is usually not applied provided that other, better estimators can be constructed. In general, no claims can be made on the unbiasedness and consistency of estimates given by the method of moments. Sometimes the moment method does not even lead to an acceptable estimator.

These negative remarks have implications in independent component analysis. Algebraic, cumulant-based methods proposed for ICA are typically based on estimating fourth-order moments and cross-moments of the components of the observation (data) vectors. Hence, one could claim that cumulant-based ICA methods inefficiently utilize, in general, the information contained in the data vectors. On the other hand, these methods have some advantages. They will be discussed in more detail in Chapter 11, and related methods can be found in Chapter 8 as well.

4.4 LEAST-SQUARES ESTIMATION

4.4.1 Linear least-squares method

The least-squares method can be regarded as a deterministic approach to the estimation problem where no assumptions on the probability distributions, etc., are necessary. However, statistical arguments can be used to justify the least-squares method, and they give further insight into its properties. Least-squares estimation is discussed in numerous books, in a more thorough fashion from estimation point-of-view, for example, in [407, 299].

In the basic linear least-squares method, the T-dimensional data vectors $\mathbf{x}_T$ are assumed to obey the following model:

$$\mathbf{x}_T = \mathbf{H}\boldsymbol{\theta} + \mathbf{v}_T \tag{4.35}$$

Here $\boldsymbol{\theta}$ is again the m-dimensional parameter vector, and $\mathbf{v}_T$ is a T-vector whose components are the unknown measurement errors $v(j)$, $j = 1, \ldots, T$. The $T \times m$ *observation matrix* $\mathbf{H}$ is assumed to be completely *known*. Furthermore, the number of measurements is assumed to be at least as large as the number of unknown parameters, so that $T \geq m$. In addition, the matrix $\mathbf{H}$ has the maximum rank m.

First, it can be noted that if $m = T$, we can set $\mathbf{v}_T = \mathbf{0}$, and get a unique solution $\boldsymbol{\theta} = \mathbf{H}^{-1}\mathbf{x}_T$. If there were more unknown parameters than measurements $(m > T)$, infinitely many solutions would exist for Eqs. (4.35) satisfying the condition $\mathbf{v} = \mathbf{0}$. However, if the measurements are noisy or contain errors, it is generally highly desirable to have much more measurements than there are parameters to be estimated, in order to obtain more reliable estimates. So, in the following we shall concentrate on the case $T > m$.

When $T > m$, equation (4.35) has no solution for which $\mathbf{v}_T = \mathbf{0}$. Because the measurement errors $\mathbf{v}_T$ are unknown, the best that we can then do is to choose an estimator $\hat{\boldsymbol{\theta}}$ that minimizes in some sense the effect of the errors. For mathematical convenience, a natural choice is to consider the *least-squares* criterion

$$\mathcal{E}_{LS} = \frac{1}{2} \parallel \mathbf{v}_T \parallel^2 = \frac{1}{2}(\mathbf{x}_T - \mathbf{H}\boldsymbol{\theta})^T(\mathbf{x}_T - \mathbf{H}\boldsymbol{\theta}) \tag{4.36}$$

Note that this differs from the error criteria in Section 4.2 in that no expectation is involved and the criterion $\mathcal{E}_{LS}$ tries to minimize the *measurement errors* **v**, and not directly the estimation error $\boldsymbol{\theta} - \hat{\boldsymbol{\theta}}$.

Minimization of the criterion (4.36) with respect to the unknown parameters $\boldsymbol{\theta}$ leads to so-called *normal equations* [407, 320, 299]

$$(\mathbf{H}^T\mathbf{H})\hat{\boldsymbol{\theta}}_{LS} = \mathbf{H}^T\mathbf{x}_T \qquad (4.37)$$

for determining the least-squares estimate $\hat{\boldsymbol{\theta}}_{LS}$ of $\boldsymbol{\theta}$. It is often most convenient to solve $\hat{\boldsymbol{\theta}}_{LS}$ from these linear equations. However, because we assumed that the matrix **H** has full rank, we can explicitly solve the normal equations, getting

$$\hat{\boldsymbol{\theta}}_{LS} = (\mathbf{H}^T\mathbf{H})^{-1}\mathbf{H}^T\mathbf{x}_T = \mathbf{H}^+\mathbf{x}_T \qquad (4.38)$$

where $\mathbf{H}^+ = (\mathbf{H}^T\mathbf{H})^{-1}\mathbf{H}^T$ is the *pseudoinverse* of **H** (assuming that **H** has maximal rank m and more rows than columns: $T > m$) [169, 320, 299].

The least-squares estimator can be analyzed statistically by assuming that the measurement errors have zero mean: $\mathrm{E}\{\mathbf{v}_T\} = \mathbf{0}$. It is easy to see that the least-squares estimator is unbiased: $\mathrm{E}\{\hat{\boldsymbol{\theta}}_{LS} \mid \boldsymbol{\theta}\} = \boldsymbol{\theta}$. Furthermore, if the covariance matrix of the measurement errors $\mathbf{C_v} = \mathrm{E}\{\mathbf{v}_T\mathbf{v}_T^T\}$ is known, one can compute the covariance matrix (4.19) of the estimation error. These simple analyses are left as an exercise to the reader.

Example 4.5 The least-squares method is commonly applied in various branches of science to linear curve fitting. The general setting here is as follows. We try to fit to the measurements the linear model

$$y(t) = \sum_{i=1}^{m} a_i\phi_i(t) + v(t) \qquad (4.39)$$

Here $\phi_i(t)$, $i = 1, 2, \ldots, m$, are m basis functions that can be generally *nonlinear* functions of the argument t — it suffices that the model (4.39) be linear with respect to the unknown parameters a_i. Assume now that there are available measurements $y(t_1), y(t_2), \ldots, y(t_T)$ at argument values $t_1, t_2, \ldots, t_T$, respectively. The linear model (4.39) can be easily written in the vector form (4.35), where now the parameter vector is given by

$$\boldsymbol{\theta} = [a_1, a_2, \ldots, a_m]^T \qquad (4.40)$$

and the data vector by

$$\mathbf{x}_T = [y(t_1), y(t_2), \ldots, y(t_T)]^T \qquad (4.41)$$

Similarly, the vector $\mathbf{v}_T = [v(t_1), v(t_2), \ldots, v(t_T)]^T$ contains the error terms $v(t_i)$. The observation matrix becomes

$$\mathbf{H} = \begin{bmatrix} \phi_1(t_1) & \phi_2(t_1) & \cdots & \phi_m(t_1) \\ \phi_1(t_2) & \phi_2(t_2) & \cdots & \phi_m(t_2) \\ \vdots & \vdots & \ddots & \vdots \\ \phi_1(t_T) & \phi_2(t_T) & \cdots & \phi_m(t_T) \end{bmatrix} \qquad (4.42)$$

Inserting the numerical values into (4.41) and (4.42) one can now determine $\mathbf{H}$ and $\mathbf{x}_T$, and then compute the least-squares estimates $\hat{a}_{i,LS}$ of the parameters a_i of the curve from the normal equations (4.37) or directly from (4.38).

The basis functions $\phi_i(t)$ are often chosen so that they satisfy the orthonormality conditions

$$\sum_{i=1}^{T} \phi_j(t_i)\phi_k(t_i) = \left\{ \begin{array}{ll} 1, & j = k \\ 0, & j \neq k \end{array} \right. \tag{4.43}$$

Now $\mathbf{H}^T\mathbf{H} = \mathbf{I}$, since Eq. (4.43) represents this condition for the elements (j, k) of the matrix $\mathbf{H}^T\mathbf{H}$. This implies that the normal equations (4.37) reduce to the simple form $\hat{\boldsymbol{\theta}}_{LS} = \mathbf{H}^T\mathbf{x}_T$. Writing out this equation for each component of $\hat{\boldsymbol{\theta}}_{LS}$ provides for the least-squares estimate of the parameter a_i

$$\hat{a}_{i,LS} = \sum_{j=1}^{T} \phi_i(t_j)y(t_j), \quad i = 1, \ldots, m \tag{4.44}$$

Note that the linear data model (4.35) employed in the least-squares method resembles closely the noisy linear ICA model $\mathbf{x} = \mathbf{As} + \mathbf{n}$ to be discussed in Chapter 15. Clearly, the observation matrix $\mathbf{H}$ in (4.35) corresponds to the mixing matrix $\mathbf{A}$, the parameter vector $\boldsymbol{\theta}$ to the source vector $\mathbf{s}$, and the error vector $\mathbf{v}$ to the noise vector $\mathbf{n}$ in the noisy ICA model. These model structures are thus quite similar, but the assumptions made on the models are clearly different. In the least-squares model the observation matrix $\mathbf{H}$ is assumed to be completely known, while in the ICA model the mixing matrix $\mathbf{A}$ is unknown. This lack of knowledge is compensated in ICA by assuming that the components of the source vector $\mathbf{s}$ are statistically independent, while in the least-squares model (4.35) no assumptions are needed on the parameter vector $\boldsymbol{\theta}$. Even though the models look the same, the different assumptions lead to quite different methods for estimating the desired quantities.

The basic least-squares method is simple and widely used. Its success in practice depends largely on how well the physical situation can be described using the linear model (4.35). If the model (4.35) is accurate for the data and the elements of the observation matrix $\mathbf{H}$ are known from the problem setting, good estimation results can be expected.

4.4.2 Nonlinear and generalized least-squares estimators *

Generalized least-squares The least-squares problem can be generalized by adding a symmetric and positive definite weighting matrix $\mathbf{W}$ to the criterion (4.36). The weighted criterion becomes [407, 299]

$$\mathcal{E}_{WLS} = (\mathbf{x}_T - \mathbf{H}\boldsymbol{\theta})^T \mathbf{W}(\mathbf{x}_T - \mathbf{H}\boldsymbol{\theta}) \tag{4.45}$$

It turns out that a natural, optimal choice for the weighting matrix $\mathbf{W}$ is the inverse of the covariance matrix of the measurement errors (noise) $\mathbf{W} = \mathbf{C}_\mathbf{v}^{-1}$. This is because for this choice the resulting generalized least-squares estimator

$$\hat{\boldsymbol{\theta}}_{WLS} = (\mathbf{H}^T \mathbf{C}_{\mathbf{v}}^{-1} \mathbf{H})^{-1} \mathbf{H}^T \mathbf{C}_{\mathbf{v}}^{-1} \mathbf{x}_T \qquad (4.46)$$

also minimizes the mean-square estimation error $\mathcal{E}_{MSE} = \mathrm{E}\{\| \boldsymbol{\theta} - \hat{\boldsymbol{\theta}} \|^2 | \boldsymbol{\theta}\}$ [407, 299]. Here it is assumed that the estimator $\hat{\boldsymbol{\theta}}$ is linear and unbiased. The estimator (4.46) is often referred to as the *best linear unbiased estimator (BLUE)* or *Gauss-Markov estimator*.

Note that (4.46) reduces to the standard least-squares solution (4.38) if $\mathbf{C}_{\mathbf{v}} = \sigma^2 \mathbf{I}$. This happens, for example, when the measurement errors $v(j)$ have zero mean and are mutually independent and identically distributed with a common variance σ^2. The choice $\mathbf{C}_{\mathbf{v}} = \sigma^2 \mathbf{I}$ also applies if we have no prior knowledge of the covariance matrix $\mathbf{C}_{\mathbf{v}}$ of the measurement errors. In these instances, the best linear unbiased estimator (BLUE) minimizing the mean-square error coincides with the standard least-squares estimator. This connection provides a strong statistical argument supporting the use of the least-squares method, because the mean-square error criterion directly measures the estimation error $\boldsymbol{\theta} - \hat{\boldsymbol{\theta}}$.

Nonlinear least-squares   The linear data model (4.35) employed in the linear least-squares methods is not adequate for describing the dependence between the parameters $\boldsymbol{\theta}$ and the measurements $\mathbf{x}_T$ in many instances. It is therefore natural to consider the following more general nonlinear data model

$$\mathbf{x}_T = \mathbf{f}(\boldsymbol{\theta}) + \mathbf{v}_T \qquad (4.47)$$

Here $\mathbf{f}$ is a vector-valued nonlinear and continuously differentiable function of the parameter vector $\boldsymbol{\theta}$. Each component $f_i(\boldsymbol{\theta})$ of $\mathbf{f}(\boldsymbol{\theta})$ is assumed to be a known scalar function of the components of $\boldsymbol{\theta}$.

Similarly to previously, the nonlinear least-squares criterion $\mathcal{E}_{NLS}$ is defined as the squared sum of the measurement (or modeling) errors $\| \mathbf{v}_T \|^2 = \sum_j [v(j)]^2$. From the model (4.47), we get

$$\mathcal{E}_{NLS} = [\mathbf{x}_T - \mathbf{f}(\boldsymbol{\theta})]^T [\mathbf{x}_T - \mathbf{f}(\boldsymbol{\theta})] \qquad (4.48)$$

The nonlinear least-squares estimator $\hat{\boldsymbol{\theta}}_{NLS}$ is the value of $\boldsymbol{\theta}$ that minimizes $\mathcal{E}_{NLS}$. The nonlinear least-squares problem is thus nothing but a nonlinear optimization problem where the goal is to find the minimum of the function $\mathcal{E}_{NLS}$. Such problems cannot usually be solved analytically, but one must resort to iterative numerical methods for finding the minimum. One can use any suitable nonlinear optimization method for finding the estimate $\hat{\boldsymbol{\theta}}_{NLS}$. These optimization procedures are discussed briefly in Chapter 3 and more thoroughly in the books referred to there.

The basic linear least-squares method can be extended in several other directions. It generalizes easily to the case where the measurements (made, for example, at different time instants) are vector-valued. Furthermore, the parameters can be time-varying, and the least-squares estimator can be computed adaptively (recursively). See, for example, the books [407, 299] for more information.

4.5 MAXIMUM LIKELIHOOD METHOD

Maximum likelihood (ML) estimator assumes that the unknown parameters $\boldsymbol{\theta}$ are constants or there is no prior information available on them. The ML estimator has several asymptotic optimality properties that make it a theoretically desirable choice especially when the number of samples is large. It has been applied to a wide variety of problems in many application areas.

The *maximum likelihood estimate* $\hat{\boldsymbol{\theta}}_{ML}$ of the parameter vector $\boldsymbol{\theta}$ is chosen to be the value $\hat{\boldsymbol{\theta}}_{ML}$ that maximizes the *likelihood function* (joint distribution)

$$p(\mathbf{x}_T \mid \boldsymbol{\theta}) = p(x(1), x(2), \ldots, x(T) \mid \boldsymbol{\theta}) \tag{4.49}$$

of the measurements $x(1), x(2), \ldots, x(T)$. The maximum likelihood estimator corresponds to the value $\hat{\boldsymbol{\theta}}_{ML}$ that makes the obtained measurements *most likely*.

Because many density functions contain an exponential function, it is often more convenient to deal with the *log likelihood function* $\ln p(\mathbf{x}_T \mid \boldsymbol{\theta})$. Clearly, the maximum likelihood estimator $\hat{\boldsymbol{\theta}}_{ML}$ also maximizes the log likelihood. The maximum likelihood estimator is usually found from the solutions of the *likelihood equation*

$$\frac{\partial}{\partial \boldsymbol{\theta}} \ln p(\mathbf{x}_T \mid \boldsymbol{\theta}) \bigg|_{\boldsymbol{\theta}=\hat{\boldsymbol{\theta}}_{ML}} = \mathbf{0} \tag{4.50}$$

The likelihood equation gives the values of $\boldsymbol{\theta}$ that maximize (or minimize) the likelihood function. If the likelihood function is complicated, having several local maxima and minima, one must choose the value $\hat{\boldsymbol{\theta}}_{ML}$ that corresponds to the absolute maximum. Sometimes the maximum likelihood estimate can be found from the endpoints of the interval where the likelihood function is nonzero.

The construction of the likelihood function (4.49) can be very difficult if the measurements depend on each other. Therefore, it is almost always assumed in applying the ML method that the observations $x(j)$ are statistically *independent* of each other. Fortunately, this holds quite often in practice. Assuming independence, the likelihood function decouples into the product

$$p(\mathbf{x}_T \mid \boldsymbol{\theta}) = \prod_{j=1}^{T} p(x(j) \mid \boldsymbol{\theta}) \tag{4.51}$$

where $p(x(i) \mid \boldsymbol{\theta})$ is the conditional pdf of a single scalar measurement $x(j)$. Note that taking the logarithm, the product (4.51) decouples to the sum of logarithms $\sum_j \ln p(x(j) \mid \boldsymbol{\theta})$.

The vector likelihood equation (4.50) consists of m scalar equations

$$\frac{\partial}{\partial \theta_i} \ln p(\mathbf{x}_T \mid \hat{\boldsymbol{\theta}}_{ML}) \bigg|_{\boldsymbol{\theta}=\hat{\boldsymbol{\theta}}_{ML}} = 0, \quad i = 1, \ldots, m \tag{4.52}$$

for the m parameter estimates $\hat{\theta}_{i,ML}, i = 1, \ldots, m$. These equations are in general coupled and nonlinear, so they can be solved only numerically except for simple cases. In several practical applications, the computational load of the maximum likelihood method can be prohibitive, and one must resort to various approximations for simplifying the likelihood equations or to some suboptimal estimation methods.

Example 4.6 Assume that we have T independent observations $x(1), \dots, x(T)$ of a scalar random variable x that is gaussian distributed with mean μ and variance σ^2. Using (4.51), the likelihood function can be written

$$p(\mathbf{x}_T \mid \mu, \sigma^2) = (2\pi\sigma^2)^{-T/2} \exp\left[-\frac{1}{2\sigma^2} \sum_{j=1}^{T} [x(j) - \mu]^2\right] \tag{4.53}$$

The log likelihood function becomes

$$\ln p(\mathbf{x}_T \mid \mu, \sigma^2) = -\frac{T}{2} \ln(2\pi\sigma^2) - \frac{1}{2\sigma^2} \sum_{j=1}^{T} [x(j) - \mu]^2 \tag{4.54}$$

The first likelihood equation (4.52) is

$$\frac{\partial}{\partial \mu} \ln p(\mathbf{x}_T \mid \hat{\mu}_{ML}, \hat{\sigma}_{ML}^2) = \frac{1}{\hat{\sigma}_{ML}^2} \sum_{j=1}^{T} [x(j) - \hat{\mu}_{ML}] = 0 \tag{4.55}$$

Solving this yields for the maximum likelihood estimate of the mean μ the sample mean

$$\hat{\mu}_{ML} = \frac{1}{T} \sum_{j=1}^{T} x(j) \tag{4.56}$$

The second likelihood equation is obtained by differentiating the log likelihood (4.54) with respect to the variance σ^2:

$$\frac{\partial}{\partial \sigma^2} \ln p(\mathbf{x}_T \mid \hat{\mu}_{ML}, \hat{\sigma}_{ML}^2) = -\frac{T}{2\hat{\sigma}_{ML}^2} + \frac{1}{2\hat{\sigma}_{ML}^4} \sum_{j=1}^{T} [x(j) - \hat{\mu}_{ML}]^2 = 0 \tag{4.57}$$

From this equation, we get for the maximum likelihood estimate of the variance σ^2 the sample variance

$$\hat{\sigma}_{ML}^2 = \frac{1}{T} \sum_{j=1}^{T} [x(j) - \hat{\mu}_{ML}]^2 \tag{4.58}$$

This is a *biased* estimator of the true variance σ^2, while the sample mean $\hat{\mu}_{ML}$ is an unbiased estimator of the mean μ. The bias of the variance estimator $\hat{\sigma}_{ML}^2$ is due to using the *estimated* mean $\hat{\mu}_{ML}$ instead of the true one in (4.58). This reduces the amount of new information that is truly available for estimation by one sample.

Hence the unbiased estimator of the variance is given by (4.6). However, the bias of the estimator (4.58) is usually small, and it is asymptotically unbiased.

The maximum likelihood estimator is important because it provides estimates that have certain very desirable theoretical properties. In the following, we list briefly the most important of them. Somewhat heuristic but illustrative proofs can be found in [407]. For more detailed analyses, see, e.g., [477].

1. If there exists an estimator that satisfies the Cramer-Rao lower bound (4.20) as an equality, it can be determined using the maximum likelihood method.

2. The maximum likelihood estimator $\hat{\theta}_{ML}$ is consistent.

3. The maximum likelihood estimator is *asymptotically efficient*. This means that it achieves asymptotically the Cramer-Rao lower bound for the estimation error.

Example 4.7 Let us determine the Cramer-Rao lower bound (4.20) for the mean μ of a single gaussian random variable. From (4.55), the derivative of the log likelihood function with respect to μ is

$$\frac{\partial}{\partial \mu} \ln p(\mathbf{x}_T \mid \mu, \sigma^2) = \frac{1}{\sigma^2} \sum_{j=1}^{T} [x(j) - \mu] \tag{4.59}$$

Because we are now considering a single parameter μ only, the Fisher information matrix reduces to the scalar quantity

$$
\begin{aligned}
J &= \mathrm{E}\left\{ \left[\frac{\partial}{\partial \mu} \ln p(\mathbf{x}_T \mid \mu, \sigma^2) \right]^2 \mid \mu, \sigma^2 \right\} \\
&= \mathrm{E}\left\{ \left[\frac{1}{\sigma^2} \sum_{j=1}^{T} [x(j) - \mu] \right]^2 \mid \mu, \sigma^2 \right\}
\end{aligned}
\tag{4.60}
$$

Since the samples $x(j)$ are assumed to be independent, all the cross covariance terms vanish, and (4.60) simplifies to

$$J = \frac{1}{\sigma^4} \sum_{j=1}^{T} \mathrm{E}\{[x(j) - \mu]^2 \mid \mu, \sigma^2\} = \frac{T\sigma^2}{\sigma^4} = \frac{T}{\sigma^2} \tag{4.61}$$

Thus the Cramer-Rao lower bound (4.20) for the mean-square error of any unbiased estimator $\hat{\mu}$ of the mean of the gaussian density is

$$\mathrm{E}\{(\mu - \hat{\mu})^2 \mid \mu\} \geq J^{-1} = \frac{\sigma^2}{T} \tag{4.62}$$

In the previous example we found that the maximum likelihood estimator $\hat{\mu}_{ML}$ of μ is the sample mean (4.56). The mean-square error $\mathrm{E}\{(\mu - \hat{\mu}_{ML})^2\}$ of the sample mean was shown earlier in Example 4.3 to be σ^2/T. Hence the sample mean satisfies the Cramer-Rao inequality as an equation and is an efficient estimator for independent gaussian measurements.

The *expectation-maximization (EM)* algorithm [419, 172, 298, 304] provides a general iterative approach for computing maximum likelihood estimates. The main advantage of the EM algorithm is that it often allows treatment of difficult maximum likelihood problems suffering from multiple parameters and highly nonlinear likelihood functions in terms of simpler maximization problems. However, the application of the EM algorithm requires care in general because it can get stuck into a local maximum or suffer from singularity problems [48]. In context with ICA methods, the EM algorithm has been used for estimating unknown densities of source signals. Any probability density function can be approximated using a mixture-of-gaussians model [48]. A popular method for finding parameters of such a model is to use the EM algorithm. This specific but important application of the EM algorithm is discussed in detail in [48]. For a more detailed discussion of the EM algorithm, see references [419, 172, 298, 304].

The maximum likelihood method has a connection with the least-squares method. Consider the nonlinear data model (4.47). Assuming that the parameters $\boldsymbol{\theta}$ are unknown constants independent of the additive noise (error) $\mathbf{v}_T$, the (conditional) distribution $p(\mathbf{x}_T \mid \boldsymbol{\theta})$ of $\mathbf{x}_T$ is the same as the distribution of $\mathbf{v}_T$ at the point $\mathbf{v}_T = \mathbf{x}_T - \mathbf{f}(\boldsymbol{\theta})$:

$$p_{\mathbf{x}|\boldsymbol{\theta}}(\mathbf{x}_T \mid \boldsymbol{\theta}) = p_{\mathbf{v}}(\mathbf{x}_T - \mathbf{f}(\boldsymbol{\theta}) \mid \boldsymbol{\theta}) \qquad (4.63)$$

If we further assume that the noise $\mathbf{v}_T$ is zero-mean and gaussian with the covariance matrix $\sigma^2\mathbf{I}$, the preceding distribution becomes

$$p(\mathbf{x}_T \mid \boldsymbol{\theta}) = \gamma \exp\left\{ -\frac{1}{2\sigma^2}[\mathbf{x}_T - \mathbf{f}(\boldsymbol{\theta})]^T[\mathbf{x} - \mathbf{f}(\boldsymbol{\theta})] \right\} \qquad (4.64)$$

where $\gamma = (2\pi)^{-T/2}\sigma^{-T}$ is the normalizing term. Clearly, this is maximized when the exponent

$$[\mathbf{x}_T - \mathbf{f}(\boldsymbol{\theta})]^T[\mathbf{x}_T - \mathbf{f}(\boldsymbol{\theta})] = \ \| \mathbf{x}_T - \mathbf{f}(\boldsymbol{\theta}) \|^2 \qquad (4.65)$$

is minimized, since γ is a constant independent of $\boldsymbol{\theta}$. But the exponent (4.65) coincides with the nonlinear least-squares criterion (4.48). Hence if in the nonlinear data model (4.47) the noise $\mathbf{v}_T$ is zero-mean, gaussian with the covariance matrix $\mathbf{C_v} = \sigma^2\mathbf{I}$, and independent of the unknown parameters $\boldsymbol{\theta}$, the maximum likelihood estimator and the nonlinear least-squares estimator yield the same results.

4.6 BAYESIAN ESTIMATION *

All the estimation methods discussed thus far in more detail, namely the moment, the least-squares, and the maximum likelihood methods, assume that the parameters θ are unknown *deterministic constants*. In Bayesian estimation methods, the parameters θ are assumed to be *random* themselves. This randomness is modeled using the *a priori* probability density function $p_\theta(\theta)$ of the parameters. In Bayesian methods, it is typically assumed that this a priori density is *known*. Taken strictly, this is a very demanding assumption. In practice we usually do not have such far-reaching information on the parameters. However, *assuming* some useful form for the a priori density $p_\theta(\theta)$ often allows the incorporation of useful prior information on the parameters into the estimation process. For example, we may know which is the most typical value of the parameter θ_i and its typical range of variation. We can then formulate this prior information for instance by assuming that θ_i is gaussian distributed with a mean m_i and variance σ_i^2. In this case the mean m_i and variance σ_i^2 contain our prior knowledge about θ_i (together with the gaussianity assumption).

The essence of Bayesian estimation methods is *the posterior density* $p_{\theta|x}(\theta|x_T)$ of the parameters θ given the data x_T. Basically, *the posterior density contains all the relevant information on the parameters* θ. Choosing a specific estimate $\hat{\theta}$ for the parameters θ among the range of values of θ where the posterior density is high or relatively high is somewhat arbitrary. The two most popular methods for doing this are based on the mean-square error criterion and choosing the maximum of the posterior density. These are discussed in the following subsections.

4.6.1 Minimum mean-square error estimator for random parameters

In the minimum mean-square error method for random parameters θ, the optimal estimator $\hat{\theta}_{MSE}$ is chosen by minimizing the mean-square error (MSE)

$$\mathcal{E}_{MSE} = \mathrm{E}\{\| \theta - \hat{\theta} \|^2\} \qquad (4.66)$$

with respect to the estimator $\hat{\theta}$. The following theorem specifies the optimal estimator.

Theorem 4.2 *Assume that the parameters* θ *and the observations* x_T *have the joint probability density function* $p_{\theta,x}(\theta, x_T)$. *The minimum mean-square estimator* $\hat{\theta}_{MSE}$ *of* θ *is given by the conditional expectation*

$$\hat{\theta}_{MSE} = \mathrm{E}\{\theta|x_T\} \qquad (4.67)$$

The theorem can be proved by first noting that the mean-square error (4.66) can be computed in two stages. First the expectation is evaluated with respect to θ only, and after this it is taken with respect to the measurement vector x:

$$\mathcal{E}_{MSE} = \mathrm{E}\{\| \theta - \hat{\theta} \|^2\} = \mathrm{E}_x\{\mathrm{E}\{\| \theta - \hat{\theta} \|^2 |x_T\}\} \qquad (4.68)$$

This expression shows that the minimization can be carried out by minimizing the conditional expectation

$$\mathrm{E}\{\| \boldsymbol{\theta} - \hat{\boldsymbol{\theta}} \|^2 \,|\mathbf{x}_T\} = \hat{\boldsymbol{\theta}}^T \hat{\boldsymbol{\theta}} - 2\hat{\boldsymbol{\theta}}^T \mathrm{E}\{\boldsymbol{\theta}|\mathbf{x}_T\} + \mathrm{E}\{\boldsymbol{\theta}^T \boldsymbol{\theta}|\mathbf{x}_T\} \qquad (4.69)$$

The right-hand side is obtained by evaluating the squared norm and noting that $\hat{\boldsymbol{\theta}}$ is a function of the observations $\mathbf{x}_T$ only, so that it can be treated as a nonrandom vector when computing the conditional expectation (4.69). The result (4.67) now follows directly by computing the gradient $2\hat{\boldsymbol{\theta}} - 2\mathrm{E}\{\boldsymbol{\theta}|\mathbf{x}_T\}$ of (4.69) with respect to $\hat{\boldsymbol{\theta}}$ and equating it to zero.

The minimum mean-square estimator $\hat{\boldsymbol{\theta}}_{MSE}$ is unbiased since

$$\mathrm{E}\{\hat{\boldsymbol{\theta}}_{MSE}\} = \mathrm{E}_{\mathbf{x}}\{\mathrm{E}\{\boldsymbol{\theta}|\mathbf{x}_T\}\} = \mathrm{E}\{\boldsymbol{\theta}\} \qquad (4.70)$$

The minimum mean-square estimator (4.67) is theoretically very significant because of its conceptual simplicity and generality. This result holds for all distributions for which the joint distribution $p_{\boldsymbol{\theta},\mathbf{x}}(\boldsymbol{\theta}, \mathbf{x})$ exists, and remains unchanged if a weighting matrix $\mathbf{W}$ is added into the criterion (4.66) [407].

However, actual computation of the minimum mean-square estimator is often very difficult. This is because in practice we only know or assume the prior distribution $p_{\boldsymbol{\theta}}(\boldsymbol{\theta})$ and the conditional distribution of the observations $p_{\mathbf{x}|\boldsymbol{\theta}}(\mathbf{x}|\boldsymbol{\theta})$ given the parameters $\boldsymbol{\theta}$. In constructing the optimal estimator (4.67), one must first compute the posterior density from Bayes' formula (see Section 2.4)

$$p_{\boldsymbol{\theta}|\mathbf{x}}(\boldsymbol{\theta}|\mathbf{x}_T) = \frac{p_{\mathbf{x}|\boldsymbol{\theta}}(\mathbf{x}_T|\boldsymbol{\theta})p_{\boldsymbol{\theta}}(\boldsymbol{\theta})}{p_{\mathbf{x}}(\mathbf{x}_T)} \qquad (4.71)$$

where the denominator is computed by integrating the numerator:

$$p_{\mathbf{x}}(\mathbf{x}_T) = \int_{-\infty}^{\infty} p_{\mathbf{x}|\boldsymbol{\theta}}(\mathbf{x}_T|\boldsymbol{\theta})p_{\boldsymbol{\theta}}(\boldsymbol{\theta})d\boldsymbol{\theta} \qquad (4.72)$$

The computation of the conditional expectation (4.67) then requires still another integration. These integrals are usually impossible to evaluate at least analytically except for special cases.

There are, however, two important special cases where the minimum mean-square estimator $\hat{\boldsymbol{\theta}}_{MSE}$ for random parameters $\boldsymbol{\theta}$ can be determined fairly easily. If the estimator $\hat{\boldsymbol{\theta}}$ is constrained to be a *linear* function of the data: $\hat{\boldsymbol{\theta}} = \mathbf{L}\mathbf{x}_T$, then it can be shown [407] that the optimal linear estimator $\hat{\boldsymbol{\theta}}_{LMSE}$ minimizing the MSE criterion (4.66) is

$$\hat{\boldsymbol{\theta}}_{LMSE} = \mathbf{m}_{\boldsymbol{\theta}} + \mathbf{C}_{\boldsymbol{\theta}\mathbf{x}}\mathbf{C}_{\mathbf{x}}^{-1}(\mathbf{x}_T - \mathbf{m}_{\mathbf{x}}) \qquad (4.73)$$

where $\mathbf{m}_{\boldsymbol{\theta}}$ and $\mathbf{m}_{\mathbf{x}}$ are the mean vectors of $\boldsymbol{\theta}$ and $\mathbf{x}_T$, respectively, $\mathbf{C}_{\mathbf{x}}$ is the covariance matrix of $\mathbf{x}_T$, and $\mathbf{C}_{\boldsymbol{\theta}\mathbf{x}}$ is the cross-covariance matrix of $\boldsymbol{\theta}$ and $\mathbf{x}_T$. The error covariance matrix corresponding to the optimum linear estimator $\hat{\boldsymbol{\theta}}_{LMSE}$ is

$$\mathrm{E}\{(\boldsymbol{\theta} - \hat{\boldsymbol{\theta}}_{LMSE})(\boldsymbol{\theta} - \hat{\boldsymbol{\theta}}_{LMSE})^T\} = \mathbf{C}_{\boldsymbol{\theta}} - \mathbf{C}_{\boldsymbol{\theta}\mathbf{x}}\mathbf{C}_{\mathbf{x}}^{-1}\mathbf{C}_{\mathbf{x}\boldsymbol{\theta}} \qquad (4.74)$$

where $\mathbf{C}_\theta$ is the covariance matrix of the parameter vector θ. We can conclude that if the minimum mean-square estimator is constrained to be linear, it suffices to know the first-order and second-order statistics of the data $\mathbf{x}$ and the parameters θ, that is, their means and covariance matrices.

If the joint probability density $p_{\theta,\mathbf{x}}(\theta, \mathbf{x}_T)$ of the parameters θ and data $\mathbf{x}_T$ is *gaussian*, the results (4.73) and (4.74) obtained by constraining the minimum mean-square estimator to be linear are quite generally optimal. This is because the conditional density $p_{\theta|\mathbf{x}}(\theta|\mathbf{x}_T)$ is also gaussian with the conditional mean (4.73) and covariance matrix (4.74); see section 2.5. This again underlines the fact that for the gaussian distribution, linear processing and knowledge of first and second order statistics are usually sufficient to obtain optimal results.

4.6.2   Wiener filtering

In this subsection, we take a somewhat different signal processing viewpoint to the linear minimum MSE estimation. Many estimation algorithms have in fact been developed in context with various signal processing problems [299, 171].

Consider the following linear *filtering problem*. Let $\mathbf{z}$ be an m-dimensional data or input vector of the form

$$\mathbf{z} = [z_1, z_2, \dots, z_m]^T \tag{4.75}$$

and

$$\mathbf{w} = [w_1, w_2, \dots, w_m]^T \tag{4.76}$$

an m-dimensional *weight vector* with adjustable weights (elements) w_i, $i = 1, \dots, m$ operating linearly on $\mathbf{z}$ so that the output of the filter is

$$y = \mathbf{w}^T \mathbf{z} \tag{4.77}$$

In *Wiener filtering*, the goal is to determine the linear filter (4.77) that minimizes the mean-square error

$$\mathcal{E}_{MSE} = \mathrm{E}\{(y - d)^2\} \tag{4.78}$$

between the *desired response* d and the output y of the filter. Inserting (4.77) into (4.78) and evaluating the expectation yields

$$\mathcal{E}_{MSE} = \mathbf{w}^T \mathbf{R_z} \mathbf{w} - 2\mathbf{w}^T \mathbf{r}_{zd} + \mathrm{E}\{d^2\} \tag{4.79}$$

Here $\mathbf{R_z} = \mathrm{E}\{\mathbf{zz}^T\}$ is the data correlation matrix, and $\mathbf{r}_{zd} = \mathrm{E}\{\mathbf{z}d\}$ is the cross-correlation vector between the data vector $\mathbf{z}$ and the desired response d. Minimizing the mean-square error (4.79) with respect to the weight vector $\mathbf{w}$ provides as the optimum solution the *Wiener filter* [168, 171, 419, 172]

$$\hat{\mathbf{w}}_{MSE} = \mathbf{R_z}^{-1} \mathbf{r}_{zd} \tag{4.80}$$

provided that $\mathbf{R}_z$ is nonsingular. This is almost always the case in practice due to the noise and the statistical nature of the problem. The Wiener filter is usually computed by directly solving the linear normal equations

$$\mathbf{R}_z \hat{\mathbf{w}}_{MSE} = \mathbf{r}_{zd} \tag{4.81}$$

In practice, the correlation matrix $\mathbf{R}_z$ and the cross-correlation vector $\mathbf{r}_{zd}$ are usually unknown. They must then be replaced by their estimates, which can be computed easily from the available finite data set. In fact the Wiener estimate then becomes a standard least-squares estimator (see exercises). In signal processing applications, the correlation matrix $\mathbf{R}_z$ is often a Toeplitz matrix, since the data vectors $\mathbf{z}(i)$ consist of subsequent samples from a single signal or time series (see Section 2.8). For this special case, various fast algorithms are available for solving the normal equations efficiently [169, 171, 419].

4.6.3 Maximum a posteriori (MAP) estimator

Instead of minimizing the mean-square error (4.66) or some other performance index, we can apply to Bayesian estimation the same principle as in the maximum likelihood method. This leads to the *maximum a posteriori (MAP) estimator* $\hat{\boldsymbol{\theta}}_{MAP}$, which is defined as the value of the parameter vector $\boldsymbol{\theta}$ that maximizes the posterior density $p_{\boldsymbol{\theta}|\mathbf{x}}(\boldsymbol{\theta}|\mathbf{x}_T)$ of $\boldsymbol{\theta}$ given the measurements $\mathbf{x}_T$. The MAP estimator can be interpreted as the most probable value of the parameter vector $\boldsymbol{\theta}$ for the available data $\mathbf{x}_T$. The principle behind the MAP estimator is intuitively well justified and appealing.

We have earlier noted that the posterior density can be computed from Bayes' formula (4.71). Note that the denominator in (4.71) is the prior density $p_{\mathbf{x}}(\mathbf{x}_T)$ of the data $\mathbf{x}_T$ which does not depend on the parameter vector $\boldsymbol{\theta}$, and merely normalizes the posterior density $p_{\boldsymbol{\theta}|\mathbf{x}}(\boldsymbol{\theta}|\mathbf{x}_T)$. Hence for finding the MAP estimator it suffices to find the value of $\boldsymbol{\theta}$ that maximizes the numerator of (4.71), which is the joint density

$$p_{\boldsymbol{\theta},\mathbf{x}}(\boldsymbol{\theta}, \mathbf{x}_T) = p_{\mathbf{x}|\boldsymbol{\theta}}(\mathbf{x}_T|\boldsymbol{\theta})p_{\boldsymbol{\theta}}(\boldsymbol{\theta}) \tag{4.82}$$

Quite similarly to the maximum likelihood method, the MAP estimator $\hat{\boldsymbol{\theta}}_{MAP}$ can usually be found by solving the (logarithmic) likelihood equation. This now has the form

$$\frac{\partial}{\partial \boldsymbol{\theta}} \ln p(\boldsymbol{\theta}, \mathbf{x}_T) = \frac{\partial}{\partial \boldsymbol{\theta}} \ln p(\mathbf{x}_T|\boldsymbol{\theta}) + \frac{\partial}{\partial \boldsymbol{\theta}} \ln p(\boldsymbol{\theta}) = 0 \tag{4.83}$$

where we have dropped the subscripts of the probability densities for notational simplicity.

A comparison with the respective likelihood equation (4.50) for the maximum likelihood method shows that these equations are otherwise the same, but the MAP likelihood equation (4.83) contains an additional term $\partial(\ln p(\boldsymbol{\theta}))/\partial \boldsymbol{\theta}$, which takes into account the prior information on the parameters $\boldsymbol{\theta}$. If the prior density $p(\boldsymbol{\theta})$ is uniform for parameter values $\boldsymbol{\theta}$ for which $p(\mathbf{x}_T|\boldsymbol{\theta})$ is markedly greater than zero, then the MAP and maximum likelihood estimators become the same. In this case, they are both obtained by finding the value $\hat{\theta}$ that maximizes the conditional density $p(\mathbf{x}_T | \boldsymbol{\theta})$. This is the case when there is no prior information about the parameters $\boldsymbol{\theta}$ available. However, when the prior density $p(\boldsymbol{\theta})$ is not uniform, the MAP and ML estimators are usually different.

Example 4.8 Assume that we have T independent observations $x(1), \ldots, x(T)$ from a scalar random quantity x that is gaussian distributed with mean μ_x and variance σ_x^2. This time the mean μ_x is itself a gaussian random variable having mean zero and the variance σ_μ^2. We assume that both the variances σ_x^2 and σ_μ^2 are known and wish to estimate μ using the MAP method.

Using the preceding information, it is straightforward to form the likelihood equation for the MAP estimator $\hat{\mu}_{MAP}$ and solve it. The solution is (the derivation is left as a exercise)

$$\hat{\mu}_{MAP} = \frac{\sigma_\mu^2}{\sigma_x^2 + T\sigma_\mu^2} \sum_{j=1}^{T} x(j) \tag{4.84}$$

The case in which we do not have any prior information on μ can be modeled by letting $\sigma_\mu^2 \to \infty$, reflecting our uncertainty about μ [407]. Then clearly

$$\hat{\mu}_{MAP} \to \frac{1}{T} \sum_{j=1}^{T} x(j) \tag{4.85}$$

so that the MAP estimator $\hat{\mu}_{MAP}$ tends to the sample mean. The same limiting value is obtained if the number of samples $T \to \infty$. This shows that the influence of the prior information, contained in the variance σ_μ^2, gradually decreases as the number of the measurements increases. Hence asymptotically the MAP estimator coincides with the maximum likelihood estimator $\hat{\mu}_{ML}$ which we found earlier in (4.56) to be the sample mean (4.85).

Note also that if we are relatively confident about the prior value 0 of the mean μ, but the samples are very noisy so that $\sigma_x^2 \gg \sigma_\mu^2$, the MAP estimator (4.84) for small T stays close to the prior value 0 of μ, and the number T of samples must grow large until the MAP estimator approaches its limiting value (4.85). In contrast, if $\sigma_\mu^2 \gg \sigma_x^2$, so that the samples are reliable compared to the prior information on μ, the MAP estimator (4.84) rapidly approaches the sample mean (4.85). Thus the MAP estimator (4.84) weights in a meaningful way the prior information and the samples according to their relative reliability.

Roughly speaking, the MAP estimator is a compromise between the general minimum mean-square error estimator (4.67) and the maximum likelihood estimator. The MAP method has the advantage over the maximum likelihood method that it takes into account the (possibly available) prior information about the parameters $\boldsymbol{\theta}$, but it is computationally somewhat more difficult to determine because a second term appears in the likelihood equation (4.83). On the other hand, both the ML and MAP estimators are obtained from likelihood equations, avoiding the generally difficult integrations needed in computing the minimum mean-square estimator. If the posterior distribution $p(\boldsymbol{\theta}|\mathbf{x}_T)$ is symmetric around its peak value, the MAP estimator and MSE estimator coincide.

There is no guarantee that the MAP estimator is unbiased. It is also generally difficult to compute the covariance matrix of the estimation error for the MAP and ML estimators. However, the MAP estimator is intuitively sensible, yields in most cases good results in practice, and it has good asymptotic properties under appropriate conditions. These desirable characteristics justify its use.

4.7 CONCLUDING REMARKS AND REFERENCES

In this chapter, we have dealt with basic concepts in estimation theory and the most widely used estimation methods. These include the maximum likelihood method, minimum mean-square error estimator, the maximum a posteriori method, and the least-squares method for both linear and nonlinear data models. We have also pointed out their interrelationships, and discussed the method of moments because of its relationship to higher-order statistics. Somewhat different estimation methods must be used depending on whether the parameters are considered to be deterministic, in which case the maximum likelihood method is the most common choice, or random, in which case Bayesian methods such as maximum a posteriori estimation can be used.

Rigorous treatment of estimation theory requires a certain mathematical background as well as a good knowledge of probability and statistics, linear algebra, and matrix differential calculus. The interested reader can find more information on estimation theory in several textbooks, including both mathematically [244, 407, 477] and signal-processing oriented treatments [242, 299, 393, 419]. There are several topics worth mentioning that we have not discussed in this introductory chapter. These include dynamic estimation methods in which the parameters and/or the data model are time dependent, for example, Kalman filtering [242, 299]. In this chapter, we have derived several estimators by minimizing error criteria or maximizing conditional probability distributions. Alternatively, optimal estimators can often be derived from the orthogonality principle, which states that the estimator and its associated estimation error must be statistically orthogonal, having a zero cross-covariance matrix.

From a theoretical viewpoint, the posterior density $p_{\boldsymbol{\theta}|\mathbf{x}}(\boldsymbol{\theta}|\mathbf{x}_T)$ contains all the information about the random parameters that the measurements $\mathbf{x}_T$ provide. Knowledge of the posterior density allows in principle the use of any suitable optimality criterion for determining an estimator. Figure 4.2 shows an example of a hypothetical posterior density $p(\theta \mid \mathbf{x})$ of a scalar parameter θ. Because of the asymmetricity of this density, different estimators yield different results. The minimum absolute error estimator $\hat{\theta}_{ABS}$ minimizes the absolute error $E\{|\,\theta - \hat{\theta}\,|\}$. The choice of a specific estimator is somewhat arbitrary, since the true value of the parameter θ is unknown, and can be anything within the range of the posterior density.

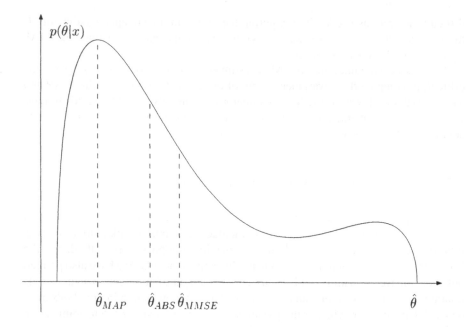

Fig. 4.2 A posterior density $p(\theta \mid \mathbf{x})$, and the respective MAP estimate $\hat{\theta}_{MAP}$, minimum MSE estimate $\hat{\theta}_{MSE}$, and the minimum absolute error estimate $\hat{\theta}_{ABS}$.

Regrettably, it is generally difficult to determine the posterior distribution in a form that allows for convenient mathematical analysis [407]. However, various advanced and approximative techniques have been developed to facilitate Bayesian estimation; see [142]. When the number of measurements increases, the importance of prior information gradually decreases, and the maximum likelihood estimator becomes asymptotically optimal.

Finally, we point out that neural networks provide in many instances a useful practical tool for nonlinear estimation, even though they lie outside the range of classic estimation theory. For example, the well-known back-propagation algorithm [48, 172, 376] is in fact a stochastic gradient algorithm for minimizing the mean-square error criterion

$$\mathcal{E}_{MSE} = \mathrm{E}\{\|\, \mathbf{d} - \mathbf{f}(\boldsymbol{\theta}, \mathbf{z})\, \|^2\} \tag{4.86}$$

Here $\mathbf{d}$ is the desired response vector and $\mathbf{z}$ the (input) data vector. The parameters $\boldsymbol{\theta}$ consist of weights that are adjusted so that the mapping error (4.86) is minimized. The nonlinear function $\mathbf{f}(\boldsymbol{\theta}, \mathbf{z})$ has enough parameters and a flexible form, so that it can actually model with sufficient accuracy any regular nonlinear function. The back-propagation algorithm learns the parameters $\boldsymbol{\theta}$ that define the estimated input-output mapping $\mathbf{f}(\boldsymbol{\theta}, \mathbf{z})$. See [48, 172, 376] for details and applications.

Problems

4.1   Show that:

4.1.1. the maximum likelihood estimator of the variance (4.58) becomes unbiased if the estimated mean $\hat{\mu}_{ML}$ is replaced in (4.58) by the true one μ.

4.1.2. if the mean is estimated from the observations, one must use the formula (4.6) for getting an unbiased estimator.

4.2   Assume that $\hat{\theta}_1$ and $\hat{\theta}_2$ are unbiased estimators of the parameter θ having variances $\mathrm{var}(\hat{\theta}_1) = \sigma_1^2$, $\mathrm{var}(\hat{\theta}_2) = \sigma_2^2$.

4.2.1. Show that for any scalar $0 \le \alpha \le 1$, the estimator $\hat{\theta}_3 = \alpha\hat{\theta}_1 + (1 - \alpha)\hat{\theta}_2$ is unbiased.

4.2.2. Determine the mean-square error of $\hat{\theta}_3$ assuming that $\hat{\theta}_1$ and $\hat{\theta}_2$ are statistically independent.

4.2.3. Find the value of α that minimizes this mean-square error.

4.3   Let the scalar random variable z be uniformly distributed on the interval $[0, \theta)$. There exist T independent samples $z(1), \ldots, z(T)$ from z. Using them, the estimate $\hat{\theta} = \max(z(i))$ is constructed for the parameter θ.

4.3.1. Compute the probability density function of $\hat{\theta}$. (*Hint:* First construct the cumulative distribution function.)

4.3.2. Is $\hat{\theta}$ unbiased or asymptotically unbiased?

4.3.3. What is the mean-square error $\mathrm{E}\{(\hat{\theta} - \theta)^2 \mid \theta\}$ of the estimate $\hat{\theta}$?

4.4   Assume that you know T independent observations of a scalar quantity that is gaussian distributed with unknown mean μ and variance σ^2. Estimate μ and σ^2 using the method of moments.

4.5   Assume that $x(1), x(2), \ldots, x(K)$ are independent gaussian random variables having all the mean 0 and variance σ_x^2. Then the sum of their squares

$$y = \sum_{j=1}^{K} [x(j)]^2$$

is χ^2-distributed with the mean $K\sigma_x^2$ and variance $2K\sigma_x^4$. Estimate the parameters K and σ_x^2 using the method of moments, assuming that there exist T measurements $y(1), y(2), \ldots, y(T)$ on the sum of squares y.

4.6   Derive the normal equations (4.37) for the least-squares criterion (4.36). Justify why these equations indeed provide the minimum of the criterion.

4.7   Assume that the measurement errors have zero mean: $\mathrm{E}\{\mathbf{v}_T\} = \mathbf{0}$, and that the covariance matrix of the measurement errors is $\mathbf{C_v} = \mathrm{E}\{\mathbf{v}_T\mathbf{v}_T^T\}$. Consider the properties of the least-squares estimator $\hat{\boldsymbol{\theta}}_{LS}$ in (4.38).

4.7.1. Show that the estimator $\hat{\boldsymbol{\theta}}_{LS}$ is unbiased.

4.7.2. Compute the error covariance matrix $\mathbf{C}_{\tilde{\theta}}$ defined in (4.19).

4.7.3. Compute $\mathbf{C}_{\tilde{\theta}}$ when $\mathbf{C_v} = \sigma^2\mathbf{I}$.

4.8 Consider line fitting using the linear least-squares method. Assume that you know T measurements $x(1), x(2), \dots, x(T)$ on the scalar quantity x made, respectively, at times (or argument values) $t(1), t(2), \dots, t(T)$. The task is to fit the line

$$x = \alpha_0 + \alpha_1 t$$

to these measurements.

4.8.1. Construct the normal equations for this problem using the standard linear least-squares method.

4.8.2. Assume that the sampling interval Δt is constant and has been scaled so that the measurement times are integers $1, 2, \dots, T$. Solve the normal equations in this important special case.

4.9 * Consider the equivalence of the generalized least-squares and linear unbiased minimum mean-square estimators. Show that

4.9.1. The optimal solution minimizing the generalized least-squares criterion (4.45) is

$$\hat{\boldsymbol{\theta}}_{WLS} = (\mathbf{H}^T \mathbf{W} \mathbf{H})^{-1} \mathbf{H}^T \mathbf{W} \mathbf{x}_T$$

4.9.2. An unbiased linear mean-square estimator $\hat{\boldsymbol{\theta}}_{MSE} = \mathbf{L}\mathbf{x}_T$ satisfies the condition $\mathbf{L}\mathbf{H} = \mathbf{I}$.

4.9.3. The mean-square error can be written in the form

$$\mathcal{E}_{MSE} = \mathrm{E}\{\|\, \boldsymbol{\theta} - \hat{\boldsymbol{\theta}} \,\|^2 |\, \boldsymbol{\theta}\} = \mathrm{trace}(\mathbf{L}\mathbf{C_v}\mathbf{L}^T)$$

4.9.4. Minimization of the preceding criterion $\mathcal{E}_{MSE}$ under the constraint $\mathbf{L}\mathbf{H} = \mathbf{I}$ leads to the BLUE estimator (4.46).

4.10 For a fixed amount of gas, the following connection holds between the pressure P and the volume V:

$$PV^\gamma = c,$$

where γ and c are constants. Assume that we know T pairs of measurements (P_i, V_i). We want to estimate the parameters γ and c using the *linear* least-squares method. Express the situation in the form of a matrix-vector model and explain how the estimates are computed (you need not compute the exact solution).

4.11 Let the probability density function of a scalar-valued random variable z be

$$p(z \mid \theta) = \theta^2 z e^{-\theta z}, \qquad z \geq 0, \quad \theta > 0$$

Determine the maximum likelihood estimate of the parameter θ. There are available T independent measurements $z(1), \dots, z(T)$ on z.

4.12 In a signal processing application five sensors placed mutually according to a cross pattern yield, respectively, the measurements x_0, x_1, x_2, x_3, and x_4, that can be collected to the measurement vector $\mathbf{x}$. The measurements are quantized with 7 bits accuracy so that their values are integers in the interval $0, \dots, 127$. The joint density $p(\mathbf{x} \mid \theta)$ of the measurements is a multinomial density that depends on the unknown parameter θ as follows:

$$p(\mathbf{x} \mid \theta) = k(\mathbf{x})(1/2)^{x_0}(\theta/4)^{x_1}(1/4 - \theta/4)^{x_2}(1/4 - \theta/4)^{x_3}(\theta/4)^{x_4}$$

where the scaling term

$$k(\mathbf{x}) = \frac{(x_0 + x_1 + x_2 + x_3 + x_4)!}{x_0! x_1! x_2! x_3! x_4!}$$

Determine the maximum likelihood estimate of the parameter θ in terms of the measurement vector $\mathbf{x}$. (Here, you can here treat the individual measurements in a similar manner as mutually independent scalar measurements.)

4.13   Consider the sum $z = x_1 + x_2 + \ldots + x_K$, where the scalar random variables x_i are statistically independent and gaussian, each having the same mean 0 and variance σ_x^2.

4.13.1. Construct the maximum likelihood estimate for the number K of the terms in the sum.

4.13.2. Is this estimate unbiased?

4.14   * Consider direct evaluation of the Wiener filter.

4.14.1. Show that the mean-square filtering error (4.78) can be evaluated to the form (4.79).

4.14.2. What is the minimum mean-square error given by the Wiener estimate?

4.15   The random variables x_1, x_2, and a third, related random variable y are jointly distributed. Define the random vector

$$\mathbf{z} = [y, x_1, x_2]^T$$

It is known that $\mathbf{z}$ has the mean vector $\mathbf{m}_z$ and the covariance matrix $\mathbf{C}_z$ given by

$$\mathbf{m}_z = \begin{bmatrix} 1/4 \\ 1/2 \\ 1/2 \end{bmatrix}, \qquad \mathbf{C}_z = \frac{1}{10} \begin{bmatrix} 7 & 1 & 1 \\ 1 & 3 & -1 \\ 1 & -1 & 3 \end{bmatrix}$$

Find the optimum linear mean-square estimate of y based on x_1 and x_2.

4.16   * Assume that you know T data vectors $\mathbf{z}(1), \mathbf{z}(2), \ldots, \mathbf{z}(T)$ and their corresponding desired responses $d(1), d(2), \ldots, d(T)$. Standard estimates of the correlation matrix and the cross-correlation vector needed in Wiener filtering are [172]

$$\hat{\mathbf{R}}_{\mathbf{z}} = \frac{1}{t}\sum_{i=1}^{T} \mathbf{z}(i)\mathbf{z}(i)^T, \qquad \hat{\mathbf{r}}_{zd} = \frac{1}{T}\sum_{i=1}^{T} \mathbf{z}(i)d(i) \tag{4.87}$$

4.16.1. Express the estimates (4.87) in matrix form and show that when they are used in the Wiener filter (4.80) instead of the true values, the filter coincides with a least-squares solution.

4.16.2. What is the discrete data model corresponding to this least-squares estimator?

4.17 * The joint density function of the random variables x and y is given by

$$p_{xy}(x,y) = 8xy, \quad 0 \leq y \leq x \leq 1,$$

and $p_{xy}(x,y) = 0$ outside the region defined above.

4.17.1. Find and sketch the conditional density $p_{y|x}(y \mid x)$.

4.17.2. Compute the MAP (maximum a posteriori) estimate of y.

4.17.3. Compute the optimal mean-square error estimate of y.

4.18 * Suppose that a scalar random variable y is of the form $y = z + v$, where the pdf of v is $p_v(t) = t/2$ on the interval $[0,2]$, and the pdf of z is $p_z(t) = 2t$ on the interval $[0,1]$. Both the densities are zero elsewhere. There is available a single measurement value $y = 2.5$.

4.18.1. Compute the maximum likelihood estimate of y.

4.18.2. Compute the MAP (maximum a posteriori) estimate of y.

4.18.3. Compute the minimum mean-square estimate of y.

4.19 * Consider the MAP estimator (4.84) of the mean μ.

4.19.1. Derive the estimator.

4.19.2. Express the estimator in recursive form.

Computer assignments

4.1 Choose a suitable set of two-dimensional data. Plenty of real-world data can be found for example using the links of the WWW page of this book, as well as in [376] and at the following Web sites:
http://ferret.wrc.noaa.gov/
http://www.ics.uci.edu/ mlearn/MLSummary.html

4.1.1. Plot the data (or part of it, if the data set is large).

4.1.2. Based on the plot, choose a suitable function (which is linear with respect to the parameters), and fit it to your data using the standard least-squares method. (Alternatively, you can use nonlinear least-squares method if the parameters of the chosen function depend nonlinearly on the data.)

4.1.3. Plot the fitted curve and the fitting error. Assess the quality of your least-squares model.

4.2 * Use the Bayesian linear minimum mean-square estimator for predicting a scalar measurement from other measurements.

4.2.1. Choose first a suitable data set in which the components of the data vectors are correlated (see the previous computer assignment for finding data).

4.2.2. Compute the linear minimum mean-square estimator.

4.2.3. Compute the variance of the measurement that you have predicted and compare it with your minimum mean-square estimation (prediction) error.

5

Information Theory

Estimation theory gives one approach to characterizing random variables. This was based on building parametric models and describing the data by the parameters.

An alternative approach is given by information theory. Here the emphasis is on *coding*. We want to code the observations. The observations can then be stored in the memory of a computer, or transmitted by a communications channel, for example. Finding a suitable code depends on the statistical properties of the data. In independent component analysis (ICA), estimation theory and information theory offer the two principal theoretical approaches.

In this chapter, the basic concepts of information theory are introduced. The latter half of the chapter deals with a more specialized topic: approximation of entropy. These concepts are needed in the ICA methods of Part II.

5.1 ENTROPY

5.1.1 Definition of entropy

Entropy is the basic concept of information theory. Entropy H is defined for a discrete-valued random variable X as

$$H(X) = - \sum_i P(X = a_i) \log P(X = a_i) \qquad (5.1)$$

where the a_i are the possible values of X. Depending on what the base of the logarithm is, different units of entropy are obtained. Usually, the logarithm with base 2 is used, in which case the unit is called a bit. In the following the base is

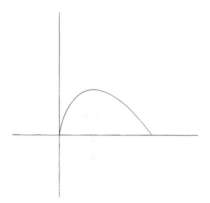

Fig. 5.1 The function f in (5.2), plotted on the interval $[0, 1]$.

not important since it only changes the measurement scale, so it is not explicitly mentioned.

Let us define the function f as

$$f(p) = -p \log p, \quad \text{for } 0 \leq p \leq 1 \qquad (5.2)$$

This is a nonnegative function that is zero for $p = 0$ and for $p = 1$, and positive for values in between; it is plotted in Fig. 5.1. Using this function, entropy can be written as

$$H(X) = \sum_i f(P(X = a_i)) \qquad (5.3)$$

Considering the shape of f, we see that the entropy is small if the probabilities $P(X = a_i)$ are close to 0 or 1, and large if the probabilities are in between.

In fact, the entropy of a random variable can be interpreted as the degree of information that the observation of the variable gives. The more "random", i.e., unpredictable and unstructured the variable is, the larger its entropy. Assume that the probabilities are all close to 0, expect for one that is close to 1 (the probabilities must sum up to one). Then there is little randomness in the variable, since it almost always takes the same value. This is reflected in its small entropy. On the other hand, if all the probabilities are equal, then they are relatively far from 0 and 1, and f takes large values. This means that the entropy is large, which reflects the fact that the variable is really random: We cannot predict which value it takes.

Example 5.1 Let us consider a random variable X that can have only two values, a and b. Denote by p the probability that it has the value a, then the probability that it is b is equal to $1 - p$. The entropy of this random variable can be computed as

$$H(X) = f(p) + f(1 - p) \qquad (5.4)$$

Thus, entropy is a simple function of p. (It does not depend on the values a and b.) Clearly, this function has the same properties as f: it is a nonnegative function that is zero for $p = 0$ and for $p = 1$, and positive for values in between. In fact, it it is maximized for $p = 1/2$ (this is left as an exercice). Thus, the entropy is largest when the values are both obtained with a probability of 50%. In contrast, if one of these values is obtained almost always (say, with a probability of 99.9%), the entropy of X is small, since there is little randomness in the variable.

5.1.2   Entropy and coding length

The connection between entropy and randomness can be made more rigorous by considering *coding length*. Assume that we want to find a binary code for a large number of observations of X, so that the code uses the minimum number of bits possible. According to the fundamental results of information theory, entropy is very closely related to the length of the code required. Under some simplifying assumptions, the length of the shortest code is bounded below by the entropy, and this bound can be approached arbitrarily close, see, e.g., [97]. So, entropy gives roughly the average minimum code length of the random variable.

Since this topic is out of the scope of this book, we will just illustrate it with two examples.

Example 5.2 Consider again the case of a random variable with two possible values, a and b. If the variable almost always takes the same value, its entropy is small. This is reflected in the fact that the variable is easy to code. In fact, assume the value a is almost always obtained. Then, one efficient code might be obtained simply by counting how many a's are found between two subsequent observations of b, and writing down these numbers. If we need to code only a few numbers, we are able to code the data very efficiently.

In the extreme case where the probability of a is 1, there is actually nothing left to code and the coding length is zero. On the other hand, if both values have the same probability, this trick cannot be used to obtain an efficient coding mechanism, and every value must be coded separately by one bit.

Example 5.3 Consider a random variable X that can have eight different values with probabilities $(1/2, 1/4, 1/8, 1/16, 1/64, 1/64, 1/64, 1/64)$. The entropy of X is 2 bits (this computation is left as an exercice to the reader). If we just coded the data in the ordinary way, we would need 3 bits for every observation. But a more intelligent way is to code frequent values with short binary strings and infrequent values with longer strings. Here, we could use the following strings for the outcomes: 0,10,110,1110,111100,111101,111110,111111. (Note that the strings can be written one after another with no spaces since they are designed so that one always knows when the string ends.) With this encoding the *average* number of bits needed for each outcome is only 2, which is in fact equal to the entropy. So we have gained a 33% reduction of coding length.

5.1.3 Differential entropy

The definition of entropy for a discrete-valued random variable can be generalized for continuous-valued random variables and vectors, in which case it is often called differential entropy.

The differential entropy H of a random variable x with density $p_x(.)$ is defined as:

$$H(x) = -\int p_x(\xi) \log p_x(\xi) d\xi = \int f(p_x(\xi)) d\xi \tag{5.5}$$

Differential entropy can be interpreted as a measure of randomness in the same way as entropy. If the random variable is concentrated on certain small intervals, its differential entropy is small.

Note that differential entropy can be negative. Ordinary entropy cannot be negative because the function f in (5.2) is nonnegative in the interval $[0, 1]$, and discrete probabilities necessarily stay in this interval. But probability densities can be larger than 1, in which case f takes negative values. So, when we speak of a "small differential entropy", it may be negative and have a large absolute value.

It is now easy to see what kind of random variables have small entropies. They are the ones whose probability densities take large values, since these give strong negative contributions to the integral in (5.8). This means that certain intervals are quite probable. Thus we again find that entropy is small when the variable is not very random, that is, it is contained in some limited intervals with high probabilities.

Example 5.4 Consider a random variable x that has a uniform probability distribution in the interval $[0, a]$. Its density is given by

$$p_x(\xi) = \begin{cases} 1/a, & \text{for } 0 \leq \xi \leq a \\ 0, & \text{otherwise} \end{cases} \tag{5.6}$$

The differential entropy can be evaluated as

$$H(x) = -\int_0^a \frac{1}{a} \log \frac{1}{a} d\xi = \log a \tag{5.7}$$

Thus we see that the entropy is large if a is large, and small if a is small. This is natural because the smaller a is, the less randomness there is in x. In the limit where a goes to 0, differential entropy goes to $-\infty$, because in the limit, x is no longer random at all: it is always 0.

The interpretation of entropy as coding length is more or less valid with differential entropy. The situation is more complicated, however, since the coding length interpretation requires that we discretize (quantize) the values of x. In this case, the coding length depends on the discretization, i.e., on the accuracy with which we want to represent the random variable. Thus the actual coding length is given by the sum of entropy and a function of the accuracy of representation. We will not go into the details here; see [97] for more information.

The definition of differential entropy can be straightforwardly generalized to the multidimensional case. Let $\mathbf{x}$ be a random vector with density $p_x(.)$. The differential entropy is then defined as:

$$H(\mathbf{x}) = -\int p_x(\boldsymbol{\xi}) \log p_x(\boldsymbol{\xi}) d\boldsymbol{\xi} = \int f(p_x(\boldsymbol{\xi})) d\boldsymbol{\xi} \qquad (5.8)$$

5.1.4 Entropy of a transformation

Consider an *invertible* transformation of the random vector $\mathbf{x}$, say

$$\mathbf{y} = \mathbf{f}(\mathbf{x}) \qquad (5.9)$$

In this section, we show the connection between the entropy of $\mathbf{y}$ and that of $\mathbf{x}$.

A short, if somewhat sloppy derivation is as follows. (A more rigorous derivation is given in the Appendix.) Denote by $J\mathbf{f}(\boldsymbol{\xi})$ the Jacobian matrix of the function $\mathbf{f}$, i.e., the matrix of the partial derivatives of $\mathbf{f}$ at point $\boldsymbol{\xi}$. The classic relation between the density p_y of $\mathbf{y}$ and the density p_x of $\mathbf{x}$, as given in Eq. (2.82), can then be formulated as

$$p_y(\boldsymbol{\eta}) = p_x(\mathbf{f}^{-1}(\boldsymbol{\eta})) |\det J\mathbf{f}(\mathbf{f}^{-1}(\boldsymbol{\eta}))|^{-1} \qquad (5.10)$$

Now, expressing the entropy as an expectation

$$H(\mathbf{y}) = -E\{\log p_y(\mathbf{y})\} \qquad (5.11)$$

we get

$$
\begin{aligned}
E\{\log p_y(\mathbf{y})\} &= E\{\log[p_x(\mathbf{f}^{-1}(\mathbf{y}))|\det J\mathbf{f}(\mathbf{f}^{-1}(\mathbf{y}))|^{-1}]\} \\
&= E\{\log[p_x(\mathbf{x})|\det J\mathbf{f}(\mathbf{x})|^{-1}]\} = E\{\log p_x(\mathbf{x})\} - E\{\log|\det J\mathbf{f}(\mathbf{x})|\} \quad (5.12)
\end{aligned}
$$

Thus we obtain the relation between the entropies as

$$H(\mathbf{y}) = H(\mathbf{x}) + E\{\log|\det J\mathbf{f}(\mathbf{x})|\} \qquad (5.13)$$

In other words, the entropy is increased in the transformation by $E\{\log|\det J\mathbf{f}(\mathbf{x})|\}$.

An important special case is *the linear transformation*

$$\mathbf{y} = \mathbf{M}\mathbf{x} \qquad (5.14)$$

in which case we obtain

$$H(\mathbf{y}) = H(\mathbf{x}) + \log|\det \mathbf{M}| \qquad (5.15)$$

This also shows that differential entropy is *not scale-invariant*. Consider a random variable x. If we multiply it by a scalar constant, α, differential entropy changes as

$$H(\alpha x) = H(x) + \log|\alpha| \qquad (5.16)$$

Thus, just by changing the scale, we can change the differential entropy. This is why the scale of x often is fixed before measuring its differential entropy.

5.2 MUTUAL INFORMATION

5.2.1 Definition using entropy

Mutual information is a measure of the information that members of a set of random variables have on the other random variables in the set. Using entropy, we can define the mutual information I between n (scalar) random variables, $x_i, i = 1, ..., n$, as follows

$$I(x_1, x_2, ..., x_n) = \sum_{i=1}^{n} H(x_i) - H(\mathbf{x}) \qquad (5.17)$$

where $\mathbf{x}$ is the vector containing all the x_i.

Mutual information can be interpreted by using the interpretation of entropy as code length. The terms $H(x_i)$ give the lengths of codes for the x_i when these are coded separately, and $H(\mathbf{x})$ gives the code length when $\mathbf{x}$ is coded as a random vector, i.e., all the components are coded in the same code. Mutual information thus shows what code length reduction is obtained by coding the whole vector instead of the separate components. In general, better codes can be obtained by coding the whole vector. However, if the x_i are independent, they give no information on each other, and one could just as well code the variables separately without increasing code length.

5.2.2 Definition using Kullback-Leibler divergence

Alternatively, mutual information can be interpreted as a distance, using what is called the Kullback-Leibler divergence. This is defined between two n-dimensional probability density functions (pdf's) p^1 and p^2 as

$$\delta(p^1, p^2) = \int p^1(\boldsymbol{\xi}) \log \frac{p^1(\boldsymbol{\xi})}{p^2(\boldsymbol{\xi})} d\boldsymbol{\xi} \qquad (5.18)$$

The Kullback-Leibler divergence can be considered as a kind of a distance between the two probability densities, because it is always nonnegative, and zero if and only if the two distributions are equal. This is a direct consequence of the (strict) convexity of the negative logarithm, and the application of the classic Jensen's inequality. Jensen's inequality (see [97]) says that for any strictly convex function f and any random variable y, we have

$$E\{f(y)\} \geq f(E\{y\}) \qquad (5.19)$$

Take $f(y) = -\log(y)$, and assume that $y = p^2(x)/p^1(x)$ where x has the distribution given by p^1. Then we have

$$\delta(p^1, p^2) = E\{f(y)\} = E\{-\log \frac{p^2(\mathbf{x})}{p^1(\mathbf{x})}\} = \int p^1(\boldsymbol{\xi})\{-\log \frac{p^2(\boldsymbol{\xi})}{p^1(\boldsymbol{\xi})}\}d\boldsymbol{\xi}$$

$$\geq f(E\{y\}) = -\log \int p^1(\boldsymbol{\xi})\{\frac{p^2(\boldsymbol{\xi})}{p^1(\boldsymbol{\xi})}\}d\boldsymbol{\xi} = -\log \int p^2(\boldsymbol{\xi})d\boldsymbol{\xi} = 0 \quad (5.20)$$

Moreover, we have equality in Jensen's inequality if and only if y is constant. In our case, it is constant if and only if the two distributions are equal, so we have proven the announced property of the Kullback-Leibler divergence.

Kullback-Leibler divergence is not a proper distance measure, though, because it is not symmetric.

To apply Kullback-Leibler divergence here, let us begin by considering that if random variables x_i were independent, their joint probability density could be factorized according to the definition of independence. Thus one might measure the independence of the x_i as the Kullback-Leibler divergence between the real density $p^1 = p_x(\xi)$ and the factorized density $p^2 = p_1(\xi_1)p_2(\xi_2)...p_n(\xi_n)$, where the $p_i(.)$ are the marginal densities of the x_i. In fact, simple algebraic manipulations show that this quantity equals the mutual information that we defined using entropy in (5.17), which is left as an exercice.

The interpretation as Kullback-Leibler divergence implies the following important property: *Mutual information is always nonnegative, and it is zero if and only if the variables are independent.* This is a direct consequence of the properties of the Kullback-Leibler divergence.

5.3   MAXIMUM ENTROPY

5.3.1   Maximum entropy distributions

An important class of methods that have application in many domains is given by the maximum entropy methods. These methods apply the concept of entropy to the task of regularization.

Assume that the information available on the density $p_x(.)$ of the scalar random variable x is of the form

$$\int p(\xi)F^i(\xi)d\xi = c_i, \text{ for } i = 1, ..., m \qquad (5.21)$$

which means in practice that we have estimated the expectations $E\{F^i(x)\}$ of m different functions F^i of x. (Note that i is here an index, not an exponent.)

The question is now: What is the probability density function p_0 that satisfies the constraints in (5.21), and has maximum entropy among such densities? (Earlier, we defined the entropy of random variable, but the definition can be used with pdf's as well.) This question can be motivated by noting that a finite number of observations cannot tell us exactly what p is like. So we might use some kind of *regularization* to obtain the most useful p compatible with these measurements. Entropy can be here considered as a regularization measure that helps us find the least structured density compatible with the measurements. In other words, the maximum entropy density can be interpreted as the density that is compatible with the measurements and makes the minimum number of assumptions on the data. This is because entropy can be interpreted as a measure of randomness, and therefore the maximum entropy density is the most random of all the pdf's that satisfy the constraints. For further details on why entropy can be used as a measure of regularity, see [97, 353].

The basic result of the maximum entropy method (see, e.g. [97, 353]) tells us that under some regularity conditions, the density $p_0(\xi)$ which satisfies the constraints (5.21) and has maximum entropy among all such densities, is of the form

$$p_0(\xi) = A \exp(\sum_i a_i F^i(\xi)) \tag{5.22}$$

Here, A and a_i are constants that are determined from the c_i, using the constraints in (5.21) (i.e., by substituting the right-hand side of (5.22) for p in (5.21)), and the constraint $\int p_0(\xi)d\xi = 1$. This leads in general to a system of $n + 1$ nonlinear equations that may be difficult to solve, and in general, numerical methods must be used.

5.3.2 Maximality property of the gaussian distribution

Now, consider the set of random variables that can take all the values on the real line, and have zero mean and a fixed variance, say 1 (thus, we have two constraints). The maximum entropy distribution for such variables is the gaussian distribution. This is because by (5.22), the distribution has the form

$$p_0(\xi) = A \exp(a_1 \xi^2 + a_2 \xi) \tag{5.23}$$

and all probability densities of this form are gaussian by definition (see Section 2.5).

Thus we have the fundamental result that *a gaussian variable has the largest entropy among all random variables of unit variance*. This means that entropy could be used as a measure of nongaussianity. In fact, this shows that the gaussian distribution is the "most random" or the least structured of all distributions. Entropy is small for distributions that are clearly concentrated on certain values, i.e., when the variable is clearly clustered, or has a pdf that is very "spiky". This property can be generalized to arbitrary variances, and what is more important, to multidimensional spaces: The gaussian distribution has maximum entropy among all distributions with a given covariance matrix.

5.4 NEGENTROPY

The maximality property given in Section 5.3.2 shows that entropy could be used to define a measure of nongaussianity. A measure that is zero for a gaussian variable and always nonnegative can be simply obtained from differential entropy, and is called negentropy. Negentropy J is defined as follows

$$J(\mathbf{x}) = H(\mathbf{x}_{gauss}) - H(\mathbf{x}) \tag{5.24}$$

where $\mathbf{x}_{gauss}$ is a gaussian random vector of the same covariance matrix $\mathbf{\Sigma}$ as $\mathbf{x}$. Its entropy can be evaluated as

$$H(\mathbf{x}_{gauss}) = \frac{1}{2} \log |\det \mathbf{\Sigma}| + \frac{n}{2}[1 + \log 2\pi] \qquad (5.25)$$

where n is the dimension of $\mathbf{x}$.

Due to the previously mentioned maximality property of the gaussian distribution, negentropy is always nonnegative. Moreover, it is zero if and only if $\mathbf{x}$ has a gaussian distribution, since the maximum entropy distribution is unique.

Negentropy has the additional interesting property that it is *invariant for invertible linear transformations*. This is because for $\mathbf{y} = \mathbf{M}\mathbf{x}$ we have $E\{\mathbf{y}\mathbf{y}^T\} = \mathbf{M}\mathbf{\Sigma}\mathbf{M}^T$, and, using preceding results, the negentropy can be computed as

$$J(\mathbf{M}\mathbf{x}) = \frac{1}{2} \log |\det(\mathbf{M}\mathbf{\Sigma}\mathbf{M}^T)| + \frac{n}{2}[1 + \log 2\pi] - (H(\mathbf{x}) + \log |\mathbf{M}|)$$

$$= \frac{1}{2} \log |\det \mathbf{\Sigma}| + 2\frac{1}{2} \log |\det \mathbf{M}| + \frac{n}{2}[1 + \log 2\pi] - H(\mathbf{x}) - \log |\det \mathbf{M}|$$

$$= \frac{1}{2} \log |\det \mathbf{\Sigma}| + \frac{n}{2}[1 + \log 2\pi] - H(\mathbf{x})$$

$$= H(\mathbf{x}_{gauss}) - H(\mathbf{x}) = J(\mathbf{x}) \quad (5.26)$$

In particular negentropy is scale-invariant, i.e., multiplication of a random variable by a constant does not change its negentropy. This was not true for differential entropy, as we saw earlier.

5.5 APPROXIMATION OF ENTROPY BY CUMULANTS

In the previous section we saw that negentropy is a principled measure of nongaussianity. The problem in using negentropy is, however, that it is computationally very difficult. To use differential entropy or negentropy in practice, we could compute the integral in the definition in (5.8). This is, however, quite difficult since the integral involves the probability density function. The density could be estimated using basic density estimation methods such as kernel estimators. Such a simple approach would be very error prone, however, because the estimator would depend on the correct choice of the kernel parameters. Moreover, it would be computationally rather complicated.

Therefore, differential entropy and negentropy remain mainly theoretical quantities. In practice, some approximations, possibly rather coarse, have to be used. In this section and the next one we discuss different approximations of negentropy that will be used in the ICA methods in Part II of this book.

5.5.1 Polynomial density expansions

The classic method of approximating negentropy is using higher-order cumulants (defined in Section 2.7). These are based on the idea of using an expansion not unlike a Taylor expansion. This expansion is taken for the pdf of a random variable, say x, in the vicinity of the gaussian density. (We only consider the case of scalar random variables here, because it seems to be sufficient in most applications.) For simplicity, let us first make x zero-mean and of unit variance. Then, we can make the technical assumption that the density $p_x(\xi)$ of x is near the *standardized* gaussian density

$$\varphi(\xi) = \exp(-\xi^2/2)/\sqrt{2\pi} \tag{5.27}$$

Two expansions are usually used in this context: the Gram-Charlier expansion and the Edgeworth expansion. They lead to very similar approximations, so we only consider the Gram-Charlier expansion here. These expansions use the so-called Chebyshev-Hermite polynomials, denoted by H_i where the index i is a nonnegative integer. These polynomials are defined by the derivatives of the standardized gaussian pdf $\varphi(\xi)$ by the equation

$$\frac{\partial^i \varphi(\xi)}{\partial \xi^i} = (-1)^i H_i(\xi)\varphi(\xi) \tag{5.28}$$

Thus, H_i is a polynomial of order i. These polynomials have the nice property of forming an orthonormal system in the following sense:

$$\int \varphi(\xi) H_i(\xi) H_j(\xi) d\xi = \begin{cases} 1, & \text{if } i = j \\ 0, & \text{if } i \neq j \end{cases} \tag{5.29}$$

The Gram-Charlier expansion of the pdf of x, truncated to include the two first nonconstant terms, is then given by

$$p_x(\xi) \approx \hat{p}_x(\xi) = \varphi(\xi)(1 + \kappa_3(x)\frac{H_3(\xi)}{3!} + \kappa_4(x)\frac{H_4(\xi)}{4!}) \tag{5.30}$$

This expansion is based on the idea that the pdf of x is very close to a gaussian one, which allows a Taylor-like approximation to be made. Thus, the nongaussian part of the pdf is directly given by the higher-order cumulants, in this case the third- and fourth-order cumulants. Recall that these are called the skewness and kurtosis, and are given by $\kappa_3(x) = E\{x^3\}$ and $\kappa_4(x) = E\{x^4\} - 3$. The expansion has an infinite number of terms, but only those given above are of interest to us. Note that the expansion starts directly from higher-order cumulants, because we standardized x to have zero mean and unit variance.

5.5.2   Using density expansions for entropy approximation

Now we could plug the density in (5.30) into the definition of entropy, to obtain

$$H(x) \approx -\int \hat{p}_x(\xi) \log \hat{p}_x(\xi) d\xi \tag{5.31}$$

This integral is not very simple to evaluate, though. But again using the idea that the pdf is very close to a gaussian one, we see that the cumulants in (5.30) are very small, and thus we can use the simple approximation

$$\log(1 + \epsilon) \approx \epsilon - \epsilon^2/2 \tag{5.32}$$

which gives

$$H(x) \approx - \int \varphi(\xi)(1 + \kappa_3(x)\frac{H_3(\xi)}{3!} + \kappa_4(x)\frac{H_4(\xi)}{4!})$$
$$[\log \varphi(\xi) + \kappa_3(x)\frac{H_3(\xi)}{3!} + \kappa_4(x)\frac{H_4(\xi)}{4!} - (\kappa_3(x)\frac{H_3(\xi)}{3!} + \kappa_4(x)\frac{H_4(\xi)}{4!})^2/2]$$

$$\tag{5.33}$$

This expression can be simplified (see exercices). Straightforward algebraic manipulations then give

$$H(x) \approx - \int \varphi(\xi) \log \varphi(\xi) d\xi - \frac{\kappa_3(x)^2}{2 \times 3!} - \frac{\kappa_4(\xi)^2}{2 \times 4!} \tag{5.34}$$

Thus we finally obtain an approximation of the negentropy of a standardized random variable as

$$J(x) \approx \frac{1}{12}E\{x^3\}^2 + \frac{1}{48}\text{kurt}(x)^2 \tag{5.35}$$

This gives a computationally very simple approximation of the nongaussianity measured by negentropy.

5.6 APPROXIMATION OF ENTROPY BY NONPOLYNOMIAL FUNCTIONS

In the previous section, we introduced cumulant-based approximations of (neg)entropy. However, such cumulant-based methods sometimes provide a rather poor approximation of entropy. There are two main reasons for this. First, finite-sample estimators of higher-order cumulants are highly sensitive to outliers: their values may depend on only a few, possibly erroneous, observations with large values. This means that outliers may completely determine the estimates of cumulants, thus making them useless. Second, even if the cumulants were estimated perfectly, they mainly measure the tails of the distribution, and are largely unaffected by structure near the center of the distribution. This is because expectations of polynomials like the fourth power are much more strongly affected by data far away from zero than by data close to zero.

In this section, we introduce entropy approximations that are based on an approximative maximum entropy method. The motivation for this approach is that the entropy of a distribution cannot be determined from a given finite number of estimated expectations as in (5.21), even if these were estimated exactly. As explained in

Section 5.3, there exist an infinite number of distributions for which the constraints in (5.21) are fulfilled, but whose entropies are very different from each other. In particular, the differential entropy reaches $-\infty$ in the limit where x takes only a finite number of values.

A simple solution to this is the maximum entropy method. This means that we compute the *maximum entropy* that is compatible with our constraints or measurements in (5.21), which is a well-defined problem. This maximum entropy, or further approximations thereof, can then be used as a meaningful approximation of the entropy of a random variable. This is because in ICA we usually want to minimize entropy. The maximum entropy method gives an upper bound for entropy, and its minimization is likely to minimize the true entropy as well.

In this section, we first derive a first-order approximation of the maximum entropy density for a continuous one-dimensional random variable, given a number of simple constraints. This results in a density expansion that is somewhat similar to the classic polynomial density expansions by Gram-Charlier and Edgeworth. Using this approximation of density, an approximation of 1-D differential entropy is derived. The approximation of entropy is both more exact and more robust against outliers than the approximations based on the polynomial density expansions, without being computationally more expensive.

5.6.1   Approximating the maximum entropy

Let us thus assume that we have observed (or, in practice, estimated) a number of expectations of x, of the form

$$\int p(\xi)F^i(\xi)d\xi = c_i, \text{ for } i = 1, ..., m \qquad (5.36)$$

The functions F^i are not, in general, polynomials. In fact, if we used simple polynomials, we would end up with something very similar to what we had in the preceding section.

Since in general the maximum entropy equations cannot be solved analytically, we make a simple approximation of the maximum entropy density p_0. This is based on the assumption that the density $p(\xi)$ is not very far from the gaussian density of the same mean and variance; this assumption is similar to the one made using polynomial density expansions.

As with the polynomial expansions, we can assume that x has zero mean and unit variance. Therefore we put two additional constraints in (5.36), defined by

$$F^{n+1}(\xi) = \xi, \quad c_{n+1} = 0 \qquad (5.37)$$

$$F^{n+2}(\xi) = \xi^2, \quad c_{n+2} = 1 \qquad (5.38)$$

To further simplify the calculations, let us make another, purely technical assumption: The functions $F^i, i = 1, ..., n$, form an orthonormal system according to the metric defined by φ in (5.27), and are orthogonal to all polynomials of second degree. In other words, for all $i, j = 1, ..., n$

$$\int \varphi(\xi) F^i(\xi) F^j(\xi) d\xi = \begin{cases} 1, & \text{if } i = j \\ 0, & \text{if } i \neq j \end{cases} \tag{5.39}$$

$$\int \varphi(\xi) F^i(\xi) \xi^k d\xi = 0, \text{ for } k = 0, 1, 2 \tag{5.40}$$

Again, these orthogonality constraints are very similar to those of Chebyshev-Hermite polynomials. For any set of linearly independent functions F^i (not containing second-order polynomials), this assumption can always be made true by ordinary Gram-Schmidt orthonormalization.

Now, note that the assumption of near-gaussianity implies that all the other a_i in (5.22) are very small compared to $a_{n+2} \approx -1/2$, since the exponential in (5.22) is not far from $\exp(-\xi^2/2)$. Thus we can make a first-order approximation of the exponential function (detailed derivations can be found in the Appendix). This allows for simple solutions for the constants in (5.22), and we obtain the *approximative maximum entropy density*, which we denote by $\hat{p}(\xi)$:

$$\hat{p}(\xi) = \varphi(\xi)(1 + \sum_{i=1}^{n} c_i F^i(\xi)) \tag{5.41}$$

where $c_i = E\{F^i(\xi)\}$.

Now we can derive an approximation of differential entropy using this density approximation. As with the polynomial density expansions, we can use (5.31) and (5.32). After some algebraic manipulations (see the Appendix), we obtain

$$J(x) \approx \frac{1}{2} \sum_{i=1}^{n} E\{F^i(x)\}^2 \tag{5.42}$$

Note that even in cases where this approximation is not very accurate, (5.42) can be used to construct a measure of nongaussianity that is consistent in the sense that (5.42) obtains its minimum value, 0, when x has a gaussian distribution. This is because according to the latter part of (5.39) with $k = 0$, we have $E\{F^i(\nu)\} = 0$.

5.6.2   Choosing the nonpolynomial functions

Now it remains to choose the "measuring" functions F^i that define the information given in (5.36). As noted in Section 5.6.1, one can take practically any set of linearly independent functions, say $G^i, i = 1, ..., m$, and then apply Gram-Schmidt orthonormalization on the set containing those functions and the monomials $\xi^k, k = 0, 1, 2$, so as to obtain the set F^i that fulfills the orthogonality assumptions in (5.39).

This can be done, in general, by numerical integration. In the practical choice of the functions G^i, the following criteria must be emphasized:

1. The practical estimation of $E\{G^i(x)\}$ should not be statistically difficult. In particular, this estimation should not be too sensitive to outliers.

2. The maximum entropy method assumes that the function p_0 in (5.22) is integrable. Therefore, to ensure that the maximum entropy distribution exists in the first place, the $G^i(x)$ must not grow faster than quadratically as a function of $|x|$, because a function growing faster might lead to the nonintegrability of p_0.

3. The G^i must capture aspects of the distribution of X that are pertinent in the computation of entropy. In particular, if the density $p(\xi)$ were known, the optimal function G^{opt} would clearly be $-\log p(\xi)$, because $-E\{\log p(x)\}$ gives the entropy directly Thus, one might use for G^i the log-densities of some known important densities.

The first two criteria are met if the $G^i(x)$ are functions that do not grow too fast (not faster than quadratically) as $|x|$ increases. This excludes, for example, the use of higher-order polynomials, which are used in the Gram-Charlier and Edgeworth expansions. One might then search, according to criterion 3, for log-densities of some well-known distributions that also fulfill the first two conditions. Examples will be given in the next subsection.

It should be noted, however, that the criteria above only delimit the space of functions that can be used. Our framework enables the use of very different functions (or just one) as G^i. However, if prior knowledge is available on the distributions whose entropy is to be estimated, criterion 3 shows how to choose the optimal function.

5.6.3 Simple special cases

A simple special case of (5.41) is obtained if one uses two functions G^1 and G^2, which are chosen so that G^1 is *odd* and G^2 is *even*. Such a system of two functions can measure the two most important features of nongaussian 1-D distributions. The odd function measures the asymmetry, and the even function measures the dimension of bimodality vs. peak at zero, closely related to sub- vs. supergaussianity. Classically, these features have been measured by skewness and kurtosis, which correspond to $G^1(x) = x^3$ and $G^2(x) = x^4$, but we do not use these functions for the reasons explained in Section 5.6.2. (In fact, with these choices, the approximation in (5.41) becomes identical to the one obtained from the Gram-Charlier expansion in (5.35).)

In this special case, the approximation in (5.42) simplifies to

$$J(x) \approx k_1 (E\{G^1(x)\})^2 + k_2 (E\{G^2(x)\} - E\{G^2(\nu)\})^2 \tag{5.43}$$

where k_1 and k_2 are positive constants (see the Appendix). Practical examples of choices of G^i that are consistent with the requirements in Section 5.6.2 are the following. First, for measuring bimodality/sparsity, one might use, according to the recommendations of Section 5.6.2, the log-density of the Laplacian distribution:

$$G^{2a}(x) = |x| \tag{5.44}$$

For computational reasons, a smoother version of G^{2a} might also be used. Another choice would be the gaussian function, which can be considered as the log-density of a distribution with infinitely heavy tails (since it stays constant when going to infinity):

$$G^{2b}(x) = \exp(-x^2/2) \tag{5.45}$$

For measuring asymmetry, one might use, on more heuristic grounds, the following function:

$$G^1(x) = x \exp(-x^2/2) \tag{5.46}$$

that is smooth and robust against outliers.

Using the preceding examples one obtains two practical examples of (5.43):

$$J_a(x) = k_1 (E\{x \exp(-x^2/2)\})^2 + k_2^a (E\{|x|\} - \sqrt{2/\pi})^2 \tag{5.47}$$

and

$$J_b(x) = k_1 (E\{x \exp(-x^2/2)\})^2 + k_2^b (E\{\exp(-x^2/2)\} - \sqrt{1/2})^2 \tag{5.48}$$

with $k_1 = 36/(8\sqrt{3} - 9)$, $k_2^a = 1/(2 - 6/\pi)$, and $k_2^b = 24/(16\sqrt{3} - 27)$. These approximations $J_a(x)$ and $J_b(x)$ can be considered more robust and accurate generalizations of the approximation derived using the Gram-Charlier expansion in Section 5.5.

Even simpler approximations of negentropy can be obtained by using only one nonquadratic function, which amounts to omitting one of the terms in the preceding approximations.

5.6.4 Illustration

Here we illustrate the differences in accuracy of the different approximations of negentropy. The expectations were here evaluated exactly, ignoring finite-sample effects. Thus these results do not illustrate the robustness of the maximum entropy approximation with respect to outliers; this is quite evident anyway.

First, we used a family of gaussian mixture densities, defined by

$$p(\xi) = \mu\varphi(x) + (1 - \mu)2\varphi(2(x - 1)) \tag{5.49}$$

where μ is a parameter that takes all the values in the interval $0 \le \mu \le 1$. This family includes asymmetric densities of both negative and positive kurtosis. The results are depicted in Fig. 5.2. One can see that both of the approximations J_a and J_b introduced in Section 5.6.3 were considerably more accurate than the cumulant-based approximation in (5.35).

Second, we considered the exponential power family of density functions:

$$p_\alpha(\xi) = C_1 \exp(-C_2|\xi|^\alpha) \tag{5.50}$$

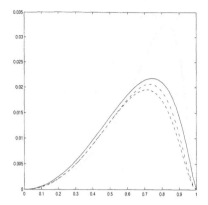

Fig. 5.2 Comparison of different approximations of negentropy for the family of mixture densities in (5.49) parametrized by μ ranging from 0 to 1 (horizontal axis). Solid curve: true negentropy. Dotted curve: cumulant-based approximation as in (5.35). Dashed curve: approximation J_a in (5.47). Dot-dashed curve: approximation J_b in (5.48). The two maximum entropy approximations were clearly better than the cumulant-based one.

where α is a positive constant, and C_1, C_2 are normalization constants that make p_α a probability density of unit variance. For different values of α, the densities in this family exhibit different shapes. For $\alpha < 2$, one obtains densities of positive kurtosis (supergaussian). For $\alpha = 2$, one obtains the gaussian density, and for $\alpha > 2$, a density of negative kurtosis. Thus the densities in this family can be used as examples of different symmetric nongaussian densities. In Fig. 5.3, the different negentropy approximations are plotted for this family, using parameter values $0.5 \le \alpha \le 3$. Since the densities used are all symmetric, the first terms in the approximations were neglected. Again, it is clear that both of the approximations J_a and J_b introduced in Section 5.6.3 were considerably more accurate than the cumulant-based approximation in (5.35). Especially in the case of supergaussian densities, the cumulant-based approximation performed very poorly; this is probably because it gives too much weight to the tails of the distribution.

5.7 CONCLUDING REMARKS AND REFERENCES

Most of the material in this chapter can be considered classic. The basic definitions of information theory and the relevant proofs can be found, e.g., in [97, 353]. The approximations of entropy are rather recent, however. The cumulant-based approximation was proposed in [222], and it is almost identical to those proposed in [12, 89]. The approximations of entropy using nonpolynomial functions were introduced in [196], and they are closely related to the measures of nongaussianity that have been proposed in the projection pursuit literature, see, e.g., [95].

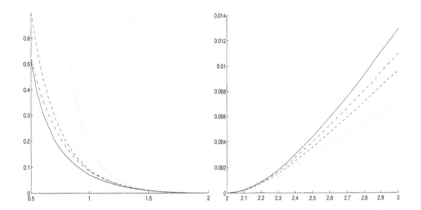

Fig. 5.3 Comparison of different approximations of negentropy for the family of densities (5.50) parametrized by α (horizontal axis). On the left, approximations for densities of positive kurtosis ($0.5 \leq \alpha < 2$) are depicted, and on the right, approximations for densities of negative kurtosis ($2 < \alpha \leq 3$). Solid curve: true negentropy. Dotted curve: cumulant-based approximation as in (5.35). Dashed curve: approximation J_a in (5.47). Dot-dashed curve: approximation J_b in (5.48). Clearly, the maximum entropy approximations were much better than the cumulant-based one, especially in the case of densities of positive kurtosis.

Problems

5.1   Assume that the random variable X can have two values, a and b, as in Example 5.1. Compute the entropy as a function of the probability of obtaining a. Show that this is maximized when the probability is $1/2$.

5.2   Compute the entropy of X in Example 5.3.

5.3   Assume x has a Laplacian distribution of arbitrary variance with pdf

$$p_x(\xi) = \frac{1}{\sqrt{2}\sigma} \exp(\frac{\sqrt{2}}{\sigma}|\xi|) \qquad (5.51)$$

Compute the differential entropy.

5.4   Prove (5.15).

5.5   Prove (5.25).

5.6   Show that the definition of mutual information using Kullback-Leibler divergence is equal to the one given by entropy.

5.7   Compute the three first Chebyshev-Hermite polynomials.

5.8   Prove (5.34). Use the orthogonality in (5.29), and in particular the fact that H_3 and H_4 are orthogonal to any second-order polynomial (prove this first!). Furthermore, use the fact that any expression involving a third-order monomial of the higher-order cumulants is infinitely smaller than terms involving only second-order monomials (due to the assumption that the pdf is very close to gaussian).

Computer assignments

5.1   Consider random variables with (1) a uniform distribution and (2) a Laplacian distribution, both with zero mean and unit variance. Compute their differential entropies with numerical integration. Then, compute the approximations given by the polynomial and nonpolynomial approximations given in this chapter. Compare the results.

Appendix proofs

First, we give a detailed proof of (5.13). We have by (5.10)

$$
H(\mathbf{y}) = - \int p_y(\boldsymbol{\eta}) \log p_y(\boldsymbol{\eta}) d\boldsymbol{\eta}
$$

$$
= - \int p_x(\mathbf{f}^{-1}(\boldsymbol{\eta})) |\det J\mathbf{f}(\mathbf{f}^{-1}(\boldsymbol{\eta}))|^{-1} \log[p_x(\mathbf{f}^{-1}(\boldsymbol{\eta})) |\det J\mathbf{f}(\mathbf{f}^{-1}(\boldsymbol{\eta}))|^{-1}] d\boldsymbol{\eta}
$$

$$
= - \int p_x(\mathbf{f}^{-1}(\boldsymbol{\eta})) \log[p_x(\mathbf{f}^{-1}(\boldsymbol{\eta}))] |\det J\mathbf{f}(\mathbf{f}^{-1}(\boldsymbol{\eta}))|^{-1} d\boldsymbol{\eta}
$$

$$
- \int p_x(\mathbf{f}^{-1}(\boldsymbol{\eta})) \log[|\det J\mathbf{f}(\mathbf{f}^{-1}(\boldsymbol{\eta}))|^{-1}] |\det J\mathbf{f}(\mathbf{f}^{-1}(\boldsymbol{\eta}))|^{-1} d\boldsymbol{\eta} \quad \text{(A.1)}
$$

Now, let us make the change of integration variable

$$
\boldsymbol{\xi} = \mathbf{f}^{-1}(\boldsymbol{\eta}) \tag{A.2}
$$

which gives us

$$
H(\mathbf{y}) = - \int p_x(\boldsymbol{\xi}) \log[p_x(\boldsymbol{\xi})] |\det J\mathbf{f}(\boldsymbol{\xi})|^{-1} |\det J\mathbf{f}(\boldsymbol{\xi})| d\boldsymbol{\xi}
$$

$$
- \int p_x(\boldsymbol{\xi}) \log[|\det J\mathbf{f}(\boldsymbol{\xi})|^{-1}] |\det J\mathbf{f}(\boldsymbol{\xi})|^{-1} |\det J\mathbf{f}(\boldsymbol{\xi})| d\boldsymbol{\xi} \quad \text{(A.3)}
$$

where the Jacobians cancel each other, and we have

$$
H(\mathbf{y}) = - \int p_x(\boldsymbol{\xi}) \log[p_x(\boldsymbol{\xi})] d\boldsymbol{\xi} + \int p_x(\boldsymbol{\xi}) \log |\det J\mathbf{f}(\boldsymbol{\xi})| d\boldsymbol{\xi} \tag{A.4}
$$

which gives (5.13).

Now follow the proofs connected with the entropy approximations. First, we prove (5.41). Due to the assumption of near-gaussianity, we can write $p_0(\xi)$ as

$$
p_0(\xi) = A \exp(-\xi^2/2 + a_{n+1}\xi + (a_{n+2} + 1/2)\xi^2 + \sum_{i=1}^{n} a_i G^i(\xi)), \tag{A.5}
$$

where in the exponential, all other terms are very small with respect to the first one. Thus, using the first-order approximation $\exp(\epsilon) \approx 1 + \epsilon$, we obtain

$$p_0(\xi) \approx \tilde{A}\varphi(\xi)(1 + a_{n+1}\xi + (a_{n+2} + 1/2)\xi^2 + \sum_{i=1}^{n} a_i G^i(\xi)), \qquad (A.6)$$

where $\varphi(\xi) = (2\pi)^{-1/2} \exp(-\xi^2/2)$ is the standardized gaussian density, and $\tilde{A} = \sqrt{2\pi}A$. Due to the orthogonality constraints in (5.39), the equations for solving $\tilde{A}$ and a_i become linear and almost diagonal:

$$\int p_0(\xi)d\xi = \tilde{A}(1 + (a_{n+2} + 1/2)) = 1 \qquad (A.7)$$

$$\int p_0(\xi)\xi d\xi = \tilde{A}a_{n+1} = 0 \qquad (A.8)$$

$$\int p_0(\xi)\xi^2 d\xi = \tilde{A}(1 + 3(a_{n+2} + 1/2)) = 1 \qquad (A.9)$$

$$\int p_0(\xi)G^i(\xi)d\xi = \tilde{A}a_i = c_i, \text{ for } i = 1, ..., n \qquad (A.10)$$

and can be easily solved to yield $\tilde{A} = 1$, $a_{n+1} = 0$, $a_{n+2} = -1/2$ and $a_i = c_i, i = 1, .., n$. This gives (5.41).

Second, we prove (5.42). Using the Taylor expansion $(1+\epsilon)\log(1+\epsilon) = \epsilon + \epsilon^2/2 + o(\epsilon^2)$, one obtains

$$-\int \hat{p}(\xi)\log\hat{p}(\xi)d\xi \qquad (A.11)$$

$$= -\int \varphi(\xi)(1 + \sum c_i G^i(\xi))(\log(1 + \sum c_i G^i(\xi)) + \log\varphi(\xi))d\xi \qquad (A.12)$$

$$= -\int \varphi(\xi)\log\varphi(\xi) - \int \varphi(\xi)\sum c_i G^i(\xi)\log\varphi(\xi) \qquad (A.13)$$

$$-\int \varphi(\xi)[\sum c_i G^i(\xi) + \frac{1}{2}(\sum c_i G^i(\xi))^2 + o((\sum c_i G^i(\xi))^2)] \qquad (A.14)$$

$$= H(\nu) - 0 - 0 - \frac{1}{2}\sum c_i^2 + o((\sum c_i)^2) \qquad (A.15)$$

due to the orthogonality relationships in (5.39).

Finally, we prove (5.43), (5.47) and (5.48). First, we must orthonormalize the two functions G^1 and G^2 according to (5.39). To do this, it is enough to determine constants $\beta_1, \delta_1, \alpha_2, \gamma_2, \delta_2$ so that the functions $F^1(x) = (G^1(x) + \beta_1 x)/\delta_1$ and $F^2(x) = (G^2(x) + \alpha_2 x^2 + \gamma_2)/\delta_2$ are orthogonal to any second degree polynomials as in (5.39), and have unit norm in the metric defined by φ. In fact, as will be seen below, this modification gives a G^1 that is odd and a

G^2 that is even, and therefore the G^i are automatically orthogonal with respect to each other. Thus, first we solve the following equations:

$$\int \varphi(\xi)\xi(G^1(\xi) + \beta_1\xi)d\xi = 0 \tag{A.16}$$

$$\int \varphi(\xi)\xi^k(G^2(\xi) + \alpha_2\xi^2 + \gamma_2)d\xi = 0, \text{ for } k = 0, 2 \tag{A.17}$$

A straightforward solution gives:

$$\beta_1 = -\int \varphi(\xi)G^1(\xi)\xi d\xi \tag{A.18}$$

$$\alpha_2 = \frac{1}{2}\left(\int \varphi(\xi)G^2(\xi)d\xi - \int \varphi(\xi)G^2(\xi)\xi^2 d\xi\right) \tag{A.19}$$

$$\gamma_2 = \frac{1}{2}\left(\int \varphi(\xi)G^2(\xi)\xi^2 d\xi - 3\int \varphi(\xi)G^2(\xi)d\xi\right) \tag{A.20}$$

Next note that together with the standardization $\int \varphi(\xi)(G^2(\xi) + \alpha_2\xi^2 + \gamma_2)d\xi = 0$ implies

$$c_i = E\{F^i(x)\} = [E\{G^i(x)\} - E\{G^i(\nu)\}]/\delta_i \tag{A.21}$$

This implies (5.43), with $k_i^2 = 1/(2\delta_i^2)$. Thus we only need to determine explicitly the δ_i for each function. We solve the two equations

$$\int \varphi(\xi)(G^1(\xi) + \beta_1\xi)^2/\delta_1 d\xi = 1 \tag{A.22}$$

$$\int \varphi(\xi)(G^2(\xi) + \alpha_2\xi^2 + \gamma_2)^2/\delta_2 d\xi = 1 \tag{A.23}$$

which, after some tedious manipulations, yield:

$$\delta_1^2 = \int \varphi(\xi)G^1(\xi)^2 d\xi - \left(\int \varphi(\xi)G^1(\xi)\xi \, d\xi\right)^2 \tag{A.24}$$

$$\delta_2^2 = \int \varphi(\xi)G^2(\xi)^2 d\xi - \left(\int \varphi(\xi)G^2(\xi)d\xi\right)^2$$
$$- \frac{1}{2}\left(\int \varphi(\xi)G^2(\xi)d\xi - \int \varphi(\xi)G^2(\xi)\xi^2 d\xi\right)^2. \tag{A.25}$$

Evaluating the δ_i for the given functions G^i, one obtains (5.47) and (5.48) by the relation $k_i^2 = 1/(2\delta_i^2)$.

6

Principal Component Analysis and Whitening

Principal component analysis (PCA) and the closely related Karhunen-Loève transform, or the Hotelling transform, are classic techniques in statistical data analysis, feature extraction, and data compression, stemming from the early work of Pearson [364]. Given a set of multivariate measurements, the purpose is to find a smaller set of variables with less redundancy, that would give as good a representation as possible. This goal is related to the goal of independent component analysis (ICA). However, in PCA the redundancy is measured by correlations between data elements, while in ICA the much richer concept of independence is used, and in ICA the reduction of the number of variables is given less emphasis. Using only the correlations as in PCA has the advantage that the analysis can be based on second-order statistics only. In connection with ICA, PCA is a useful preprocessing step.

The basic PCA problem is outlined in this chapter. Both the closed-form solution and on-line learning algorithms for PCA are reviewed. Next, the related linear statistical technique of factor analysis is discussed. The chapter is concluded by presenting how data can be preprocessed by whitening, removing the effect of first- and second-order statistics, which is very helpful as the first step in ICA.

6.1 PRINCIPAL COMPONENTS

The starting point for PCA is a random vector $\mathbf{x}$ with n elements. There is available a sample $\mathbf{x}(1), ..., \mathbf{x}(T)$ from this random vector. No explicit assumptions on the probability density of the vectors are made in PCA, as long as the first- and second-order statistics are known or can be estimated from the sample. Also, no generative model is assumed for vector **x**. Typically the elements of **x** are measurements like pixel gray levels or values of a signal at different time instants. It is essential in PCA that the elements are mutually correlated, and there is thus some redundancy in **x**, making compression possible. If the elements are independent, nothing can be achieved by PCA.

In the PCA transform, the vector **x** is first centered by subtracting its mean:

$$\mathbf{x} \leftarrow \mathbf{x} - E\{\mathbf{x}\}$$

The mean is in practice estimated from the available sample $\mathbf{x}(1), ..., \mathbf{x}(T)$ (see Chapter 4). Let us assume in the following that the centering has been done and thus $E\{\mathbf{x}\} = 0$. Next, **x** is linearly transformed to another vector **y** with m elements, $m < n$, so that the redundancy induced by the correlations is removed. This is done by finding a rotated orthogonal coordinate system such that the elements of **x** in the new coordinates become uncorrelated. At the same time, the variances of the projections of **x** on the new coordinate axes are maximized so that the first axis corresponds to the maximal variance, the second axis corresponds to the maximal variance in the direction orthogonal to the first axis, and so on.

For instance, if **x** has a gaussian density that is constant over ellipsoidal surfaces in the n-dimensional space, then the rotated coordinate system coincides with the principal axes of the ellipsoid. A two-dimensional example is shown in Fig. 2.7 in Chapter 2. The principal components are now the projections of the data points on the two principal axes, $\mathbf{e}_1$ and $\mathbf{e}_2$. In addition to achieving uncorrelated components, the variances of the components (projections) also will be very different in most applications, with a considerable number of the variances so small that the corresponding components can be discarded altogether. Those components that are left constitute the vector **y**.

As an example, take a set of 8×8 pixel windows from a digital image, an application that is considered in detail in Chapter 21. They are first transformed, e.g., using row-by-row scanning, into vectors **x** whose elements are the gray levels of the 64 pixels in the window. In real-time digital video transmission, it is essential to reduce this data as much as possible without losing too much of the visual quality, because the total amount of data is very large. Using PCA, a compressed representation vector **y** can be obtained from **x**, which can be stored or transmitted. Typically, **y** can have as few as 10 elements, and a good replica of the original 8×8 image window can still be reconstructed from it. This kind of compression is possible because neighboring elements of **x**, which are the gray levels of neighboring pixels in the digital image, are heavily correlated. These correlations are utilized by PCA, allowing almost the same information to be represented by a much smaller vector **y**. PCA is a linear technique, so computing **y** from **x** is not heavy, which makes real-time processing possible.

6.1.1 PCA by variance maximization

In mathematical terms, consider a linear combination

$$y_1 = \sum_{k=1}^{n} w_{k1}x_k = \mathbf{w}_1^T\mathbf{x}$$

of the elements $x_1, ..., x_n$ of the vector $\mathbf{x}$. The $w_{11}, ..., w_{n1}$ are scalar coefficients or weights, elements of an n-dimensional vector $\mathbf{w}_1$, and $\mathbf{w}_1^T$ denotes the transpose of $\mathbf{w}_1$.

The factor y_1 is called the first principal component of $\mathbf{x}$, if the variance of y_1 is maximally large. Because the variance depends on both the norm and orientation of the weight vector $\mathbf{w}_1$ and grows without limits as the norm grows, we impose the constraint that the norm of $\mathbf{w}_1$ is constant, in practice equal to 1. Thus we look for a weight vector $\mathbf{w}_1$ maximizing the PCA criterion

$$J_1^{PCA}(\mathbf{w}_1) = E\{y_1^2\} = E\{(\mathbf{w}_1^T\mathbf{x})^2\} = \mathbf{w}_1^T E\{\mathbf{x}\mathbf{x}^T\}\mathbf{w}_1 = \mathbf{w}_1^T\mathbf{C}_\mathbf{x}\mathbf{w}_1 \quad (6.1)$$
$$\text{so that } \|\mathbf{w}_1\| = 1 \quad (6.2)$$

There $E\{.\}$ is the expectation over the (unknown) density of input vector $\mathbf{x}$, and the norm of $\mathbf{w}_1$ is the usual Euclidean norm defined as

$$\|\mathbf{w}_1\| = (\mathbf{w}_1^T\mathbf{w}_1)^{1/2} = [\sum_{k=1}^{n} w_{k1}^2]^{1/2}$$

The matrix $\mathbf{C}_\mathbf{x}$ in Eq. (6.1) is the $n \times n$ covariance matrix of $\mathbf{x}$ (see Chapter 4) given for the zero-mean vector $\mathbf{x}$ by the correlation matrix

$$\mathbf{C}_\mathbf{x} = E\{\mathbf{x}\mathbf{x}^T\} \quad (6.3)$$

It is well known from basic linear algebra (see, e.g., [324, 112]) that the solution to the PCA problem is given in terms of the unit-length eigenvectors $\mathbf{e}_1, ..., \mathbf{e}_n$ of the matrix $\mathbf{C}_\mathbf{x}$. The ordering of the eigenvectors is such that the corresponding eigenvalues $d_1, ..., d_n$ satisfy $d_1 \geq d_2 \geq ... \geq d_n$. The solution maximizing (6.1) is given by

$$\mathbf{w}_1 = \mathbf{e}_1$$

Thus the first principal component of $\mathbf{x}$ is $y_1 = \mathbf{e}_1^T\mathbf{x}$.

The criterion J_1^{PCA} in eq. (6.1) can be generalized to m principal components, with m any number between 1 and n. Denoting the m-th $(1 \leq m < n)$ principal component by $y_m = \mathbf{w}_m^T\mathbf{x}$, with $\mathbf{w}_m$ the corresponding unit norm weight vector, the variance of y_m is now maximized under the constraint that y_m is uncorrelated with all the previously found principal components:

$$E\{y_m y_k\} = 0, \; k < m. \quad (6.4)$$

Note that the principal components y_m have zero means because

$$E\{y_m\} = \mathbf{w}_m^T E\{\mathbf{x}\} = 0$$

The condition (6.4) yields:

$$E\{y_m y_k\} = E\{(\mathbf{w}_m^T \mathbf{x})(\mathbf{w}_k^T \mathbf{x})\} = \mathbf{w}_m^T \mathbf{C}_{\mathbf{x}} \mathbf{w}_k = 0 \tag{6.5}$$

For the second principal component, we have the condition that

$$\mathbf{w}_2^T \mathbf{C} \mathbf{w}_1 = d_1 \mathbf{w}_2^T \mathbf{e}_1 = 0 \tag{6.6}$$

because we already know that $\mathbf{w}_1 = \mathbf{e}_1$. We are thus looking for maximal variance $E\{y_2^2\} = E\{(\mathbf{w}_2^T \mathbf{x})^2\}$ in the subspace orthogonal to the first eigenvector of $\mathbf{C}_{\mathbf{x}}$. The solution is given by

$$\mathbf{w}_2 = \mathbf{e}_2$$

Likewise, recursively it follows that

$$\mathbf{w}_k = \mathbf{e}_k$$

Thus the kth principal component is $y_k = \mathbf{e}_k^T \mathbf{x}$.

Exactly the same result for the $\mathbf{w}_i$ is obtained if the variances of y_i are maximized under the constraint that the principal component vectors are orthonormal, or $\mathbf{w}_i^T \mathbf{w}_j = \delta_{ij}$. This is left as an exercise.

6.1.2   PCA by minimum mean-square error compression

In the preceding subsection, the principal components were defined as weighted sums of the elements of $\mathbf{x}$ with maximal variance, under the constraints that the weights are normalized and the principal components are uncorrelated with each other. It turns out that this is strongly related to minimum mean-square error compression of $\mathbf{x}$, which is another way to pose the PCA problem. Let us search for a set of m orthonormal basis vectors, spanning an m-dimensional subspace, such that the mean-square error between $\mathbf{x}$ and its projection on the subspace is minimal. Denoting again the basis vectors by $\mathbf{w}_1, ..., \mathbf{w}_m$, for which we assume

$$\mathbf{w}_i^T \mathbf{w}_j = \delta_{ij}$$

the projection of $\mathbf{x}$ on the subspace spanned by them is $\sum_{i=1}^{m}(\mathbf{w}_i^T \mathbf{x})\mathbf{w}_i$. The mean-square error (MSE) criterion, to be minimized by the orthonormal basis $\mathbf{w}_1, ..., \mathbf{w}_m$, becomes

$$J_{MSE}^{PCA} = E\{\|\mathbf{x} - \sum_{i=1}^{m}(\mathbf{w}_i^T \mathbf{x})\mathbf{w}_i\|^2\} \tag{6.7}$$

It is easy to show (see exercises) that due to the orthogonality of the vectors $\mathbf{w}_i$, this criterion can be further written as

$$J_{MSE}^{PCA} = E\{\|\mathbf{x}\|^2\} - E\{\sum_{j=1}^{m}(\mathbf{w}_j^T \mathbf{x})^2\} \tag{6.8}$$

$$= \text{trace}(\mathbf{C}_{\mathbf{x}}) - \sum_{j=1}^{m} \mathbf{w}_j^T \mathbf{C}_{\mathbf{x}} \mathbf{w}_j \tag{6.9}$$

It can be shown (see, e.g., [112]) that the minimum of (6.9) under the orthonormality condition on the $\mathbf{w}_i$ is given by any orthonormal basis of the PCA subspace spanned by the m first eigenvectors $\mathbf{e}_1, ..., \mathbf{e}_m$. However, the criterion does not specify the basis of this subspace at all. Any orthonormal basis of the subspace will give the same optimal compression. While this ambiguity can be seen as a disadvantage, it should be noted that there may be some other criteria by which a certain basis in the PCA subspace is to be preferred over others. Independent component analysis is a prime example of methods in which PCA is a useful preprocessing step, but once the vector $\mathbf{x}$ has been expressed in terms of the first m eigenvectors, a further rotation brings out the much more useful independent components.

It can also be shown [112] that the value of the minimum mean-square error of (6.7) is

$$J_{MSE}^{PCA} = \sum_{i=m+1}^{n} d_i \qquad (6.10)$$

the sum of the eigenvalues corresponding to the discarded eigenvectors $\mathbf{e}_{m+1}, ..., \mathbf{e}_n$.

If the orthonormality constraint is simply changed to

$$\mathbf{w}_j^T \mathbf{w}_k = \omega_k \delta_{jk} \qquad (6.11)$$

where all the numbers ω_k are positive and different, then the mean-square error problem will have a unique solution given by scaled eigenvectors [333].

6.1.3 Choosing the number of principal components

From the result that the principal component basis vectors $\mathbf{w}_i$ are eigenvectors $\mathbf{e}_i$ of $\mathbf{C_x}$, it follows that

$$E\{y_m^2\} = E\{\mathbf{e}_m^T \mathbf{x}\mathbf{x}^T \mathbf{e}_m\} = \mathbf{e}_m^T \mathbf{C_x} \mathbf{e}_m = d_m \qquad (6.12)$$

The variances of the principal components are thus directly given by the eigenvalues of $\mathbf{C_x}$. Note that, because the principal components have zero means, a small eigenvalue (a small variance) d_m indicates that the value of the corresponding principal component y_m is mostly close to zero.

An important application of PCA is data compression. The vectors $\mathbf{x}$ in the original data set (that have first been centered by subtracting the mean) are approximated by the truncated PCA expansion

$$\hat{\mathbf{x}} = \sum_{i=1}^{m} y_i \mathbf{e}_i \qquad (6.13)$$

Then we know from (6.10) that the mean-square error $E\{\|\mathbf{x} - \hat{\mathbf{x}}\|^2\}$ is equal to $\sum_{i=m+1}^{n} d_i$. As the eigenvalues are all positive, the error decreases when more and more terms are included in (6.13), until the error becomes zero when $m = n$ or all the principal components are included. A very important practical problem is how to choose m in (6.13); this is a trade-off between error and the amount of data needed for the expansion. Sometimes a rather small number of principal components are sufficient.

Fig. 6.1 Leftmost column: some digital images in a 32×32 grid. Second column: means of the samples. Remaining columns: reconstructions by PCA when 1, 2, 5, 16, 32, and 64 principal components were used in the expansion.

Example 6.1 In digital image processing, the amount of data is typically very large, and data compression is necessary for storage, transmission, and feature extraction. PCA is a simple and efficient method. Fig. 6.1 shows 10 handwritten characters that were represented as binary 32×32 matrices (left column) [183]. Such images, when scanned row by row, can be represented as 1024-dimensional vectors. For each of the 10 character classes, about 1700 handwritten samples were collected, and the sample means and covariance matrices were computed by standard estimation methods. The covariance matrices were 1024×1024 matrices. For each class, the first 64 principal component vectors or eigenvectors of the covariance matrix were computed. The second column in Fig. 6.1 shows the sample means, and the other columns show the reconstructions (6.13) for various values of m. In the reconstructions, the sample means have been added again to scale the images for visual display. Note how a relatively small percentage of the 1024 principal components produces reasonable reconstructions.

The condition (6.12) can often be used in advance to determine the number of principal components m, if the eigenvalues are known. The eigenvalue sequence $d_1, d_2, ..., d_n$ of a covariance matrix for real-world measurement data is usually sharply decreasing, and it is possible to set a limit below which the eigenvalues, hence principal components, are insignificantly small. This limit determines how many principal components are used.

Sometimes the threshold can be determined from some prior information on the vectors $\mathbf{x}$. For instance, assume that $\mathbf{x}$ obeys a signal-noise model

$$\mathbf{x} = \sum_{i=1}^{m} \mathbf{a}_i s_i + \mathbf{n} \tag{6.14}$$

where $m < n$. There $\mathbf{a}_i$ are some fixed vectors and the coefficients s_i are random numbers that are zero mean and uncorrelated. We can assume that their variances have been absorbed in vectors $\mathbf{a}_i$ so that they have unit variances. The term $\mathbf{n}$ is white noise, for which $E\{\mathbf{n}\mathbf{n}^T\} = \sigma^2 \mathbf{I}$. Then the vectors $\mathbf{a}_i$ span a subspace, called the *signal subspace*, that has lower dimensionality than the whole space of vectors $\mathbf{x}$. The subspace orthogonal to the signal subspace is spanned by pure noise and it is called the noise subspace.

It is easy to show (see exercises) that in this case the covariance matrix of $\mathbf{x}$ has a special form:

$$\mathbf{C_x} = \sum_{i=1}^{m} \mathbf{a}_i \mathbf{a}_i^T + \sigma^2 \mathbf{I} \tag{6.15}$$

The eigenvalues are now the eigenvalues of $\sum_{i=1}^{m} \mathbf{a}_i \mathbf{a}_i^T$, added by the constant σ^2. But the matrix $\sum_{i=1}^{m} \mathbf{a}_i \mathbf{a}_i^T$ has at most m nonzero eigenvalues, and these correspond to eigenvectors that span the signal subspace. When the eigenvalues of $\mathbf{C_x}$ are computed, the first m form a decreasing sequence and the rest are small constants, equal to σ^2:

$$d_1 > d_2 > ... > d_m > d_{m+1} = d_{m+2} = ... = d_n = \sigma^2$$

It is usually possible to detect where the eigenvalues become constants, and putting a threshold at this index, m, cuts off the eigenvalues and eigenvectors corresponding to pure noise. Then only the signal part remains.

A more disciplined approach to this problem was given by [453]; see also [231]. They give formulas for two well-known information theoretic modeling criteria, Akaike's information criterion (AIC) and the minimum description length criterion (MDL), as functions of the signal subspace dimension m. The criteria depend on the length T of the sample $\mathbf{x}(1), ..., \mathbf{x}(T)$ and on the eigenvalues $d_1, ..., d_n$ of the matrix $\mathbf{C_x}$. Finding the minimum point gives a good value for m.

6.1.4   Closed-form computation of PCA

To use the closed-form solution $\mathbf{w}_i = \mathbf{e}_i$ given earlier for the PCA basis vectors, the eigenvectors of the covariance matrix $\mathbf{C_x}$ must be known. In the conventional use of

PCA, there is a sufficiently large sample of vectors $\mathbf{x}$ available, from which the mean and the covariance matrix $\mathbf{C_x}$ can be estimated by standard methods (see Chapter 4). Solving the eigenvector–eigenvalue problem for $\mathbf{C_x}$ gives the estimate for $\mathbf{e_1}$. There are several efficient numerical methods available for solving the eigenvectors, e.g., the QR algorithm with its variants [112, 153, 320].

However, it is not always feasible to solve the eigenvectors by standard numerical methods. In an on-line data compression application like image or speech coding, the data samples $\mathbf{x}(t)$ arrive at high speed, and it may not be possible to estimate the covariance matrix and solve the eigenvector–eigenvalue problem once and for all. One reason is computational: the eigenvector problem is numerically too demanding if the dimensionality n is large and the sampling rate is high. Another reason is that the covariance matrix $\mathbf{C_x}$ may not be stationary, due to fluctuating statistics in the sample sequence $\mathbf{x}(t)$, so the estimate would have to be incrementally updated. Therefore, the PCA solution is often replaced by suboptimal nonadaptive transformations like the discrete cosine transform [154].

6.2   PCA BY ON-LINE LEARNING

Another alternative is to derive gradient ascent algorithms or other on-line methods for the preceding maximization problems. The algorithms will then converge to the solutions of the problems, that is, to the eigenvectors. The advantage of this approach is that such algorithms work on-line, using each input vector $\mathbf{x}(t)$ once as it becomes available and making an incremental change to the eigenvector estimates, without computing the covariance matrix at all. This approach is the basis of the PCA neural network learning rules.

Neural networks provide a novel way for parallel on-line computation of the PCA expansion. The PCA network [326] is a layer of parallel linear artificial neurons shown in Fig. 6.2. The output of the ith unit ($i = 1, ..., m$) is $y_i = \mathbf{w}_i^T \mathbf{x}$, with $\mathbf{x}$ denoting the n-dimensional input vector of the network and $\mathbf{w}_i$ denoting the weight vector of the ith unit. The number of units, m, will determine how many principal components the network will compute. Sometimes this can be determined in advance for typical inputs, or m can be equal to n if all principal components are required.

The PCA network learns the principal components by unsupervised learning rules, by which the weight vectors are gradually updated until they become orthonormal and tend to the theoretically correct eigenvectors. The network also has the ability to track slowly varying statistics in the input data, maintaining its optimality when the statistical properties of the inputs do not stay constant. Due to their parallelism and adaptivity to input data, such learning algorithms and their implementations in neural networks are potentially useful in feature detection and data compression tasks.

In ICA, where decorrelating the mixture variables is a useful preprocessing step, these learning rules can be used in connection to on-line ICA.

Input vector **x**

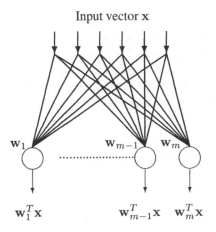

Fig. 6.2 The basic linear PCA layer

6.2.1 The stochastic gradient ascent algorithm

In this learning rule, the gradient of y_1^2 is taken with respect to $\mathbf{w}_1$ and the normalizing constraint $\|\mathbf{w}_1\| = 1$ is taken into account. The learning rule is

$$\mathbf{w}_1(t+1) = \mathbf{w}_1(t) + \gamma(t)[y_1(t)\mathbf{x}(t) - y_1^2(t)\mathbf{w}_1(t)]$$

with $y_1(t) = \mathbf{w}_1(t)^T\mathbf{x}(t)$. This is iterated over the training set $\mathbf{x}(1), \mathbf{x}(2), \dots..$ The parameter $\gamma(t)$ is the learning rate controlling the speed of convergence.

In this chapter we will use the shorthand notation introduced in Chapter 3 and write the learning rule as

$$\Delta\mathbf{w}_1 = \gamma(y_1\mathbf{x} - y_1^2\mathbf{w}_1) \tag{6.16}$$

The name stochastic gradient ascent (SGA) is due to the fact that the gradient is not with respect to the variance $\mathrm{E}\{y_1^2\}$ but with respect to the instantaneous random value y_1^2. In this way, the gradient can be updated every time a new input vector becomes available, contrary to batch mode learning. Mathematically, this is a stochastic approximation type of algorithm (for details, see Chapter 3). Convergence requires that the learning rate is decreased during learning at a suitable rate. For tracking nonstationary statistics, the learning rate should remain at a small constant value. For a derivation of this rule, as well as for the mathematical details of its convergence, see [323, 324, 330]. The algorithm (6.16) is often called Oja's rule in the literature.

Likewise, taking the gradient of y_j^2 with respect to the weight vector $\mathbf{w}_j$ and using the normalization and orthogonality constraints, we end up with the learning rule

$$\Delta\mathbf{w}_j = \gamma y_j[\mathbf{x} - y_j\mathbf{w}_j - 2\sum_{i<j} y_i\mathbf{w}_i] \tag{6.17}$$

On the right-hand side there is a term $y_j\mathbf{x}$, which is a so-called Hebbian term, product of the output y_j of the jth neuron and the input $\mathbf{x}$ to it. The other terms are implicit orthonormality constraints. The case $j = 1$ gives the one-unit learning rule (6.16) of the basic PCA neuron. The convergence of the vectors $\mathbf{w}_1, ..., \mathbf{w}_m$ to the eigenvectors $\mathbf{e}_1, ..., \mathbf{e}_m$ was established in [324, 330]. A modification called the generalized Hebbian algorithm (GHA) was later presented by Sanger [391], who also applied it to image coding, texture segmentation, and the development of receptive fields.

6.2.2 The subspace learning algorithm

The following algorithm [324, 458]

$$\Delta\mathbf{w}_j = \gamma y_j[\mathbf{x} - \sum_{i=1}^{m} y_i\mathbf{w}_i] \tag{6.18}$$

is obtained as a constrained gradient ascent maximization of $\sum_{j=1}^{m}(\mathbf{w}_j^T\mathbf{x})^2$, the mean of which gives criterion (6.9). The regular structure allows this algorithm to be written in a simple matrix form: denoting by $\mathbf{W} = (\mathbf{w}_1...\mathbf{w}_m)^T$ the $m \times n$ matrix whose rows are the weight vectors $\mathbf{w}_j$, we have the update rule

$$\Delta\mathbf{W} = \gamma[\mathbf{W}\mathbf{x}\mathbf{x}^T - (\mathbf{W}\mathbf{x}\mathbf{x}^T\mathbf{W}^T)\mathbf{W}]. \tag{6.19}$$

The network implementation of (6.18) is analogous to the SGA algorithm but still simpler because the normalizing feedback term, depending on the other weight vectors, is the same for all neuron units. The convergence was studied by Williams [458], who showed that the weight vectors $\mathbf{w}_1, ..., \mathbf{w}_m$ will not tend to the eigenvectors $\mathbf{e}_1, ..., \mathbf{e}_m$ but only to some rotated basis in the subspace spanned by them, in analogy with the minimum mean-square criterion of Section 6.1.2. For this reason, this learning rule is called the subspace algorithm. A global convergence analysis was given in [465, 75].

A variant of the subspace algorithm (6.18) is the weighted subspace algorithm

$$\Delta\mathbf{w}_j = \gamma y_j[\mathbf{x} - \theta_j \sum_{i=1}^{m} y_i\mathbf{w}_i] \tag{6.20}$$

Algorithm (6.20) is similar to (6.18) except for the scalar parameters $\theta_1, ..., \theta_m$, which are inverses of the parameters $\omega_1, ..., \omega_m$ in criterion (6.11). If all of them are chosen different and positive, then it was shown by [333] that the vectors $\mathbf{w}_1, ..., \mathbf{w}_m$ will tend to the true PCA eigenvectors $\mathbf{e}_1, ..., \mathbf{e}_m$ multiplied by scalars. The algorithm is appealing because it produces the true eigenvectors but can be computed in a fully parallel way in a homogeneous network. It can be easily presented in a matrix form, analogous to (6.19).

Other related on-line algorithms have been introduced in [136, 388, 112, 450]. Some of them, like the APEX algorithm by Diamantaras and Kung [112], are based on a feedback neural network. Also minor components defined by the eigenvectors corresponding to the smallest eigenvalues can be computed by similar algorithms [326]. Overviews of these and related neural network realizations of signal processing algorithms are given by [83, 112].

6.2.3 Recursive least-squares approach: the PAST algorithm *

The on-line algorithms reviewed in preceding sections typically suffer from slow convergence. The learning rate $\gamma(t)$ would have to be tuned optimally to speed up the convergence. One way of doing this is to use the recursive least squares (RLS) principle.

Recursive least squares methods have a long history in statistics, adaptive signal processing, and control; see [171, 299]. For example in adaptive signal processing, it is well known that RLS methods converge much faster than the standard stochastic gradient based least-mean-square (LMS) algorithm at the expense of somewhat greater computational cost [171].

Consider the mean-square error criterion (6.7). The cost function is in practice estimated from a fixed sample $\mathbf{x}(1), ..., \mathbf{x}(T)$ as

$$\hat{J}_{MSE}^{PCA} = \frac{1}{T} \sum_{j=1}^{T} [\| \mathbf{x}(j) - \sum_{i=1}^{m} (\mathbf{w}_i^T \mathbf{x}(j)) \mathbf{w}_i \|^2] \tag{6.21}$$

For simplicity of notation, let us write this in matrix form: denoting again $\mathbf{W} = (\mathbf{w}_1 ... \mathbf{w}_m)^T$ we have

$$\hat{J}_{MSE}^{PCA} = \frac{1}{T} \sum_{j=1}^{T} [\| \mathbf{x}(j) - \mathbf{W}^T \mathbf{W} \mathbf{x}(j) \|^2] \tag{6.22}$$

In [466], the following exponentially weighted sum was considered instead:

$$J_{MSE}(t) = \sum_{j=1}^{t} \beta^{t-j} [\| \mathbf{x}(j) - \mathbf{W}(t)^T \mathbf{W}(t) \mathbf{x}(j) \|^2] \tag{6.23}$$

The fixed multiplier $\frac{1}{T}$ has now been replaced by an exponential smoother β^{t-j}, where the "forgetting factor" β is between 0 and 1. If $\beta = 1$, all the samples are given the same weight, and no forgetting of old data takes place. Choosing $\beta < 1$ is especially useful in tracking nonstationary changes in the sources. The solution is denoted by $\mathbf{W}(t)$ to indicate that it depends on the sample $\mathbf{x}(1), ..., \mathbf{x}(t)$ up to time index t. The problem is to solve $\mathbf{W}(t)$ recursively: knowing $\mathbf{W}(t-1)$, we compute $\mathbf{W}(t)$ from an update rule.

Note that the cost function (6.23) is fourth order in the elements of $\mathbf{W}(t)$. It can be simplified by approximating the vector $\mathbf{W}(t)\mathbf{x}(j)$ in the sum (6.23) by the vector $\mathbf{y}(j) = \mathbf{W}(j-1)\mathbf{x}(j)$. These vectors can be easily computed because the estimated weight matrices $\mathbf{W}(j-1)$ for the previous iteration steps $j = 1, ..., t$ are already known at step t. The approximation error is usually rather small after initial convergence. This approximation yields the modified least-squares-type criterion

$$J'_{MSE}(t) = \sum_{j=1}^{t} \beta^{t-j} [\|\mathbf{x}(j) - \mathbf{W}^T(t)\mathbf{y}(j)\|]^2 \tag{6.24}$$

The cost function (6.24) is now of the standard form used in recursive least-squares methods. Any of the available algorithms [299] can be used for solving the weight matrix $\mathbf{W}(t)$ iteratively. The algorithm proposed by Yang [466], which he calls the Projection Approximation Subspace Tracking (PAST) algorithm, is as follows:

$$
\begin{aligned}
\mathbf{y}(t) &= \mathbf{W}(t-1)\mathbf{x}(t) \\
\mathbf{h}(t) &= \mathbf{P}(t-1)\mathbf{y}(t) \\
\mathbf{m}(t) &= \mathbf{h}(t)/(\beta + \mathbf{y}^T(t)\mathbf{h}(t)) \\
\mathbf{P}(t) &= \frac{1}{\beta}\mathrm{Tri}\left[\mathbf{P}(t-1) - \mathbf{m}(t)\mathbf{h}^T(t)\right] \\
\mathbf{e}(t) &= \mathbf{x}(t) - \mathbf{W}^T(t-1)\mathbf{y}(t) \\
\mathbf{W}(t) &= \mathbf{W}(t-1) + \mathbf{m}(t)\mathbf{e}^T(t)
\end{aligned}
\tag{6.25}
$$

The notation Tri means that only the upper triangular part of the argument is computed and its transpose is copied to the lower triangular part, making thus the matrix $\mathbf{P}(t)$ symmetric. The simplest way to choose the initial values is to set both $\mathbf{W}(0)$ and $\mathbf{P}(0)$ to $n \times n$ unit matrices.

The PAST algorithm (6.25) can be regarded either as a neural network learning algorithm or adaptive signal processing algorithm. It does not require any matrix inversions, because the most complicated operation is division by a scalar. The computational cost is thus low. The convergence of the algorithm is relatively fast, as shown in [466].

6.2.4 PCA and back-propagation learning in multilayer perceptrons *

Another possibility for PCA computation in neural networks is the multilayer perceptron (MLP) network, which learns using the back-propagation algorithm (see [172]) in unsupervised autoassociative mode. The network is depicted in Fig. 6.3.

The input and output layers have n units and the hidden layer has $m < n$ units. The outputs of the hidden layer are given by

$$\mathbf{h} = \sigma(\mathbf{W}_1\mathbf{x} + \mathbf{b}_1) \tag{6.26}$$

where $\mathbf{W}_1$ is the input-to-hidden-layer weight matrix, $\mathbf{b}_1$ is the corresponding bias vector, and σ is the activation function, to be applied elementwise. The output $\mathbf{y}$ of the network is an affine linear function of the hidden-layer outputs:

$$\mathbf{y} = \mathbf{W}_2\mathbf{h} + \mathbf{b}_2 \tag{6.27}$$

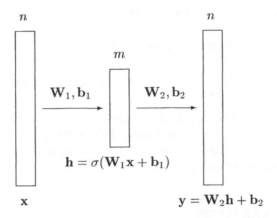

Fig. 6.3   The three-layer MLP in autoassociative mode.

with obvious notation.

In autoassociative mode, the same vectors **x** are used both as inputs and as desired outputs in back-propagation learning. If σ is *linear*, then the hidden layer outputs will become the principal components of **x** [23]. For the linear network, back-propagation learning is especially feasible because it can be shown that the "energy" function has no local minima.

This network with a nonlinear hidden layer was suggested for data compression by [96], and it was shown to be closely connected to the theoretical PCA by [52]. It is not equivalent to PCA, however, as shown by [220], unless the hidden layer is linear.

6.2.5   Extensions of PCA to nonquadratic criteria *

In the on-line learning rules reviewed earlier, the explicit computation of the eigenvectors of the covariance matrix has been replaced by gradient ascent. This makes it possible to widely extend the PCA criteria. In fact, any criterion $J(\mathbf{w}_1, ..., \mathbf{w}_n)$ such that its maximum over the constraint set $\mathbf{w}_i^T \mathbf{w}_j = \delta_{ij}$ coincides with the dominant eigenvectors of $\mathbf{C_x}$, or is a basis of the subspace spanned by them, could now be used instead. The criterion does not have to be quadratic anymore, like all the criteria presented in Section 6.1. An advantage might then be faster convergence to the PCA basis.

Recently, Miao and Hua [300] proposed such a criterion: again denoting by $\mathbf{W} = (\mathbf{w}_1...\mathbf{w}_m)^T$ the matrix whose rows are the weight vectors, this "novel information criterion" (NIC) is

$$J_{NIC}(\mathbf{W}) = \frac{1}{2}\{\text{trace}[\log(\mathbf{W}\mathbf{C_x}\mathbf{W}^T)] - \text{trace}(\mathbf{W}\mathbf{W}^T)\}. \tag{6.28}$$

It was shown in [300] that the matrix gradient is

$$\frac{\partial J_{NIC}(\mathbf{W})}{\partial \mathbf{W}} = \mathbf{C_x}\mathbf{W}^T(\mathbf{W}\mathbf{C_x}\mathbf{W}^T)^{-1} - \mathbf{W}^T \tag{6.29}$$

Setting this to zero gives the eigenvalue–eigenvector equation for $\mathbf{C_x}$, so the stationary points of this criterion are given by bases of PCA subspaces.

Other extensions and analysis of both PCA and minor component learning rules have been given by [340, 480].

6.3   FACTOR ANALYSIS

The PCA model was discussed above as a distribution-free method with no underlying statistical model. However, PCA can also be derived from a generative latent variable model: assume

$$\mathbf{x} = \mathbf{A}\mathbf{y} + \mathbf{n} \tag{6.30}$$

where $\mathbf{y}$ is gaussian, zero-mean and white, so that $E\{\mathbf{y}\mathbf{y}^T\} = \mathbf{I}$, and $\mathbf{n}$ is zero-mean gaussian white noise. It is now easy to formulate the likelihood function, because the density of $\mathbf{x}$, given $\mathbf{y}$, is gaussian. Scaled eigenvectors of $\mathbf{C_x}$ are obtained as the rows of $\mathbf{A}$ in the maximum likelihood solution, in the limiting case when the noise tends to zero.

This approach is one of the methods for the classic statistical technique of factor analysis (FA). It is called principal factor analysis [166]. Generally, the goal in factor analysis is different from PCA. Factor analysis was originally developed in social sciences and psychology. In these disciplines, the researchers want to find relevant and meaningful factors that explain observed results [166, 243, 454]. The model has the form of (6.30), with the interpretation that the elements of $\mathbf{y}$ are the unobservable factors. The elements a_{ij} of matrix $\mathbf{A}$ are called *factor loadings*. The elements of the additive term $\mathbf{n}$ are called *specific factors*, instead of noise. Let us make the simplifying assumption that the data has been normalized to zero mean.

In FA we assume that the elements of $\mathbf{y}$ (the factors) are uncorrelated and gaussian, and their variances can be absorbed into the unknown matrix $\mathbf{A}$ so that we can assume

$$E\{\mathbf{y}\mathbf{y}^T\} = \mathbf{I}. \tag{6.31}$$

The elements of $\mathbf{n}$ are uncorrelated with each other and also with the factors y_i; denote $\mathbf{Q} = E\{\mathbf{n}\mathbf{n}^T\}$. It is a diagonal matrix, but the variances of the noise elements are generally not assumed to be equal or infinitely small, as in the special case of principal FA. We can write the covariance matrix of the observations from (6.30) as

$$E\{\mathbf{x}\mathbf{x}^T\} = \mathbf{C_x} = \mathbf{A}\mathbf{A}^T + \mathbf{Q} \tag{6.32}$$

In practice, we have a good estimate of $\mathbf{C_x}$ available, given by the sample co-variance matrix. The main problem is then to solve the matrix $\mathbf{A}$ of factor loadings and the diagonal noise covariance matrix $\mathbf{Q}$ such that they will explain the observed covariances from (6.32). There is no closed-form analytic solution for $\mathbf{A}$ and $\mathbf{Q}$.

Assuming $\mathbf{Q}$ is known or can be estimated, we can attempt to solve $\mathbf{A}$ from $\mathbf{A}\mathbf{A}^T = \mathbf{C_x} - \mathbf{Q}$. The number of factors is usually constrained to be much smaller than the number of dimensions in the data, so this equation cannot be exactly solved; something similar to a least-squares solution should be used instead. Clearly, this problem does not have a unique solution: any orthogonal transform or rotation of $\mathbf{A} \rightarrow \mathbf{A}\mathbf{T}$, with $\mathbf{T}$ an orthogonal matrix (for which $\mathbf{T}\mathbf{T}^T = \mathbf{I}$), will produce exactly the same left-hand side. We need some extra constraints to make the problem more unique.

Now, looking for a factor-based interpretation of the observed variables, FA typically tries to solve the matrix $\mathbf{A}$ in such a way that the variables would have high loadings on a small number of factors, and very low loadings on the remaining factors. The results are then easier to interpret. This principle has been used in such techniques as varimax, quartimax, and oblimin rotations. Several classic techniques for such *factor rotations* are covered by Harman [166].

There are some important differences between PCA, FA, and ICA. Principal component analysis is not based on a generative model, although it can be derived from one. It is a linear transformation that is based either on variance maximization or minimum mean-square error representation. The PCA model is invertible in the (theoretical) case of no compression, i.e., when all the principal components are retained. Once the principal components y_i have been found, the original observations can be readily expressed as their linear functions as $\mathbf{x} = \sum_{i=1}^{n} y_i \mathbf{w}_i$, and also the principal components are simply obtained as linear functions of the observations: $y_i = \mathbf{w}_i^T \mathbf{x}$.

The FA model is a generative latent variable model; the observations are expressed in terms of the factors, but the values of the factors cannot be directly computed from the observations. This is due to the additive term of specific factors or noise which is considered important in some application fields. Further, the rows of matrix $\mathbf{A}$ are generally not (proportional to) eigenvectors of $\mathbf{C_x}$; several different estimation methods exist.

FA, as well as PCA, is a purely second-order statistical method: only covariances between the observed variables are used in the estimation, which is due to the assumption of gaussianity of the factors. The factors are further assumed to be uncorrelated, which also implies independence in the case of gaussian data. ICA is a similar generative latent variable model, but now the factors or independent components are assumed to be statistically independent and nongaussian — a much stronger assumption that removes the rotational redundancy of the FA model. In fact, ICA can be considered as one particular method of determining the factor rotation.

The noise term is usually omitted in the ICA model; see Chapter 15 for a detailed discussion on this point.

6.4 WHITENING

As already discussed in Chapter 1, the ICA problem is greatly simplified if the observed mixture vectors are first whitened or sphered. A zero-mean random vector $z = (z_1...z_n)^T$ is said to be *white* if its elements z_i are *uncorrelated* and have *unit variances*:

$$E\{z_i z_j\} = \delta_{ij}$$

In terms of the covariance matrix, this obviously means that $E\{zz^T\} = I$, with I the unit matrix. The best-known example is white noise; then the elements z_i would be the intensities of noise at consequent time points $i = 1, 2, ...$ and there are no temporal correlations in the noise process. The term "white" comes from the fact that the power spectrum of white noise is constant over all frequencies, somewhat like the spectrum of white light contains all colors.

A synonym for white is *sphered*. If the density of the vector z is radially symmetric and suitably scaled, then it is sphered. An example is the multivariate gaussian density that has zero mean and unit covariance matrix. The opposite does not hold: the density of a sphered vector does not have to be radially symmetric. An example is a two-dimensional uniform density that has the shape of a rotated square; see Fig. 7.10. It is easy to see that in this case both the variables z_1 and z_2 on the coordinate axes have unit variance (if the side of the square has length $2\sqrt{3}$) and they are uncorrelated, independently of the rotation angle. Thus vector z is sphered, even if the density is highly nonsymmetric. Note that the densities of the elements z_i of a sphered random vector need not be the same.

Because whitening is essentially decorrelation followed by scaling, the technique of PCA can be used. This implies that whitening can be done with a linear operation. The problem of whitening is now: Given a random vector x with n elements, find a linear transformation V into another vector z such that

$$z = Vx$$

is white (sphered).

The problem has a straightforward solution in terms of the PCA expansion. Let $E = (e_1...e_n)$ be the matrix whose columns are the unit-norm eigenvectors of the covariance matrix $C_x = E\{xx^T\}$. These can be computed from a sample of the vectors x either directly or by one of the on-line PCA learning rules. Let $D = \text{diag}(d_1...d_n)$ be the diagonal matrix of the eigenvalues of C. Then a linear whitening transform is given by

$$V = D^{-1/2}E^T \tag{6.33}$$

This matrix always exists when the eigenvalues d_i are positive; in practice, this is not a restriction. Remember (see Chapter 4) that C_x is positive semidefinite, in practice positive definite for almost any natural data, so its eigenvalues will be positive.

It is easy to show that the matrix $\mathbf{V}$ of Eq. (6.33) is indeed a whitening transformation. Recalling that $\mathbf{C_x}$ can be written in terms of its eigenvector and eigenvalue matrices $\mathbf{E}$ and $\mathbf{D}$ as $\mathbf{C_x} = \mathbf{EDE}^T$, with $\mathbf{E}$ an orthogonal matrix satisfying $\mathbf{E}^T\mathbf{E} = \mathbf{EE}^T = \mathbf{I}$, it holds:

$$E\{\mathbf{zz}^T\} = \mathbf{V}E\{\mathbf{xx}^T\}\mathbf{V}^T = \mathbf{D}^{-1/2}\mathbf{E}^T\mathbf{EDE}^T\mathbf{ED}^{-1/2} = \mathbf{I}$$

The covariance of $\mathbf{z}$ is the unit matrix, hence $\mathbf{z}$ is white.

The linear operator $\mathbf{V}$ of (6.33) is by no means the only unique whitening matrix. It is easy to see that *any* matrix $\mathbf{UV}$, with $\mathbf{U}$ an orthogonal matrix, is also a whitening matrix. This is because for $\mathbf{z} = \mathbf{UVx}$ it holds:

$$E\{\mathbf{zz}^T\} = \mathbf{UV}E\{\mathbf{xx}^T\}\mathbf{V}^T\mathbf{U}^T = \mathbf{UIU}^T = \mathbf{I}$$

An important instance is the matrix $\mathbf{ED}^{-1/2}\mathbf{E}^T$. This is a whitening matrix because it is obtained by multiplying $\mathbf{V}$ of Eq. (6.33) from the left by the orthogonal matrix $\mathbf{E}$. This matrix is called the inverse square root of $\mathbf{C_x}$, and denoted by $\mathbf{C_x}^{-1/2}$, because it comes from the standard extension of square roots to matrices.

It is also possible to perform whitening by on-line learning rules, similar to the PCA learning rules reviewed earlier. One such direct rule is

$$\Delta\mathbf{V} = \gamma(\mathbf{I} - \mathbf{Vxx}^T\mathbf{V}^T)\mathbf{V} = \gamma(\mathbf{I} - \mathbf{zz}^T)\mathbf{V} \tag{6.34}$$

It can be seen that at a stationary point, when the change in the value of $\mathbf{V}$ is zero on the average, it holds

$$(\mathbf{I} - E\{\mathbf{zz}^T\})\mathbf{V} = 0$$

for which a whitened $\mathbf{z} = \mathbf{Vx}$ is a solution. It can be shown (see, e.g., [71]) that the algorithm will indeed converge to a whitening transformation $\mathbf{V}$.

6.5 ORTHOGONALIZATION

In some PCA and ICA algorithms, we know that in theory the solution vectors (PCA basis vectors or ICA basis vectors) are orthogonal or orthonormal, but the iterative algorithms do not always automatically produce orthogonality. Then it may be necessary to orthogonalize the vectors after each iteration step, or at some suitable intervals. In this subsection, we look into some basic orthogonalization methods.

Simply stated, the problem is as follows: given a set of n-dimensional linearly independent vectors $\mathbf{a}_1, ..., \mathbf{a}_m$, with $m \leq n$, compute another set of m vectors $\mathbf{w}_1, ..., \mathbf{w}_m$ that are *orthogonal* or *orthonormal* (i.e., orthogonal and having unit Euclidean norm) and that span the same subspace as the original vectors. This means that each $\mathbf{w}_i$ is some linear combination of the $\mathbf{a}_j$.

The classic approach is the Gram-Schmidt orthogonalization (GSO) method [284]:

$$\mathbf{w}_1 = \mathbf{a}_1 \tag{6.35}$$

$$\mathbf{w}_j = \mathbf{a}_j - \sum_{i=1}^{j-1} \frac{\mathbf{w}_i^T\mathbf{a}_j}{\mathbf{w}_i^T\mathbf{w}_i}\mathbf{w}_i \tag{6.36}$$

As a result, $\mathbf{w}_i^T \mathbf{w}_j = 0$ for $i \neq j$, as is easy to show by induction. Assume that the first $j - 1$ basis vectors are already orthogonal; from (6.36) it then follows for any $k < j$ that $\mathbf{w}_k^T \mathbf{w}_j = \mathbf{w}_k^T \mathbf{a}_j - \sum_{i=1}^{j-1} \frac{\mathbf{w}_i^T \mathbf{a}_j}{\mathbf{w}_i^T \mathbf{w}_i} (\mathbf{w}_k^T \mathbf{w}_i)$. In the sum, all the inner products $\mathbf{w}_k^T \mathbf{w}_i$ are zero except the one where $i = k$. This term becomes equal to $\frac{\mathbf{w}_k^T \mathbf{a}_j}{\mathbf{w}_k^T \mathbf{w}_k} (\mathbf{w}_k^T \mathbf{w}_k) = \mathbf{w}_k^T \mathbf{a}_j$, and thus the inner product $\mathbf{w}_k^T \mathbf{w}_j$ is zero, too.

If in the GSO each $\mathbf{w}_j$ is further divided by its norm, the set will be orthonormal. The GSO is a *sequential* orthogonalization procedure. It is the basis of *deflation* approaches to PCA and ICA. A problem with sequential orthogonalization is the cumulation of errors.

In *symmetric* orthonormalization methods, none of the original vectors $\mathbf{a}_i$ is treated differently from the others. If it is sufficient to find any orthonormal basis for the subspace spanned by the original vectors, without other constraints on the new vectors, then this problem does not have a unique solution. This can be accomplished for instance by first forming the matrix $\mathbf{A} = (\mathbf{a}_1 ... \mathbf{a}_m)$ whose columns are the vectors to be orthogonalized, then computing $(\mathbf{A}^T \mathbf{A})^{-1/2}$ using the eigendecomposition of the symmetric matrix $(\mathbf{A}^T \mathbf{A})$, and finally putting

$$\mathbf{W} = \mathbf{A}(\mathbf{A}^T \mathbf{A})^{-1/2} \tag{6.37}$$

Obviously, for matrix $\mathbf{W}$ it holds $\mathbf{W}^T \mathbf{W} = \mathbf{I}$, and its columns $\mathbf{w}_1, ..., \mathbf{w}_m$ span the same subspace as the columns of matrix $\mathbf{A}$. These vectors are thus a suitable orthonormalized basis. This solution to the symmetric orthonormalization problem is by no means unique; again, any matrix $\mathbf{W}\mathbf{U}$ with $\mathbf{U}$ an orthogonal matrix will do quite as well.

However, among these solutions, there is one specific orthogonal matrix that is *closest* to matrix $\mathbf{A}$ (in an appropriate matrix norm). Then this matrix is the orthogonal projection of $\mathbf{A}$ onto the set of orthogonal matrices [284]. This is somewhat analogous to the normalization of one vector $\mathbf{a}$; the vector $\mathbf{a}/\|\mathbf{a}\|$ is the projection of $\mathbf{a}$ onto the set of unit-norm vectors (the unit sphere). For matrices, it can be shown that the matrix $\mathbf{A}(\mathbf{A}^T \mathbf{A})^{-1/2}$ in Eq. (6.37) is in fact the unique orthogonal projection of $\mathbf{A}$ onto this set.

This orthogonalization should be preferred in gradient algorithms that minimize a function $\mathcal{J}(\mathbf{W})$ under the constraint $\mathbf{W}^T \mathbf{W} = \mathbf{I}$. As explained in Chapter 3, one iteration step consists of two parts: first, the matrix $\mathbf{W}$ is updated by the usual gradient descent, and second, the updated matrix is projected orthogonally onto the constraint set. For this second stage, the form given in (6.37) for orthogonalizing the updated matrix should be used.

There are iterative methods for symmetric orthonormalization that avoid the matrix eigendecomposition and inversion. An example is the following iterative algorithm [197], starting from a nonorthogonal matrix $\mathbf{W}(0)$:

$$\mathbf{W}(1) = \mathbf{W}(0)/\|\mathbf{W}(0)\|, \tag{6.38}$$

$$\mathbf{W}(t+1) = \frac{3}{2}\mathbf{W}(t) - \frac{1}{2}\mathbf{W}(t)\mathbf{W}(t)^T \mathbf{W}(t) \tag{6.39}$$

The iteration is continued until $\mathbf{W}(t)^T\mathbf{W}(t) \approx \mathbf{I}$. The convergence of this iteration can be proven as follows [197]: matrices $\mathbf{W}(t)^T\mathbf{W}(t)$ and $\mathbf{W}(t+1)^T\mathbf{W}(t+1) = \frac{9}{4}\mathbf{W}(t)^T\mathbf{W}(t) - \frac{3}{2}[\mathbf{W}(t)^T\mathbf{W}(t)]^2 + \frac{1}{4}[\mathbf{W}(t)^T\mathbf{W}(t)]^3$ have clearly the same eigenvectors, and the relation between the eigenvalues is

$$d(t+1) = \frac{9}{4}d(t) - \frac{3}{2}d^2(t) + \frac{1}{4}d^3(t) \tag{6.40}$$

This nonlinear scalar iteration will converge on the interval $[0, 1]$ to 1 (see exercises). Due to the original normalization, all the eigenvalues are on this interval, assuming that the norm in the normalization is appropriately chosen (it must be a proper norm in the space of matrices; most conventional norms, except for the Frobenius norm, have this property). Because the eigenvalues tend to 1, the matrix itself tends to the unit matrix.

6.6 CONCLUDING REMARKS AND REFERENCES

Good general discussions on PCA are [14, 109, 324, 112]. The variance maximization criterion of PCA covered in Section 6.1.1 is due to Hotelling [185], while in the original work by Pearson [364], the starting point was minimizing the squared reconstruction error (Section 6.1.2). These are not the only criteria leading to the PCA solution; yet another information-theoretic approach is maximization of mutual information between the inputs and outputs in a linear gaussian channel [112]. An expansion closely related to PCA is the Karhunen-Loève expansion for continuous second-order stochastic processes, whose autocovariance function can be expanded in terms of its eigenvalues and orthonormal eigenfunctions in a convergent series [237, 283].

The on-line algorithms of Section 6.2 are especially suitable for neural network implementations. In numerical analysis and signal processing, many other adaptive algorithms of varying complexity have been reported for different computing hardware. A good review is given by Comon and Golub [92]. Experimental results on PCA algorithms both for finding the eigenvectors of stationary training sets, and for tracking the slowly changing eigenvectors of nonstationary input data streams, have been reported in [324, 391, 350]. An obvious extension of PCA neural networks would be to use nonlinear units, e.g., perceptrons, instead of the linear units. It turns out that such "nonlinear PCA" networks will in some cases give the independent components of the input vectors, instead of just uncorrelated components [232, 233] (see Chapter 12).

Good general texts on factor analysis are [166, 243, 454]. The principal FA model has been recently discussed by [421] and [387].

Problems

6.1   Consider the problem of maximizing the variance of $y_m = \mathbf{w}_m^T \mathbf{x}$ ($m = 1, ..., n$) under the constraint that $\mathbf{w}_m$ must be of unit Euclidean norm and orthogonal to all the previously-found principal vectors $\mathbf{w}_i$, $i < m$. Show that the solution is given by $\mathbf{w}_m = \mathbf{e}_m$ with $\mathbf{e}_m$ the eigenvector of $\mathbf{C_x}$ corresponding to the mth largest eigenvalue.

6.2   Show that the criterion (6.9) is equivalent to the mean-square error (6.7). Show that at the optimum, if $\mathbf{w}_i = \mathbf{e}_i$, the value of (6.7) is given by (6.10).

6.3   Given the data model (6.14), show that the covariance matrix has the form (6.15).

6.4   The learning rule for a PCA neuron is based on maximization of $y = (\mathbf{w}^T \mathbf{x})^2$ under constraint $\|\mathbf{w}\| = 1$. (We have now omitted the subscript 1 because only one neuron is involved.)

   6.4.1. Show that an unlimited gradient ascent method would compute the new vector $\mathbf{w}$ from

$$\mathbf{w} \leftarrow \mathbf{w} + \gamma(\mathbf{w}^T \mathbf{x})\mathbf{x}$$

with γ the learning rate. Show that the norm of the weight vector always grows in this case.

   6.4.2. Thus the norm must be bounded. A possibility is the following update rule:

$$\mathbf{w} \leftarrow [\mathbf{w} + \gamma(\mathbf{w}^T \mathbf{x})\mathbf{x}]/\|\mathbf{w} + \gamma(\mathbf{w}^T \mathbf{x})\mathbf{x}\|$$

Now the norm will stay equal to 1. Derive an approximation to this update rule for a small value of γ, by taking a Taylor expansion of the right-hand side with respect to γ and dropping all higher powers of γ. Leave only terms linear in γ. Show that the result is

$$\mathbf{w} \leftarrow \mathbf{w} + \gamma[(\mathbf{w}^T \mathbf{x})\mathbf{x} - (\mathbf{w}^T \mathbf{x})^2 \mathbf{w}]$$

which is the basic PCA learning rule of Eq. (6.16).

   6.4.3. Take averages with respect to the random input vector $\mathbf{x}$ and show that in a stationary point of the iteration, where there is no change on the average in the value of $\mathbf{w}$, it holds: $\mathbf{C_x}\mathbf{w} = (\mathbf{w}^T \mathbf{C_x}\mathbf{w})\mathbf{w}$ with $\mathbf{C_x} = E\{\mathbf{x}\mathbf{x}^T\}$.

   6.4.4. Show that the only possible solutions will be the eigenvectors of $\mathbf{C_x}$.

6.5   The covariance matrix of vector $\mathbf{x}$ is

$$\mathbf{C_x} = \begin{pmatrix} 2.5 & 1.5 \\ 1.5 & 2.5 \end{pmatrix} \tag{6.41}$$

Compute a whitening transformation for $\mathbf{x}$.

6.6   * Based on the first step (6.38) of the orthogonalization algorithm, show that $0 < d(1) < 1$ where $d(1)$ is any eigenvalue of $\mathbf{W}(1)$. Next consider iteration (6.40). Write $d(t + 1) - 1$ in terms of $d(t) - 1$. Show that $d(t)$ converges to 1. How fast is the convergence?

Part II

BASIC INDEPENDENT COMPONENT ANALYSIS

7

What is Independent
Component Analysis?

In this chapter, the basic concepts of independent component analysis (ICA) are defined. We start by discussing a couple of practical applications. These serve as motivation for the mathematical formulation of ICA, which is given in the form of a statistical estimation problem. Then we consider under what conditions this model can be estimated, and what exactly can be estimated.

After these basic definitions, we go on to discuss the connection between ICA and well-known methods that are somewhat similar, namely principal component analysis (PCA), decorrelation, whitening, and sphering. We show that these methods do something that is weaker than ICA: they estimate essentially one half of the model. We show that because of this, ICA is not possible for gaussian variables, since little can be done in addition to decorrelation for gaussian variables. On the positive side, we show that whitening is a useful thing to do before performing ICA, because it does solve one-half of the problem and it is very easy to do.

In this chapter we do not yet consider how the ICA model can actually be estimated. This is the subject of the next chapters, and in fact the rest of Part II.

7.1 MOTIVATION

Imagine that you are in a room where three people are speaking simultaneously. (The number three is completely arbitrary, it could be anything larger than one.) You also have three microphones, which you hold in different locations. The microphones give you three recorded time signals, which we could denote by $x_1(t), x_2(t)$ and $x_3(t)$, with x_1, x_2 and x_3 the amplitudes, and t the time index. Each of these recorded

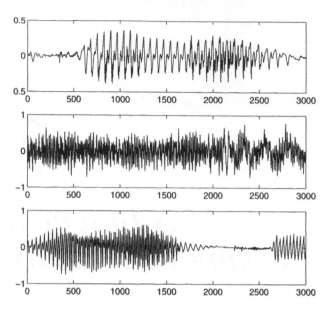

Fig. 7.1 The original audio signals.

signals is a weighted sum of the speech signals emitted by the three speakers, which we denote by $s_1(t)$, $s_2(t)$, and $s_3(t)$. We could express this as a linear equation:

$$x_1(t) = a_{11}s_1(t) + a_{12}s_2(t) + a_{13}s_3(t) \qquad (7.1)$$

$$x_2(t) = a_{21}s_1(t) + a_{22}s_2(t) + a_{23}s_3(t) \qquad (7.2)$$

$$x_3(t) = a_{31}s_1(t) + a_{32}s_2(t) + a_{33}s_3(t) \qquad (7.3)$$

where the a_{ij} with $i, j = 1, ..., 3$ are some parameters that depend on the distances of the microphones from the speakers. It would be very useful if you could now estimate the original speech signals $s_1(t)$, $s_2(t)$, and $s_3(t)$, using only the recorded signals $x_i(t)$. This is called the *cocktail-party problem*. For the time being, we omit any time delays or other extra factors from our simplified mixing model. A more detailed discussion of the cocktail-party problem can be found later in Section 24.2.

As an illustration, consider the waveforms in Fig. 7.1 and Fig. 7.2. The original speech signals could look something like those in Fig. 7.1, and the mixed signals could look like those in Fig. 7.2. The problem is to recover the "source" signals in Fig. 7.1 using only the data in Fig. 7.2.

Actually, if we knew the mixing parameters a_{ij}, we could solve the linear equation in (7.1) simply by inverting the linear system. The point is, however, that here we know *neither* the a_{ij} *nor* the $s_i(t)$, so the problem is considerably more difficult.

One approach to solving this problem would be to use some information on the statistical properties of the signals $s_i(t)$ to estimate both the a_{ij} and the $s_i(t)$. Actually, and perhaps surprisingly, it turns out that it is enough to assume that

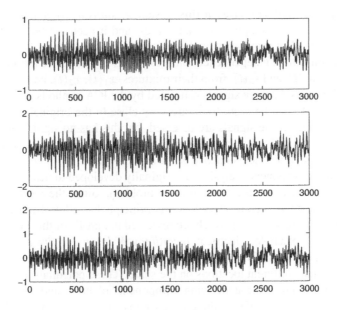

Fig. 7.2 The observed mixtures of the original signals in Fig. 7.1.

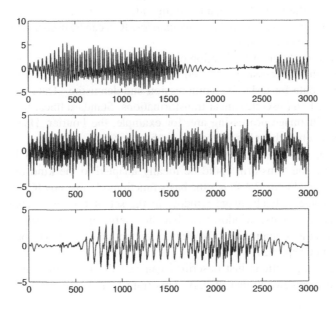

Fig. 7.3 The estimates of the original signals, obtained using only the observed signals in Fig. 7.2. The original signals were very accurately estimated, up to multiplicative signs.

$s_1(t)$, $s_2(t)$, and $s_3(t)$ are, at each time instant t, *statistically independent.* This is not an unrealistic assumption in many cases, and it need not be exactly true in practice. Independent component analysis can be used to estimate the a_{ij} based on the information of their independence, and this allows us to separate the three original signals, $s_1(t)$, $s_2(t)$, and $s_3(t)$, from their mixtures, $x_1(t)$, $x_2(t)$, and $x_2(t)$.

Figure 7.3 gives the three signals estimated by the ICA methods discussed in the next chapters. As can be seen, these are very close to the original source signals (the signs of some of the signals are reversed, but this has no significance.) These signals were estimated using only the mixtures in Fig. 7.2, together with the very weak assumption of the independence of the source signals.

Independent component analysis was originally developed to deal with problems that are closely related to the cocktail-party problem. Since the recent increase of interest in ICA, it has become clear that this principle has a lot of other interesting applications as well, several of which are reviewed in Part IV of this book.

Consider, for example, electrical *recordings of brain activity* as given by an electroencephalogram (EEG). The EEG data consists of recordings of electrical potentials in many different locations on the scalp. These potentials are presumably generated by mixing some underlying components of brain and muscle activity. This situation is quite similar to the cocktail-party problem: we would like to find the original components of brain activity, but we can only observe mixtures of the components. ICA can reveal interesting information on brain activity by giving access to its independent components. Such applications will be treated in detail in Chapter 22. Furthermore, finding underlying independent causes is a central concern in the social sciences, for example, *econometrics*. ICA can be used as an econometric tool as well; see Section 24.1.

Another, very different application of ICA is *feature extraction*. A fundamental problem in signal processing is to find suitable representations for image, audio or other kind of data for tasks like compression and denoising. Data representations are often based on (discrete) linear transformations. Standard linear transformations widely used in image processing are, for example, the Fourier, Haar, and cosine transforms. Each of them has its own favorable properties.

It would be most useful to estimate the linear transformation from the data itself, in which case the transform could be ideally adapted to the kind of data that is being processed. Figure 7.4 shows the basis functions obtained by ICA from patches of natural images. Each image window in the set of training images would be a superposition of these windows so that the coefficient in the superposition are independent, at least approximately. Feature extraction by ICA will be explained in more detail in Chapter 21.

All of the applications just described can actually be formulated in a unified mathematical framework, that of ICA. This framework will be defined in the next section.

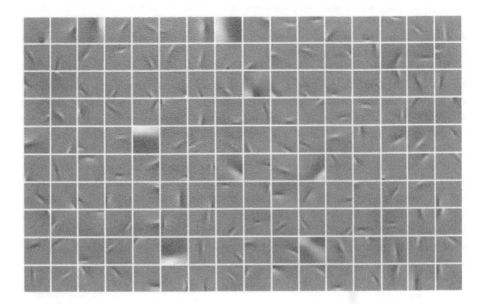

Fig. 7.4 Basis functions in ICA of natural images. These basis functions can be considered as the independent features of images. Every image window is a linear sum of these windows.

7.2   DEFINITION OF INDEPENDENT COMPONENT ANALYSIS

7.2.1   ICA as estimation of a generative model

To rigorously define ICA, we can use a statistical "latent variables" model. We observe n random variables $x_1, ..., x_n$, which are modeled as linear combinations of n random variables $s_1, ..., s_n$:

$$x_i = a_{i1}s_1 + a_{i2}s_2 + ... + a_{in}s_n, \quad \text{for all } i = 1, ..., n \qquad (7.4)$$

where the $a_{ij}, i, j = 1, ..., n$ are some real coefficients. By definition, the s_i are statistically mutually independent.

This is the basic ICA model. The ICA model is a generative model, which means that it describes how the observed data are generated by a process of mixing the components s_j. The independent components s_j (often abbreviated as ICs) are latent variables, meaning that they cannot be directly observed. Also the mixing coefficients a_{ij} are assumed to be unknown. All we observe are the random variables x_i, and we must estimate *both* the mixing coefficients a_{ij} *and* the ICs s_i using the $\mathbf{x}_i$. This must be done under as general assumptions as possible.

Note that we have here dropped the time index t that was used in the previous section. This is because in this basic ICA model, we assume that each mixture x_i as well as each independent component s_j is a random variable, instead of a proper time signal or time series. The observed values $x_i(t)$, e.g., the microphone signals in the cocktail party problem, are then a sample of this random variable. We also neglect any time delays that may occur in the mixing, which is why this basic model is often called the *instantaneous* mixing model.

ICA is very closely related to the method called *blind source separation* (BSS) or blind signal separation. A "source" means here an original signal, i.e., independent component, like the speaker in the cocktail-party problem. "Blind" means that we know very little, if anything, of the mixing matrix, and make very weak assumptions on the source signals. ICA is one method, perhaps the most widely used, for performing blind source separation.

It is usually more convenient to use vector-matrix notation instead of the sums as in the previous equation. Let us denote by x the random vector whose elements are the mixtures $x_1, ..., x_n$, and likewise by s the random vector with elements $s_1, ..., s_n$. Let us denote by $\mathbf{A}$ the matrix with elements a_{ij}. (Generally, bold lowercase letters indicate vectors and bold uppercase letters denote matrices.) All vectors are understood as column vectors; thus x^T, or the transpose of x, is a row vector. Using this vector-matrix notation, the mixing model is written as

$$x = \mathbf{A}s \tag{7.5}$$

Sometimes we need the columns of matrix $\mathbf{A}$; if we denote them by $\mathbf{a}_j$ the model can also be written as

$$x = \sum_{i=1}^{n} \mathbf{a}_i s_i \tag{7.6}$$

The definition given here is the most basic one, and in Part II of this book, we will essentially concentrate on this basic definition. Some generalizations and modifications of the definition will be given later (especially in Part III), however. For example, in many applications, it would be more realistic to assume that there is some *noise* in the measurements, which would mean adding a noise term in the model (see Chapter 15). For simplicity, we omit any noise terms in the basic model, since the estimation of the noise-free model is difficult enough in itself, and seems to be sufficient for many applications. Likewise, in many cases the *number of ICs and observed mixtures may not be equal*, which is treated in Section 13.2 and Chapter 16, and the mixing might be *nonlinear*, which is considered in Chapter 17. Furthermore, let us note that an *alternative definition* of ICA that does not use a generative model will be given in Chapter 10.

7.2.2   Restrictions in ICA

To make sure that the basic ICA model just given can be estimated, we have to make certain assumptions and restrictions.

1. The independent components are assumed statistically *independent*.

This is the principle on which ICA rests. Surprisingly, not much more than this assumption is needed to ascertain that the model can be estimated. This is why ICA is such a powerful method with applications in many different areas.

Basically, random variables $y_1, y_2, ..., y_n$ are said to be independent if information on the value of y_i does not give any information on the value of y_j for $i \neq j$. Technically, independence can be defined by the probability densities. Let us denote by $p(y_1, y_2, ..., y_n)$ the joint probability density function (pdf) of the y_i, and by $p_i(y_i)$ the marginal pdf of y_i, i.e., the pdf of y_i when it is considered alone. Then we say that the y_i are *independent* if and only if the joint pdf is factorizable in the following way:

$$p(y_1, y_2, ..., y_n) = p_1(y_1)p_2(y_2)...p_n(y_n). \tag{7.7}$$

For more details, see Section 2.3.

2. The independent components must have *nongaussian* distributions.

Intuitively, one can say that the gaussian distributions are "too simple". The higher-order cumulants are zero for gaussian distributions, but such higher-order information is essential for estimation of the ICA model, as will be seen in Section 7.4.2. Thus, ICA is essentially impossible if the observed variables have gaussian distributions. The case of gaussian components is treated in more detail in Section 7.5 below. Note that in the basic model we do *not* assume that we know what the nongaussian distributions of the ICs look like; if they are known, the problem will be considerably simplified. Also, note that a completely different class of ICA methods, in which the assumption of nongaussianity is replaced by some assumptions on the *time structure* of the signals, will be considered later in Chapter 18.

3. For simplicity, we assume that the unknown mixing matrix is *square*.

In other words, the number of independent components is equal to the number of observed mixtures. This assumption can sometimes be relaxed, as explained in Chapters 13 and 16. We make it here because it simplifies the estimation very much. Then, after estimating the matrix $\mathbf{A}$, we can compute its inverse, say $\mathbf{B}$, and obtain the independent components simply by

$$\mathbf{s} = \mathbf{Bx} \tag{7.8}$$

It is also assumed here that the mixing matrix is *invertible*. If this is not the case, there are redundant mixtures that could be omitted, in which case the matrix would not be square; then we find again the case where the number of mixtures is not equal to the number of ICs.

Thus, under the preceding three assumptions (or at the minimum, the two first ones), the ICA model is identifiable, meaning that the mixing matrix and the ICs can be estimated up to some trivial indeterminacies that will be discussed next. We will not prove the identifiability of the ICA model here, since the proof is quite complicated; see the end of the chapter for references. On the other hand, in the next chapter we develop estimation methods, and the developments there give a kind of a nonrigorous, constructive proof of the identifiability.

7.2.3 Ambiguities of ICA

In the ICA model in Eq. (7.5), it is easy to see that the following ambiguities or indeterminacies will necessarily hold:

1. We cannot determine the variances (energies) of the independent components.

The reason is that, both s and **A** being unknown, any scalar multiplier in one of the sources s_i could always be canceled by dividing the corresponding column $\mathbf{a}_i$ of **A** by the same scalar, say α_i:

$$\mathbf{x} = \sum_i (\frac{1}{\alpha_i}\mathbf{a}_i)(s_i\alpha_i) \tag{7.9}$$

As a consequence, we may quite as well fix the magnitudes of the independent components. Since they are random variables, the most natural way to do this is to assume that each has unit variance: $E\{s_i^2\} = 1$. Then the matrix **A** will be adapted in the ICA solution methods to take into account this restriction. Note that this still leaves the *ambiguity of the sign*: we could multiply an independent component by -1 without affecting the model. This ambiguity is, fortunately, insignificant in most applications.

2. We cannot determine the order of the independent components.

The reason is that, again both s and **A** being unknown, we can freely change the order of the terms in the sum in (7.6), and call any of the independent components the first one. Formally, a permutation matrix **P** and its inverse can be substituted in the model to give $\mathbf{x} = \mathbf{AP}^{-1}\mathbf{Ps}$. The elements of **Ps** are the original independent variables s_j, but in another order. The matrix $\mathbf{AP}^{-1}$ is just a new unknown mixing matrix, to be solved by the ICA algorithms.

7.2.4 Centering the variables

Without loss of generality, we can assume that both the mixture variables and the independent components have zero mean. This assumption simplifies the theory and algorithms quite a lot; it is made in the rest of this book.

If the assumption of zero mean is not true, we can do some preprocessing to make it hold. This is possible by *centering* the observable variables, i.e., subtracting their sample mean. This means that the original mixtures, say $\mathbf{x}'$ are preprocessed by

$$\mathbf{x} = \mathbf{x}' - E\{\mathbf{x}'\} \tag{7.10}$$

before doing ICA. Thus the independent components are made zero mean as well, since

$$E\{\mathbf{s}\} = \mathbf{A}^{-1}E\{\mathbf{x}\} \tag{7.11}$$

The mixing matrix, on the other hand, remains the same after this preprocessing, so we can always do this without affecting the estimation of the mixing matrix. After

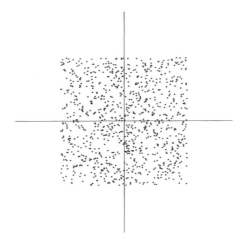

Fig. 7.5 The joint distribution of the independent components s_1 and s_2 with uniform distributions. Horizontal axis: s_1, vertical axis: s_2.

estimating the mixing matrix and the independent components for the zero-mean data, the subtracted mean can be simply reconstructed by adding $\mathbf{A}^{-1}E\{\mathbf{x}'\}$ to the zero-mean independent components.

7.3 ILLUSTRATION OF ICA

To illustrate the ICA model in statistical terms, consider two independent components that have the following uniform distributions:

$$p(s_i) = \begin{cases} \frac{1}{2\sqrt{3}}, & \text{if } |s_i| \leq \sqrt{3} \\ 0, & \text{otherwise} \end{cases} \tag{7.12}$$

The range of values for this uniform distribution were chosen so as to make the mean zero and the variance equal to one, as was agreed in the previous section. The joint density of s_1 and s_2 is then uniform on a square. This follows from the basic definition that the joint density of two independent variables is just the product of their marginal densities (see Eq. (7.7)): we simply need to compute the product. The joint density is illustrated in Fig. 7.5 by showing data points randomly drawn from this distribution.

Now let us mix these two independent components. Let us take the following mixing matrix:

$$\mathbf{A}_0 = \begin{pmatrix} 5 & 10 \\ 10 & 2 \end{pmatrix} \tag{7.13}$$

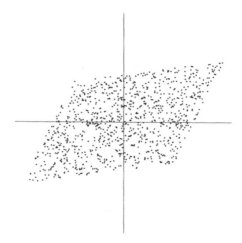

Fig. 7.6 The joint distribution of the observed mixtures x_1 and x_2. Horizontal axis: x_1, vertical axis: x_2. (Not in the same scale as Fig. 7.5.)

This gives us two mixed variables, x_1 and x_2. It is easily computed that the mixed data has a uniform distribution on a parallelogram, as shown in Fig. 7.6. Note that the random variables x_1 and x_2 are not independent anymore; an easy way to see this is to consider whether it is possible to predict the value of one of them, say x_2, from the value of the other. Clearly, if x_1 attains one of its maximum or minimum values, then this completely determines the value of x_2. They are therefore not independent. (For variables s_1 and s_2 the situation is different: from Fig. 7.5 it can be seen that knowing the value of s_1 does not in any way help in guessing the value of s_2.)

The problem of estimating the data model of ICA is now to estimate the mixing matrix $\mathbf{A}$ using only information contained in the mixtures x_1 and x_2. Actually, from Fig. 7.6 you can see an intuitive way of estimating $\mathbf{A}$: The *edges* of the parallelogram are in the directions of the columns of $\mathbf{A}$. This means that we could, in principle, estimate the ICA model by first estimating the joint density of x_1 and x_2, and then locating the edges. So, the problem seems to have a solution.

On the other hand, consider a mixture of ICs with a different type of distribution, called supergaussian (see Section 2.7.1). Supergaussian random variables typically have a pdf with a peak a zero. The marginal distribution of such an IC is given in Fig. 7.7. The joint distribution of the original independent components is given in Fig. 7.8, and the mixtures are shown in Fig. 7.9. Here, we see some kind of edges, but in very different places this time.

In practice, however, locating the edges would be a very poor method because it only works with variables that have very special distributions. For most distributions, such edges cannot be found; we use only for illustration purposes distributions that visually show edges. Moreover, methods based on finding edges, or other similar heuristic methods, tend to be computationally quite complicated, and unreliable.

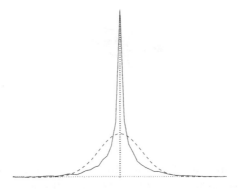

Fig. 7.7 The density of one supergaussian independent component. The gaussian density if give by the dashed line for comparison.

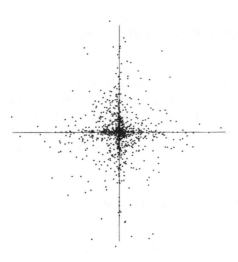

Fig. 7.8 The joint distribution of the independent components s_1 and s_2 with supergaussian distributions. Horizontal axis: s_1, vertical axis: s_2.

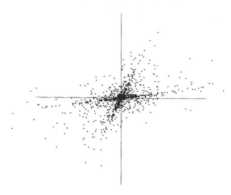

Fig. 7.9   The joint distribution of the observed mixtures x_1 and x_2, obtained from super-gaussian independent components. Horizontal axis: x_1, vertical axis: x_2.

What we need is a method that works for any distributions of the independent components, and works fast and reliably. Such methods are the main subject of this book, and will be presented in Chapters 8–12. In the rest of this chapter, however, we discuss the connection between ICA and whitening.

7.4   ICA IS STRONGER THAT WHITENING

Given some random variables, it is straightforward to linearly transform them into uncorrelated variables. Therefore, it would be tempting to try to estimate the independent components by such a method, which is typically called whitening or sphering, and often implemented by principal component analysis. In this section, we show that this is not possible, and discuss the relation between ICA and decorrelation methods. It will be seen that whitening is, nevertheless, a useful preprocessing technique for ICA.

7.4.1   Uncorrelatedness and whitening

A weaker form of independence is uncorrelatedness. Here we review briefly the relevant definitions that were already encountered in Chapter 2.

Two random variables y_1 and y_2 are said to be *uncorrelated*, if their covariance is zero:

$$\text{cov}(y_1, y_2) = E\{y_1 y_2\} - E\{y_1\}E\{y_2\} = 0 \qquad (7.14)$$

In this book, all random variables are assumed to have zero mean, unless otherwise mentioned. Thus, covariance is equal to correlation $\text{corr}(y_1, y_2) = E\{y_1 y_2\}$, and uncorrelatedness is the same thing as zero correlation (see Section 2.2).[1]

If random variables are independent, they are uncorrelated. This is because if the y_1 and y_2 are independent, then for any two functions, h_1 and h_2, we have

$$E\{h_1(y_1)h_2(y_2)\} = E\{h_1(y_1)\}E\{h_2(y_2)\} \tag{7.15}$$

see Section 2.3. Taking $h_1(y_1) = y_1$ and $h_2(y_2) = y_2$, we see that this implies uncorrelatedness.

On the other hand, uncorrelatedness does *not* imply independence. For example, assume that (y_1, y_2) are discrete valued and follow such a distribution that the pair are with probability $1/4$ equal to any of the following values: $(0, 1), (0, -1), (1, 0)$, and $(-1, 0)$. Then y_1 and y_2 are uncorrelated, as can be simply calculated. On the other hand,

$$E\{y_1^2 y_2^2\} = 0 \neq \frac{1}{4} = E\{y_1^2\}E\{y_2^2\} \tag{7.16}$$

so the condition in Eq. (7.15) is violated, and the variables cannot be independent.

A slightly stronger property than uncorrelatedness is *whiteness*. Whiteness of a zero-mean random vector, say $\mathbf{y}$, means that its components are uncorrelated and their variances equal unity. In other words, the covariance matrix (as well as the correlation matrix) of $\mathbf{y}$ equals the identity matrix:

$$E\{\mathbf{y}\mathbf{y}^T\} = \mathbf{I} \tag{7.17}$$

Consequently, whitening means that we linearly transform the observed data vector $\mathbf{x}$ by linearly multiplying it with some matrix $\mathbf{V}$

$$\mathbf{z} = \mathbf{V}\mathbf{x} \tag{7.18}$$

so that we obtain a new vector $\mathbf{z}$ that is white. Whitening is sometimes called sphering.

A whitening transformation is always possible. Some methods were reviewed in Chapter 6. One popular method for whitening is to use the eigenvalue decomposition (EVD) of the covariance matrix

$$E\{\mathbf{x}\mathbf{x}^T\} = \mathbf{E}\mathbf{D}\mathbf{E}^T \tag{7.19}$$

where $\mathbf{E}$ is the orthogonal matrix of eigenvectors of $E\{\mathbf{x}\mathbf{x}^T\}$ and $\mathbf{D}$ is the diagonal matrix of its eigenvalues, $\mathbf{D} = \text{diag}(d_1, ..., d_n)$. Whitening can now be done by the whitening matrix

$$\mathbf{V} = \mathbf{E}\mathbf{D}^{-1/2}\mathbf{E}^T \tag{7.20}$$

---

[1]In statistical literature, correlation is often defined as a normalized version of covariance. Here, we use this simpler definition that is more widely spread in signal processing. In any case, the concept of uncorrelatedness is the same.

where the matrix $\mathbf{D}^{-1/2}$ is computed by a simple componentwise operation as $\mathbf{D}^{-1/2} = \mathrm{diag}(d_1^{-1/2}, ..., d_n^{-1/2})$. A whitening matrix computed this way is denoted by $E\{\mathbf{xx}^T\}^{-1/2}$ or $\mathbf{C}^{-1/2}$. Alternatively, whitening can be performed in connection with principal component analysis, which gives a related whitening matrix. For details, see Chapter 6.

7.4.2   Whitening is only half ICA

Now, suppose that the data in the ICA model is whitened, for example, by the matrix given in (7.20). Whitening transforms the mixing matrix into a new one, $\tilde{\mathbf{A}}$. We have from (7.5) and (7.18)

$$\mathbf{z} = \mathbf{VAs} = \tilde{\mathbf{A}}\mathbf{s} \tag{7.21}$$

One could hope that whitening solves the ICA problem, since whiteness or uncorrelatedness is related to independence. This is, however, not so. Uncorrelatedness is weaker than independence, and is not in itself sufficient for estimation of the ICA model. To see this, consider an *orthogonal* transformation $\mathbf{U}$ of $\mathbf{z}$:

$$\mathbf{y} = \mathbf{U}\mathbf{z} \tag{7.22}$$

Due to the orthogonality of $\mathbf{U}$, we have

$$E\{\mathbf{yy}^T\} = E\{\mathbf{U}\mathbf{yy}^T\mathbf{U}^T\} = \mathbf{U}\mathbf{I}\mathbf{U}^T = \mathbf{I} \tag{7.23}$$

In other words, $\mathbf{y}$ is white as well. Thus, we cannot tell if the independent components are given by $\mathbf{z}$ or $\mathbf{y}$ using the whiteness property alone. Since $\mathbf{y}$ could be any orthogonal transformation of $\mathbf{z}$, *whitening gives the ICs only up to an orthogonal transformation*. This is not sufficient in most applications.

On the other hand, whitening is useful as a preprocessing step in ICA. The utility of whitening resides in the fact that the *new mixing matrix $\tilde{\mathbf{A}} = \mathbf{VA}$ is orthogonal*. This can be seen from

$$E\{\mathbf{zz}^T\} = \tilde{\mathbf{A}}E\{\mathbf{ss}^T\}\tilde{\mathbf{A}}^T = \tilde{\mathbf{A}}\tilde{\mathbf{A}}^T = \mathbf{I} \tag{7.24}$$

This means that we can restrict our search for the mixing matrix to the space of orthogonal matrices. Instead of having to estimate the n^2 parameters that are the elements of the original matrix $\mathbf{A}$, we only need to estimate an orthogonal mixing matrix $\tilde{\mathbf{A}}$. An orthogonal matrix contains $n(n-1)/2$ degrees of freedom. For example, in two dimensions, an orthogonal transformation is determined by a single angle parameter. In larger dimensions, an orthogonal matrix contains only about half of the number of parameters of an arbitrary matrix.

Thus one can say that whitening solves half of the problem of ICA. Because whitening is a very simple and standard procedure, much simpler than any ICA algorithms, it is a good idea to reduce the complexity of the problem this way. The remaining half of the parameters has to be estimated by some other method; several will be introduced in the next chapters.

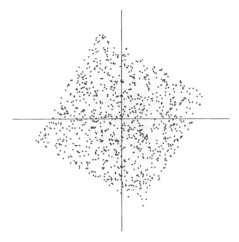

Fig. 7.10 The joint distribution of the whitened mixtures of uniformly distributed independent components.

A graphical illustration of the effect of whitening can be seen in Fig. 7.10, in which the data in Fig. 7.6 has been whitened. The square defining the distribution is now clearly a rotated version of the original square in Fig. 7.10. All that is left is the estimation of a single angle that gives the rotation.

In many chapters of this book, we assume that the data has been preprocessed by whitening, in which case we denote the data by z. Even in cases where whitening is not explicitly required, it is recommended, since it reduces the number of free parameters and considerably increases the performance of the methods, especially with high-dimensional data.

7.5 WHY GAUSSIAN VARIABLES ARE FORBIDDEN

Whitening also helps us understand why gaussian variables are forbidden in ICA. Assume that the joint distribution of two ICs, s_1 and s_2, is gaussian. This means that their joint pdf is given by

$$p(s_1, s_2) = \frac{1}{2\pi} \exp(-\frac{s_1^2 + s_2^2}{2}) = \frac{1}{2\pi} \exp(-\frac{\|s\|^2}{2}) \qquad (7.25)$$

(For more information on the gaussian distribution, see Section 2.5.) Now, assume that the mixing matrix A is orthogonal. For example, we could assume that this is so because the data has been whitened. Using the classic formula of transforming pdf's in (2.82), and noting that for an orthogonal matrix $A^{-1} = A^T$ holds, we get

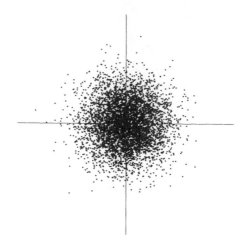

Fig. 7.11   The multivariate distribution of two independent gaussian variables.

the joint density of the mixtures x_1 and x_2 as density is given by

$$p(x_1, x_2) = \frac{1}{2\pi} \exp(-\frac{\|\mathbf{A}^T\mathbf{x}\|^2}{2})|\det \mathbf{A}^T| \tag{7.26}$$

Due to the orthogonality of $\mathbf{A}$, we have $\|\mathbf{A}^T\mathbf{x}\|^2 = \|\mathbf{x}\|^2$ and $|\det \mathbf{A}| = 1$; note that if $\mathbf{A}$ is orthogonal, so is $\mathbf{A}^T$. Thus we have

$$p(x_1, x_2) = \frac{1}{2\pi} \exp(-\frac{\|\mathbf{x}\|^2}{2}) \tag{7.27}$$

and we see that the orthogonal mixing matrix does not change the pdf, since it does not appear in this pdf at all. The original and mixed distributions are identical. Therefore, there is no way how we could infer the mixing matrix from the mixtures.

The phenomenon that the orthogonal mixing matrix cannot be estimated for gaussian variables is related to the property that uncorrelated jointly gaussian variables are necessarily independent (see Section 2.5). Thus, the information on the independence of the components does not get us any further than whitening.

Graphically, we can see this phenomenon by plotting the distribution of the orthogonal mixtures, which is in fact the same as the distribution of the ICs. This distribution is illustrated in Fig. 7.11. The figure shows that the density is rotationally symmetric. Therefore, it does not contain any information on the directions of the columns of the mixing matrix $\mathbf{A}$. This is why $\mathbf{A}$ cannot be estimated.

Thus, in the case of gaussian independent components, we can only estimate the ICA model up to an orthogonal transformation. In other words, the matrix $\mathbf{A}$ is not identifiable for gaussian independent components. With gaussian variables, all we can do is whiten the data. There is some choice in the whitening procedure, however; PCA is the classic choice.

What happens if we try to estimate the ICA model and *some of the components are gaussian, some nongaussian*? In this case, we *can* estimate all the nongaussian components, but the gaussian components cannot be separated from each other. In other words, some of the estimated components will be arbitrary linear combinations of the gaussian components. Actually, this means that in the case of just one gaussian component, we can estimate the model, because the single gaussian component does not have any other gaussian components that it could be mixed with.

7.6 CONCLUDING REMARKS AND REFERENCES

ICA is a very general-purpose statistical technique in which observed random data are expressed as a linear transform of components that are statistically independent from each other. In this chapter, we formulated ICA as the estimation of a generative model, with independent latent variables. Such a decomposition is identifiable, i.e., well defined, if the independent components are nongaussian (except for perhaps one). To simplify the estimation problem, we can begin by whitening the data. This estimates part of the parameters, but leaves an orthogonal transformation unspecified. Using the higher-order information contained in nongaussian variables, we can estimate this orthogonal transformation as well.

Practical methods for estimating the ICA model will be treated in the rest of Part II. A simple approach based on finding the maxima of nongaussianity is presented first in Chapter 8. Next, the classic maximum likelihood estimation method is applied on ICA in Chapter 9. An information-theoretic framework that also shows a connection between the previous two is given by mutual information in Chapter 10. Some further methods are considered in Chapters 11 and 12. Practical considerations on the application of ICA methods, in particular on the preprocessing of the data, are treated in Chapter 13. The different ICA methods are compared with each other, and the choice of the "best" method is considered in Chapter 14, which concludes Part II.

The material that we treated in this chapter can be considered classic. The ICA model was first defined as herein in [228]; somewhat related developments were given in [24]. The identifiability is treated in [89, 423]. Whitening was proposed in [61] as well. In addition to this research in signal processing, a parallel neuroscientific line of research developed ICA independently. This was started by [26, 27, 28], being more qualitative in nature. The first quantitative results in this area were proposed in [131], and in [335], a model that is essentially equivalent to the noisy version of the ICA model (see Chapter 15) was proposed. More on the history of ICA can be found in Chapter 1, as well as in [227]. For recent reviews on ICA, see [10, 65, 201, 267, 269, 149]. A shorter tutorial text is in [212].

Problems

7.1   Show that given a random vector $\mathbf{x}$, there is only one *symmetric* whitening matrix for $\mathbf{x}$, given by (7.20).

7.2   Show that two (zero-mean) random variables that have a jointly gaussian distribution are independent if and only if they are uncorrelated. (Hint: The pdf can be found in (2.68). Uncorrelatedness means that the covariance matrix is diagonal. Show that this implies that the joint pdf can be factorized.)

7.3   If both $\mathbf{x}$ and $\mathbf{s}$ could be observed, how would you estimate the ICA model? (Assume there is some noise in the data as well.)

7.4   Assume that the data $\mathbf{x}$ is multiplied by a matrix $\mathbf{M}$. Does this change the independent components?

7.5   In our definition, the signs of the independent components are left undetermined. How could you complement the definition so that they are determined as well?

7.6   Assume that there are more independent components than observed mixtures. Assume further that we have been able to estimate the mixing matrix. Can we recover the values of the independent components? What if there are more observed mixtures than ICs?

Computer assignments

7.1   Generate samples of two independent components that follow a Laplacian distribution (see Eq. 2.96). Mix them with three different random mixing matrices. Plot the distributions of the independent components. Can you see the matrix $\mathbf{A}$ in the plots? Do the same for ICs that are obtained by taking absolute values of gaussian random variables.

7.2   Generate samples of two independent gaussian random variables. Mix them with a random mixing matrix. Compute a whitening matrix. Compute the product of the whitening matrix and the mixing matrix. Show that this is almost orthogonal. Why is it not exactly orthogonal?

8

ICA by Maximization of Nongaussianity

In this chapter, we introduce a simple and intuitive principle for estimating the model of independent component analysis (ICA). This is based on maximization of nongaussianity.

Nongaussianity is actually of paramount importance in ICA estimation. Without nongaussianity the estimation is not possible at all, as shown in Section 7.5. Therefore, it is not surprising that nongaussianity could be used as a leading principle in ICA estimation. This is at the same time probably the main reason for the rather late resurgence of ICA research: In most of classic statistical theory, random variables are assumed to have gaussian distributions, thus precluding methods related to ICA. (A completely different approach may then be possible, though, using the time structure of the signals; see Chapter 18.)

We start by intuitively motivating the maximization of nongaussianity by the central limit theorem. As a first practical measure of nongaussianity, we introduce the fourth-order cumulant, or kurtosis. Using kurtosis, we derive practical algorithms by gradient and fixed-point methods. Next, to solve some problems associated with kurtosis, we introduce the information-theoretic quantity called negentropy as an alternative measure of nongaussianity, and derive the corresponding algorithms for this measure. Finally, we discuss the connection between these methods and the technique called projection pursuit.

8.1   "NONGAUSSIAN IS INDEPENDENT"

The central limit theorem is a classic result in probability theory that was presented in Section 2.5.2. It says that the distribution of a sum of independent random variables tends toward a gaussian distribution, under certain conditions. Loosely speaking, a sum of two independent random variables usually has a distribution that is closer to gaussian than any of the two original random variables.

Let us now assume that the data vector $\mathbf{x}$ is distributed according to the ICA data model:

$$\mathbf{x} = \mathbf{As} \tag{8.1}$$

i.e., it is a mixture of independent components. For pedagogical purposes, let us assume in this motivating section that all the independent components have identical distributions. Estimating the independent components can be accomplished by finding the right linear combinations of the mixture variables, since we can invert the mixing as

$$\mathbf{s} = \mathbf{A}^{-1}\mathbf{x} \tag{8.2}$$

Thus, to estimate one of the independent components, we can consider a linear combination of the x_i. Let us denote this by $y = \mathbf{b}^T\mathbf{x} = \sum_i b_i x_i$, where $\mathbf{b}$ is a vector to be determined. Note that we also have $y = \mathbf{b}^T\mathbf{As}$. Thus, y is a certain linear combination of the s_i, with coefficients given by $\mathbf{b}^T\mathbf{A}$. Let us denote this vector by $\mathbf{q}$. Then we have

$$y = \mathbf{b}^T\mathbf{x} = \mathbf{q}^T\mathbf{s} = \sum_i q_i s_i \tag{8.3}$$

If $\mathbf{b}$ were one of the rows of the inverse of $\mathbf{A}$, this linear combination $\mathbf{b}^T\mathbf{x}$ would actually equal one of the independent components. In that case, the corresponding $\mathbf{q}$ would be such that just one of its elements is 1 and all the others are zero.

The question is now: *How could we use the central limit theorem to determine* $\mathbf{b}$ *so that it would equal one of the rows of the inverse of* $\mathbf{A}$? In practice, we cannot determine such a $\mathbf{b}$ exactly, because we have no knowledge of matrix $\mathbf{A}$, but we can find an estimator that gives a good approximation.

Let us vary the coefficients in $\mathbf{q}$, and see how the distribution of $y = \mathbf{q}^T\mathbf{s}$ changes. The fundamental idea here is that since a sum of even two independent random variables is more gaussian than the original variables, $y = \mathbf{q}^T\mathbf{s}$ is usually more gaussian than any of the s_i and becomes least gaussian when it in fact equals one of the s_i. (Note that this is strictly true only if the s_i have identical distributions, as we assumed here.) In this case, obviously only one of the elements q_i of $\mathbf{q}$ is nonzero.

We do not in practice know the values of $\mathbf{q}$, but we do not need to, because $\mathbf{q}^T\mathbf{s} = \mathbf{b}^T\mathbf{x}$ by the definition of $\mathbf{q}$. We can just let $\mathbf{b}$ vary and look at the distribution of $\mathbf{b}^T\mathbf{x}$.

Therefore, we could take as $\mathbf{b}$ a vector that *maximizes the nongaussianity* of $\mathbf{b}^T\mathbf{x}$. Such a vector would necessarily correspond to a $\mathbf{q} = \mathbf{A}^T\mathbf{b}$, which has only one nonzero component. This means that $y = \mathbf{b}^T\mathbf{x} = \mathbf{q}^T\mathbf{s}$ equals one of the independent components! Maximizing the nongaussianity of $\mathbf{b}^T\mathbf{x}$ thus gives us one of the independent components.

In fact, the optimization landscape for nongaussianity in the n-dimensional space of vectors $\mathbf{b}$ has $2n$ local maxima, two for each independent component, corresponding to s_i and $-s_i$ (recall that the independent components can be estimated only up to a multiplicative sign).

We can illustrate the principle of maximizing nongaussianity by simple examples. Let us consider two independent components that have uniform densities. (They also have zero mean, as do all the random variables in this book.) Their joint distribution is illustrated in Fig. 8.1, in which a sample of the independent components is plotted on the two-dimensional (2-D) plane. Figure 8.2 also shows a histogram estimate of the uniform densities. These variables are then linearly mixed, and the mixtures are whitened as a preprocessing step. Whitening is explained in Section 7.4; let us recall briefly that it means that $\mathbf{x}$ is linearly transformed into a random vector

$$\mathbf{z} = \mathbf{V}\mathbf{x} = \mathbf{V}\mathbf{A}\mathbf{s} \tag{8.4}$$

whose correlation matrix equals unity: $E\{\mathbf{z}\mathbf{z}^T\} = \mathbf{I}$. Thus the ICA model still holds, though with a different mixing matrix. (Even without whitening, the situation would be similar.) The joint density of the whitened mixtures is given in Fig. 8.3. It is a rotation of the original joint density, as explained in Section 7.4.

Now, let us look at the densities of the two linear mixtures z_1 and z_2. These are estimated in Fig. 8.4. One can clearly see that the densities of the mixtures are closer to a gaussian density than the densities of the independent components shown in Fig. 8.2. Thus we see that the mixing makes the variables closer to gaussian. Finding the rotation that rotates the square in Fig. 8.3 back to the original ICs in Fig. 8.1 would give us the two maximally nongaussian linear combinations with uniform distributions.

A second example with very different densities shows the same result. In Fig. 8.5, the joint distribution of very supergaussian independent components is shown. The marginal density of a component is estimated in Fig. 8.6. The density has a large peak at zero, as is typical of supergaussian densities (see Section 2.7.1 or below). Whitened mixtures of the independent components are shown in Fig. 8.7. The densities of two linear mixtures are given in Fig. 8.8. They are clearly more gaussian than the original densities, as can be seen from the fact that the peak is much lower. Again, we see that mixing makes the distributions more gaussian.

To recapitulate, we have formulated ICA estimation as the search for directions that are maximally nongaussian: Each local maximum gives one independent component. Our approach here is somewhat heuristic, but it will be seen in the next section and Chapter 10 that it has a perfectly rigorous justification. From a practical point of view, we now have to answer the following questions: How can the nongaussianity of $\mathbf{b}^T\mathbf{x}$ be measured? And how can we compute the values of $\mathbf{b}$ that maximize (locally) such a measure of nongaussianity? The rest of this chapter is devoted to answering these questions.

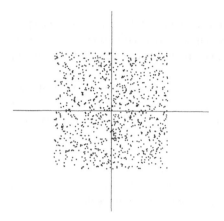

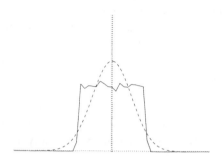

Fig. 8.1 The joint distribution of two independent components with uniform densities.

Fig. 8.2 The estimated density of one uniform independent component, with the gaussian density (dashed curve) given for comparison.

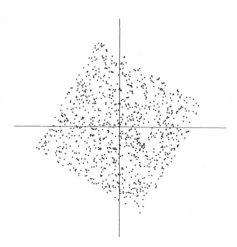

Fig. 8.3 The joint density of two whitened mixtures of independent components with uniform densities.

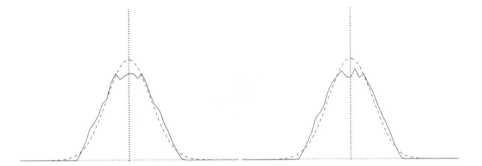

Fig. 8.4 The marginal densities of the whitened mixtures. They are closer to the gaussian density (given by the dashed curve) than the densities of the independent components.

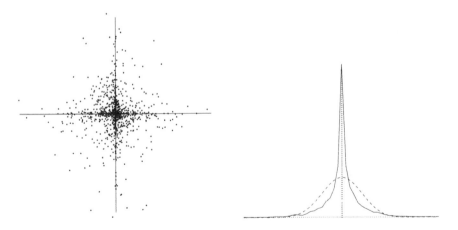

Fig. 8.5 The joint distribution of the two independent components with supergaussian densities.

Fig. 8.6 The estimated density of one supergaussian independent component.

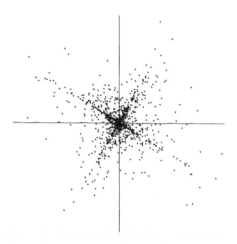

Fig. 8.7 The joint distribution of two whitened mixtures of independent components with supergaussian densities.

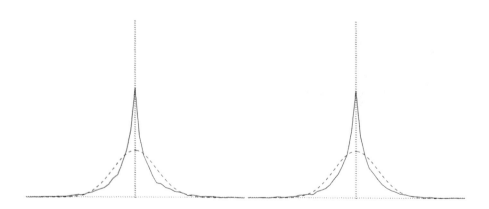

Fig. 8.8 The marginal densities of the whitened mixtures in Fig. 8.7. They are closer to the gaussian density (given by dashed curve) than the densities of the independent components.

8.2 MEASURING NONGAUSSIANITY BY KURTOSIS

8.2.1 Extrema of kurtosis give independent components

Kurtosis and its properties To use nongaussianity in ICA estimation, we must have a quantitative measure of nongaussianity of a random variable, say y. In this section, we show how to use kurtosis, a classic measure of nongaussianity, for ICA estimation. Kurtosis is the name given to the fourth-order cumulant of a random variable; for a general discussion of cumulants; see Section 2.7. Thus we obtain an estimation method that can be considered a variant of the classic method of moments; see Section 4.3.

The kurtosis of y, denoted by kurt(y), is defined by

$$\text{kurt}(y) = E\{y^4\} - 3(E\{y^2\})^2 \tag{8.5}$$

Remember that all the random variables here have zero mean; in the general case, the definition of kurtosis is slightly more complicated. To simplify things, we can further assume that y has been normalized so that its variance is equal to one: $E\{y^2\} = 1$. Then the right-hand side simplifies to $E\{y^4\} - 3$. This shows that kurtosis is simply a normalized version of the fourth moment $E\{y^4\}$. For a gaussian y, the fourth moment equals $3(E\{y^2\})^2$. Thus, kurtosis is zero for a gaussian random variable. For most (but not quite all) nongaussian random variables, kurtosis is nonzero.

Kurtosis can be both positive or negative. Random variables that have a negative kurtosis are called subgaussian, and those with positive kurtosis are called supergaussian. In statistical literature, the corresponding expressions platykurtic and leptokurtic are also used. For details, see Section 2.7.1. Supergaussian random variables have typically a "spiky" probability density function (pdf) with heavy tails, i.e., the pdf is relatively large at zero and at large values of the variable, while being small for intermediate values. A typical example is the Laplacian distribution, whose pdf is given by

$$p(y) = \frac{1}{\sqrt{2}} \exp(\sqrt{2}|y|) \tag{8.6}$$

Here we have normalized the variance to unity; this pdf is illustrated in Fig. 8.9. Subgaussian random variables, on the other hand, have typically a "flat" pdf, which is rather constant near zero, and very small for larger values of the variable. A typical example is the uniform distribution, whose density is given by

$$p(y) = \begin{cases} \frac{1}{2\sqrt{3}}, & \text{if } |y| \leq \sqrt{3} \\ 0, & \text{otherwise} \end{cases} \tag{8.7}$$

which is normalized to unit variance as well; it is illustrated in Fig. 8.10.

Typically nongaussianity is measured by the absolute value of kurtosis. The square of kurtosis can also be used. These measures are zero for a gaussian variable, and greater than zero for most nongaussian random variables. There are nongaussian random variables that have zero kurtosis, but they can be considered to be very rare.

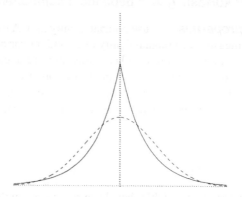

Fig. 8.9 The density function of the Laplacian distribution, which is a typical supergaussian distribution. For comparison, the gaussian density is given by a dashed curve. Both densities are normalized to unit variance.

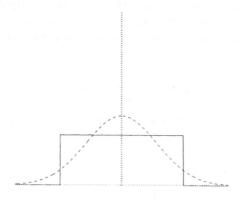

Fig. 8.10 The density function of the uniform distribution, which is a typical *sub*gaussian distribution. For comparison, the gaussian density is given by a dashed line. Both densities are normalized to unit variance.

Kurtosis, or rather its absolute value, has been widely used as a measure of nongaussianity in ICA and related fields. The main reason is its simplicity, both computational and theoretical. Computationally, kurtosis can be estimated simply by using the fourth moment of the sample data (if the variance is kept constant). Theoretical analysis is simplified because of the following linearity property: If x_1 and x_2 are two independent random variables, it holds

$$\text{kurt}(x_1 + x_2) = \text{kurt}(x_1) + \text{kurt}(x_2) \tag{8.8}$$

and

$$\text{kurt}(\alpha x_1) = \alpha^4 \text{kurt}(x_1) \tag{8.9}$$

where α is a constant. These properties can be easily proven using the general definition of cumulants, see Section 2.7.2.

Optimization landscape in ICA To illustrate in a simple example what the optimization landscape for kurtosis looks like, and how independent components could be found by kurtosis minimization or maximization, let us look at a 2-D model $\mathbf{x} = \mathbf{As}$. Assume that the independent components s_1, s_2 have kurtosis values $\text{kurt}(s_1), \text{kurt}(s_2)$, respectively, both different from zero. Recall that they have unit variances by definition. We look for one of the independent components as $y = \mathbf{b}^T\mathbf{x}$.

Let us again consider the transformed vector $\mathbf{q} = \mathbf{A}^T\mathbf{b}$. Then we have $y = \mathbf{b}^T\mathbf{x} = \mathbf{b}^T\mathbf{As} = \mathbf{q}^T\mathbf{s} = q_1 s_1 + q_2 s_2$. Now, based on the additive property of kurtosis, we have

$$\text{kurt}(y) = \text{kurt}(q_1 s_1) + \text{kurt}(q_2 s_2) = q_1^4 \text{kurt}(s_1) + q_2^4 \text{kurt}(s_2) \tag{8.10}$$

On the other hand, we made the constraint that the variance of y is equal to 1, based on the same assumption concerning s_1, s_2. This implies a constraint on $\mathbf{q}$: $E\{y^2\} = q_1^2 + q_2^2 = 1$. Geometrically, this means that vector $\mathbf{q}$ is constrained to the unit circle on the 2-D plane.

The optimization problem is now: What are the maxima of the function $|\text{kurt}(y)| = |q_1^4 \text{kurt}(s_1) + q_2^4 \text{kurt}(s_2)|$ on the unit circle? To begin with, we may assume for simplicity that the kurtoses are equal to 1. In this case, we are simply considering the function

$$F(\mathbf{q}) = q_1^4 + q_2^4 \tag{8.11}$$

Some contours of this function, i.e., curves in which this function is constant, are shown in Fig. 8.11. The unit sphere, i.e., the set where $q_1^2 + q_2^2 = 1$, is shown as well. This gives the "optimization landscape" for the problem.

It is not hard to see that the maxima are at those points where exactly one of the elements of vector $\mathbf{q}$ is zero and the other nonzero; because of the unit circle constraint, the nonzero element must be equal to 1 or -1. But these points are exactly

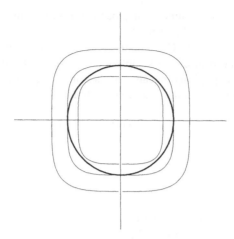

Fig. 8.11 The optimization landscape of kurtosis. The thick curve is the unit sphere, and the thin curves are the contours where F in (8.11) is constant.

the ones when y equals one of the independent components $\pm s_i$, and the problem has been solved.

If the kurtoses are both equal to -1, the situation is similar, because taking the absolute values, we get exactly the same function to maximize. Finally, if the kurtoses are completely arbitrary, as long as they are nonzero, more involved algebraic manipulations show that the absolute value of kurtosis is still maximized when $y = \mathbf{b}^T\mathbf{x}$ equals one of the independent components. A proof is given in the exercises.

Now we see the *utility of preprocessing by whitening*. For whitened data $\mathbf{z}$, we seek for a linear combination $\mathbf{w}^T\mathbf{z}$ that maximizes nongaussianity. This simplifies the situation here, since we have $\mathbf{q} = (\mathbf{VA})^T\mathbf{w}$ and therefore

$$\|\mathbf{q}\|^2 = (\mathbf{w}^T\mathbf{VA})(\mathbf{A}^T\mathbf{V}^T\mathbf{w}) = \|\mathbf{w}\|^2 \qquad (8.12)$$

This means that constraining $\mathbf{q}$ to lie on the unit sphere is equivalent to constraining $\mathbf{w}$ to be on the unit sphere. Thus we maximize the absolute value of kurtosis of $\mathbf{w}^T\mathbf{z}$ under the simpler constraint that $\|\mathbf{w}\| = 1$. Also, after whitening, the linear combinations $\mathbf{w}^T\mathbf{z}$ can be interpreted as *projections* on the line (that is, a 1-D subspace) spanned by the vector $\mathbf{w}$. Each point on the unit sphere corresponds to one projection.

As an example, let us consider the whitened mixtures of uniformly distributed independent components in Fig. 8.3. We search for a vector $\mathbf{w}$ such that the linear combination or projection $\mathbf{w}^T\mathbf{x}$ has maximum nongaussianity, as illustrated in Fig. 8.12. In this two-dimensional case, we can parameterize the points on the unit sphere by the *angle* that the corresponding vector $\mathbf{w}$ makes with the horizontal axis. Then, we can plot the kurtosis of $\mathbf{w}^T\mathbf{z}$ as a function of this angle, which is given in

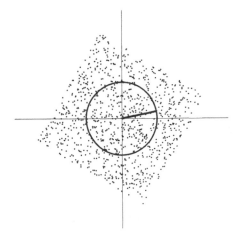

Fig. 8.12 We search for projections (which correspond to points on the unit circle) that maximize nongaussianity, using whitened mixtures of uniformly distributed independent components. The projections can be parameterized by the angle.

Fig. 8.13. The plot shows kurtosis is always negative, and is minimized at approximately 1 and 2.6 radians. These directions are thus such that the absolute value of kurtosis is maximized. They can be seen in Fig. 8.12 to correspond to the directions given by the edges of the square, and thus they do give the independent components.

In the second example, we see the same phenomenon for whitened mixtures of supergaussian independent components. Again, we search for a vector $\mathbf{w}$ such that the projection in that direction has maximum nongaussianity, as illustrated in Fig. 8.14. We can plot the kurtosis of $\mathbf{w}^T\mathbf{z}$ as a function of the angle in which $\mathbf{w}$ points, as given in Fig. 8.15. The plot shows kurtosis is always positive, and is maximized in the directions of the independent components. These angles are the same as in the preceding example because we used the same mixing matrix. Again, they correspond to the directions in which the absolute value of kurtosis is maximized.

8.2.2 Gradient algorithm using kurtosis

In practice, to maximize the absolute value of kurtosis, we would start from some vector $\mathbf{w}$, compute the direction in which the absolute value of the kurtosis of $y = \mathbf{w}^T\mathbf{z}$ is growing most strongly, based on the available sample $\mathbf{z}(1), ..., \mathbf{z}(T)$ of mixture vector $\mathbf{z}$, and then move the vector $\mathbf{w}$ in that direction. This idea is implemented in gradient methods and their extensions.

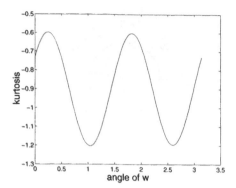

Fig. 8.13 The values of kurtosis for projections as a function of the angle as in Fig. 8.12. Kurtosis is minimized, and its absolute value maximized, in the directions of the independent components.

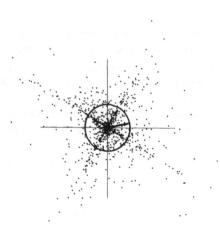

Fig. 8.14 Again, we search for projections that maximize nongaussianity, this time with whitened mixtures of supergaussian independent components. The projections can be parameterized by the angle.

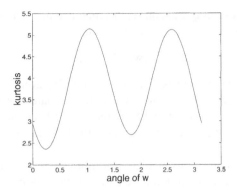

Fig. 8.15 The values of kurtosis for projections in different angles as in Fig. 8.14. Kurtosis, as well as its absolute value, is maximized in the directions of the independent components.

Using the principles in Chapter 3, the gradient of the absolute value of kurtosis of $\mathbf{w}^T\mathbf{z}$ can be simply computed as

$$\frac{\partial |\text{kurt}(\mathbf{w}^T\mathbf{x})|}{\partial \mathbf{w}} = 4\,\text{sign}(\text{kurt}(\mathbf{w}^T\mathbf{z}))[E\{\mathbf{z}(\mathbf{w}^T\mathbf{z})^3\} - 3\mathbf{w}\|\mathbf{w}\|^2] \tag{8.13}$$

since for whitened data we have $E\{(\mathbf{w}^T\mathbf{z})^2\} = \|\mathbf{w}\|^2$. Since we are optimizing this function on the unit sphere $\|\mathbf{w}\|^2 = 1$, the gradient method must be complemented by projecting $\mathbf{w}$ on the unit sphere after every step. This can be done simply by dividing $\mathbf{w}$ by its norm.

To further simplify the algorithm, note that since the latter term in brackets in (8.13) would simply be changing the norm of $\mathbf{w}$ in the gradient algorithm, and not its direction, it can be omitted. This is because only the direction of $\mathbf{w}$ is interesting, and any change in the norm is insignificant because the norm is normalized to unity anyway.

Thus we obtain the following gradient algorithm:

$$\Delta\mathbf{w} \propto \text{sign}(\text{kurt}(\mathbf{w}^T\mathbf{z}))E\{\mathbf{z}(\mathbf{w}^T\mathbf{z})^3\} \tag{8.14}$$

$$\mathbf{w} \leftarrow \mathbf{w}/\|\mathbf{w}\| \tag{8.15}$$

An on-line (or adaptive) version of this algorithm can be obtained as well. This is possible by omitting the second expectation operation in the algorithm, yielding:

$$\Delta\mathbf{w} \propto \text{sign}(\text{kurt}(\mathbf{w}^T\mathbf{z}))\mathbf{z}(\mathbf{w}^T\mathbf{z})^3 \tag{8.16}$$

$$\mathbf{w} \leftarrow \mathbf{w}/\|\mathbf{w}\| \tag{8.17}$$

Then every observation $\mathbf{z}(t)$ can be used in the algorithm at once. However, it must be noted that when computing $\text{sign}(\text{kurt}(\mathbf{w}^T\mathbf{x}))$, the expectation operator in the definition of kurtosis cannot be omitted. Instead, the kurtosis must be properly estimated from a time-average; of course, this time-average can be estimated on-line. Denoting by γ the estimate of the kurtosis, we could use

$$\Delta\gamma \propto ((\mathbf{w}^T\mathbf{z})^4 - 3) - \gamma \tag{8.18}$$

This gives the estimate of kurtosis as a kind of a running average.

Actually, in many cases one knows in advance the nature of the distributions of the independent components, i.e., whether they are subgaussian or supergaussian. Then one can simply plug the correct sign of kurtosis in the algorithm, and avoid its estimation.

More general versions of this gradient algorithm are introduced in Section 8.3.4. In the next subsection we shall introduce an algorithm that maximizes the absolute value of kurtosis much more efficiently than the gradient method.

8.2.3 A fast fixed-point algorithm using kurtosis

In the previous subsection, we derived a gradient method for maximizing nongaussianity as measured by the absolute value of kurtosis. The advantage of such gradient methods, closely connected to learning in neural networks, is that the inputs $\mathbf{z}(t)$ can be used in the algorithm at once, thus enabling fast adaptation in a nonstationary environment. A resulting trade-off, however, is that the convergence is slow, and depends on a good choice of the learning rate sequence. A bad choice of the learning rate can, in practice, destroy convergence. Therefore, some ways to make the learning radically faster and more reliable may be needed. The fixed-point iteration algorithms are such an alternative.

To derive a more efficient fixed-point iteration, we note that at a stable point of the gradient algorithm, the gradient must point in the direction of $\mathbf{w}$, that is, the gradient must be equal to $\mathbf{w}$ multiplied by some scalar constant. Only in such a case, adding the gradient to $\mathbf{w}$ does not change its direction, and we can have convergence (this means that after normalization to unit norm, the value of $\mathbf{w}$ is not changed except perhaps by changing its sign). This can be proven more rigorously using the technique of Lagrange multipliers; see Exercice 3.9. Equating the gradient of kurtosis in (8.13) with $\mathbf{w}$, this means that we should have

$$\mathbf{w} \propto [E\{\mathbf{z}(\mathbf{w}^T\mathbf{z})^3\} - 3\|\mathbf{w}\|^2\mathbf{w}] \tag{8.19}$$

This equation immediately suggests a fixed-point algorithm where we first compute the right-hand side, and give this as the new value for $\mathbf{w}$:

$$\mathbf{w} \leftarrow E\{\mathbf{z}(\mathbf{w}^T\mathbf{z})^3\} - 3\mathbf{w} \tag{8.20}$$

After every fixed-point iteration, $\mathbf{w}$ is divided by its norm to remain on the constraint set. (Thus $\|\mathbf{w}\| = 1$ always, which is why it can be omitted from (8.19).) The final vector $\mathbf{w}$ gives one of the independent components as the linear combination $\mathbf{w}^T\mathbf{z}$. In practice, the expectations in (8.20) must be replaced by their estimates.

Note that convergence of this fixed-point iteration means that the old and new values of $\mathbf{w}$ point in the same direction, i.e., their dot-product is (almost) equal to

1. It is not necessary that the vector converges to a single point, since $\mathbf{w}$ and $-\mathbf{w}$ define the same direction. This is again because the independent components can be defined only up to a multiplicative sign.

Actually, it turns out that such an algorithm works very well, converging very fast and reliably. This algorithm is called FastICA [210]. The FastICA algorithm has a couple of properties that make it clearly superior to the gradient-based algorithms in most cases. First of all, it can be shown (see Appendix), that the convergence of this algorithm is *cubic*. This means very fast convergence. Second, contrary to gradient-based algorithms, there is no learning rate or other adjustable parameters in the algorithm, which makes it easy to use, and more reliable. Gradient algorithms seem to be preferable only in cases where fast adaptation in a changing environment is necessary.

More sophisticated versions of FastICA are introduced in Section 8.3.5.

8.2.4 Examples

Here we show what happens when we run the FastICA algorithm that maximizes the absolute value of kurtosis, using the two example data sets used in this chapter. First we take a mixture of two uniformly distributed independent components. The mixtures are whitened, as always in this chapter. The goal is now to find a direction in the data that maximizes the absolute value of kurtosis, as illustrated in Fig. 8.12.

We initialize, for purposes of the illustration, the vector $\mathbf{w}$ as $\mathbf{w} = (1,0)^T$. Running the FastICA iteration just two times, we obtain convergence. In Fig. 8.16, the obtained vectors $\mathbf{w}$ are shown. The dashed line gives the direction of $\mathbf{w}$ after the first iteration, and the solid line gives the direction of $\mathbf{w}$ after the second iteration. The third iteration did not significantly change the direction of $\mathbf{w}$, which means that the algorithm converged. (The corresponding vector is not plotted.) The figure shows that the value of $\mathbf{w}$ may change drastically during the iteration, because the values $\mathbf{w}$ and $-\mathbf{w}$ are considered as equivalent. This is because the sign of the vector cannot be determined in the ICA model.

The kurtoses of the projections $\mathbf{w}^T\mathbf{z}$ obtained in the iterations are plotted in Fig. 8.17, as a function of iteration count. The plot shows that the algorithm steadily increased the *absolute value* of the kurtosis of the projection, until it reached convergence at the third iteration.

Similar experiments were performed for the whitened mixtures of two *supergaussian* independent components, as illustrated in Fig. 8.14. The obtained vectors are shown in Fig. 8.18. Again, convergence was obtained after two iterations. The kurtoses of the projections $\mathbf{w}^T\mathbf{z}$ obtained in the iterations are plotted in Fig. 8.19, as a function of iteration count. As in the preceding experiment, the absolute value of the kurtosis of the projection steadily increased, until it reached convergence at the third iteration.

In these examples, we only estimated one independent component. Of course, one often needs more than one component. Figures 8.12 and 8.14 indicate how this can be done: The directions of the independent components are orthogonal in the whitened space, so the second independent component can be found as the direction

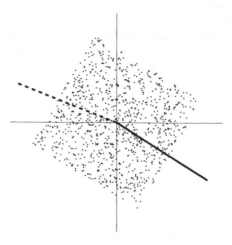

Fig. 8.16 Result of FastICA using kurtosis, for ICs with uniform distributions. Dashed line: **w** after the first iteration (plotted longer than actual size). Solid line: **w** after the second iteration.

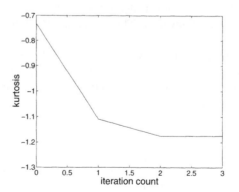

Fig. 8.17 The convergence of FastICA using kurtosis, for ICs with uniform distributions. The value of kurtosis shown as function of iteration count.

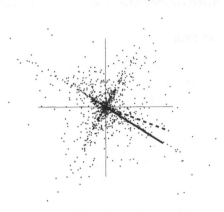

Fig. 8.18   Result of FastICA with kurtosis, this time for supergaussian ICs. Dash-dotted line: **w** after the first iteration (plotted longer than actual size). Solid line: **w** after the second iteration.

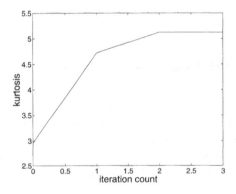

Fig. 8.19   The convergence of FastICA using kurtosis, for supergaussian ICs. The value of kurtosis shown as a function of iteration count.

orthogonal to the **w** corresponding to the estimated independent component. For more dimensions, we need to rerun the algorithm, always constraining the current **w** to be orthogonal to the previously estimated vectors **w**. This will be explained in more detail in Section 8.4.

8.3 MEASURING NONGAUSSIANITY BY NEGENTROPY

8.3.1 Critique of kurtosis

In the preceding section, we showed how to measure nongaussianity by kurtosis, thus obtaining a simple ICA estimation method. However, kurtosis also has some drawbacks in practice, when its value has to be estimated from a measured sample. The main problem is that kurtosis can be very sensitive to outliers. Assume, for example, that a sample of 1000 values of a random variable (with zero mean and unit variance, say) contains one value equal to 10. Then the kurtosis equals at least $10^4/1000 - 3 = 7$, which means that the single value makes kurtosis large. Thus we see that the value of kurtosis may depend on only a few observations in the tails of the distribution, which may be erroneous or irrelevant observations. In other words, kurtosis is not a robust measure of nongaussianity.

Thus, other measures of nongaussianity might be better than kurtosis in some situations. In this section, we shall consider negentropy, which is the second important measure of nongaussianity. Its properties are in many ways opposite to those of kurtosis: It is robust but computationally complicated. We also introduce computationally simple approximations of negentropy that more or less combine the good properties of both measures.

8.3.2 Negentropy as nongaussianity measure

Negentropy is based on the information-theoretic quantity of differential entropy, which we here call simply entropy. Entropy is the basic concept of information theory; for a more detailed discussion, see Chapter 5. The entropy of a random variable is related to the information that the observation of the variable gives. The more "random", i.e., unpredictable and unstructured the variable is, the larger its entropy. The (differential) entropy H of a random vector $\mathbf{y}$ with density $p_y(\boldsymbol{\eta})$ is defined as

$$H(\mathbf{y}) = -\int p_y(\boldsymbol{\eta}) \log p_y(\boldsymbol{\eta}) \mathrm{d}\boldsymbol{\eta} \tag{8.21}$$

A fundamental result of information theory is that *a gaussian variable has the largest entropy* among all random variables of equal variance (see Section 5.3.2). This means that entropy could be used as a measure of nongaussianity. In fact, this shows that the gaussian distribution is the "most random" or the least structured of all distributions. Entropy is small for distributions that are clearly concentrated on certain values, i.e., when the variable is clearly clustered, or has a pdf that is very "spiky".

To obtain a measure of nongaussianity that is zero for a gaussian variable and always nonnegative, one often uses a normalized version of differential entropy, called negentropy. Negentropy J is defined as follows

$$J(\mathbf{y}) = H(\mathbf{y}_{gauss}) - H(\mathbf{y}) \tag{8.22}$$

where y_{gauss} is a gaussian random variable of the same correlation (and covariance) matrix as y. Due to the above-mentioned properties, negentropy is always nonnegative, and it is zero if and only if y has a gaussian distribution. Negentropy has the additional interesting property that it is invariant for invertible linear transformations (see Section 5.4).

The advantage of using negentropy, or equivalently, differential entropy, as a measure of nongaussianity is that it is well justified by statistical theory. In fact, negentropy is in some sense the optimal estimator of nongaussianity, as far as the statistical performance is concerned, as will be seen in Section 14.3. The problem in using negentropy is, however, that it is computationally very difficult. Estimating negentropy using the definition would require an estimate (possibly nonparametric) of the pdf. Therefore, simpler approximations of negentropy are very useful, as will be discussed next. These will be used to derive an efficient method for ICA.

8.3.3 Approximating negentropy

In practice, we only need approximation of 1-D (neg)entropies, so we only consider the scalar case here.

The classic method of approximating negentropy is using higher-order cumulants, using the polynomial density expansions as explained in Section 5.5. This gives the approximation

$$J(y) \approx \frac{1}{12}E\{y^3\}^2 + \frac{1}{48}\text{kurt}(y)^2 \tag{8.23}$$

The random variable y is assumed to be of zero mean and unit variance. Actually, this approximation often leads to the use of kurtosis as in the preceding section. This is because the first term on the right-hand side of (8.23) is zero in the case of random variables with (approximately) symmetric distributions, which is quite common. In this case, the approximation in (8.23) is equivalent to the square of kurtosis. Maximization of the square of kurtosis is of course equivalent to maximization of its absolute value. Thus this approximation leads more or less to the method in Section 8.2. In particular, this approximation suffers from the nonrobustness encountered with kurtosis. Therefore, we develop here more sophisticated approximations of negentropy.

One useful approach is to generalize the higher-order cumulant approximation so that it uses expectations of general nonquadratic functions, or "nonpolynomial moments". This was described in Section 5.6. In general we can replace the polynomial functions y^3 and y^4 by any other functions G^i (where i is an index, not a power), possibly more than two. The method then gives a simple way of approximating the negentropy based on the expectations $E\{G^i(y)\}$. As a simple special case, we can take any two nonquadratic functions G^1 and G^2 so that G^1 is odd and G^2 is even, and we obtain the following approximation:

$$J(y) \approx k_1 (E\{G^1(y)\})^2 + k_2 (E\{G^2(y)\} - E\{G^2(\nu)\})^2 \tag{8.24}$$

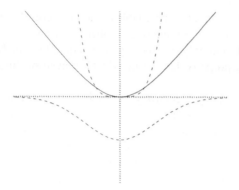

Fig. 8.20   The functions G_1 in Eq. (8.26), G_2 in Eq. (8.27), given by the solid curve and the dashed curve, respectively. The fourth power, as used in kurtosis, is given for comparison by the dash-dotted curve.

where k_1 and k_2 are positive constants, and ν is a gaussian variable of zero mean and unit variance (i.e., standardized). The variable y is assumed to have zero mean and unit variance. Note that even in cases where this approximation is not very accurate, (8.24) can be used to construct a measure of nongaussianity that is consistent in the sense that it is always nonnegative, and equal to zero if y has a gaussian distribution. This is a generalization of the moment-based approximation in (8.23), which is obtained by taking $G^1(y) = y^3$ and $G^2(y) = y^4$.

In the case where we use only one nonquadratic function G, the approximation becomes

$$J(y) \propto [E\{G(y)\} - E\{G(\nu)\}]^2 \qquad (8.25)$$

for practically any nonquadratic function G. This is a generalization of the moment-based approximation in (8.23) if y has a symmetric distribution, in which case the first term in (8.23) vanishes. Indeed, taking $G(y) = y^4$, one then obtains a kurtosis-based approximation.

But the point here is that by choosing G wisely, one obtains approximations of negentropy that are better than the one given by (8.23). In particular, choosing a G that does not grow too fast, one obtains more robust estimators. The following choices of G have proved very useful:

$$G_1(y) = \frac{1}{a_1} \log \cosh a_1 y, \qquad (8.26)$$

$$G_2(y) = - \exp(-y^2/2) \qquad (8.27)$$

where $1 \leq a_1 \leq 2$ is some suitable constant, often taken equal to one. The functions in (8.26)–(8.27) are illustrated in Fig. 8.20.

Thus we obtain approximations of negentropy that give a very good compromise between the properties of the two classic nongaussianity measures given by kurtosis and negentropy. They are conceptually simple, fast to compute, yet have appealing statistical properties, especially robustness. Therefore, we shall use these objective functions in our ICA methods. Interestingly, kurtosis can be expressed in this same framework.

8.3.4 Gradient algorithm using negentropy

Gradient algorithm As with kurtosis, we can derive a simple gradient algorithm for maximizing negentropy. Taking the gradient of the approximation of negentropy in (8.25) with respect to $\mathbf{w}$, and taking the normalization $E\{(\mathbf{w}^T\mathbf{z})^2\} = \|\mathbf{w}\|^2 = 1$ into account, one obtains the following algorithm

$$\Delta\mathbf{w} \propto \gamma E\{\mathbf{z}g(\mathbf{w}^T\mathbf{z})\} \tag{8.28}$$

$$\mathbf{w} \leftarrow \mathbf{w}/\|\mathbf{w}\| \tag{8.29}$$

where $\gamma = E\{G(\mathbf{w}^T\mathbf{z})\} - E\{G(\nu)\}$, ν being a standardized gaussian random variable. The normalization is necessary to project $\mathbf{w}$ on the unit sphere to keep the variance of $\mathbf{w}^T\mathbf{z}$ constant. The function g is the derivative of the function G used in the approximation of negentropy. The expectation could be omitted to obtain an on-line (adaptive) stochastic gradient algorithm.

The constant, γ, which gives the algorithm a kind of "self-adaptation" quality, can be easily estimated on-line as follows:

$$\Delta\gamma \propto (G(\mathbf{w}^T\mathbf{z}) - E\{G(\nu)\}) - \gamma \tag{8.30}$$

This constant corresponds to the sign of kurtosis in (8.13).

As for the function g, we can use the derivatives of the functions in (8.26)– (8.27) that give robust approximations of negentropy. Alternatively, we could use the derivative corresponding to the fourth power as in kurtosis, which leads to the method that was already described in the previous section. Thus we can choose from:

$$g_1(y) = \tanh(a_1 y) \tag{8.31}$$

$$g_2(y) = y\exp(-y^2/2) \tag{8.32}$$

$$g_3(y) = y^3 \tag{8.33}$$

where $1 \leq a_1 \leq 2$ is some suitable constant, often taken as $a_1 = 1$. These functions are illustrated in Fig. 8.21.

The final form of the on-line stochastic gradient algorithm is summarized on Table 8.1.

This algorithm can be further simplified. First note that the constant γ does not change the stationary points of the learning rule. Its sign does affect their stability, though. Therefore, one can replace the γ by its sign without essentially affecting the behavior of the learning rule. This is useful, for example, in cases where we have some a priori information on the distributions of the independent components. For example, speech signals are usually highly supergaussian. One might thus roughly

---

1. Center the data to make its mean zero.

2. Whiten the data to give $\mathbf{z}$.

3. Choose an initial (e.g., random) vector $\mathbf{w}$ of unit norm, and an initial value for γ.

4. Update $\Delta\mathbf{w} \propto \gamma\mathbf{z}g(\mathbf{w}^T\mathbf{z})$, where g is defined e.g. as in (8.31)-(8.33)

5. Normalize $\mathbf{w} \leftarrow \mathbf{w}/\|\mathbf{w}\|$

6. If the sign of γ is not known a priori, update $\Delta\gamma \propto (G(\mathbf{w}^T\mathbf{z}) - E\{G(\nu)\}) - \gamma$.

7. If not converged, go back to step 4.

Table 8.1   The on-line stochastic gradient algorithm for finding one maximally nongaussian direction, i.e., estimating one independent component.

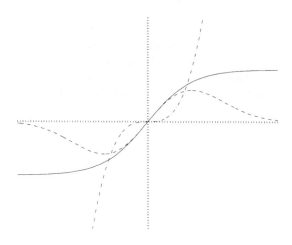

Fig. 8.21   The robust nonlinearities g_1 in Eq. (8.31), g_2 in Eq. (8.32), given by the solid line and the dashed line, respectively. The third power in (8.33), as used in kurtosis-based methods, is given by the dash-dotted line.

evaluate $E\{G(s_i) - G(\nu)\}$ for some supergaussian independent components and then take this, or its sign, as the value of γ. For example, if g is the tanh function, then $\gamma = -1$ works for supergaussian independent components.

Stability analysis * This section contains a theoretical analysis that can be skipped at first reading.

Since the approximation of negentropy in (8.25) may be rather crude, one may wonder if the estimator obtained from (8.28) really converges to the direction of one of the independent components, assuming the ICA data model. It can be proven that this is so, under rather mild conditions. The key to this proof is the following theorem, proven in the Appendix:

Theorem 8.1 *Assume that the input data follows the ICA model with whitened data:* $\mathbf{z} = \mathbf{VA}s$ *where* $\mathbf{V}$ *is the whitening matrix, and that* G *is a sufficiently smooth even function. Then the local maxima (resp. minima) of* $E\{G(\mathbf{w}^T\mathbf{z})\}$ *under the constraint* $\|\mathbf{w}\| = 1$ *include those rows of the mixing matrix* $\mathbf{VA}$ *such that the corresponding independent components* s_i *satisfy*

$$E\{s_i g(s_i) - g'(s_i)\} > 0 \ (resp. \ < 0) \tag{8.34}$$

where $g(.)$ *is the derivative of* $G(.)$, *and* $g'(.)$ *is the derivative of* $g(.)$.

This theorem shows that practically any nonquadratic function G may be used to perform ICA. More precisely, any function G divides the space of probability distributions into two half-spaces, depending on whether the nonpolynomial moment in the theorem is positive or negative. Independent components whose distribution is in one of the half-spaces can be estimated by maximizing $E\{G(\mathbf{w}^T\mathbf{z})\}$, and ICs whose distribution is in the other half-space can be estimated by minimizing the same function. The theorem gives the exact boundary between these two half-spaces.

In particular, this theorem implies the following:

Theorem 8.2 *Assume that the input data follows the ICA data model in (8.1), and that* G *is a sufficiently smooth even function. Then the asymptotically stable points of algorithm in (8.28) include the* ith *row of the inverse of the whitened mixing matrix* $\mathbf{VA}$ *such that the corresponding independent component* s_i *fulfills*

$$E\{s_i g(s_i) - g'(s_i)\}[E\{G(s_i)\} - E\{G(\nu)\}] > 0 \tag{8.35}$$

where $g(.)$ *is the derivative of* $G(.)$, *and* ν *is a standardized gaussian variable.*

Note that if $\mathbf{w}$ equals the ith row of $(\mathbf{VA})^{-1}$, the linear combination equals the ith independent component: $\mathbf{w}^T\mathbf{z} = \pm s_i$.

This theorem simply says that the question of stability of the gradient learning rule of the approximation of negentropy boils down to the question: Does the division into two half-spaces as given by Theorem 8.1 give the same division that is given by the sign of $E\{G(s_i) - G(\nu)\}$? This seems to be approximately true for most reasonable choices of G, and distributions of the s_i. In particular, if $G(y) = y^4$, we encounter the kurtosis-based criterion, and the condition is fulfilled for any distribution of nonzero kurtosis.

Theorem 8.1 also shows how to modify the algorithm in (8.28) so that it is (practically) always stable. This is possible by defining the self-adaptation constant γ as

$$\gamma = \text{sign}(E\{yg(y) - g'(y)\}) \tag{8.36}$$

The drawback with this definition is that now the algorithm cannot be interpreted as optimization of an objective function.

8.3.5   A fast fixed-point algorithm using negentropy

As with kurtosis, a much faster method for maximizing negentropy than that given by the gradient method, can be found using a fixed-point algorithm. The resulting FastICA algorithm [197] finds a direction, i.e., a unit vector $\mathbf{w}$, such that the projection $\mathbf{w}^T\mathbf{z}$ maximizes nongaussianity. Nongaussianity is here measured by the approximation of negentropy $J(\mathbf{w}^T\mathbf{z})$ given in (8.25). Recall that the variance of $\mathbf{w}^T\mathbf{z}$ must here be constrained to unity; for whitened data, this is equivalent to constraining the norm of $\mathbf{w}$ to be unity.

FastICA is based on a fixed-point iteration scheme for finding a maximum of the nongaussianity of $\mathbf{w}^T\mathbf{z}$, as measured in (8.25). More rigorously, it can be derived as an approximative Newton iteration. The FastICA algorithm using negentropy combines the superior algorithmic properties resulting from the fixed-point iteration with the preferable statistical properties due to negentropy.

Derivation of algorithm *   In this subsection, we derive the fixed-point algorithm using negentropy. This can be skipped by the reader not interested in mathematical details.

Looking at the gradient method in (8.28) immediately suggests the following fixed-point iteration:

$$\mathbf{w} \leftarrow E\{\mathbf{z}g(\mathbf{w}^T\mathbf{z})\} \tag{8.37}$$

which would of course be followed by normalization of $\mathbf{w}$. The coefficient γ can be omitted because it would be eliminated by the normalization anyway.

The iteration in (8.37) does not, however, have the good convergence properties of the FastICA using kurtosis, because the nonpolynomial moments do not have the same nice algebraic properties as real cumulants like kurtosis. Therefore, the iteration in (8.37) has to be modified. This is possible because we can add $\mathbf{w}$, multiplied by some constant α, on both sides of (8.37) without modifying the fixed points. In fact, we have

$$\mathbf{w} = E\{\mathbf{z}g(\mathbf{w}^T\mathbf{z})\} \tag{8.38}$$

$$\Leftrightarrow$$

$$(1 + \alpha)\mathbf{w} = E\{\mathbf{z}g(\mathbf{w}^T\mathbf{z})\} + \alpha\mathbf{w} \tag{8.39}$$

and because of the subsequent normalization of $\mathbf{w}$ to unit norm, the latter equation (8.39) gives a fixed-point iteration that has the same fixed points. Thus, by choosing α wisely, it may be possible to obtain an algorithm that converges as fast as the fixed-point algorithm using kurtosis. In fact, such a α *can* be found, as we show here.

The suitable coefficient α, and thus the FastICA algorithm, can be found using an approximative Newton method. The Newton method is a fast method for solving equations; see Chapter 3. When it is applied on the gradient, it gives an optimization method that usually converges in a small number of steps. The problem with the Newton method, however, is that it usually requires a matrix inversion at every step. Therefore, the total computational load may not be smaller than with gradient methods. What is quite surprising is that using the special properties of the ICA problem, we can find an approximation of the Newton method that does *not* need a matrix inversion but still converges roughly with the same number of iterations as the real Newton method (at least in theory). This approximative Newton method gives a fixed-point algorithm of the form (8.39).

To derive the approximative Newton method, first note that the maxima of the approximation of the negentropy of $\mathbf{w}^T\mathbf{z}$ are typically obtained at certain optima of $E\{G(\mathbf{w}^T\mathbf{z})\}$. According to the Lagrange conditions (see Chapter 3), the optima of $E\{G(\mathbf{w}^T\mathbf{z})\}$ under the constraint $E\{(\mathbf{w}^T\mathbf{z})^2\} = \|\mathbf{w}\|^2 = 1$ are obtained at points where the gradient of the Lagrangian is zero:

$$E\{\mathbf{z}g(\mathbf{w}^T\mathbf{z})\} + \beta\mathbf{w} = 0 \tag{8.40}$$

Now let us try to solve this equation by Newton's method, which is equivalent to finding the optima of the Lagrangian by Newton's method. Denoting the function on the left-hand side of (8.40) by F, we obtain its gradient (which is the second derivative of the Lagrangian) as

$$\frac{\partial F}{\partial \mathbf{w}} = E\{\mathbf{z}\mathbf{z}^T g'(\mathbf{w}^T\mathbf{z})\} + \beta\mathbf{I} \tag{8.41}$$

To simplify the inversion of this matrix, we decide to approximate the first term in (8.41). Since the data is sphered, a reasonable approximation seems to be $E\{\mathbf{z}\mathbf{z}^T g'(\mathbf{w}^T\mathbf{z})\} \approx E\{\mathbf{z}\mathbf{z}^T\}E\{g'(\mathbf{w}^T\mathbf{z})\} = E\{g'(\mathbf{w}^T\mathbf{z})\}\mathbf{I}$. Thus the gradient becomes diagonal, and can easily be inverted. Thus we obtain the following approximative Newton iteration:

$$\mathbf{w} \leftarrow \mathbf{w} - [E\{\mathbf{z}g(\mathbf{w}^T\mathbf{z})\} + \beta\mathbf{w}]/[E\{g'(\mathbf{w}^T\mathbf{z})\} + \beta] \tag{8.42}$$

This algorithm can be further simplified by multiplying both sides of (8.42) by $\beta + E\{g'(\mathbf{w}^T\mathbf{z})\}$. This gives, after straightforward algebraic simplification:

$$\mathbf{w} \leftarrow E\{\mathbf{z}g(\mathbf{w}^T\mathbf{z}) - E\{g'(\mathbf{w}^T\mathbf{z})\}\mathbf{w}\} \tag{8.43}$$

This is the basic fixed-point iteration in FastICA.

1. Center the data to make its mean zero.

2. Whiten the data to give $\mathbf{z}$.

3. Choose an initial (e.g., random) vector $\mathbf{w}$ of unit norm.

4. Let $\mathbf{w} \leftarrow E\{\mathbf{z}g(\mathbf{w}^T\mathbf{z})\} - E\{g'(\mathbf{w}^T\mathbf{z})\}\mathbf{w}$, where g is defined, e.g., as in (8.31)–(8.33).

5. Let $\mathbf{w} \leftarrow \mathbf{w}/\|\mathbf{w}\|$.

6. If not converged, go back to step 4.

Table 8.2 The FastICA algorithm for finding one maximally nongaussian direction, i.e., estimating one independent component. The expectations are estimated in practice as an average over the available data sample.

The fixed-point algorithm   The preceding derivation gives us the FastICA algorithm that can be described as follows.

First, we choose a nonlinearity g, which is the derivative of the nonquadratic function G used in (8.25). For example, we can use the derivatives of the functions in (8.26)–(8.27) that give robust approximations of negentropy. Alternatively, we could use the derivative corresponding to the fourth power as in kurtosis, which leads to the method that was already described in the previous section. Thus we can choose from the same functions in (8.31)–(8.33) as with the gradient algorithm, illustrated in Fig. 8.21.

Then we use the iteration in (8.43), followed by normalization. Thus, the basic form of the FastICA algorithm is then as described in Table 8.2.

The functions g' can be computed as

$$g_1'(y) = a_1(1 - \tanh^2(a_1 y)) \tag{8.44}$$

$$g_2'(y) = (1 - y^2)\exp(-y^2/2) \tag{8.45}$$

$$g_3'(y) = 3y^2 \tag{8.46}$$

Note that since we have constrained $E\{y^2\} = 1$, the derivative in (8.46) is essentially reduced to the constant 3.

Note that as above, convergence means that the old and new values of $\mathbf{w}$ point in the same direction, i.e., the absolute value of their dot-product is (almost) equal to 1. It is not necessary that the vector converges to a single point, since $\mathbf{w}$ and $-\mathbf{w}$ define the same direction.

As already discussed in connection with kurtosis, FastICA has properties that make it clearly superior to the gradient-based algorithms when fast adaptation to a changing environment is not needed. Even when using general approximations of negentropy, convergence is at least quadratic, which means much faster than the linear convergence obtained by gradient methods. Moreover, there is no learning rate or other adjustable parameters in the algorithm, which makes it easy to use, and more

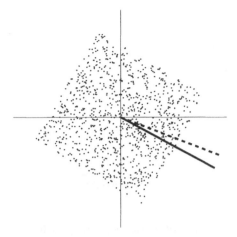

Fig. 8.22 Results with FastICA using negentropy, for ICs with uniform distributions. Dashed line: **w** after the first iteration (plotter longer than actual size). Solid line: **w** after the second iteration.

reliable. Using robust approximations of negentropy instead of kurtosis enhances the statistical properties of the resulting estimator, as discussed in Section 8.3.1.

The algorithm just given estimates only one independent component. To estimate more independent components, different kinds of decorrelation schemes should be used; see Section 8.4.

Examples   Here we show what happens when we run this version of the FastICA algorithm that maximizes the negentropy, using the two example data sets used in this chapter. First we take a mixture of two uniformly distributed independent components. The mixtures are whitened, as always in this chapter. The goal is now to find a direction in the data that maximizes the negentropy, as illustrated in 8.12.

For purposes of the illustration, we initialize the vector **w** as $\mathbf{w} = (1, 0)^T$. Running the FastICA iteration just two times (using the tanh nonlinearity in 8.31)), we obtain convergence. In Fig. 8.22, the obtained vectors **w** are shown. The dashed line gives the direction of **w** after the first iteration, and the solid line gives the direction of **w** after the second iteration. The third iteration didn't make any significant change in the direction of **w**, which means that the algorithm converged. (The corresponding vector in not plotted.) The figure shows that the value of **w** may change drastically during the iteration, because the values **w** and $-\mathbf{w}$ are considered as equivalent. This is because the sign of the vector cannot be determined in the ICA model.

The negentropies of the projections $\mathbf{w}^T\mathbf{z}$ obtained in the iterations are plotted in Fig. 8.23, as a function of iteration count. The plot shows that the algorithm steadily increased the negentropy of the projection, until it reached convergence at the third iteration.

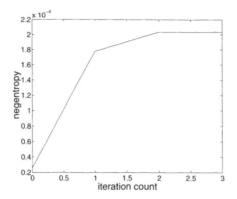

Fig. 8.23   The convergence of FastICA using negentropy, for ICs with uniform distributions. The value of negentropy shown as a function of iteration count. (Note that the value of negentropy was not properly scaled because a multiplying constant was omitted.)

Similar experiments were performed for the whitened mixtures of two *supergaussian* independent components, as illustrated in Fig. 8.14. In Fig. 8.24, the obtained vectors are shown. Convergence was obtained after three iterations. The negentropies of the projections $\mathbf{w}^T \mathbf{z}$ obtained in the iterations are plotted in Fig. 8.25, as a function of iteration count. As earlier, the negentropy of the projection steadily increased, until it reached convergence at the third iteration.

In these examples, we only estimated one independent component. In practice, we have many more dimensions and, therefore, we usually want to estimate more than one independent component. This can be done using a decorrelation scheme, as will be discussed next.

8.4   ESTIMATING SEVERAL INDEPENDENT COMPONENTS

8.4.1   Constraint of uncorrelatedness

In this chapter, we have so far estimated only one independent component. This is why these algorithms are sometimes called "one-unit" algorithms. In principle, we could find more independent components by running the algorithm many times and using different initial points. This would not be a reliable method of estimating many independent components, however.

The key to extending the method of maximum nongaussianity to estimate more independent component is based on the following property: The vectors $\mathbf{w}_i$ corresponding to different independent components are orthogonal in the whitened space, as shown in Chapter 7. To recapitulate, the independence of the components requires that they are uncorrelated, and in the whitened space we have $E\{(\mathbf{w}_i^T \mathbf{z})(\mathbf{w}_j^T \mathbf{z})\} = \mathbf{w}_i^T \mathbf{w}_j$, and therefore uncorrelatedness in equivalent to orthog-

Fig. 8.24 Results with FastICA using negentropy, second experiment. Dashed line: **w** after the first iteration (plotted longer than actual size). Solid line: **w** after the second iteration.

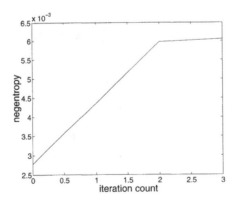

Fig. 8.25 The convergence of FastICA using negentropy, for supergaussian ICs. The value of negentropy shown as function of iteration count. (Again, note that the value of negentropy was not properly scaled.)

onality. This property is a direct consequence of the fact that after whitening, the mixing matrix can be taken to be orthogonal. The $\mathbf{w}_i$ are in fact by definition the rows of the inverse of the mixing matrix, and these are equal to the columns of the mixing matrix, because by orthogonality $\mathbf{A}^{-1} = \mathbf{A}^T$.

Thus, to estimate several independent components, we need to run any of the one-unit algorithms several times (possibly using several units) with vectors $\mathbf{w}_1, ..., \mathbf{w}_n$, and to prevent different vectors from converging to the same maxima we must *orthogonalize* the vectors $\mathbf{w}_1, ..., \mathbf{w}_n$ after every iteration. We present in the following different methods for achieving decorrelation.

8.4.2 Deflationary orthogonalization

A simple way of orthogonalization is deflationary orthogonalization using the Gram-Schmidt method. This means that we estimate the independent components one by one. When we have estimated p independent components, or p vectors $\mathbf{w}_1, ..., \mathbf{w}_p$, we run any one-unit algorithm for $\mathbf{w}_{p+1}$, and after every iteration step subtract from $\mathbf{w}_{p+1}$ the projections $(\mathbf{w}_{p+1}^T \mathbf{w}_j)\mathbf{w}_j, j = 1, ..., p$ of the previously estimated p vectors, and then renormalize $\mathbf{w}_{p+1}$. More precisely, we alternate the following steps:

1. Choose m, the number of ICs to estimate. Set $p \leftarrow 1$.

2. Initialize $\mathbf{w}_p$ (e.g. randomly)

3. Do an iteration of a one-unit algorithm on $\mathbf{w}_p$.

4. Do the following orthogonalization:

$$\mathbf{w}_p \leftarrow \mathbf{w}_p - \sum_{j=1}^{p-1} (\mathbf{w}_p^T \mathbf{w}_j)\mathbf{w}_j \tag{8.47}$$

5. Normalize $\mathbf{w}_p$ by dividing it by its norm.

6. If $\mathbf{w}_p$ has not converged, go back to step 3.

7. Set $p \leftarrow p + 1$. If p is not greater than the desired number of ICs, go back to step 2.

In particular, we give the FastICA algorithm with deflationary orthogonalization in Table 8.3.

8.4.3 Symmetric orthogonalization

In certain applications, it may be desirable to use a symmetric decorrelation, in which no vectors are "privileged" over others. This means that the vectors $\mathbf{w}_i$ are not estimated one by one; instead, they are estimated in parallel. One motivation for this is that the deflationary method has the drawback that estimation errors in the first vectors are cumulated in the subsequent ones by the orthogonalization. Another one is that the symmetric orthogonalization methods enable parallel computation of ICs.

Symmetric orthogonalization is done by first doing the iterative step of the one-unit algorithm on every vector $\mathbf{w}_i$ in parallel, and afterwards orthogonalizing all the $\mathbf{w}_i$ by special symmetric methods. In other words:

1. Choose the number of independent components to estimate, say m.

2. Initialize the $\mathbf{w}_i, i = 1, ..., m$ (e.g., randomly).

3. Do an iteration of a one-unit algorithm on every $\mathbf{w}_i$ in parallel.

4. Do a symmetric orthogonalization of the matrix $\mathbf{W} = (\mathbf{w}_1, ..., \mathbf{w}_m)^T$.

5. If not converged, go back to step 3.

In Chapter 6, methods for symmetric orthogonalization were discussed. The symmetric orthogonalization of $\mathbf{W}$ can be accomplished, e.g., by the classic method involving matrix square roots,

$$\mathbf{W} \leftarrow (\mathbf{W}\mathbf{W}^T)^{-1/2}\mathbf{W} \tag{8.48}$$

The inverse square root $(\mathbf{W}\mathbf{W}^T)^{-1/2}$ is obtained from the eigenvalue decomposition of $\mathbf{W}\mathbf{W}^T = \mathbf{E} \operatorname{diag}(d_1, ..., d_m) \mathbf{E}^T$ as

$$(\mathbf{W}\mathbf{W}^T)^{-1/2} = \mathbf{E} \operatorname{diag}(d_1^{-1/2}, ..., d_m^{-1/2}) \mathbf{E}^T \tag{8.49}$$

A simpler alternative is the following iterative algorithm:

1. Let $\mathbf{W} \leftarrow \mathbf{W}/\|\mathbf{W}\|$.

2. Let $\mathbf{W} \leftarrow \frac{3}{2}\mathbf{W} - \frac{1}{2}\mathbf{W}\mathbf{W}^T\mathbf{W}$.

3. If $\mathbf{W}\mathbf{W}^T$ is not close enough to identity, go back to step 2.

The norm in step 1 can be almost any ordinary matrix norm, e.g., the 2-norm or the largest absolute row or column sum (but not the Frobenius norm); see Section 6.5 for details.

We give a detailed version of the FastICA algorithm that uses the symmetric orthogonalization in Table 8.4.

1. Center the data to make its mean zero.

2. Whiten the data to give $\mathbf{z}$.

3. Choose m, the number of ICs to estimate. Set counter $p \leftarrow 1$.

4. Choose an initial value of unit norm for $\mathbf{w}_p$, e.g., randomly.

5. Let $\mathbf{w}_p \leftarrow E\{\mathbf{z}g(\mathbf{w}_p^T\mathbf{z})\} - E\{g'(\mathbf{w}_p^T\mathbf{z})\}\mathbf{w}$, where g is defined, e.g., as in (8.31)–(8.33).

6. Do the following orthogonalization:

$$\mathbf{w}_p \leftarrow \mathbf{w}_p - \sum_{j=1}^{p-1}(\mathbf{w}_p^T\mathbf{w}_j)\mathbf{w}_j \qquad (8.50)$$

7. Let $\mathbf{w}_p \leftarrow \mathbf{w}_p/\|\mathbf{w}_p\|$.

8. If $\mathbf{w}_p$ has not converged, go back to step 5.

9. Set $p \leftarrow p + 1$. If $p \leq m$, go back to step 4.

Table 8.3   The FastICA algorithm for estimating several ICs, with *deflationary* orthogonalization. The expectations are estimated in practice as sample averages.

1. Center the data to make its mean zero.

2. Whiten the data to give $\mathbf{z}$.

3. Choose m, the number of independent components to estimate.

4. Choose initial values for the $\mathbf{w}_i, i = 1, ..., m$, each of unit norm. Orthogonalize the matrix $\mathbf{W}$ as in step 6 below.

5. For every $i = 1, ..., m$, let $\mathbf{w}_i \leftarrow E\{\mathbf{z}g(\mathbf{w}_i^T\mathbf{z})\} - E\{g'(\mathbf{w}_i^T\mathbf{z})\}\mathbf{w}$, where g is defined, e.g., as in (8.31)–(8.33).

6. Do a symmetric orthogonalization of the matrix $\mathbf{W} = (\mathbf{w}_1, ..., \mathbf{w}_m)^T$ by

$$\mathbf{W} \leftarrow (\mathbf{W}\mathbf{W}^T)^{-1/2}\mathbf{W}, \qquad (8.51)$$

or by the iterative algorithm in Sec. 8.4.3.

7. If not converged, go back to step 5.

Table 8.4   The FastICA algorithm for estimating several ICs, with *symmetric* orthogonalization. The expectations are estimated in practice as sample averages.

8.5 ICA AND PROJECTION PURSUIT

It is interesting to note how the approach to ICA described in this Chapter makes explicit the connection between ICA and another technique: projection pursuit.

8.5.1 Searching for interesting directions

Projection pursuit is a technique developed in statistics for finding "interesting" projections of multidimensional data. Such projections can then be used for optimal visualization of the data, and for such purposes as density estimation and regression.

When projection pursuit is used for exploratory data analysis, we usually compute a couple of the most interesting 1-D projections. (The definition of interestingness will be treated in the next section.) Some structure of the data can then be visualized by showing the distribution of the data in the 1-D subspaces, or on 2-D planes spanned by two of the projection pursuit directions. This method is en extension of the classic method of using principal component analysis (PCA) for visualization, in which the distribution of the data is shown on the plane spanned by the two first principal components.

An example of the problem can be seen in Fig. 8.26. In reality, projection pursuit is of course used in situations where the number of dimensions is very large, but for purposes of illustration, we use here a trivial 2-D example. In the figure, the interesting projection of the data would be on the horizontal axis. This is because that projection shows the clustering structure of the data. In contrast, projections in very different directions (here, projection on the vertical axis) would show only an ordinary gaussian distribution. It would thus be useful to have a method that automatically finds the horizontal projection in this example.

8.5.2 Nongaussian is interesting

The basic question in projection pursuit is thus to define what kind of projections are interesting.

It is usually argued that the gaussian distribution is the least interesting one, and that the most interesting directions are those that show the least gaussian distribution. One motivation for this is that distributions that are multimodal, i.e., show some clustering structure, are far from gaussian.

An information-theoretic motivation for nongaussianity is that entropy is maximized by the gaussian distribution, and entropy can be considered as a measure of the lack of structure (see Chapter 5). This is related to the interpretation of entropy as code length: a variable that has a clear structure is usually easy to code. Thus, since the gaussian distribution has the largest entropy, it is the most difficult to code, and therefore it can be considered as the least structured.

The usefulness of using the most nongaussian projections for visualization can be seen in Fig. 8.26. Here the most nongaussian projection is on the horizontal axis; this is also the projection that most clearly shows the clustered structure of the data. On

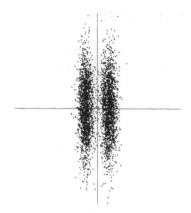

Fig. 8.26 An illustration of projection pursuit and the "interesting" directions. The data in this figure is clearly divided into two clusters. The goal in projection pursuit is to find the projection (here, on the horizontal axis) that reveals the clustering or other structure of the data.

the other hand, the projection on the vertical direction, which is also the direction of the first principal component, fails to show this structure. This also shows that PCA does not use the clustering structure. In fact, clustering structure is not visible in the covariance or correlation matrix on which PCA is based.

Thus projection pursuit is usually performed by finding the most nongaussian projections of the data. This is the same thing that we did in this chapter to estimate the ICA model. This means that all the nongaussianity measures and the corresponding ICA algorithms presented in this chapter could also be called projection pursuit "indices" and algorithms.

It should be noted that in the formulation of projection pursuit, no data model or assumption about independent components is made. If the ICA model holds, optimizing the ICA nongaussianity measures produce independent components; if the model does not hold, then what we get are the projection pursuit directions.

8.6 CONCLUDING REMARKS AND REFERENCES

A fundamental approach to ICA is given by the principle of nongaussianity. The independent components can be found by finding directions in which the data is maximally nongaussian. Nongaussianity can be measured by entropy-based measures or cumulant-based measures like kurtosis. Estimation of the ICA model can then be performed by maximizing such nongaussianity measures; this can be done by gradient methods or by fixed-point algorithms. Several independent components can be found by finding several directions of maximum nongaussianity under the constraint of decorrelation.

This approach is closely connected to projection pursuit, in which the maximally nongaussian directions are considered interesting from the viewpoint of visualization and exploratory data analysis [222, 137, 138, 95, 160, 151, 316, 414, 139]; a useful review is provided in [189]. From a modeling viewpoint, this approach was first developed in the context of blind deconvolution [114, 399]. Blind deconvolution, discussed in more detail in Chapter 19, is a linear model not unlike the ICA model, but the mixing is done by convolution of a one-dimensional signal. In the context of ICA, the principle of nongaussianity was probably first used in [107], where the maximality property of kurtosis was proven rigorously, and further developed in [197, 210, 211, 291], on which this chapter is based.

Problems

8.1 Prove (8.8) and (8.9)

   8.1.1. by algebraically manipulating the formulas

   8.1.2. by using the general definition of cumulants (see Section 2.7.2).

8.2 Derive the gradient in (8.13).

8.3 Derive the gradient in (8.28).

8.4 What happens to the ICA algorithms in (8.28) and (8.43) if the nonlinearity g is taken to be linear?

8.5 How does the behavior of the algorithms in (8.28) and (8.43) change if

   8.5.1. a linear function is added to g?

   8.5.2. a constant is added to g?

8.6 Derive a fixed-point algorithm for the third-order cumulant: $E\{y^3\}$. When could this algorithm be useful? Why is kurtosis preferred in most applications?

8.7 In this exercice, we prove the fundamental maximality property of kurtosis. More precisely, we prove that the maxima of the function

$$F(\mathbf{q}) = |\text{kurt}(\mathbf{q}^T \mathbf{s})| = |q_1^4 \text{kurt}(s_1) + q_2^4 \text{kurt}(s_2)| \qquad (8.52)$$

in the constraint set $\|\mathbf{q}\|^2 = 1$ are obtained when only one of the components of $\mathbf{q}$ is nonzero. For simplicity, we consider here the 2-D case first.

   8.7.1. Make the change of variables $t_i = q_i^2$. What is the geometrical form of the constraint set of $\mathbf{t} = (t_1, t_2)$? Note that the objective function is now quadratic.

   8.7.2. Assume that both of the kurtoses are positive. What is the geometrical shape of the sets $F(\mathbf{t}) = \text{const.}$? By a geometrical argument, show that the maximum of $F(\mathbf{t})$ is obtained when one of the t_i is one and the other one is zero. Show how this proves the maximality property if the kurtoses are both positive.

   8.7.3. Assume that both kurtoses are negative. Using exactly the same logic as in the preceding point, show that the maximality property holds if the kurtoses are both negative.

8.7.4. Assume that the kurtoses have different signs. What is the geometrical shape of the sets $F(t) = $ const. now? By geometrical arguments, show the maximality property holds even in this case.

8.7.5. Let us redo the proof algebraically. Express t_2 as a function of t_1, and reformulate the problem. Solve it explicitly.

8.8 * Now we extend the preceding geometric proof the n dimensions. We will need some basic concepts of convex analysis [284].

8.8.1. Make the same change of variables. Prove that the constraint set is convex (in fact, it is what is called a simplex).

8.8.2. Assume that the kurtoses of the s_i are all positive. Show that the objective function is strictly convex.

8.8.3. Show that a strictly convex function defined on a simplex obtains its maxima in the extremal points.

8.8.4. Show that our objective function is maximized at the points where just one of the t_i is 1 and the others are zero, and that these correspond to the independent components.

8.8.5. Show the same when all the kurtoses are negative.

8.8.6. In the case of kurtoses of different signs, first show by a simple argument that if t_i and t_j corresponding to components with kurtoses of different signs are nonzero, the objective function can be increased by reducing one of them and increasing the other by the same amount. Conclude that in the maximum, only two of the t_i, corresponding to two different signs of kurtoses, can be nonzero. Show that the problem then reduces to what was already shown in the 2-D case.

Computer assignments

8.1 In this computer assignment, the central limit theorem is studied experimentally. Let $x(t)$, $t = 1, \ldots, T$, be T independent random numbers distributed uniformly on the interval $[-1, 1]$, and

$$y = \sum_{t=1}^{T} x(t)$$

their sum. Generate 5000 different realizations of the random variable y for the following numbers of terms in the sum: $T = 2, T = 4$, and $T = 12$.

8.1.1. Plot the experimental pdf's of y, and compare it with the gaussian pdf having the same (zero) mean and variance. (*Hint:* you can here estimate the pdf from the generated samples simply by dividing their value range into small bins of width 0.1 or 0.05, count the number of samples falling into each bin, and divide by the total number of samples. You can compute the difference between the respective gaussian and the estimated density to get a better idea of the similarity of the two distributions.)

8.1.2. Plot the kurtoses in each case. Note that you must normalize all the variables to unit variance. What if you don't normalize?

8.2 Program the FastICA algorithm in (8.4) in some computer environment.

8.2.1. Take the data $x(t)$ in the preceding assignment as two independent components by splitting the sample in two. Mix them using a random mixing matrix, and estimate the model, using one of the nonlinearity in (8.31).

8.2.2. Reduce the sample size to 100. Estimate the mixing matrix again. What do you see?

8.2.3. Try the different nonlinearities in (8.32)–(8.33). Do you see any difference?

8.2.4. Try the nonlinearity $g(u) = u^2$. Why does this not work?

Appendix proofs

Proof of Theorem 8.1 Denote by $H(\mathbf{w})$ the function to be minimized/maximized, $E\{G(\mathbf{w}^T\mathbf{z})\}$. Make the orthogonal change of coordinates $\mathbf{q} = \mathbf{A}^T\mathbf{V}^T\mathbf{w}$. Then we can calculate the gradient as $\frac{\partial H(\mathbf{q})}{\partial \mathbf{q}} = E\{\mathbf{s}g(\mathbf{q}^T\mathbf{s})\}$ and the Hessian as $\frac{\partial^2 H(\mathbf{q})}{\partial \mathbf{q}^2} = E\{\mathbf{s}\mathbf{s}^T g'(\mathbf{q}^T\mathbf{s})\}$. Without loss of generality, it is enough to analyze the stability of the point $\mathbf{q} = \mathbf{e}_1$, where $\mathbf{e}_1 = (1, 0, 0, 0, ...)$. Evaluating the gradient and the Hessian at point $\mathbf{q} = \mathbf{e}_1$, we get using the independence of the s_i,

$$\frac{\partial H(\mathbf{e}_1)}{\partial \mathbf{q}} = \mathbf{e}_1 E\{s_1 g(s_1)\} \tag{A.1}$$

and

$$\frac{\partial^2 H(\mathbf{e}_1)}{\partial \mathbf{q}^2} = \mathrm{diag}(E\{s_1^2 g'(s_1)\}, E\{g'(s_1)\}, E\{g'(s_1)\}, ...). \tag{A.2}$$

Making a small perturbation $\epsilon = (\epsilon_1, \epsilon_2, ...)$, we obtain

$$H(\mathbf{e}_1 + \epsilon) = H(\mathbf{e}_1) + \epsilon^T \frac{\partial H(\mathbf{e}_1)}{\partial \mathbf{q}} + \frac{1}{2}\epsilon^T \frac{\partial^2 H(\mathbf{e}_1)}{\partial \mathbf{q}^2} \epsilon + o(\|\epsilon\|^2)$$

$$= H(\mathbf{e}_1) + E\{s_1 g(s_1)\}\epsilon_1 + \frac{1}{2}[E\{s_1^2 g'(s_1)\}\epsilon_1^2 + E\{g'(s_1)\}\sum_{i>1}\epsilon_i^2] + o(\|\epsilon\|^2) \tag{A.3}$$

Due to the constraint $\|\mathbf{w}\| = 1$ we get $\epsilon_1 = \sqrt{1 - \epsilon_2^2 - \epsilon_3^2 - ...} - 1$. Due to the fact that $\sqrt{1 - \gamma} = 1 - \gamma/2 + o(\gamma)$, the term of order ϵ_1^2 in (A.3) is $o(\|\epsilon\|^2)$, i.e., of higher order, and can be neglected. Using the aforementioned first-order approximation for ϵ_1 we obtain $\epsilon_1 = -\sum_{i>1}\epsilon_i^2/2 + o(\|\epsilon\|^2)$, which finally gives

$$H(\mathbf{e}_1 + \epsilon) = H(\mathbf{e}_1) + \frac{1}{2}[E\{g'(s_1)\} - s_1 g(s_1)\}]\sum_{i>1}\epsilon_i^2 + o(\|\epsilon\|^2) \tag{A.4}$$

which clearly proves $\mathbf{q} = \mathbf{e}_1$ is an extremum, and of the type implied by the condition of the theorem.

Proof of convergence of FastICA The convergence is proven under the assumptions that first, the data follows the ICA data model (8.1) and second, that the expectations are evaluated exactly.

Let g be the nonlinearity used in the algorithm. In the case of the kurtosis-based algorithm in Section 8.2.3, this is the cubic function, so we obtain that algorithm as a special case of the following proof for a general g. We must also make the following technical assumption:

$$E\{s_i g(s_i) - g'(s_i)\} \neq 0, \quad \text{for any } i \tag{A.5}$$

which can be considered a generalization of the condition valid when we use kurtosis, that the kurtosis of the independent components must be nonzero. If (A.5) is true for a subset of independent components, we can estimate just those independent components.

To begin with, make the change of variable $\mathbf{q} = \mathbf{A}^T \mathbf{V}^T \mathbf{w}$, as earlier, and assume that $\mathbf{q}$ is in the neighborhood of a solution (say, $q_1 \approx 1$ as before). As shown in the proof of Theorem 8.1, the change in q_1 is then of a lower order than the change in the other coordinates, due to the constraint $\|\mathbf{q}\| = 1$. Then we can expand the terms in (8.43) using a Taylor approximation for g and g', first obtaining

$$g(\mathbf{q}^T\mathbf{s}) = \quad g(q_1 s_1) + g'(q_1 s_1)\mathbf{q}_{-1}^T \mathbf{s}_{-1} + \tfrac{1}{2}g''(q_1 s_1)(\mathbf{q}_{-1}^T \mathbf{s}_{-1})^2$$
$$+ \tfrac{1}{6}g'''(q_1 s_1)(\mathbf{q}_{-1}^T \mathbf{s}_{-1})^3 + O(\|\mathbf{q}_{-1}\|^4) \tag{A.6}$$

and then

$$g'(\mathbf{q}^T\mathbf{s}) = \quad g'(q_1 s_1) + g''(q_1 s_1)\mathbf{q}_{-1}^T \mathbf{s}_{-1}$$
$$+ \tfrac{1}{2}g'''(q_1 s_1)(\mathbf{q}_{-1}^T \mathbf{s}_{-1})^2 + O(\|\mathbf{q}_{-1}\|^3) \tag{A.7}$$

where $\mathbf{q}_{-1}$ and $\mathbf{s}_{-1}$ are the vectors $\mathbf{q}$ and $\mathbf{s}$ without their first components. Denote by $\mathbf{q}^+$ the new value of $\mathbf{q}$ (after one iteration). Thus we obtain, using the independence of the s_i and doing some tedious but straightforward algebraic manipulations,

$$q_1^+ = E\{s_1 g(q_1 s_1) - g'(q_1 s_1)\} + O(\|q_{-1}\|^2) \tag{A.8}$$

$$q_i^+ = \quad \tfrac{1}{2}E\{s_i^3\}E\{g''(s_1)\}q_i^2$$
$$+ \tfrac{1}{6}\mathrm{kurt}(s_i)E\{g'''(s_1)\}q_i^3 + O(\|q_{-1}\|^4), \quad \text{for } i > 1 \tag{A.9}$$

We obtain also

$$\mathbf{q}^* = \mathbf{q}^+/\|\mathbf{q}^+\| \tag{A.10}$$

This shows clearly that under assumption (A.5), the algorithm converges (locally) to such a vector $\mathbf{q}$ that $q_1 = \pm 1$ and $q_i = 0$ for $i > 1$. This means that $\mathbf{w} = ((\mathbf{VA})^T)^{-1}\mathbf{q}$ converges, up to the sign, to one of the rows of the inverse of the mixing matrix $\mathbf{VA}$, which implies that $\mathbf{w}^T\mathbf{z}$ converges to one of the s_i. Moreover, if $E\{g''(s_1)\} = 0$, i.e., if the s_i has a symmetric distribution, as is usually the case, (A.9) shows that the convergence is cubic. In other cases, the convergence is quadratic. If kurtosis is used, however, we always have $E\{g''(s_1)\} = 0$ and thus cubic convergence. In addition, if $G(y) = y^4$, the local approximations are exact, and the convergence is global.

9

ICA by Maximum Likelihood Estimation

A very popular approach for estimating the independent component analysis (ICA) model is maximum likelihood (ML) estimation. Maximum likelihood estimation is a fundamental method of statistical estimation; a short introduction was provided in Section 4.5. One interpretation of ML estimation is that we take those parameter values as estimates that give the highest probability for the observations. In this section, we show how to apply ML estimation to ICA estimation. We also show its close connection to the neural network principle of maximization of information flow (infomax).

9.1 THE LIKELIHOOD OF THE ICA MODEL

9.1.1 Deriving the likelihood

It is not difficult to derive the likelihood in the noise-free ICA model. This is based on using the well-known result on the density of a linear transform, given in (2.82). According to this result, the density p_x of the mixture vector

$$\mathbf{x} = \mathbf{As} \tag{9.1}$$

can be formulated as

$$p_x(\mathbf{x}) = |\det \mathbf{B}| p_s(\mathbf{s}) = |\det \mathbf{B}| \prod_i p_i(s_i) \tag{9.2}$$

where $\mathbf{B} = \mathbf{A}^{-1}$, and the p_i denote the densities of the independent components. This can be expressed as a function of $\mathbf{B} = (\mathbf{b}_1, ..., \mathbf{b}_n)^T$ and $\mathbf{x}$, giving

$$p_x(\mathbf{x}) = |\det \mathbf{B}| \prod_i p_i(\mathbf{b}_i^T \mathbf{x}) \tag{9.3}$$

Assume that we have T observations of $\mathbf{x}$, denoted by $\mathbf{x}(1), \mathbf{x}(2), ..., \mathbf{x}(T)$. Then the likelihood can be obtained (see Section 4.5) as the product of this density evaluated at the T points. This is denoted by L and considered as a function of $\mathbf{B}$:

$$L(\mathbf{B}) = \prod_{t=1}^{T} \prod_{i=1}^{n} p_i(\mathbf{b}_i^T \mathbf{x}(t)) |\det \mathbf{B}| \tag{9.4}$$

Very often it is more practical to use the logarithm of the likelihood, since it is algebraically simpler. This does not make any difference here since the maximum of the logarithm is obtained at the same point as the maximum of the likelihood. The log-likelihood is given by

$$\log L(\mathbf{B}) = \sum_{t=1}^{T} \sum_{i=1}^{n} \log p_i(\mathbf{b}_i^T \mathbf{x}(t)) + T \log |\det \mathbf{B}| \tag{9.5}$$

The basis of the logarithm makes no difference, though in the following the natural logarithm is used.

To simplify notation and to make it consistent to what was used in the previous chapter, we can denote the sum over the sample index t by an expectation operator, and divide the likelihood by T to obtain

$$\frac{1}{T} \log L(\mathbf{B}) = E\{\sum_{i=1}^{n} \log p_i(\mathbf{b}_i^T \mathbf{x})\} + \log |\det \mathbf{B}| \tag{9.6}$$

The expectation here is not the theoretical expectation, but an average computed from the observed sample. Of course, in the algorithms the expectations are eventually replaced by sample averages, so the distinction is purely theoretical.

9.1.2 Estimation of the densities

Problem of semiparametric estimation In the preceding, we have expressed the likelihood as a function of the parameters of the model, which are the elements of the mixing matrix. For simplicity, we used the elements of the inverse $\mathbf{B}$ of the mixing matrix. This is allowed since the mixing matrix can be directly computed from its inverse.

There is another thing to estimate in the ICA model, though. This is the densities of the independent components. Actually, the likelihood is a function of these densities as well. This makes the problem much more complicated, because the estimation of densities is, in general, a nonparametric problem. Nonparametric means that it cannot be reduced to the estimation of a finite parameter set. In fact the number of parameters to be estimated is infinite, or in practice, very large. Thus the estimation of the ICA model has also a nonparametric part, which is why the estimation is sometimes called "semiparametric".

Nonparametric estimation of densities is known to be a difficult problem. Many parameters are always more difficult to estimate than just a few; since nonparametric problems have an infinite number of parameters, they are the most difficult to estimate. This is why we would like to avoid the nonparametric density estimation in the ICA. There are two ways to avoid it.

First, in some cases we might know the densities of the independent components in advance, using some prior knowledge on the data at hand. In this case, we could simply use these prior densities in the likelihood. Then the likelihood would really be a function of $\mathbf{B}$ only. If reasonably small errors in the specification of these prior densities have little influence on the estimator, this procedure will give reasonable results. In fact, it will be shown below that this is the case.

A second way to solve the problem of density estimation is to approximate the densities of the independent components by a family of densities that are specified by a limited number of parameters. If the number of parameters in the density family needs to be very large, we do not gain much from this approach, since the goal was to reduce the number of parameters to be estimated. However, if it is possible to use a very simple family of densities to estimate the ICA model for any densities p_i, we will get a simple solution. Fortunately, this turns out to be the case. We can use an extremely simple parameterization of the p_i, consisting of the choice between two densities, i.e., a single binary parameter.

A simple density family   It turns out that in maximum likelihood estimation, it is enough to use just *two* approximations of the density of an independent component. For each independent component, we just need to determine which one of the two approximations is better. This shows that, first, we can make small errors when we fix the densities of the independent components, since it is enough that we use a density that is in the same half of the space of probability densities. Second, it shows that we can estimate the independent components using very simple models of their densities, in particular, using models consisting of only two densities.

This situation can be compared with the one encountered in Section 8.3.4, where we saw that any nonlinearity can be seen to divide the space of probability distributions in half. When the distribution of an independent component is in one of the halves, the nonlinearity can be used in the gradient method to estimate that independent component. When the distribution is in the other half, the *negative* of the nonlinearity must be used in the gradient method. In the ML case, a nonlinearity corresponds to a density approximation.

The validity of these approaches is shown in the following theorem, whose proof can be found in the appendix. This theorem is basically a corollary of the stability theorem in Section 8.3.4.

Theorem 9.1 *Denote by $\tilde{p}_i$ the assumed densities of the independent components, and*

$$g_i(s_i) = \frac{\partial}{\partial s_i} \log \tilde{p}_i(s_i) = \frac{\tilde{p}_i'(s_i)}{\tilde{p}_i(s_i)} \tag{9.7}$$

Constrain the estimates of the independent components $y_i = \mathbf{b}_i^T \mathbf{x}$ to be uncorrelated and to have unit variance. Then the ML estimator is locally consistent, if the assumed densities $\tilde{p}_i$ fulfill

$$E\{s_i g_i(s_i) - g'(s_i)\} > 0 \tag{9.8}$$

for all i.

This theorem shows rigorously that small misspecifications in the densities p_i do not affect the local consistency of the ML estimator, since sufficiently small changes do not change the sign in (9.8).

Moreover, the theorem shows how to construct families consisting of only two densities, so that the condition in (9.8) is true for one of these densities. For example, consider the following log-densities:

$$\log \tilde{p}_i^+(s) = \alpha_1 - 2 \log \cosh(s) \tag{9.9}$$
$$\log \tilde{p}_i^-(s) = \alpha_2 - [s^2/2 - \log \cosh(s)] \tag{9.10}$$

where α_1, α_2 are positive parameters that are fixed so as to make these two functions logarithms of probability densities. Actually, these constants can be ignored in the following. The factor 2 in (9.9) is not important, but it is usually used here; also, the factor $1/2$ in (9.10) could be changed.

The motivation for these functions is that $\tilde{p}_i^+$ is a *supergaussian* density, because the log cosh function is close to the absolute value that would give the Laplacian density. The density given by $\tilde{p}_i^-$ is *subgaussian*, because it is like a gaussian log-density, $-s^2/2$ plus a constant, that has been somewhat "flattened" by the log cosh function.

Simple computations show that the value of the nonpolynomial moment in (9.8) is for $\tilde{p}_i^+$

$$2E\{-\tanh(s_i)s_i + (1 - \tanh(s_i)^2)\} \tag{9.11}$$

and for $\tilde{p}_i^-$ it is

$$E\{\tanh(s_i)s_i - (1 - \tanh(s_i)^2)\} \tag{9.12}$$

since the derivative of $\tanh(s)$ equals $1 - \tanh(s)^2$, and $E\{s_i^2\} = 1$ by definition. We see that the signs of these expressions are always opposite. Thus, for practically any distributions of the s_i, one of these functions fulfills the condition, i.e., has the desired sign, and estimation is possible. Of course, for some distribution of the s_i the nonpolynomial moment in the condition could be zero, which corresponds to the case of zero kurtosis in cumulant-based estimation; such cases can be considered to be very rare.

Thus we can just compute the nonpolynomial moments for the two prior distributions in (9.9) and (9.10), and choose the one that fulfills the stability condition in (9.8). This can be done on-line during the maximization of the likelihood. This always provides a (locally) consistent estimator, and solves the problem of semiparametric estimation.

In fact, the nonpolynomial moment in question measures the shape of the density function in much the same way as kurtosis. For $g(s) = -s^3$, we would actually obtain kurtosis. Thus, the choice of nonlinearity could be compared with the choice whether to minimize or maximize kurtosis, as previously encountered in Section 8.2. That choice was based on the value of the sign of kurtosis; here we use the sign of a nonpolynomial moment.

Indeed, the nonpolynomial moment of this chapter is the same as the one encountered in Section 8.3 when using more general measures of nongaussianity. However, it must be noted that the set of nonlinearities that we can use here is more restricted than those used in Chapter 8. This is because the nonlinearities g_i used must correspond to the derivative of the logarithm of a probability density function (pdf). For example, we cannot use the function $g(s) = s^3$ because the corresponding pdf would be of the form $\exp(s^4/4)$, and this is not integrable, i.e., it is not a pdf at all.

9.2 ALGORITHMS FOR MAXIMUM LIKELIHOOD ESTIMATION

To perform maximum likelihood estimation in practice, we need an algorithm to perform the numerical maximization of likelihood. In this section, we discuss different methods to this end. First, we show how to derive simple gradient algorithms, of which especially the natural gradient algorithm has been widely used. Then we show how to derive a fixed-point algorithm, a version of FastICA, that maximizes the likelihood faster and more reliably.

9.2.1 Gradient algorithms

The Bell-Sejnowski algorithm The simplest algorithms for maximizing likelihood are obtained by gradient methods. Using the well-known results in Chapter 3, one can easily derive the stochastic gradient of the log-likelihood in (9.6) as:

$$\frac{1}{T}\frac{\partial \log L}{\partial \mathbf{B}} = [\mathbf{B}^T]^{-1} + E\{\mathbf{g}(\mathbf{Bx})\mathbf{x}^T\} \tag{9.13}$$

Here, $\mathbf{g}(\mathbf{y}) = (g_i(y_i), ..., g_n(y_n))$ is a component-wise vector function that consists of the so-called (negative) score functions g_i of the distributions of s_i, defined as

$$g_i = (\log p_i)' = \frac{p_i'}{p_i}. \tag{9.14}$$

This immediately gives the following algorithm for ML estimation:

$$\Delta \mathbf{B} \propto [\mathbf{B}^T]^{-1} + E\{\mathbf{g}(\mathbf{B}\mathbf{x})\mathbf{x}^T\} \tag{9.15}$$

A stochastic version of this algorithm could be used as well. This means that the expectation is omitted, and in each step of the algorithm, only one data point is used:

$$\Delta \mathbf{B} \propto [\mathbf{B}^T]^{-1} + \mathbf{g}(\mathbf{B}\mathbf{x})\mathbf{x}^T. \tag{9.16}$$

This algorithm is often called the Bell-Sejnowski algorithm. It was first derived in [36], though from a different approach using the infomax principle that is explained in Section 9.3 below.

The algorithm in Eq. (9.15) converges very slowly, however, especially due to the inversion of the matrix **B** that is needed in every step. The convergence can be improved by whitening the data, and especially by using the natural gradient.

The natural gradient algorithm   The natural (or relative) gradient method simplifies the maximization of the likelihood considerably, and makes it better conditioned. The principle of the natural gradient is based on the geometrical structure of the parameter space, and is related to the principle of relative gradient, which uses the Lie group structure of the ICA problem. See Chapter 3 for more details. In the case of basic ICA, both of these principles amount to multiplying the right-hand side of (9.15) by $\mathbf{B}^T\mathbf{B}$. Thus we obtain

$$\Delta \mathbf{B} \propto (\mathbf{I} + E\{\mathbf{g}(\mathbf{y})\mathbf{y}^T\})\mathbf{B} \tag{9.17}$$

Interestingly, this algorithm can be interpreted as *nonlinear decorrelation*. This principle will be treated in more detail in Chapter 12. The idea is that the algorithm converges when $E\{\mathbf{g}(\mathbf{y})\mathbf{y}^T\} = \mathbf{I}$, which means that the y_i and $g_j(y_j)$ are uncorrelated for $i \neq j$. This is a nonlinear extension of the ordinary requirement of uncorrelatedness, and, in fact, this algorithm is a special case of the nonlinear decorrelation algorithms to be introduced in Chapter 12.

In practice, one can use, for example, the two densities described in Section 9.1.2. For supergaussian independent components, the pdf defined by (9.9) is usually used. This means that the component-wise nonlinearity g is the tanh function:

$$g^+(y) = -2\tanh(y) \tag{9.18}$$

For subgaussian independent components, other functions must be used. For example, one could use the pdf in (9.10), which leads to

$$g^-(y) = \tanh(y) - y \tag{9.19}$$

(Another possibility is to use $g(y) = -y^3$ for subgaussian components.) These nonlinearities are illustrated in Fig. 9.1.

The choice between the two nonlinearities in (9.18) and (9.19) can be made by computing the nonpolynomial moment:

$$E\{-\tanh(s_i)s_i + (1 - \tanh(s_i)^2)\} \tag{9.20}$$

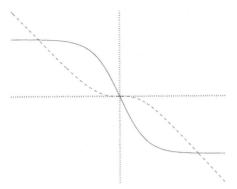

Fig. 9.1 The functions g^+ in Eq. (9.18) and g^- in Eq. (9.19), given by the solid line and the dashed line, respectively.

using some estimates of the independent components. If this nonpolynomial moment is positive, the nonlinearity in (9.18) should be used, otherwise the nonlinearity in (9.19) should be used. This is because of the condition in Theorem 9.1.

The choice of nonlinearity can be made while running the gradient algorithm, using the running estimates of the independent components to estimate the nature of the independent components (that is, the sign of the nonpolynomial moment). Note that the use of the polynomial moment requires that the estimates of the independent components are first scaled properly, constraining them to unit variance, as in the theorem. Such normalizations are often omitted in practice, which may in some cases lead to situations in which the wrong nonlinearity is chosen.

The resulting algorithm is recapitulated in Table 9.1. In this version, whitening and the above-mentioned normalization in the estimation of the nonpolynomial moments are omitted; in practice, these may be very useful.

9.2.2 A fast fixed-point algorithm

Likelihood can be maximized by a fixed-point algorithm as well. The fixed-point algorithm given by FastICA is a very fast and reliable maximization method that was introduced in Chapter 8 to maximize the measures of nongaussianity used for ICA estimation. Actually, the FastICA algorithm can be directly applied to maximization of the likelihood.

The FastICA algorithm was derived in Chapter 8 for optimization of $E\{G(\mathbf{w}^T\mathbf{z})\}$ under the constraint of the unit norm of $\mathbf{w}$. In fact, maximization of likelihood gives us an almost identical optimization problem, if we constrain the estimates of the independent components to be white (see Chapter 7). In particular, this implies that the term $\log |\det \mathbf{W}|$ is constant, as proven in the Appendix, and thus the likelihood basically consists of the sum of n terms of the form optimized by FastICA. Thus

1. Center the data to make its mean zero

2. Choose an initial (e.g., random) separating matrix **B**. Choose initial values of $\gamma_i, i = 1, ..., n$, either randomly or using prior information. Choose the learning rates μ and μ_γ.

3. Compute $\mathbf{y} = \mathbf{Bx}$.

4. If the nonlinearities are not fixed a priori:

   (a) update $\gamma_i = (1 - \mu_\gamma)\gamma_i + \mu_\gamma E\{-\tanh(y_i)y_i + (1 - \tanh(y_i)^2)\}$.

   (b) if $\gamma_i > 0$, define g_i as in (9.18), otherwise define it as in (9.19).

5. Update the separating matrix by

$$\mathbf{B} \leftarrow \mathbf{B} + \mu[\mathbf{I} + \mathbf{g}(\mathbf{y})\mathbf{y}^T]\mathbf{B} \tag{9.21}$$

where $\mathbf{g}(y) = (g_1(y_1), ..., g_n(y_n))^T$.

6. If not converged, go back to step 3.

Table 9.1   The on-line stochastic natural gradient algorithm for maximum likelihood estimation. Preliminary whitening is not shown here, but in practice it is highly recommended.

we could use directly the same kind of derivation of fixed-point iteration as used in Chapter 8.

In Eq. (8.42) in Chapter 8 we had the following form of the FastICA algorithm (for whitened data):

$$\mathbf{w} \leftarrow \mathbf{w} - [E\{\mathbf{z}g(\mathbf{w}^T\mathbf{z})\} + \beta\mathbf{w}]/[E\{g'(\mathbf{w}^T\mathbf{z})\} + \beta] \tag{9.22}$$

where β can be computed from (8.40) as $\beta = -E\{y_i g(y_i)\}$. If we write this in matrix form, we obtain:

$$\mathbf{W} \leftarrow \mathbf{W} + \mathrm{diag}(\alpha_i)[\mathrm{diag}(\beta_i) + E\{\mathbf{g}(\mathbf{y})\mathbf{y}^T\}]\mathbf{W} \tag{9.23}$$

where α_i is defined as $-1/(E\{g'(\mathbf{w}^T\mathbf{z}) + \beta_i\})$, and $\mathbf{y} = \mathbf{Wz}$. To express this using nonwhitened data, as we have done in this chapter, it is enough to multiply both sides of (9.23) from the right by the whitening matrix. This means simply that we replace the **W** by **B**, since we have $\mathbf{Wz} = \mathbf{WVx}$ which implies $\mathbf{B} = \mathbf{WV}$.

Thus, we obtain the basic iteration of FastICA as:

$$\mathbf{B} \leftarrow \mathbf{B} + \mathrm{diag}(\alpha_i)[\mathrm{diag}(\beta_i) + E\{\mathbf{g}(\mathbf{y})\mathbf{y}^T\}]\mathbf{B} \tag{9.24}$$

where $\mathbf{y} = \mathbf{Bx}$, $\beta_i = -E\{y_i g(y_i)\}$, and $\alpha_i = -1/(\beta_i + E\{g'(y_i)\})$.

After every step, the matrix **B** must be projected on the set of whitening matrices. This can be accomplished by the classic method involving matrix square roots,

$$\mathbf{B} \leftarrow (\mathbf{BCB}^T)^{-1/2}\mathbf{B} \tag{9.25}$$

where $\mathbf{C} = E\{\mathbf{x}\mathbf{x}^T\}$ is the correlation matrix of the data (see exercises). The inverse square root is obtained as in (7.20). For alternative methods, see Section 8.4 and Chapter 6, but note that those algorithms require that the data is prewhitened, since they simply orthogonalize the matrix.

This version of FastICA is recapitulated in Table 9.2. FastICA could be compared with the natural gradient method for maximizing likelihood given in (9.17). Then we see that FastICA can be considered as a *computationally optimized version* of the gradient algorithm. In FastICA, convergence speed is optimized by the choice of the matrices diag(α_i) and diag(β_i). These two matrices give an optimal step size to be used in the algorithm.

Another advantage of FastICA is that it can estimate both sub- and supergaussian independent components without any additional steps: We can fix the nonlinearity to be equal to the tanh nonlinearity for all the independent components. The reason is clear from (9.24): The matrix diag(α_i) contains estimates on the nature (sub- or supergaussian) of the independent components. These estimates are used as in the gradient algorithm in the previous subsection. On the other hand, the matrix diag(β_i) can be considered as a scaling of the nonlinearities, since we could reformulate $[\text{diag}(\beta_i) + E\{\mathbf{g}(\mathbf{y})\mathbf{y}^T\}] = \text{diag}(\beta_i)[\mathbf{I} + \text{diag}(\beta_i^{-1})E\{\mathbf{g}(\mathbf{y})\mathbf{y}^T\}]$. Thus we can say that FastICA uses a richer parameterization of the densities than that used in Section 9.1.2: a parameterized family instead of just two densities.

Note that in FastICA, the outputs y_i are decorrelated and normalized to unit variance after every step. No such operations are needed in the gradient algorithm. FastICA is not stable if these additional operations are omitted. Thus the optimization space is slightly reduced.

In the version given here, no preliminary whitening is done. In practice, it is often highly recommended to do prewhitening, possibly combined with PCA dimension reduction.

9.3 THE INFOMAX PRINCIPLE

An estimation principle for ICA that is very closely related to maximum likelihood is the infomax principle [282, 36]. This is based on maximizing the output entropy, or information flow, of a neural network with nonlinear outputs. Hence the name infomax.

Assume that $\mathbf{x}$ is the input to the neural network whose outputs are of the form

$$y_i = \phi_i(\mathbf{b}_i^T\mathbf{x}) + \mathbf{n} \tag{9.31}$$

where the ϕ_i are some nonlinear scalar functions, and the $\mathbf{b}_i$ are the weight vectors of the neurons. The vector $\mathbf{n}$ is additive gaussian white noise. One then wants to maximize the entropy of the outputs:

$$H(\mathbf{y}) = H(\phi_1(\mathbf{b}_1^T\mathbf{x}), ..., \phi_n(\mathbf{b}_n^T\mathbf{x})) \tag{9.32}$$

This can be motivated by considering information flow in a neural network. Efficient information transmission requires that we maximize the mutual information between

---

1. Center the data to make its mean zero. Compute correlation matrix $\mathbf{C} = E\{\mathbf{xx}^T\}$.

2. Choose an initial (e.g., random) separating matrix $\mathbf{B}$.

3. Compute

$$\mathbf{y} = \mathbf{Bx} \tag{9.26}$$

$$\beta_i = -E\{y_i g(y_i)\}, \text{for } i = 1, ..., n \tag{9.27}$$

$$\alpha_i = -1/(\beta_i + E\{g'(y_i)\}), \text{for } i = 1, ..., n \tag{9.28}$$

4. Update the separating matrix by

$$\mathbf{B} \leftarrow \mathbf{B} + \text{diag}(\alpha_i)[\text{diag}(\beta_i) + E\{\mathbf{g}(\mathbf{y})\mathbf{y}^T\}]\mathbf{B} \tag{9.29}$$

5. Decorrelate and normalize by

$$\mathbf{B} \leftarrow (\mathbf{BCB}^T)^{-1/2}\mathbf{B} \tag{9.30}$$

6. If not converged, go back to step 3.

Table 9.2 The FastICA algorithm for maximum likelihood estimation. This is a version without whitening; in practice, whitening combined with PCA may often be useful. The nonlinear function g is typically the tanh function.

the inputs $\mathbf{x}$ and the outputs $\mathbf{y}$. This problem is meaningful only if there is some information loss in the transmission. Therefore, we assume that there is some noise in the network. It can then be shown (see exercices) that in the limit of no noise (i.e., with infinitely weak noise), maximization of this mutual information is equivalent to maximization of the output entropy in (9.32). For simplicity, we therefore assume in the following that the noise is of zero variance.

Using the classic formula of the entropy of a transformation (see Eq. (5.13) we have

$$H(\phi_1(\mathbf{b}_1^T \mathbf{x}), ..., \phi_n(\mathbf{b}_n^T \mathbf{x})) = H(\mathbf{x}) + E\{\log | \det \frac{\partial \mathbf{F}}{\partial \mathbf{B}}(\mathbf{x})|\}$$

(9.33)

where $\mathbf{F}(\mathbf{x}) = (\phi_1(\mathbf{w}_1^T \mathbf{x}), ..., \phi_n(\mathbf{w}_n^T \mathbf{x}))$ denotes the function defined by the neural network. We can simply calculate the derivative to obtain

$$E\{\log | \det \frac{\partial \mathbf{F}}{\partial \mathbf{B}}(\mathbf{x})|\} = \sum_i E\{\log \phi_i'(\mathbf{b}_i^T \mathbf{x})\} + \log | \det \mathbf{B}|$$

(9.34)

Now we see that the output entropy is of the same form as the expectation of the likelihood as in Eq. 9.6. The pdf's of the independent components are here replaced by the functions ϕ_i'. Thus, if the nonlinearities ϕ_i used in the neural network are chosen as the cumulative distribution functions corresponding to the densities p_i, i.e., $\phi_i'(.) = p_i(.)$, the output entropy is actually equal to the likelihood. This means that infomax is equivalent to maximum likelihood estimation.

9.4 EXAMPLES

Here we show the results of applying maximum likelihood estimation to the two mixtures introduced in Chapter 7. Here, we use whitened data. This is not strictly necessary, but the algorithms converge much better with whitened data. The algorithms were always initialized so that $\mathbf{B}$ was the identity matrix.

First, we used the natural gradient ML algorithm in Table 9.1. In the first example, we used the data consisting of two mixtures of two subgaussian (uniformly distributed) independent components, and took the nonlinearity to be the one in (9.18), corresponding to the density in (9.9). The algorithm did *not* converge properly, as shown in Fig. 9.2. This is because the nonlinearity was not correctly chosen. Indeed, computing the nonpolynomial moment (9.20), we saw that it was negative, which means that the nonlinearity in (9.19) should have been used. Using the correct nonlinearity, we obtained correct convergence, as in Fig. 9.3. In both cases, several hundred iterations were performed.

Next we did the corresponding estimation for two mixtures of two *supergaussian* independent components. This time, the nonlinearity in (9.18) was the correct one, and gave the estimates in Fig. 9.4. This could be checked by computing the nonpolynomial moment in (9.20): It was positive. In contrast, using the nonlinearity in (9.19) gave completely wrong estimates, as seen in Fig. 9.5.

In contrast to the gradient algorithm, FastICA effortlessly finds the independent components in both cases. In Fig. 9.6, the results are shown for the subgaussian data, and in Fig. 9.7, the results are shown for the supergaussian data. In both cases the algorithm converged correctly, in a couple of iterations.

9.5 CONCLUDING REMARKS AND REFERENCES

Maximum likelihood estimation, perhaps the most commonly used statistical estimation principle, can be used to estimate the ICA model as well. It is closely related to the infomax principle used in neural network literature. If the densities of the independent components are known in advance, a very simple gradient algorithm can be derived. To speed up convergence, the natural gradient version and especially the FastICA fixed-point algorithm can be used. If the densities of the independent components are not known, the situation is somewhat more complicated. Fortunately, however, it is enough to use a very rough density approximation. In the extreme case, a family that contains just two densities to approximate the densities of the independent components is enough. The choice of the density can then be based on the information whether the independent components are sub- or supergaussian. Such an estimate can be simply added to the gradient methods, and it is automatically done in FastICA.

The first approaches to using maximum likelihood estimation for ICA were in [140, 372]; see also [368, 371]. This approach became very popular after the introduction of the algorithm in (9.16) by Bell and Sejnowski, who derived it using the infomax principle [36]; see also [34]. The connection between these two approaches was later proven by [64, 322, 363]. The natural gradient algorithm in (9.17) is sometimes called the Bell-Sejnowski algorithm as well. However, the natural gradient extension was actually introduced only in [12, 3]; for the underlying theory, see [4, 11, 118]. This algorithm is actually almost identical to those introduced previously [85, 84] based on nonlinear decorrelation, and quite similar to the one in [255, 71] (see Chapter 12). In particular, [71] used the relative gradient approach, which in this case is closely related to the natural gradient; see Chapter 14 for more details. Our two-density family is closely related to those in [148, 270]; for alternative approaches on modeling the distributions of the ICs, see [121, 125, 133, 464].

The stability criterion in Theorem 9.1 has been presented in different forms by many authors [9, 71, 67, 69, 211]. The different forms are mainly due to the complication of different normalizations, as discussed in [67]. We chose to normalize the components to unit variance, which gives a simple theorem and is in line with the approach of the other chapters. Note that in [12], it was proposed that a single very high-order polynomial nonlinearity could be used as a universal nonlinearity. Later research has shown that this is not possible, since we need at least two different nonlinearities, as discussed in this chapter. Moreover, a high-order polynomial leads to very nonrobust estimators.

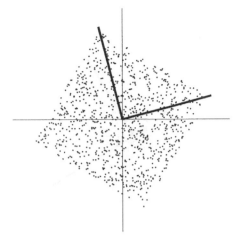

Fig. 9.2 Problems of convergence with the (natural) gradient method for maximum likelihood estimation. The data was two whitened mixtures of subgaussian independent components. The nonlinearity was the one in (9.18), which was not correct in this case. The resulting estimates of the columns of the whitened mixing matrix are shown in the figure: they are not aligned with the edges of the square, as they should be.

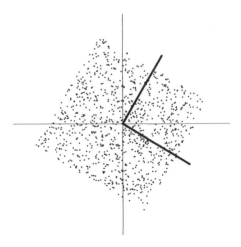

Fig. 9.3 The same as in Fig. 9.2, but with the correct nonlinearity, given by (9.19). This time, the natural gradient algorithm gave the right result. The estimated vectors *are* aligned with the edges of the square.

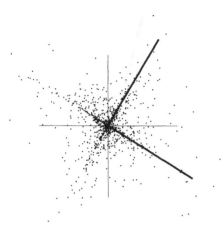

Fig. 9.4 In this experiment, data was two whitened mixtures of supergaussian independent components. The nonlinearity was the one in (9.18). The natural gradient algorithm converged correctly.

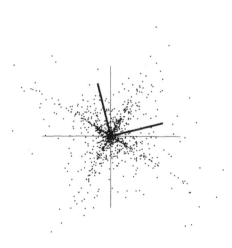

Fig. 9.5 Again, problem of convergence with the natural gradient method for maximum likelihood estimation. The nonlinearity was the one in (9.19), which was not correct in this case.

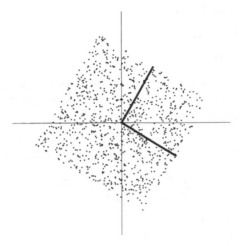

Fig. 9.6 FastICA automatically estimates the nature of the independent components, and converges fast to the maximum likelihood solution. Here, the solution was found in 2 iterations for subgaussian independent components.

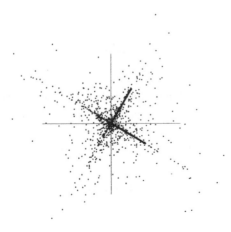

Fig. 9.7 FastICA this time applied on supergaussian mixtures. Again, the solution was found in 2 iterations.

Problems

9.1   Derive the likelihood in (9.4).

9.2   Derive (9.11) and (9.12).

9.3   Derive the gradient in (9.13).

9.4   Instead of the function in (9.19), one could use the function $g(y) = -y^3$. Show that this corresponds to a subgaussian distribution by computing the kurtosis of the distribution. Note the normalization constants involved.

9.5   After the preceding problem, one might be tempted to use $g(y) = y^3$ for super-gaussian variables. Why is this not correct in the maximum likelihood framework?

9.6   Take a linear $g(y) = -y$. What is the interpretation of this in the ML framework? Conclude (once again) that g must be nonlinear.

9.7   Assume that you use the general function family $g(y) = -2\tanh(a_1 y)$ instead the simple $-\tanh$ function in (9.18), where $a_1 \geq 1$ is a constant. What is the interpretation of a_1 in the likelihood framework?

9.8   Show that for a gaussian random variable, the nonpolynomial moment in (9.8) is zero for any g.

9.9   The difference between using $-\tanh(y)$ or $-2\tanh(y)$ in the nonlinearity (9.18) is a matter of normalization. Does it make any difference in the algorithms? Consider separately the natural gradient algorithm and the FastICA algorithm.

9.10   Show that maximizing the mutual information of inputs and outputs in a network of the form

$$y_i = \phi_i(\mathbf{b}_i^T \mathbf{x}) + n_i \tag{9.35}$$

where n_i is gaussian noise, and the output and input spaces have the same dimension, as in (9.31), is equivalent to maximizing the entropy of the outputs, in the limit of no zero noise level. (You can compute the joint entropy of inputs and outputs using the entropy transformation formula. Show that it is constant. Then set the noise level to infinitely small.)

9.11   Show that after (9.25), $\mathbf{y} = \mathbf{Bx}$ is white.

Computer assignments

9.1   Take random variables of (1) uniform and (2) Laplacian distribution. Compute the values of the nonpolynomial moment in (9.8), for different nonlinearities g. Are the moments of different signs for any nonlinearity?

9.2   Reproduce the experiments in Section 9.4.

9.3   The densities of the independent components could be modeled by a density family given by

$$p(\xi) = c_1 \exp(c_2|\xi|^{\alpha}) \tag{9.36}$$

where c_1 and c_2 are normalization constants to make this a pdf of unit variance. For different values of α, ranging from 0 to infinity, we get distributions of different properties.

9.3.1. What happens when we have $\alpha = 2$?

9.3.2. Plot the pdf's, the logarithms of the pdf's, and the corresponding score functions for the following values of α: 0.2,1,2,4,10.

9.3.3. Conclude that for $\alpha < 2$, we have subgaussian densities, and for $\alpha > 2$, we have supergaussian densities.

Appendix proofs

Here we prove Theorem 9.1. Looking at the expectation of log-likelihood, using assumed densities $\tilde{p}_i$:

$$\frac{1}{T}\log L(\mathbf{B}) = \sum_{i=1}^{n} E\{\log \tilde{p}_i(\mathbf{b}_i^T \mathbf{x})\} + \log|\det \mathbf{B}| \tag{A.1}$$

we see that the first term on the right-hand side is a sum of terms of the form $E\{G(\mathbf{b}_i^T\mathbf{x})\}$, as in the stability theorem in Section 8.3.4. Using that theorem, we see immediately that the first term is maximized when $\mathbf{y} = \mathbf{Bx}$ gives independent components.

Thus if we prove that the second term remains constant under the conditions of the theorem, the theorem is proven. Now, uncorrelatedness and unit variance of the $\mathbf{y}$ means $E\{\mathbf{yy}^T\} = \mathbf{W}E\{\mathbf{xx}^T\}\mathbf{W}^T = \mathbf{I}$, which implies

$$\det \mathbf{I} = 1 = (\det \mathbf{W}E\{\mathbf{xx}^T\}\mathbf{W}^T) = (\det \mathbf{W})(\det E\{\mathbf{xx}^T\})(\det \mathbf{W}^T) \tag{A.2}$$

and this implies that $\det \mathbf{W}$ must be constant. Thus the theorem is proven.

10

ICA by Minimization of Mutual Information

An important approach for independent component analysis (ICA) estimation, inspired by information theory, is minimization of mutual information.

The motivation of this approach is that it may not be very realistic in many cases to assume that the data follows the ICA model. Therefore, we would like to develop an approach that does not assume anything about the data. What we want to have is a general-purpose measure of the dependence of the components of a random vector. Using such a measure, we could define ICA as a linear decomposition that minimizes that dependence measure. Such an approach can be developed using mutual information, which is a well-motivated information-theoretic measure of statistical dependence.

One of the main utilities of mutual information is that it serves as a unifying framework for many estimation principles, in particular maximum likelihood (ML) estimation and maximization of nongaussianity. In particular, this approach gives a rigorous justification for the heuristic principle of nongaussianity.

10.1 DEFINING ICA BY MUTUAL INFORMATION

10.1.1 Information-theoretic concepts

The information-theoretic concepts needed in this chapter were explained in Chapter 5. Readers not familiar with information theory are advised to read that chapter before this one.

We recall here very briefly the basic definitions of information theory. The differential entropy H of a random vector $\mathbf{y}$ with density $p(\mathbf{y})$ is defined as:

$$H(\mathbf{y}) = -\int p(\mathbf{y}) \log p(\mathbf{y}) d\mathbf{y} \qquad (10.1)$$

Entropy is closely related to the code length of the random vector. A normalized version of entropy is given by negentropy J, which is defined as follows

$$J(\mathbf{y}) = H(\mathbf{y}_{gauss}) - H(\mathbf{y}) \qquad (10.2)$$

where $\mathbf{y}_{gauss}$ is a gaussian random vector of the same covariance (or correlation) matrix as $\mathbf{y}$. Negentropy is always nonnegative, and zero only for gaussian random vectors. Mutual information I between m (scalar) random variables, $y_i, i = 1...m$ is defined as follows

$$I(y_1, y_2, ..., y_m) = \sum_{i=1}^{m} H(y_i) - H(\mathbf{y}) \qquad (10.3)$$

10.1.2   Mutual information as measure of dependence

We have seen earlier (Chapter 5) that mutual information is a natural measure of the dependence between random variables. It is always nonnegative, and zero if and only if the variables are statistically independent. Mutual information takes into account the whole dependence structure of the variables, and not just the covariance, like principal component analysis (PCA) and related methods.

Therefore, we can use mutual information as the criterion for finding the ICA representation. This approach is an alternative to the model estimation approach. We define the ICA of a random vector $\mathbf{x}$ as an invertible transformation:

$$\mathbf{s} = \mathbf{B}\mathbf{x} \qquad (10.4)$$

where the matrix $\mathbf{B}$ is determined so that the mutual information of the transformed components s_i is minimized. If the data follows the ICA model, this allows estimation of the data model. On the other hand, in this definition, we do not need to assume that the data follows the model. In any case, minimization of mutual information can be interpreted as giving the maximally independent components.

10.2 MUTUAL INFORMATION AND NONGAUSSIANITY

Using the formula for the differential entropy of a transformation as given in (5.13) of Chapter 5, we obtain a corresponding result for mutual information. We have for an invertible linear transformation $\mathbf{y} = \mathbf{Bx}$:

$$I(y_1, y_2, ..., y_n) = \sum_i H(y_i) - H(\mathbf{x}) - \log|\det \mathbf{B}| \tag{10.5}$$

Now, let us consider what happens if we constrain the y_i to be *uncorrelated* and of unit variance. This means $E\{\mathbf{yy}^T\} = \mathbf{B}E\{\mathbf{xx}^T\}\mathbf{B}^T = \mathbf{I}$, which implies

$$\det \mathbf{I} = 1 = \det(\mathbf{B}E\{\mathbf{xx}^T\}\mathbf{B}^T) = (\det \mathbf{B})(\det E\{\mathbf{xx}^T\})(\det \mathbf{B}^T) \tag{10.6}$$

and this implies that $\det \mathbf{B}$ must be constant since $\det E\{\mathbf{xx}^T\}$ does not depend on $\mathbf{B}$. Moreover, for y_i of unit variance, entropy and negentropy differ only by a constant and the sign, as can be seen in (10.2). Thus we obtain,

$$I(y_1, y_2, ..., y_n) = \text{const.} - \sum_i J(y_i) \tag{10.7}$$

where the constant term does not depend on $\mathbf{B}$. This shows the fundamental relation between negentropy and mutual information.

We see in (10.7) that finding an invertible linear transformation $\mathbf{B}$ that minimizes the mutual information is roughly equivalent to finding directions in which the negentropy is maximized. We have seen previously that negentropy is a measure of nongaussianity. Thus, (10.7) shows that *ICA estimation by minimization of mutual information is equivalent to maximizing the sum of nongaussianities of the estimates of the independent components*, when the estimates are constrained to be uncorrelated.

Thus, we see that the formulation of ICA as minimization of mutual information gives another rigorous justification of our more heuristically introduced idea of finding maximally nongaussian directions, as used in Chapter 8.

In practice, however, there are also some important differences between these two criteria.

1. Negentropy, and other measures of nongaussianity, enable the deflationary, i.e., one-by-one, estimation of the independent components, since we can look for the maxima of nongaussianity of a single projection $\mathbf{b}^T\mathbf{x}$. This is not possible with mutual information or most other criteria, like the likelihood.

2. A smaller difference is that in using nongaussianity, we force the estimates of the independent components to be uncorrelated. This is not necessary when using mutual information, because we could use the form in (10.5) directly, as will be seen in the next section. Thus the optimization space is slightly reduced.

10.3 MUTUAL INFORMATION AND LIKELIHOOD

Mutual information and likelihood are intimately connected. To see the connection between likelihood and mutual information, consider the expectation of the log-likelihood in (9.5):

$$\frac{1}{T}E\{\log L(\mathbf{B})\} = \sum_{i=1}^{n} E\{\log p_i(\mathbf{b}_i^T \mathbf{x})\} + \log|\det \mathbf{B}| \qquad (10.8)$$

If the p_i were equal to the actual pdf's of $\mathbf{b}_i^T \mathbf{x}$, the first term would be equal to $-\sum_i H(\mathbf{b}_i^T \mathbf{x})$. Thus the likelihood would be equal, up to an additive constant given by the total entropy of $\mathbf{x}$, to the negative of mutual information as given in Eq. (10.5).

In practice, the connection may be just as strong, or even stronger. This is because in practice we do not know the distributions of the independent components that are needed in ML estimation. A reasonable approach would be to estimate the density of $\mathbf{b}_i^T \mathbf{x}$ as part of the ML estimation method, and use this as an approximation of the density of s_i. This is what we did in Chapter 9. Then, the p_i in this approximation of likelihood are indeed equal to the actual pdf's $\mathbf{b}_i^T \mathbf{x}$. Thus, the equivalency would really hold.

Conversely, to approximate mutual information, we could take a fixed approximation of the densities y_i, and plug this in the definition of entropy. Denote the pdf's by $G_i(y_i) = \log p_i(y_i)$. Then we could approximate (10.5) as

$$I(y_1, y_2, ..., y_n) = -\sum_i E\{G_i(y_i)\} - \log|\det \mathbf{B}| - H(\mathbf{x}) \qquad (10.9)$$

Now we see that this approximation is equal to the approximation of the likelihood used in Chapter 9 (except, again, for the global sign and the additive constant given by $H(\mathbf{x})$). This also gives an alternative method of approximating mutual information that is different from the approximation that uses the negentropy approximations.

10.4 ALGORITHMS FOR MINIMIZATION OF MUTUAL INFORMATION

To use mutual information in practice, we need some method of estimating or approximating it from real data. Earlier, we saw two methods for approximating mutual entropy. The first one was based on the negentropy approximations introduced in Section 5.6. The second one was based on using more or less fixed approximations for the densities of the ICs in Chapter 9.

Thus, using mutual information leads essentially to the same algorithms as used for maximization of nongaussianity in Chapter 8, or for maximum likelihood estimation in Chapter 9. In the case of maximization of nongaussianity, the corresponding algorithms are those that use symmetric orthogonalization, since we are maximizing the sum of nongaussianities, so that no order exists between the components. Thus, we do not present any new algorithms in this chapter; the reader is referred to the two preceding chapters.

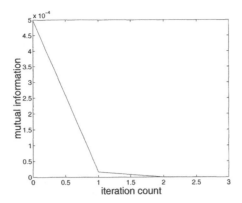

Fig. 10.1 The convergence of FastICA for ICs with uniform distributions. The value of mutual information shown as function of iteration count.

10.5 EXAMPLES

Here we show the results of applying minimization of mutual information to the two mixtures introduced in Chapter 7. We use here the whitened mixtures, and the FastICA algorithm (which is essentially identical whichever approximation of mutual information is used). For illustration purposes, the algorithm was always initialized so that $\mathbf{W}$ was the identity matrix. The function G was chosen as G_1 in (8.26).

First, we used the data consisting of two mixtures of two subgaussian (uniformly distributed) independent components. To demonstrate the convergence of the algorithm, the mutual information of the components at each iteration step is plotted in Fig. 10.1. This was obtained by the negentropy-based approximation. At convergence, after two iterations, mutual information was practically equal to zero. The corresponding results for two supergaussian independent components are shown in Fig. 10.2. Convergence was obtained after three iterations, after which mutual information was practically zero.

10.6 CONCLUDING REMARKS AND REFERENCES

A rigorous approach to ICA that is different from the maximum likelihood approach is given by minimization of mutual information. Mutual information is a natural information-theoretic measure of dependence, and therefore it is natural to estimate the independent components by minimizing the mutual information of their estimates. Mutual information gives a rigorous justification of the principle of searching for maximally nongaussian directions, and in the end turns out to be very similar to the likelihood as well.

Mutual information can be approximated by the same methods that negentropy is approximated. Alternatively, is can be approximated in the same way as likelihood.

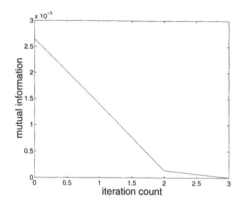

Fig. 10.2 The convergence of FastICA for ICs with supergaussian distributions. The value of mutual information shown as function of iteration count.

Therefore, we find here very much the same objective functions and algorithms as in maximization of nongaussianity and maximum likelihood. The same gradient and fixed-point algorithms can be used to optimize mutual information.

Estimation of ICA by minimization of mutual information was probably first proposed in [89], who derived an approximation based on cumulants. The idea has, however, a longer history in the context of neural network research, where it has been proposed as a sensory coding strategy. It was proposed in [26, 28, 30, 18], that decomposing sensory data into features that are maximally independent is useful as a preprocessing step. Our approach follows that of [197] for the negentropy approximations.

A nonparametric algorithm for minimization of mutual information was proposed in [175], and an approach based on order statistics was proposed in [369]. See [322, 468] for a detailed analysis of the connection between mutual information and infomax or maximum likelihood. A more general framework was proposed in [377].

Problems

10.1 Derive the formula in (10.5).

10.2 Compute the constant in (10.7).

10.3 If the variances of the y_i are not constrained to unity, does this constant change?

10.4 Compute the mutual information for a gaussian random vector with covariance matrix **C**.

Computer assignments

10.1 Create a sample of 2-D gaussian data with the two covariance matrices

$$\begin{pmatrix} 3 & 0 \\ 0 & 2 \end{pmatrix} \text{ and } \begin{pmatrix} 3 & 1 \\ 1 & 2 \end{pmatrix} \tag{10.10}$$

Estimate numerically the mutual information using the definition. (Divide the data into bins, i.e., boxes of fixed size, and estimate the density at each bin by computing the number of data points that belong to that bin and dividing it by the size of the bin. This elementary density approximation can then be used in the definition.)

11

ICA by Tensorial Methods

One approach for estimation of independent component analysis (ICA) consists of using higher-order cumulant tensor. Tensors can be considered as generalization of matrices, or linear operators. Cumulant tensors are then generalizations of the covariance matrix. The covariance matrix is the second-order cumulant tensor, and the fourth order tensor is defined by the fourth-order cumulants $\text{cum}(x_i, x_j, x_k, x_l)$. For an introduction to cumulants, see Section 2.7.

As explained in Chapter 6, we can use the eigenvalue decomposition of the covariance matrix to whiten the data. This means that we transform the data so that second-order correlations are zero. As a generalization of this principle, we can use the fourth-order cumulant tensor to make the fourth-order cumulants zero, or at least as small as possible. This kind of (approximative) higher-order decorrelation gives one class of methods for ICA estimation.

11.1 DEFINITION OF CUMULANT TENSOR

We shall here consider only the fourth-order cumulant tensor, which we call for simplicity the cumulant tensor. The cumulant tensor is a four-dimensional array whose entries are given by the fourth-order cross-cumulants of the data: $\text{cum}(x_i, x_j, x_k, x_l)$, where the indices i, j, k, l are from 1 to n. This can be considered as a "four-dimensional matrix", since it has four different indices instead of the usual two. For a definition of cross-cumulants, see Eq. (2.106).

In fact, all fourth-order cumulants of linear combinations of x_i can be obtained as linear combinations of the cumulants of x_i. This can be seen using the additive properties of the cumulants as discussed in Section 2.7. The kurtosis of a linear combination is given by

$$\text{kurt} \sum_i w_i x_i = \text{cum}(\sum_i w_i x_i, \sum_j w_j x_j, \sum_k w_k x_k, \sum_l w_l x_l)$$

$$= \sum_{ijkl} w_i^4 w_j^4 w_k^4 w_l^4 \text{cum}(x_i, x_j, x_k, x_l) \quad (11.1)$$

Thus the (fourth-order) cumulants contain all the fourth-order information of the data, just as the covariance matrix gives all the second-order information on the data. Note that if the x_i are independent, all the cumulants with at least two different indices are zero, and therefore we have the formula that was already widely used in Chapter 8: $\text{kurt} \sum_i q_i s_i = \sum_i q_i^4 \text{kurt}(s_i)$.

The cumulant tensor is a linear operator defined by the fourth-order cumulants $\text{cum}(x_i, x_j, x_k, x_l)$. This is analogous to the case of the covariance matrix with elements $\text{cov}(x_i, x_j)$, which defines a linear operator just as any matrix defines one. In the case of the tensor we have a linear transformation in the space of $n \times n$ *matrices*, instead of the space of n-dimensional vectors. The space of such matrices is a linear space of dimension $n \times n$, so there is nothing extraordinary in defining the linear transformation. The i, jth element of the matrix given by the transformation, say $\mathbf{F}_{ij}$, is defined as

$$\mathbf{F}_{ij}(\mathbf{M}) = \sum_{kl} m_{kl} \, \text{cum}(x_i, x_j, x_k, x_l) \quad (11.2)$$

where m_{kl} are the elements in the matrix $\mathbf{M}$ that is transformed.

11.2   TENSOR EIGENVALUES GIVE INDEPENDENT COMPONENTS

As any symmetric linear operator, the cumulant tensor has an eigenvalue decomposition (EVD). An eigenmatrix of the tensor is, by definition, a matrix $\mathbf{M}$ such that

$$\mathbf{F}(\mathbf{M}) = \lambda \mathbf{M} \quad (11.3)$$

i.e., $\mathbf{F}_{ij}(\mathbf{M}) = \lambda \mathbf{M}_{ij}$, where λ is a scalar eigenvalue.

The cumulant tensor is a symmetric linear operator, since in the expression $\text{cum}(x_i, x_j, x_k, x_l)$, the order of the variables makes no difference. Therefore, the tensor has an eigenvalue decomposition.

Let us consider the case where the data follows the ICA model, with whitened data:

$$\mathbf{z} = \mathbf{V}\mathbf{A}\mathbf{s} = \mathbf{W}^T \mathbf{s} \quad (11.4)$$

where we denote the whitened mixing matrix by $\mathbf{W}^T$. This is because it is orthogonal, and thus it is the transpose of the separating matrix $\mathbf{W}$ for whitened data.

The cumulant tensor of $\mathbf{z}$ has a special structure that can be seen in the eigenvalue decomposition. In fact, every matrix of the form

$$\mathbf{M} = \mathbf{w}_m \mathbf{w}_m^T \tag{11.5}$$

for $m = 1, ..., n$ is an eigenmatrix. The vector $\mathbf{w}_m$ is here one of the rows of the matrix $\mathbf{W}$, and thus one of the columns of the whitened mixing matrix $\mathbf{W}^T$. To see this, we calculate by the linearity properties of cumulants

$$\mathbf{F}_{ij}(\mathbf{w}_m \mathbf{w}_m^T) = \sum_{kl} w_{mk} w_{ml} \text{cum}(z_i, z_j, z_k, z_l)$$

$$= \sum_{kl} w_{mk} w_{ml} \text{cum}(\sum_q w_{qi} s_q, \sum_{q'} w_{q'j} s_{q'}, \sum_r w_{rk} s_r, \sum_{r'} w_{r'l} s_{r'})$$

$$= \sum_{klqq'rr'} w_{mk} w_{ml} w_{qi} w_{q'j} w_{rk} w_{r'l} \text{cum}(s_q, s_{q'}, s_r, s_{r'}) \tag{11.6}$$

Now, due to the independence of the s_i, only those cumulants where $q = q' = r = r'$ are nonzero. Thus we have

$$\mathbf{F}_{ij}(\mathbf{w}_m \mathbf{w}_m^T) = \sum_{klq} w_{mk} w_{ml} w_{qi} w_{qj} w_{qk} w_{ql} \text{kurt}(s_q) \tag{11.7}$$

Due to the orthogonality of the rows of $\mathbf{W}$, we have $\sum_k w_{mk} w_{qk} = \delta_{mq}$, and similarly for index l. Thus we can take the sum first with respect to k, and then with respect to l, which gives

$$\mathbf{F}_{ij}(\mathbf{w}_m \mathbf{w}_m^T) = \sum_{lq} w_{ml} w_{qi} w_{qj} \delta_{mq} w_{ql} \text{kurt}(s_q)$$

$$= \sum_q w_{qi} w_{qj} \delta_{mq} \delta_{mq} \text{kurt}(s_q) = w_{mi} w_{mj} \text{kurt}(s_m) \tag{11.8}$$

This proves that matrices of the form in (11.5) are eigenmatrices of the tensor. The corresponding eigenvalues are given by the kurtoses of the independent components. Moreover, it can be proven that all other eigenvalues of the tensor are zero.

Thus we see that if we knew the eigenmatrices of the cumulant tensor, we could easily obtain the independent components. If the eigenvalues of the tensor, i.e., the kurtoses of the independent components, are distinct, every eigenmatrix corresponds to a nonzero eigenvalue of the form $\mathbf{w}_m \mathbf{w}_m^T$, giving one of the columns of the whitened mixing matrix.

If the eigenvalues are not distinct, the situation is more problematic: The eigenmatrices are no longer uniquely defined, since any linear combinations of the matrices $\mathbf{w}_m \mathbf{w}_m^T$ corresponding to the same eigenvalue are eigenmatrices of the tensor as well. Thus, every k-fold eigenvalue corresponds to k matrices $\mathbf{M}_i, i = 1, ..., k$ that are different linear combinations of the matrices $\mathbf{w}_{i(j)} \mathbf{w}_{i(j)}^T$ corresponding to the k ICs whose indices are denoted by $i(j)$. The matrices $\mathbf{M}_i$ can be thus expressed as:

$$\mathbf{M}_i = \sum_{j=1}^{k} \alpha_j \mathbf{w}_{i(j)} \mathbf{w}_{i(j)}^T \tag{11.9}$$

Now, vectors that can be used to construct the matrix in this way can be computed by the eigenvalue decomposition of the matrix: The $\mathbf{w}_{i(j)}$ are the (dominant) eigenvectors of $\mathbf{M}_i$.

Thus, after finding the eigenmatrices $\mathbf{M}_i$ of the cumulant tensor, we can decompose them by ordinary EVD, and the eigenvectors give the columns of the mixing matrix $\mathbf{w}_i$. Of course, it could turn out that the eigenvalues in this latter EVD are equal as well, in which case we have to figure out something else. In the algorithms given below, this problem will be solved in different ways.

This result leaves the problem of how to compute the eigenvalue decomposition of the tensor in practice. This will be treated in the next section.

11.3 COMPUTING THE TENSOR DECOMPOSITION BY A POWER METHOD

In principle, using tensorial methods is simple. One could take any method for computing the EVD of a symmetric matrix, and apply it on the cumulant tensor.

To do this, we must first consider the tensor as a matrix in the space of $n \times n$ matrices. Let q be an index that goes though all the $n \times n$ couples (i, j). Then we can consider the elements of an $n \times n$ matrix $\mathbf{M}$ as a vector. This means that we are simply vectorizing the matrices. Then the tensor can be considered as a $q \times q$ symmetric matrix $\mathbf{F}$ with elements $f_{qq'} = \mathrm{cum}(z_i, z_j, z_{i'}, z_{j'})$, where the indices (i, j) corresponds to q, and similarly for (i', j') and q'. It is on this matrix that we could apply ordinary EVD algorithms, for example the well-known QR methods. The special symmetricity properties of the tensor could be used to reduce the complexity. Such algorithms are out of the scope of this book; see e.g. [62].

The problem with the algorithm in this category, however, is that the memory requirements may be prohibitive, because often the coefficients of the fourth-order tensor must be stored in memory, which requires $O(n^4)$ units of memory. The computational load also grows quite fast. Thus these algorithms cannot be used in high-dimensional spaces. In addition, equal eigenvalues may give problems.

In the following we discuss a simple modification of the power method, that circumvents the computational problems with the tensor EVD. In general, the power method is a simple way of computing the eigenvector corresponding to the largest eigenvalue of a matrix. This algorithm consists of multiplying the matrix with the running estimate of the eigenvector, and taking the product as the new value of the vector. The vector is then normalized to unit length, and the iteration is continued until convergence. The vector then gives the desired eigenvector.

We can apply the power method quite simply to the case of the cumulant tensor. Starting from a random matrix $\mathbf{M}$, we compute $\mathbf{F}(\mathbf{M})$ and take this as the new value of $\mathbf{M}$. Then we normalize $\mathbf{M}$ and go back to the iteration step. After convergence, $\mathbf{M}$ will be of the form $\sum_k \alpha_k \mathbf{w}_{i(k)} \mathbf{w}_{i(k)}^T$. Computing its eigenvectors gives one or more of the independent components. (In practice, though, the eigenvectors will not be exactly of this form due to estimation errors.) To find several independent components, we could simply project the matrix after every step on the space of matrices that are orthogonal to the previously found ones.

In fact, in the case of ICA, such an algorithm can be considerably simplified. Since we know that the matrices $\mathbf{w}_i \mathbf{w}_i^T$ are eigenmatrices of the cumulant tensor, we can apply the power method inside that set of matrices $\mathbf{M} = \mathbf{w}\mathbf{w}^T$ only. After every computation of the product with the tensor, we must then project the obtained matrix back to the set of matrices of the form $\mathbf{w}\mathbf{w}^T$. A very simple way of doing this is to multiply the new matrix $\mathbf{M}^*$ by the old vector to obtain the new vector $\mathbf{w}^* = \mathbf{M}^*\mathbf{w}$ (which will be normalized as necessary). This can be interpreted as another power method, this time applied on the eigenmatrix to compute its eigenvectors. Since the best way of approximating the matrix $\mathbf{M}^*$ in the space of matrices of the form $\mathbf{w}\mathbf{w}^T$ is by using the dominant eigenvector, a single step of this ordinary power method will at least take us closer to the dominant eigenvector, and thus to the optimal vector.

Thus we obtain an iteration of the form

$$\mathbf{w} \leftarrow \mathbf{w}^T \mathbf{F}(\mathbf{w}\mathbf{w}^T) \tag{11.10}$$

or

$$w_i \leftarrow \sum_j w_j \sum_{kl} w_k w_l \mathrm{cum}(z_i, z_j, z_k, z_l) \tag{11.11}$$

In fact, this can be manipulated algebraically to give much simpler forms. We have equivalently

$$w_i \leftarrow \mathrm{cum}(z_i, \sum_j w_j z_j, \sum_k w_k z_k, \sum_l w_l z_l) = \mathrm{cum}(z_i, y, y, y) \tag{11.12}$$

where we denote by $y = \sum_i w_i z_i$ the estimate of an independent component. By definition of the cumulants, we have

$$\mathrm{cum}(z_i, y, y, y) = E\{z_i y^3\} - 3E\{z_i y\}E\{y^2\} \tag{11.13}$$

We can constrain y to have unit variance, as usual. Moreover, we have $E\{z_i y\} = w_i$. Thus we have

$$\mathbf{w} \leftarrow E\{\mathbf{z}y^3\} - 3\mathbf{w} \tag{11.14}$$

where $\mathbf{w}$ is normalized to unit norm after every iteration. To find several independent components, we can actually just constrain the $\mathbf{w}$ corresponding to different independent components to be orthogonal, as is usual for whitened data.

Somewhat surprisingly, (11.14) is exactly the *FastICA algorithm* that was derived as a fixed-point iteration for finding the maxima of the absolute value of kurtosis in Chapter 8, see (8.20). We see that these two methods lead to the same algorithm.

11.4   JOINT APPROXIMATE DIAGONALIZATION OF EIGENMATRICES

Joint approximate diagonalization of eigenmatrices (JADE) refers to one principle of solving the problem of equal eigenvalues of the cumulant tensor. In this algorithm, the tensor EVD is considered more as a preprocessing step.

Eigenvalue decomposition can be viewed as diagonalization. In our case, the developments in Section 11.2 can be rephrased as follows: The matrix $\mathbf{W}$ diagonalizes $\mathbf{F}(\mathbf{M})$ for any $\mathbf{M}$. In other words, $\mathbf{W}\mathbf{F}(\mathbf{M})\mathbf{W}^T$ is diagonal. This is because the matrix $\mathbf{F}$ is of a linear combination of terms of the form $\mathbf{w}_i\mathbf{w}_i^T$, assuming that the ICA model holds.

Thus, we could take a set of different matrices $\mathbf{M}_i, i = 1, ..., k$, and try to make the matrices $\mathbf{W}\mathbf{F}(\mathbf{M}_i)\mathbf{W}$ as diagonal as possible. In practice, they cannot be made exactly diagonal because the model does not hold exactly, and there are sampling errors.

The diagonality of a matrix $\mathbf{Q} = \mathbf{W}\mathbf{F}(\mathbf{M}_i)\mathbf{W}^T$ can be measured, for example, as the sum of the squares of off-diagonal elements: $\sum_{k \neq l} q_{kl}^2$. Equivalently, since an orthogonal matrix $\mathbf{W}$ does not change the total sum of squares of a matrix, minimization of the sum of squares of off-diagonal elements is equivalent to the maximization of the sum of squares of diagonal elements. Thus, we could formulate the following measure:

$$\mathcal{J}_{JADE}(\mathbf{W}) = \sum_i \|\mathrm{diag}(\mathbf{W}\mathbf{F}(\mathbf{M}_i)\mathbf{W}^T)\|^2 \qquad (11.15)$$

where $\|\mathrm{diag}(.)\|^2$ means the sum of squares of the diagonal. Maximization of $\mathcal{J}_{JADE}$ is then one method of joint approximate diagonalization of the $\mathbf{F}(\mathbf{M}_i)$.

How do we choose the matrices $\mathbf{M}_i$? A natural choice is to take the eigenmatrices of the cumulant tensor. Thus we have a set of just n matrices that give all the relevant information on the cumulants, in the sense that they span the same subspace as the cumulant tensor. This is the basic principle of the JADE algorithm.

Another benefit associated with this choice of the $\mathbf{M}_i$ is that the joint diagonalization criterion is then a function of the distributions of the $\mathbf{y} = \mathbf{W}\mathbf{z}$ and a clear link can be made to methods of previous chapters. In fact, after quite complicated algebraic manipulations, we can obtain

$$\mathcal{J}_{JADE}(\mathbf{W}) = \sum_{ijkl \neq iikl} \mathrm{cum}(y_i, y_j, y_k, y_l)^2 \qquad (11.16)$$

in other words, when we minimize $\mathcal{J}_{JADE}$ we also minimize a sum of the squared cross-cumulants of the y_i. Thus, we can interpret the method as minimizing nonlinear correlations.

JADE suffers from the same problems as all methods using an explicit tensor EVD. Such algorithms cannot be used in high-dimensional spaces, which pose no problem for the gradient or fixed-point algorithm of Chapters 8 and 9. In problems of low dimensionality (small scale), however, JADE offers a competitive alternative.

11.5   WEIGHTED CORRELATION MATRIX APPROACH

A method closely related to JADE is given by the eigenvalue decomposition of the weighted correlation matrix. For historical reasons, the basic method is simply called fourth-order blind identification (FOBI).

11.5.1   The FOBI algorithm

Consider the matrix

$$\Omega = E\{\mathbf{z}\mathbf{z}^T \|\mathbf{z}\|^2\} \tag{11.17}$$

Assuming that the data follows the whitened ICA model, we have

$$\Omega = E\{\mathbf{V}\mathbf{A}\mathbf{s}\mathbf{s}^T(\mathbf{V}\mathbf{A})^T\|\mathbf{V}\mathbf{A}\mathbf{s}\|^2\} = \mathbf{W}^T E\{\mathbf{s}\mathbf{s}^T\|\mathbf{s}\|^2\}\mathbf{W} \tag{11.18}$$

where we have used the orthogonality of $\mathbf{V}\mathbf{A}$, and denoted the separating matrix by $\mathbf{W} = (\mathbf{V}\mathbf{A})^T$. Using the independence of the s_i, we obtain (see exercises)

$$\Omega = \mathbf{W}^T \mathrm{diag}(E\{s_i^2\|\mathbf{s}\|^2\})\mathbf{W} = \mathbf{W}^T \mathrm{diag}(E\{s_i^4\} + n - 1)\mathbf{W} \tag{11.19}$$

Now we see that this is in fact the eigenvalue decomposition of Ω. It consists of the orthogonal separating matrix $\mathbf{W}$ and the diagonal matrix whose entries depend on the fourth-order moments of the s_i. Thus, if the eigenvalue decomposition is unique, which is the case if the diagonal matrix has distinct elements, we can simply compute the decomposition on Ω, and the separating matrix is obtained immediately.

FOBI is probably the simplest method for performing ICA. FOBI allows the computation of the ICA estimates using standard methods of linear algebra on matrices of reasonable complexity ($n \times n$). In fact, the computation of the eigenvalue decomposition of the matrix Ω is of the same complexity as whitening the data. Thus, this method is computationally very efficient: It is probably the most efficient ICA method that exists.

However, FOBI works only under the restriction that the kurtoses of the ICs are all different. (If only some of the ICs have identical kurtoses, those that have distinct kurtoses can still be estimated). This restricts the applicability of the method considerably. In many cases, the ICs have identical distributions, and this method fails completely.

11.5.2   From FOBI to JADE

Now we show how we can generalize FOBI to get rid of its limitations, which actually leads us to JADE.

First, note that for whitened data, the definition of the cumulant can be written as

$$\mathbf{F}(\mathbf{M}) = E\{(\mathbf{z}^T\mathbf{M}\mathbf{z})\mathbf{z}\mathbf{z}^T\} - 2\mathbf{M} - \mathrm{tr}(\mathbf{M})\mathbf{I} \tag{11.20}$$

which is left as an exercice. Thus, we could alternatively define the weighted correlation matrix using the tensor as

$$\Omega = F(I) \tag{11.21}$$

because we have

$$F(I) = E\{\|z\|^2 zz^T\} - (n+2)I \tag{11.22}$$

and the identity matrix does not change the EVD in any significant way.

Thus we could take some matrix M and use the matrix $F(M)$ in FOBI instead of $F(I)$. This matrix would have as its eigenvalues some linear combinations of the cumulants of the ICs. If we are lucky, these linear combinations could be distinct, and FOBI works. But the more powerful way to utilize this general definition is to take several matrices $F(M_i)$ and jointly (approximately) diagonalize them. But this is what JADE is doing, for its particular set of matrices! Thus we see how JADE is a generalization of FOBI.

11.6 CONCLUDING REMARKS AND REFERENCES

An approach to ICA estimation that is rather different from those in the previous chapters is given by tensorial methods. The fourth-order cumulants of mixtures give all the fourth-order information inherent in the data. They can be used to define a tensor, which is a generalization of the covariance matrix. Then we can apply eigenvalue decomposition on this matrix. The eigenvectors more or less directly give the mixing matrix for whitened data. One simple way of computing the eigenvalue decomposition is to use the power method that turns out to be the same as the FastICA algorithm with the cubic nonlinearity. Joint approximate diagonalization of eigenmatrices (JADE) is another method in this category that has been successfully used in low-dimensional problems. In the special case of distinct kurtoses, a computationally very simple method (FOBI) can be devised.

The tensor methods were probably the first class of algorithms that performed ICA successfully. The simple FOBI algorithm was introduced in [61], and the tensor structure was first treated in [62, 94]. The most popular algorithm in this category is probably the JADE algorithm as proposed in [72]. The power method given by FastICA, another popular algorithm, is not usually interpreted from the tensor viewpoint, as we have seen in preceding chapters. For an alternative form of the power method, see [262]. A related method was introduced in [306]. An in-depth overview of the tensorial method is given in [261]; see also [94]. An accessible and fundamental paper is [68] that also introduces sophisticated modifications of the methods. In [473], a kind of a variant of the cumulant tensor approach was proposed by evaluating the second derivative of the characteristic function at arbitrary points.

The tensor methods, however, have become less popular recently. This is because methods that use the whole EVD (like JADE) are restricted, for computational reasons, to small dimensions. Moreover, they have statistical properties inferior to those methods using nonpolynomial cumulants or likelihood. With low-dimensional data, however, they can offer an interesting alternative, and the power method that boils down to FastICA can be used in higher dimensions as well.

Problems

11.1 Prove that $\mathbf{W}$ diagonalizes $\mathbf{F}(\mathbf{M})$ as claimed in Section 11.4.

11.2 Prove (11.19)

11.3 Prove (11.20).

Computer assignments

11.1 Compute the eigenvalue decomposition of random fourth-order tensors of size $2 \times 2 \times 2 \times 2$ and $5 \times 5 \times 5 \times 5$. Compare the computing times. What about a tensor of size $100 \times 100 \times 100 \times 100$?

11.2 Generate 2-D data according to the ICA model. First, with ICs of different distributions, and second, with identical distributions. Whiten the data, and perform the FOBI algorithm in Section 11.5. Compare the two cases.

12

ICA by Nonlinear Decorrelation and Nonlinear PCA

This chapter starts by reviewing some of the early research efforts in independent component analysis (ICA), especially the technique based on nonlinear decorrelation, that was successfully used by Jutten, Hérault, and Ans to solve the first ICA problems. Today, this work is mainly of historical interest, because there exist several more efficient algorithms for ICA.

Nonlinear decorrelation can be seen as an extension of second-order methods such as whitening and principal component analysis (PCA). These methods give components that are uncorrelated linear combinations of input variables, as explained in Chapter 6. We will show that independent components can in some cases be found as *nonlinearly* uncorrelated linear combinations. The nonlinear functions used in this approach introduce higher order statistics into the solution method, making ICA possible.

We then show how the work on nonlinear decorrelation eventually lead to the Cichocki-Unbehauen algorithm, which is essentially the same as the algorithm that we derived in Chapter 9 using the natural gradient. Next, the criterion of nonlinear decorrelation is extended and formalized to the theory of estimating functions, and the closely related EASI algorithm is reviewed.

Another approach to ICA that is related to PCA is the so-called nonlinear PCA. A nonlinear representation is sought for the input data that minimizes a least mean-square error criterion. For the linear case, it was shown in Chapter 6 that principal components are obtained. It turns out that in some cases the nonlinear PCA approach gives independent components instead. We review the nonlinear PCA criterion and show its equivalence to other criteria like maximum likelihood (ML). Then, two typical learning rules introduced by the authors are reviewed, of which the first one is a stochastic gradient algorithm and the other one a recursive least mean-square algorithm.

12.1 NONLINEAR CORRELATIONS AND INDEPENDENCE

The correlation between two random variables y_1 and y_2 was discussed in detail in Chapter 2. Here we consider zero-mean variables only, so correlation and covariance are equal. Correlation is related to independence in such a way that independent variables are always uncorrelated. The opposite is not true, however: the variables can be uncorrelated, yet dependent. An example is a uniform density in a rotated square centered at the origin of the (y_1, y_2) space, see e.g. Fig. 8.3. Both y_1 and y_2 are zero mean and uncorrelated, no matter what the orientation of the square, but they are independent only if the square is aligned with the coordinate axes. In some cases uncorrelatedness does imply independence, though; the best example is the case when the density of (y_1, y_2) is constrained to be jointly gaussian.

Extending the concept of correlation, we here define the *nonlinear correlation* of the random variables y_1 and y_2 as $\mathrm{E}\{f(y_1)g(y_2)\}$. Here, $f(y_1)$ and $g(y_2)$ are two functions, of which at least one is nonlinear. Typical examples might be polynomials of degree higher than 1, or more complex functions like the hyperbolic tangent. This means that one or both of the random variables are first transformed nonlinearly to new variables $f(y_1), g(y_2)$ and then the usual linear correlation between these new variables is considered.

The question now is: Assuming that y_1 and y_2 are nonlinearly decorrelated in the sense

$$\mathrm{E}\{f(y_1)g(y_2)\} = 0 \tag{12.1}$$

can we say something about their independence? We would hope that by making this kind of nonlinear correlation zero, independence would be obtained under some additional conditions to be specified.

There is a general theorem (see, e.g., [129]) stating that y_1 and y_2 are independent *if and only if*

$$\mathrm{E}\{f(y_1)g(y_2)\} = \mathrm{E}\{f(y_1)\}\mathrm{E}\{g(y_2)\} \tag{12.2}$$

for *all* continuous functions f and g that are zero outside a finite interval. Based on this, it seems very difficult to approach independence rigorously, because the functions f and g are almost arbitrary. Some kind of approximations are needed.

This problem was considered by Jutten and Hérault [228]. Let us assume that $f(y_1)$ and $g(y_2)$ are smooth functions that have derivatives of all orders in a neighborhood of the origin. They can be expanded in Taylor series:

$$
\begin{aligned}
f(y_1) &= f(0) + f'(0)y_1 + \frac{1}{2}f''(0)y_1^2 + \dots \\
&= \sum_{i=0}^{\infty} f_i y_1^i \\
g(y_2) &= g(0) + g'(0)y_2 + \frac{1}{2}g''(0)y_2^2 + \dots \\
&= \sum_{i=0}^{\infty} g_i y_2^i
\end{aligned}
$$

where f_i, g_i is shorthand for the coefficients of the ith powers in the series.
The product of the functions is then

$$
f(y_1)g(y_2) = \sum_{i=1}^{\infty} \sum_{j=1}^{\infty} f_i g_j y_1^i y_2^j \tag{12.3}
$$

and condition (12.1) is equivalent to

$$
E\{f(y_1)g(y_2)\} = \sum_{i=1}^{\infty} \sum_{j=1}^{\infty} f_i g_j E\{y_1^i y_2^j\} = 0 \tag{12.4}
$$

Obviously, a sufficient condition for this equation to hold is

$$
E\{y_1^i y_2^j\} = 0 \tag{12.5}
$$

for all indices i, j appearing in the series expansion (12.4). There may be other solutions in which the higher order correlations are *not* zero, but the coefficients f_i, g_j happen to be just suitable to cancel the terms and make the sum in (12.4) exactly equal to zero. For nonpolynomial functions that have infinite Taylor expansions, such spurious solutions can be considered unlikely (we will see later that such spurious solutions do exist but they can be avoided by the theory of ML estimation).

Again, a sufficient condition for (12.5) to hold is that the variables y_1 and y_2 are *independent* and one of $E\{y_1^i\}$, $E\{y_2^j\}$ is zero. Let us require that $E\{y_1^i\} = 0$ for all powers i appearing in its series expansion. But this is only possible if $f(y_1)$ is an *odd* function; then the Taylor series contains only odd powers $1, 3, 5, \dots$, and the powers i in Eq. (12.5) will also be odd. Otherwise, we have the case that even moments of y_1 like the variance are zero, which is impossible unless y_1 is constant.

To conclude, a sufficient (but not necessary) condition for the nonlinear uncorrelatedness (12.1) to hold is that y_1 and y_2 are independent, and for one of them, say y_1, the nonlinearity is an odd function such that $f(y_1)$ has zero mean.

The preceding discussion is informal but should make it credible that nonlinear correlations are useful as a possible general criterion for independence. Several things have to be decided in practice: the first one is how to actually choose the functions f, g. Is there some natural optimality criterion that can tell us that some functions

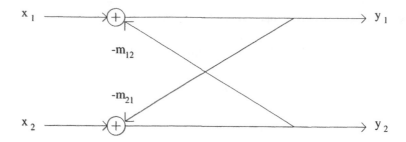

Fig. 12.1 The basic feedback circuit for the Hérault-Jutten algorithm. The element marked with + is a summation

are better than some other ones? This will be answered in Sections 12.3 and 12.4. The second problem is how we could solve Eq. (12.1), or nonlinearly decorrelate two variables y_1, y_2. This is the topic of the next section.

12.2 THE HÉRAULT-JUTTEN ALGORITHM

Consider the ICA model $\mathbf{x} = \mathbf{As}$. Let us first look at a 2×2 case, which was considered by Hérault, Jutten and Ans [178, 179, 226] in connection with the blind separation of two signals from two linear mixtures. The model is then

$$
\begin{aligned}
x_1 &= a_{11}s_1 + a_{12}s_2 \\
x_2 &= a_{21}s_1 + a_{22}s_2
\end{aligned}
$$

Hérault and Jutten proposed the feedback circuit shown in Fig. 12.1 to solve the problem. The initial outputs are fed back to the system, and the outputs are recomputed until an equilibrium is reached.

From Fig. 12.1 we have directly

$$
\begin{aligned}
y_1 &= x_1 - m_{12}y_2 & \quad (12.6) \\
y_2 &= x_2 - m_{21}y_1 & \quad (12.7)
\end{aligned}
$$

Before inputting the mixture signals x_1, x_2 to the network, they were normalized to zero mean, which means that the outputs y_1, y_2 also will have zero means. Defining a matrix $\mathbf{M}$ with off-diagonal elements m_{12}, m_{21} and diagonal elements equal to zero, these equations can be compactly written as

$$
\mathbf{y} = \mathbf{x} - \mathbf{My}
$$

Thus the input-output mapping of the network is

$$
\mathbf{y} = (\mathbf{I} + \mathbf{M})^{-1}\mathbf{x} \quad (12.8)
$$

Note that from the original ICA model we have $s = A^{-1}x$, provided that A is invertible. If $I + M = A$, then y becomes equal to s. However, the problem in blind separation is that the matrix A is unknown.

The solution that Jutten and Hérault introduced was to adapt the two feedback coefficients m_{12}, m_{21} so that the outputs of the network y_1, y_2 become independent. Then the matrix A has been implicitly inverted and the original sources have been found. For independence, they used the criterion of nonlinear correlations. They proposed the following learning rules:

$$\Delta m_{12} = \mu f(y_1) g(y_2) \qquad (12.9)$$

$$\Delta m_{21} = \mu f(y_2) g(y_1) \qquad (12.10)$$

with μ the learning rate. Both functions $f(.), g(.)$ are odd functions; typically, the functions

$$f(y) = y^3, \quad g(y) = \arctan(y)$$

were used, although the method also seems to work for $g(y) = y$ or $g(y) = \text{sign}(y)$. Now, if the learning converges, then the right-hand sides must be zero on average, implying

$$E\{f(y_1)g(y_2)\} = E\{f(y_2)g(y_1)\} = 0$$

Thus independence has hopefully been attained for the outputs y_1, y_2. A stability analysis for the Hérault-Jutten algorithm was presented by [408].

In the numerical computation of the matrix M according to algorithm (12.9,12.10), the outputs y_1, y_2 on the right-hand side must also be updated at each step of the iteration. By Eq. (12.8), they too depend on M, and solving them requires the inversion of matrix $I + M$. As noted by Cichocki and Unbehauen [84], this matrix inversion may be computationally heavy, especially if this approach is extended to more than two sources and mixtures. One way to circumvent this problem is to make a rough approximation

$$y = (I + M)^{-1} x \approx (I - M) x$$

that seems to work in practice.

Although the Hérault-Jutten algorithm was a very elegant pioneering solution to the ICA problem, we know now that it has some drawbacks in practice. The algorithm may work poorly or even fail to separate the sources altogether if the signals are badly scaled or the mixing matrix is ill-conditioned. The number of sources that the method can separate is severely limited. Also, although the local stability was shown in [408], good global convergence behavior is not guaranteed.

12.3 THE CICHOCKI-UNBEHAUEN ALGORITHM

Starting from the Hérault-Jutten algorithm Cichocki, Unbehauen, and coworkers [82, 85, 84] derived an extension that has a much enhanced performance and reliability. Instead of a feedback circuit like the Hérault-Jutten network in Fig. 12.1, Cichocki and Unbehauen proposed a *feedforward* network with weight matrix $\mathbf{B}$, with the mixture vector $\mathbf{x}$ for input and with output $\mathbf{y} = \mathbf{Bx}$. Now the dimensionality of the problem can be higher than 2. The goal is to adapt the $m \times m$ matrix $\mathbf{B}$ so that the elements of $\mathbf{y}$ become independent.   The learning algorithm for $\mathbf{B}$ is as follows:

$$\Delta\mathbf{B} = \mu[\mathbf{\Lambda} - f(\mathbf{y})g(\mathbf{y}^T)]\mathbf{B} \tag{12.11}$$

where μ is the learning rate, $\mathbf{\Lambda}$ is a diagonal matrix whose elements determine the amplitude scaling for the elements of $\mathbf{y}$ (typically, $\mathbf{\Lambda}$ could be chosen as the unit matrix $\mathbf{I}$), and f and g are two nonlinear scalar functions; the authors proposed a polynomial and a hyperbolic tangent. The notation $f(\mathbf{y})$ means a column vector with elements $f(y_1), \ldots, f(y_n)$.

The argumentation showing that this algorithm will give independent components, too, is based on nonlinear decorrelations. Consider the stationary solution of this learning rule defined as the matrix for which $\mathrm{E}\{\Delta\mathbf{B}\} = 0$, with the expectation taken over the density of the mixtures $\mathbf{x}$. For this matrix, the update is on the average zero. Because this is a stochastic-approximation-type algorithm (see Chapter 3), such stationarity is a necessary condition for convergence. Excluding the trivial solution $\mathbf{B} = 0$, we must have

$$\mathbf{\Lambda} - \mathrm{E}\{f(\mathbf{y})g(\mathbf{y}^T)\} = 0$$

Especially, for the off-diagonal elements, this implies

$$\mathrm{E}\{f(y_i)g(y_j)\} = 0 \tag{12.12}$$

which is exactly our definition of nonlinear decorrelation in Eq. (12.1) extended to n output signals $y_1, ..., y_n$. The diagonal elements satisfy

$$\mathrm{E}\{f(y_i)g(y_i)\} = \mathbf{\Lambda}_{ii}$$

showing that the diagonal elements $\mathbf{\Lambda}_{ii}$ of matrix $\mathbf{\Lambda}$ only control the amplitude scaling of the outputs.

The conclusion is that if the learning rule converges to a nonzero matrix $\mathbf{B}$, then the outputs of the network must become nonlinearly decorrelated, and hopefully independent. The convergence analysis has been performed in [84]; for general principles of analyzing stochastic iteration algorithms like (12.11), see Chapter 3.

The justification for the Cichocki-Unbehauen algorithm (12.11) in the original articles was based on nonlinear decorrelations, not on any rigorous cost functions that would be minimized by the algorithm. However, it is interesting to note that this algorithm, first appearing in the early 1990's, is in fact the same as the popular natural gradient algorithm introduced later by Amari, Cichocki, and Young [12] as an extension to the original Bell-Sejnowski algorithm [36]. All we have to do is choose $\mathbf{\Lambda}$ as the unit matrix, the function $g(\mathbf{y})$ as the linear function $g(\mathbf{y}) = \mathbf{y}$, and the function $f(\mathbf{y})$ as a sigmoidal related to the true density of the sources. The Amari-Cichocki-Young algorithm and the Bell-Sejnowski algorithm were reviewed in Chapter 9 and it was shown how the algorithms are derived from the rigorous maximum likelihood criterion. The maximum likelihood approach also tells us what kind of nonlinearities should be used, as discussed in Chapter 9.

12.4 THE ESTIMATING FUNCTIONS APPROACH *

Consider the criterion of nonlinear decorrelations being zero, generalized to n random variables $y_1, ..., y_n$, shown in Eq. (12.12). Among the possible roots $y_1, ..., y_n$ of these equations are the source signals $s_1, ..., s_n$. When solving these in an algorithm like the Hérault-Jutten algorithm or the Cichocki-Unbehauen algorithm, one in fact solves the separating matrix $\mathbf{B}$.

This notion was generalized and formalized by Amari and Cardoso [8] to the case of *estimating functions*. Again, consider the basic ICA model $\mathbf{x} = \mathbf{As}$, $\mathbf{s} = \mathbf{B}^*\mathbf{x}$ where $\mathbf{B}^*$ is a true separating matrix (we use this special notation here to avoid any confusion). An estimation function is a matrix-valued function $\mathbf{F}(\mathbf{x}, \mathbf{B})$ such that

$$\mathrm{E}\{\mathbf{F}(\mathbf{x}, \mathbf{B}^*)\} = 0. \tag{12.13}$$

This means that, taking the expectation with respect to the density of $\mathbf{x}$, the *true separating matrices are roots of the equation*. Once these are solved from Eq. (12.13), the independent components are directly obtained.

Example 12.1 Given a set of nonlinear functions $f_1(y_1), ..., f_n(y_n)$, with $\mathbf{y} = \mathbf{Bx}$, and defining a vector function $\mathbf{f}(\mathbf{y}) = [f_1(y_1), ..., f_n(y_n)]^T$, a suitable estimating function for ICA is

$$\mathbf{F}(\mathbf{x}, \mathbf{B}) = \boldsymbol{\Lambda} - \mathbf{f}(\mathbf{y})\mathbf{y}^T = \boldsymbol{\Lambda} - \mathbf{f}(\mathbf{Bx})(\mathbf{Bx})^T \tag{12.14}$$

because obviously $\mathrm{E}\{\mathbf{f}(\mathbf{y})\mathbf{y}^T\}$ becomes diagonal when $\mathbf{B}$ is a true separating matrix $\mathbf{B}^*$ and $y_1, ..., y_n$ are independent and zero mean. Then the off-diagonal elements become $\mathrm{E}\{f_i(y_i)y_j\} = \mathrm{E}\{f_i(y_i)\}\mathrm{E}\{y_j\} = 0$. The diagonal matrix $\boldsymbol{\Lambda}$ determines the scales of the separated sources. Another estimating function is the right-hand side of the learning rule (12.11),

$$\mathbf{F}(\mathbf{x}, \mathbf{B}) = [\boldsymbol{\Lambda} - f(\mathbf{y})g(\mathbf{y}^T)]\mathbf{B}$$

There is a fundamental difference in the estimating function approach compared to most of the other approaches to ICA: the usual starting point in ICA is a cost function that somehow measures how independent or nongaussian the outputs y_i are, and the independent components are solved by minimizing the cost function. In contrast, there is no such cost function here. The estimation function need not be the gradient of any other function. In this sense, the theory of estimating functions is very general and potentially useful for finding ICA algorithms. For a discussion of this approach in connection with neural networks, see [328].

It is not a trivial question how to design in practice an estimation function so that we can solve the ICA model. Even if we have two estimating functions that both have been shaped in such a way that separating matrices are their roots, what is a relevant measure to compare them? Statistical considerations are helpful here. Note that in practice, the densities of the sources s_i and the mixtures x_j are unknown in the ICA model. It is impossible in practice to solve Eq. (12.13) as such, because the expectation cannot be formed. Instead, it has to be estimated using a finite sample of $\mathbf{x}$. Denoting this sample by $\mathbf{x}(1), ..., \mathbf{x}(T)$, we use the sample function

$$E\{\mathbf{F}(\mathbf{x}, \mathbf{B})\} \approx \frac{1}{T} \sum_{t=1}^{T} \mathbf{F}(\mathbf{x}(t), \mathbf{B})$$

Its root $\hat{\mathbf{B}}$ is then an estimator for the true separating matrix. Obviously (see Chapter 4), the root $\hat{\mathbf{B}} = \hat{\mathbf{B}}[\mathbf{x}(1), ..., \mathbf{x}(T)]$ is a function of the training sample, and it is meaningful to consider its statistical properties like bias and variance. This gives a measure of goodness for the comparison of different estimation functions. The best estimating function is one that gives the smallest error between the true separating matrix $\mathbf{B}^*$ and the estimate $\hat{\mathbf{B}}$.

A particularly relevant measure is (Fisher) efficiency or asymptotic variance, as the size T of the sample $\mathbf{x}(1), ..., \mathbf{x}(T)$ grows large (see Chapter 4). The goal is to design an estimating function that gives the smallest variance, given the set of observations $\mathbf{x}(t)$. Then the optimal amount of information is extracted from the training set.

The general result provided by Amari and Cardoso [8] is that estimating functions of the form (12.14) are optimal in the sense that, given *any* estimating function $\mathbf{F}$, one can always find a better or at least equally good estimating function (in the sense of efficiency) having the form

$$\begin{aligned} \mathbf{F}(\mathbf{x}, \mathbf{B}) &= \mathbf{\Lambda} - f(\mathbf{y})\mathbf{y}^T & (12.15) \\ &= \mathbf{\Lambda} - f(\mathbf{B}\mathbf{x})(\mathbf{B}\mathbf{x})^T & (12.16) \end{aligned}$$

where $\mathbf{\Lambda}$ is a diagonal matrix. Actually, the diagonal matrix $\mathbf{\Lambda}$ has no effect on the off-diagonal elements of $\mathbf{F}(\mathbf{x}, \mathbf{B})$ which are the ones determining the independence between y_i, y_j; the diagonal elements are simply scaling factors.

The result shows that it is unnecessary to use a nonlinear function $g(\mathbf{y})$ instead of $\mathbf{y}$ as the other one of the two functions in nonlinear decorrelation. Only one nonlinear function $f(\mathbf{y})$, combined with $\mathbf{y}$, is sufficient. It is interesting that functions of exactly the type $f(\mathbf{y})\mathbf{y}^T$ naturally emerge as gradients of cost functions such as likelihood; the question of how to choose the nonlinearity $f(\mathbf{y})$ is also answered in that case. A further example is given in the following section.

The preceding analysis is not related in any way to the practical methods for finding the roots of estimating functions. Due to the nonlinearities, closed-form solutions do not exist and numerical algorithms have to be used. The simplest iterative stochastic approximation algorithm for solving the roots of $\mathbf{F}(\mathbf{x}, \mathbf{B})$ has the form

$$\Delta \mathbf{B} = -\mu \mathbf{F}(\mathbf{x}, \mathbf{B}), \qquad (12.17)$$

with μ an appropriate learning rate. In fact, we now discover that the learning rules (12.9), (12.10) and (12.11) are examples of this more general framework.

12.5   EQUIVARIANT ADAPTIVE SEPARATION VIA INDEPENDENCE

In most of the proposed approaches to ICA, the learning rules are gradient descent algorithms of cost (or contrast) functions. Many cases have been covered in previous chapters. Typically, the cost function has the form $J(\mathbf{B}) = \mathrm{E}\{G(\mathbf{y})\}$, with G some scalar function, and usually some additional constraints are used. Here again $\mathbf{y} = \mathbf{Bx}$, and the form of the function G and the probability density of $\mathbf{x}$ determine the shape of the contrast function $J(\mathbf{B})$.

It is easy to show (see the definition of matrix and vector gradients in Chapter 3) that

$$\frac{\partial J(\mathbf{B})}{\partial \mathbf{B}} = \mathrm{E}\{(\frac{\partial G(\mathbf{y})}{\partial \mathbf{y}})\mathbf{x}^T\} = \mathrm{E}\{\mathbf{g}(\mathbf{y})\mathbf{x}^T\} \tag{12.18}$$

where $\mathbf{g}(\mathbf{y})$ is the gradient of $G(\mathbf{y})$. If $\mathbf{B}$ is square and invertible, then $\mathbf{x} = \mathbf{B}^{-1}\mathbf{y}$ and we have

$$\frac{\partial J(\mathbf{B})}{\partial \mathbf{B}} = \mathrm{E}\{\mathbf{g}(\mathbf{y})\mathbf{y}^T\}(\mathbf{B}^T)^{-1} \tag{12.19}$$

For appropriate nonlinearities $G(\mathbf{y})$, these gradients are estimating functions in the sense that the elements of $\mathbf{y}$ must be statistically independent when the gradient becomes zero. Note also that in the form $\mathrm{E}\{\{\mathbf{g}(\mathbf{y})\mathbf{y}^T\}\}(\mathbf{B}^T)^{-1}$, the first factor $\mathbf{g}(\mathbf{y})\mathbf{y}^T$ has the shape of an *optimal* estimating function (except for the diagonal elements); see eq. (12.15). Now we also know how the nonlinear function $\mathbf{g}(\mathbf{y})$ can be determined: it is directly the gradient of the function $G(\mathbf{y})$ appearing in the original cost function.

Unfortunately, the matrix inversion $(\mathbf{B}^T)^{-1}$ in (12.19) is cumbersome. Matrix inversion can be avoided by using the so-called *natural gradient* introduced by Amari [4]. This is covered in Chapter 3. The natural gradient is obtained in this case by multiplying the usual matrix gradient (12.19) from the right by matrix $\mathbf{B}^T\mathbf{B}$, which gives $\mathrm{E}\{\mathbf{g}(\mathbf{y})\mathbf{y}^T\}\mathbf{B}$. The ensuing stochastic gradient algorithm to minimize the cost function $J(\mathbf{B})$ is then

$$\Delta\mathbf{B} = -\mu\mathbf{g}(\mathbf{y})\mathbf{y}^T\mathbf{B} \tag{12.20}$$

This learning rule again has the form of nonlinear decorrelations. Omitting the diagonal elements in matrix in $\mathbf{g}(\mathbf{y})\mathbf{y}^T$, the off-diagonal elements have the same form as in the Cichocki-Unbehauen algorithm (12.11), with the two functions now given by the linear function $\mathbf{y}$ and the gradient $\mathbf{g}(\mathbf{y})$.

This gradient algorithm can also be derived using the relative gradient introduced by Cardoso and Hvam Laheld [71]. This approach is also reviewed in Chapter 3. Based on this, the authors developed their *equivariant adaptive separation via independence* (EASI) learning algorithm. To proceed from (12.20) to the EASI learning rule, an extra step must be taken. In EASI, as in many other learning rules for ICA, a whitening preprocessing is considered for the mixture vectors $\mathbf{x}$ (see Chapter 6). We first transform $\mathbf{x}$ linearly to $\mathbf{z} = \mathbf{Vx}$ whose elements z_i have unit variances and zero covariances: $E\{\mathbf{z}\mathbf{z}^T\} = \mathbf{I}$. As also shown in Chapter 6, an appropriate adaptation rule for whitening is

$$\Delta\mathbf{V} = \mu(\mathbf{I} - \mathbf{z}\mathbf{z}^T)\mathbf{V} \qquad (12.21)$$

The ICA model using these whitened vectors instead of the original ones becomes $\mathbf{z} = \mathbf{V}\mathbf{A}\mathbf{s}$, and it is easily seen that the matrix $\mathbf{V}\mathbf{A}$ is an orthogonal matrix (a rotation). Thus its inverse which gives the separating matrix is also orthogonal. As in earlier chapters, let us denote the orthogonal separating matrix by $\mathbf{W}$.

Basically, the learning rule for $\mathbf{W}$ would be the same as (12.20). However, as noted by [71], certain constraints must hold in any updating of $\mathbf{W}$ if the orthogonality is to be preserved at each iteration step. Let us denote the serial update for $\mathbf{W}$ using the learning rule (12.20), briefly, as $\mathbf{W} \leftarrow \mathbf{W} + \mathbf{D}\mathbf{W}$, where now $\mathbf{D} = -\mu\mathbf{g}(\mathbf{y})\mathbf{y}^T$. The orthogonality condition for the updated matrix becomes

$$(\mathbf{W} + \mathbf{D}\mathbf{W})(\mathbf{W} + \mathbf{D}\mathbf{W})^T = \mathbf{I} + \mathbf{D} + \mathbf{D}^T + \mathbf{D}\mathbf{D}^T = \mathbf{I}$$

where $\mathbf{W}\mathbf{W}^T = \mathbf{I}$ has been substituted. Assuming $\mathbf{D}$ small, the first-order approximation gives the condition that $\mathbf{D} = -\mathbf{D}^T$, or $\mathbf{D}$ must be skew-symmetric. Applying this condition to the relative gradient learning rule (12.20) for $\mathbf{W}$, we have

$$\Delta\mathbf{W} = -\mu[\mathbf{g}(\mathbf{y})\mathbf{y}^T - \mathbf{y}\mathbf{g}(\mathbf{y})^T]\mathbf{W} \qquad (12.22)$$

where now $\mathbf{y} = \mathbf{W}\mathbf{z}$. Contrary to the learning rule (12.20), this learning rule also takes care of the diagonal elements of $\mathbf{g}(\mathbf{y})\mathbf{y}^T$ in a natural way, without imposing any conditions on them.

What is left now is to combine the two learning rules (12.21) and (12.22) into just one learning rule for the global system separation matrix. Because $\mathbf{y} = \mathbf{W}\mathbf{z} = \mathbf{W}\mathbf{V}\mathbf{x}$, this global separation matrix is $\mathbf{B} = \mathbf{W}\mathbf{V}$. Assuming the same learning rates for the two algorithms, a first order approximation gives

$$
\begin{aligned}
\Delta\mathbf{B} &= \Delta\mathbf{W}\mathbf{V} + \mathbf{W}\Delta\mathbf{V} \\
&= -\mu[\mathbf{g}(\mathbf{y})\mathbf{y}^T - \mathbf{y}\mathbf{g}(\mathbf{y})^T]\mathbf{W}\mathbf{V} + \mu[\mathbf{W}\mathbf{V} - \mathbf{W}\mathbf{z}\mathbf{z}^T\mathbf{W}^T\mathbf{W}\mathbf{V}] \\
&= -\mu[\mathbf{y}\mathbf{y}^T - \mathbf{I} + \mathbf{g}(\mathbf{y})\mathbf{y}^T - \mathbf{y}\mathbf{g}(\mathbf{y})^T]\mathbf{B} \qquad (12.23)
\end{aligned}
$$

This is the EASI algorithm. It has the nice feature of combining both whitening and separation into a single algorithm. A convergence analysis as well as some experimental results are given in [71]. One can easily see the close connection to the nonlinear decorrelation algorithm introduced earlier.

The concept of *equivariance* that forms part of the name of the EASI algorithm is a general concept in statistical estimation; see, e.g., [395]. Equivariance of an estimator means, roughly, that its performance does not depend on the actual value of the parameter. In the context of the basic ICA model, this means that the ICs can be estimated with the same performance what ever the mixing matrix may be. EASI was one of the first ICA algorithms which was explicitly shown to be equivariant. In fact, most estimators of the basic ICA model are equivariant. For a detailed discussion, see [69].

12.6 NONLINEAR PRINCIPAL COMPONENTS

One of the basic definitions of PCA was optimal least mean-square error compression, as explained in more detail in Chapter 6. Assuming a random m-dimensional zero-mean vector $\mathbf{x}$, we search for a lower dimensional subspace such that the residual error between $\mathbf{x}$ and its orthogonal projection on the subspace is minimal, averaged over the probability density of $\mathbf{x}$. Denoting an orthonormal basis of this subspace by $\mathbf{w}_1, ..., \mathbf{w}_n$, the projection of $\mathbf{x}$ on the subspace spanned by the basis is $\sum_{i=1}^{n} (\mathbf{w}_i^T \mathbf{x}) \mathbf{w}_i$. Now n is the dimension of the subspace. The minimum mean-square criterion for PCA is

$$\text{minimize } \mathrm{E}\{\|\mathbf{x} - \sum_{i=1}^{n} (\mathbf{w}_i^T \mathbf{x}) \mathbf{w}_i\|^2\} \tag{12.24}$$

A solution (although not the unique one) of this optimization problem is given by the eigenvectors $\mathbf{e}_1, ..., \mathbf{e}_n$ of the data covariance matrix $\mathbf{C_x} = \mathrm{E}\{\mathbf{x}\mathbf{x}^T\}$. Then the linear factors $\mathbf{w}_i^T \mathbf{x}$ in the sum become the principal components $\mathbf{e}_i^T \mathbf{x}$.

For instance, if $\mathbf{x}$ is two-dimensional with a gaussian density, and we seek for a one-dimensional subspace (a straight line passing through the center of the density), then the solution is given by the principal axis of the elliptical density.

We now pose the question how this criterion and its solution are changed if a nonlinearity is included in the criterion. Perhaps the simplest nontrivial nonlinear extension is provided as follows. Assuming $g_1(.), ..., g_n(.)$ a set of scalar functions, as yet unspecified, let us look at a modified criterion to be minimized with respect to the basis vectors [232]:

$$J(\mathbf{w}_1 ... \mathbf{w}_n) = \mathrm{E}\{\|\mathbf{x} - \sum_{i=1}^{n} g_i(\mathbf{w}_i^T \mathbf{x}) \mathbf{w}_i\|^2\} \tag{12.25}$$

This criterion was first considered by Xu [461] who called it the "least mean-square error reconstruction" (LMSER) criterion.

The only change with respect to (12.24) is that instead of the linear factors $\mathbf{w}_i^T \mathbf{x}$, we now have nonlinear functions of them in the expansion that gives the approximation to $\mathbf{x}$. In the optimal solution that minimizes the criterion $J(\mathbf{w}_1, ..., \mathbf{w}_n)$, such factors might be termed nonlinear principal components. Therefore, the technique of finding the basis vectors $\mathbf{w}_i$ is here called "nonlinear principal component analysis" (NLPCA).

It should be emphasized that practically always when a well-defined linear problem is extended into a nonlinear one, many ambiguities and alternative definitions arise. This is the case here, too. The term "nonlinear PCA" is by no means unique. There are several other techniques, like the method of principal curves [167, 264] or the nonlinear autoassociators [252, 325] that also give "nonlinear PCA". In these methods, the approximating subspace is a curved manifold, while the solution to the problem posed earlier is still a linear subspace. Only the coefficients corresponding to the principal components are nonlinear functions of $\mathbf{x}$. It should be noted that minimizing the criterion (12.25) does not give a smaller least mean square error than standard PCA. Instead, the virtue of this criterion is that it introduces higher-order statistics in a simple manner via the nonlinearities g_i.

Before going into any deeper analysis of (12.25), it may be instructive to see in a simple special case how it differs from linear PCA and how it is in fact related to ICA.

If the functions $g_i(y)$ were linear, as in the standard PCA technique, and the number n of terms in the sum were equal to m or the dimension of $\mathbf{x}$, then the representation error always would be zero, as long as the weight vectors are chosen orthonormal. For nonlinear functions $g_i(y)$, however, this is usually not true. Instead, in some cases, at least, it turns out that the optimal basis vectors $\mathbf{w}_i$ minimizing (12.25) will be aligned with the independent components of the input vectors.

Example 12.2 Assume that $\mathbf{x}$ is a two-dimensional random vector that has a uniform density in a unit square that is *not* aligned with the coordinate axes x_1, x_2, according to Fig. 12.2. Then it is easily shown that the elements x_1, x_2 are uncorrelated and have equal variances (equal to $1/3$), and the covariance matrix of $\mathbf{x}$ is therefore equal to $1/3\mathbf{I}$. Thus, except for the scaling by $1/3$, vector $\mathbf{x}$ is whitened (sphered). However, the elements are not independent. The problem is to find a rotation $\mathbf{s} = \mathbf{W}\mathbf{x}$ of $\mathbf{x}$ such that the elements of the rotated vector $\mathbf{s}$ are statistically independent. It is obvious from Fig. 12.2 that the elements of $\mathbf{s}$ must be aligned with the orientation of the square, because then and only then the joint density is separable into the product of the two marginal uniform densities.

Because of the whitening, we know that the rows of the separating matrix $\mathbf{W}$ must be orthogonal. This is seen by writing

$$\mathrm{E}\{\mathbf{s}\mathbf{s}^T\} = \mathbf{W}\mathrm{E}\{\mathbf{x}\mathbf{x}^T\}\mathbf{W}^T = \frac{1}{3}\mathbf{W}\mathbf{W}^T \qquad (12.26)$$

Because the elements s_1 and s_2 are uncorrelated, it must hold that $\mathbf{w}_1^T\mathbf{w}_2 = 0$.

The solution minimizing the criterion (12.25), with $\mathbf{w}_1$, $\mathbf{w}_2$ orthogonal two-dimensional vectors and $g_1(.) = g_2(.) = g(.)$ a suitable nonlinearity, provides now a rotation into independent components. This can be seen as follows. Assume that g is a very sharp sigmoid, e.g., $g(y) = \tanh(10y)$, which is approximately the sign function. The term $\sum_{i=1}^{2} g(\mathbf{w}_i^T\mathbf{x})\mathbf{w}_i$ in criterion (12.25) becomes

$$\mathbf{w}_1 g(\mathbf{w}_1^T\mathbf{x}) + \mathbf{w}_2 g(\mathbf{w}_2^T\mathbf{x})$$
$$\approx \mathbf{w}_1 \mathrm{sign}(\mathbf{w}_1^T\mathbf{x}) + \mathbf{w}_2 \mathrm{sign}(\mathbf{w}_2^T\mathbf{x})$$

Thus according to (12.25), each $\mathbf{x}$ should be optimally represented by one of the four possible points $(\pm\mathbf{w}_1, \pm\mathbf{w}_2)$, with the signs depending on the angles between $\mathbf{x}$ and the basis vectors. Each choice of the two orthogonal basis vectors divides the square of Fig. 12.2 into four quadrants, and by criterion (12.25), all the points in a given quadrant must be represented in the least mean square sense by just one point; e.g., in the first quadrant where the angles between $\mathbf{x}$ and the basis vectors are positive, by the point $\mathbf{w}_1 + \mathbf{w}_2$. From Fig. 12.2, it can be seen that the optimal fit is obtained

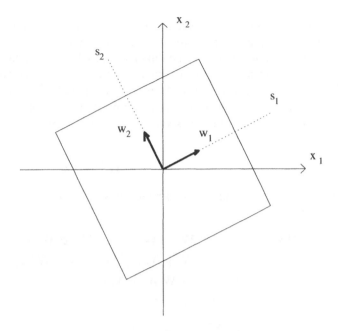

Fig. 12.2   A rotated uniform density

when the basis vectors are aligned with the axes s_1, s_2, and the point $\mathbf{w}_1 + \mathbf{w}_2$ is the center of the smaller square bordered by the positive s_1, s_2 axes.

For further confirmation, it is easy to compute the theoretical value of the cost function $J(\mathbf{w}_1, \mathbf{w}_2)$ of Eq. (12.25) when the basis vectors $\mathbf{w}_1$ and $\mathbf{w}_2$ are arbitrary orthogonal vectors [327]. Denoting the angle between $\mathbf{w}_1$ and the s_1 axis in Fig. 12.2 by θ, we then have the minimal value of $J(\mathbf{w}_1, \mathbf{w}_2)$ for the rotation $\theta = 0$, and then the lengths of the orthogonal vectors are equal to 0.5. These are the vectors shown in Fig. 12.2.

In the preceding example, it was assumed that the density of $\mathbf{x}$ is uniform. For some other densities, the same effect of rotation into independent directions would not be achieved. Certainly, this would not take place for gaussian densities with equal variances, for which the criterion $J(\mathbf{w}_1, ..., \mathbf{w}_n)$ would be independent of the orientation. Whether the criterion results in independent components, depends strongly on the nonlinearities $g_i(y)$. A more detailed analysis of the criterion (12.25) and its relation to ICA is given in the next section.

12.7   THE NONLINEAR PCA CRITERION AND ICA

Interestingly, for *prewhitened data*, it can be shown [236] that the original nonlinear PCA criterion of Eq. (12.25) has an exact relationship with other contrast functions like kurtosis maximization/minimization, maximum likelihood, or the so-called Bussgang criteria. In the prewhitened case, we use $\mathbf{z}$ instead of $\mathbf{x}$ to denote the input vector. Also, assume that in whitening, the dimension of $\mathbf{z}$ has been reduced to that of $\mathbf{s}$. We denote this dimension by n. In this case, it has been shown before (see Chapter 13) that matrix $\mathbf{W}$ is $n \times n$ and orthogonal: it holds $\mathbf{WW}^T = \mathbf{W}^T\mathbf{W} = \mathbf{I}$.

First, it is convenient to change to matrix formulation. Denoting by $\mathbf{W} = (\mathbf{w}_1...\mathbf{w}_n)^T$ the matrix that has the basis vectors $\mathbf{w}_i$ as rows, criterion (12.25) becomes

$$J(\mathbf{w}_1, ..., \mathbf{w}_n) = J(\mathbf{W}) = \mathrm{E}\{\|\mathbf{z} - \mathbf{W}^T\mathbf{g}(\mathbf{Wz})\|^2\}. \quad (12.27)$$

The function $\mathbf{g}(\mathbf{Wz})$ is a column vector with elements $g_1(\mathbf{w}_1^T\mathbf{z}), ..., g_n(\mathbf{w}_n^T\mathbf{z})$. We can write now [236]

$$
\begin{aligned}
\|\mathbf{z} - \mathbf{W}^T\mathbf{g}(\mathbf{Wz})\|^2 &= [\mathbf{z} - \mathbf{W}^T\mathbf{g}(\mathbf{Wz})]^T[\mathbf{z} - \mathbf{W}^T\mathbf{g}(\mathbf{Wz})] \\
&= [\mathbf{z} - \mathbf{W}^T\mathbf{g}(\mathbf{Wz})]^T\mathbf{W}^T\mathbf{W}[\mathbf{z} - \mathbf{W}^T\mathbf{g}(\mathbf{Wz})] \\
&= \|\mathbf{Wz} - \mathbf{WW}^T\mathbf{g}(\mathbf{Wz})\|^2 \\
&= \|\mathbf{y} - \mathbf{g}(\mathbf{y})\|^2 \\
&= \sum_{i=1}^{n}[y_i - g_i(y_i)]^2,
\end{aligned}
$$

with $\mathbf{y} = \mathbf{Wz}$. Therefore the criterion $J(\mathbf{W})$ becomes

$$J_{NLPCA}(\mathbf{W}) = \sum_{i=1}^{n}\mathrm{E}\{[y_i - g_i(y_i)]^2\} \quad (12.28)$$

This formulation of the NLPCA criterion can now be related to several other contrast functions.

As the first case, choose $g_i(y)$ as the odd quadratic function (the same for all i)

$$g_i(y) = \begin{cases} y^2 + y, & \text{if } y \geq 0 \\ -y^2 + y, & \text{if } y < 0 \end{cases}$$

Then the criterion (12.28) becomes

$$J_{\text{kurt}}(\mathbf{W}) = \sum_{i=1}^{n}\mathrm{E}\{(y_i - y_i \pm y_i^2)^2\} = \sum_{i=1}^{n}\mathrm{E}\{y_i^4\} \quad (12.29)$$

This statistic was discussed in Chapter 8. Note that because the input data has been whitened, the variance $\mathrm{E}\{y_i^2\} = \mathbf{w}_i^T\mathrm{E}\{\mathbf{zz}\}\mathbf{w}_i = \mathbf{w}_i^T\mathbf{w}_i = 1$, so in the kurtosis $\mathrm{kurt}(y_i) = \mathrm{E}\{y_i^4\} - 3(\mathrm{E}\{y_i^2\})^2$ the second term is a constant and can be dropped in kurtosis maximization/minimization. What remains is criterion (12.29). For this function, minimizing the NLPCA criterion is exactly equivalent to minimizing the sum of the kurtoses of y_i.

As a second case, consider the maximum likelihood solution of the ICA model. The maximum likelihood solution starts from the assumption that the density of $\mathbf{s}$, due to independence, is factorizable: $p(\mathbf{s}) = p_1(s_1)p_2(s_2)...p_n(s_n)$. Suppose we have a large sample $\mathbf{x}(1), ..., \mathbf{x}(T)$ of input vectors $\mathbf{x}$ available. It was shown in Chapter 9 that the log-likelihood becomes

$$\log L(\mathbf{B}) = \sum_{t=1}^{T} \sum_{i=1}^{n} \log p_i(\mathbf{b}_i^T \mathbf{x}(t)) + T \log |\det \mathbf{B}| \tag{12.30}$$

where the vectors $\mathbf{b}_i$ are the rows of matrix $\mathbf{B} = \mathbf{A}^{-1}$. In the case of a whitened sample $\mathbf{z}(1), ..., \mathbf{z}(T)$, the separating matrix will be orthogonal. Let us denote it again by $\mathbf{W}$. We have

$$\log L(\mathbf{W}) = \sum_{t=1}^{T} \sum_{i=1}^{n} \log p_i(\mathbf{w}_i^T \mathbf{z}(t)) \tag{12.31}$$

The second term in (12.30) is zero, because the determinant of the orthogonal matrix $\mathbf{W}$ is equal to one.

Because this is a maximization problem, we can multiply the cost function (12.31) by the constant $1/T$. For large T, this function tends to

$$J_{ML}(\mathbf{W}) = \sum_{i=1}^{n} \mathrm{E}\{\log p_i(y_i)\} \tag{12.32}$$

with $y_i = \mathbf{w}_i^T \mathbf{z}$.

From this, we can easily derive the connection between the NLPCA criterion (12.28) and the ML criterion (12.32). In minimizing the sum (12.28), an arbitrary additive constant and a positive multiplicative constant can be trivially added. Therefore, in the equivalence between the two criteria, we can consider the relation (dropping the subscript i from y_i for convenience)

$$\log p_i(y) = \alpha - \beta[y - g_i(y)]^2 \tag{12.33}$$

where α and $\beta > 0$ are some constants, yielding

$$p_i(y) \propto \exp[-\beta[y - g_i(y)]^2] \tag{12.34}$$

This shows how to choose the function $g_i(y)$ for any given density $p_i(y)$.

As the third case, the form of (12.28) is quite similar to the so-called Bussgang cost function used in blind equalization (see [170, 171]). We use Lambert's notation and approach [256]. He chooses only one nonlinearity $g_1(y) = \ldots = g_n(y) = g(y)$:

$$g(y) = \frac{-\mathrm{E}\{y^2\}p'(y)}{p(y)} \tag{12.35}$$

The function $p(y)$ is the density of y and $p'(y)$ its derivative. Lambert [256] also gives several algorithms for minimizing this cost function. Note that now in the whitened data case the variance of y is equal to one and the function (12.35) is simply the score function

$$-\frac{p'(y)}{p(y)} = -\frac{d}{dy}\log p(y)$$

Due to the equivalence of the maximum likelihood criterion to several other criteria like infomax or entropic criteria, further equivalences of the NLPCA criterion with these can be established. More details are given in [236] and Chapter 14.

12.8   LEARNING RULES FOR THE NONLINEAR PCA CRITERION

Once the nonlinearities $g_i(y)$ have been chosen, it remains to actually solve the minimization problem in the nonlinear PCA criterion. Here we present the simplest learning algorithms for minimizing either the original NLPCA criterion (12.25) or the prewhitened criterion (12.28). The first algorithm, the nonlinear subspace rule, is of the stochastic gradient descent type; this means that the expectation in the criterion is dropped and the gradient of the sample function that only depends on the present sample of the input vector ($\mathbf{x}$ or $\mathbf{z}$, respectively) is taken. This allows on-line learning in which each input vector is used when it comes available and then discarded; see Chapter 3 for more details. This algorithm is a nonlinear generalization of the subspace rule for PCA, covered in Chapter 6. The second algorithm reviewed in this section, the recursive least-squares learning rule, is likewise a nonlinear generalization of the PAST algorithm for PCA covered in Chapter 6.

12.8.1   The nonlinear subspace rule

Let us first consider a stochastic gradient algorithm for the original cost function (12.25), which in matrix form can be written as $J(\mathbf{W}) = \mathrm{E}\{\|\mathbf{x} - \mathbf{W}^T\mathbf{g}(\mathbf{W}\mathbf{x})\|^2\}$. This problem was considered by one of the authors [232, 233] as well as by Xu [461].
  It was shown that the stochastic gradient algorithm is

$$\Delta\mathbf{W} = \mu[\mathbf{F}(\mathbf{W}\mathbf{x})\mathbf{W}\mathbf{r}\mathbf{x}^T + \mathbf{g}(\mathbf{W}\mathbf{x})\mathbf{r}^T] \tag{12.36}$$

where

$$\mathbf{r} = \mathbf{x} - \mathbf{W}^T\mathbf{g}(\mathbf{W}\mathbf{x}) \tag{12.37}$$

is the residual error term and

$$\mathbf{F}(\mathbf{W}\mathbf{x}) = \mathrm{diag}[g'(\mathbf{w}_1^T\mathbf{x}), ..., g'(\mathbf{w}_n^T\mathbf{x})]. \tag{12.38}$$

There $g'(y)$ denotes the derivative of the function $g(y)$. We have made the simplifying assumption here that all the functions $g_1(.), ..., g_n(.)$ are equal; a generalization to the case of different functions would be straightforward but the notation would become more cumbersome.
  As motivated in more detail in [232], writing the update rule (12.36) for an individual weight vector $\mathbf{w}_i$ shows that the first term in the brackets on the right-hand

side of (12.36) affects the update of weight vectors much less than the second term, if the error **r** is relatively small in norm compared to the input vector **x**. If the first term is omitted, then we obtain the following learning rule:

$$\Delta \mathbf{W} = \mu \mathbf{g}(\mathbf{y})[\mathbf{x}^T - \mathbf{g}(\mathbf{y}^T)\mathbf{W}] \tag{12.39}$$

with **y** the vector

$$\mathbf{y} = \mathbf{W}\mathbf{x} \tag{12.40}$$

Comparing this learning rule to the subspace rule for ordinary PCA, eq. (6.19) in Chapter 6, we see that the algorithms are formally similar and become the same if g is a linear function. Both rules can be easily implemented in the one-layer PCA network shown in Fig. 6.2 of Chapter 6; the linear outputs $y_i = \mathbf{w}_i^T \mathbf{x}$ must only be changed to nonlinear versions $g(y_i) = g(\mathbf{w}_i^T \mathbf{x})$. This was the way the nonlinear PCA learning rule was first introduced in [332] as one of the extensions to numerical on-line PCA computation.

Originally, in [232] the criterion (12.25) and the learning scheme (12.39) were suggested for signal separation, but the exact relation to ICA was not clear. Even without prewhitening of the inputs **x**, the method can separate signals to a certain degree. However, if the inputs **x** are whitened first, the separation performance is greatly improved. The reason is that for whitened inputs, the criterion (12.25) and the consequent learning rule are closely connected to well-known ICA objective (cost) functions, as was shown in Section 12.7.

12.8.2 Convergence of the nonlinear subspace rule *

Let us consider the convergence and behavior of the learning rule in the case that the ICA model holds for the data. This is a very specialized section that may be skipped.

The prewhitened form of the learning rule is

$$\Delta \mathbf{W} = \mu \mathbf{g}(\mathbf{y})[\mathbf{z}^T - \mathbf{g}(\mathbf{y}^T)\mathbf{W}] \tag{12.41}$$

with **y** now the vector

$$\mathbf{y} = \mathbf{W}\mathbf{z} \tag{12.42}$$

and **z** white, thus $E\{\mathbf{z}\mathbf{z}^T\} = \mathbf{I}$. We also have to assume that the ICA model holds, i.e., there exists an orthogonal separating matrix **M** such that

$$\mathbf{s} = \mathbf{M}\mathbf{z} \tag{12.43}$$

where the elements of **s** are statistically independent. With whitening, the dimension of **z** has been reduced to that of **s**; thus both **M** and **W** are $n \times n$ matrices.

To make further analysis easier, we proceed by making a linear transformation to the learning rule (12.41): we multiply both sides by the orthogonal separating matrix $\mathbf{M}^T$, giving

$$
\begin{aligned}
\Delta(\mathbf{W}\mathbf{M}^T) &= \mu \mathbf{g}(\mathbf{W}\mathbf{z})[\mathbf{z}^T\mathbf{M}^T - \mathbf{g}(\mathbf{z}^T\mathbf{W}^T)\mathbf{W}\mathbf{M}^T] \tag{12.44}\\
&= \mu \mathbf{g}(\mathbf{W}\mathbf{M}^T\mathbf{M}\mathbf{z})[\mathbf{z}^T\mathbf{M}^T - \mathbf{g}(\mathbf{z}^T\mathbf{M}^T\mathbf{M}\mathbf{W}^T)\mathbf{W}\mathbf{M}^T]
\end{aligned}
$$

where we have used the fact that $\mathbf{M}^T\mathbf{M} = \mathbf{I}$. Denoting for the moment $\mathbf{H} = \mathbf{WM}^T$ and using (12.43), we have

$$\Delta\mathbf{H} = \mu\mathbf{g}(\mathbf{Hs})[\mathbf{s}^T - \mathbf{g}(\mathbf{s}^T\mathbf{H}^T)\mathbf{H}] \qquad (12.45)$$

This equation has exactly the same form as the original one (12.41). Geometrically the transformation by the orthogonal matrix $\mathbf{M}^T$ simply means a coordinate change to a new set of coordinates such that the elements of the input vector expressed in these coordinates are statistically independent.

The goal in analyzing the learning rule (12.41) is to show that, starting from some initial value, the matrix $\mathbf{W}$ will tend to the separating matrix $\mathbf{M}$. For the transformed weight matrix $\mathbf{H} = \mathbf{WM}^T$ in (12.45), this translates into the requirement that $\mathbf{H}$ should tend to the unit matrix or a permutation matrix. Then $\mathbf{y} = \mathbf{Hs}$ also would tend to the vector $\mathbf{s}$, or a permuted version, with independent components.

However, it turns out that in the learning rule (12.45), the unit matrix or a permutation matrix generally cannot be the asymptotic or steady state solution. This is due to the scaling given by the nonlinearity $\mathbf{g}$. Instead, we can make the more general requirement that $\mathbf{H}$ tends to a diagonal matrix or a diagonal matrix times a permutation matrix. In this case the elements of $\mathbf{y} = \mathbf{Hs}$ will become the elements of the original source vector $\mathbf{s}$, in some order, multiplied by some numbers. In view of the original problem, in which the amplitudes of the signals s_i remain unknown, this is actually no restriction, as independence is again attained.

To proceed, the difference equation (12.45) can be further analyzed by writing down the corresponding averaged differential equation; for a discussion of the technique, see Chapter 3. The limit of convergence of (12.45) is among the asymptotically stable solutions of the averaged differential equation. In practice, this also requires that the learning rate μ is decreasing to zero at a suitable rate.

Now taking averages in (12.45) and also using the same symbol $\mathbf{H} = \mathbf{H}(t)$ for the continuous-time counterpart of the transformed weight matrix $\mathbf{H}$, we obtain

$$d\mathbf{H}/dt = E\{\mathbf{g}(\mathbf{Hs})\mathbf{s}^T\} - E\{\mathbf{g}(\mathbf{Hs})\mathbf{g}(\mathbf{s}^T\mathbf{H}^T)\}\mathbf{H} \qquad (12.46)$$

The expectations are over the (unknown) density of vector $\mathbf{s}$. We are ready to state the main result of this section:

Theorem 12.1 *In the matrix differential equation (12.46), assume the following:*

1. *The random vector $\mathbf{s}$ has a symmetric density with $E\{\mathbf{s}\} = 0$;*

2. *The elements of $\mathbf{s}$, denoted $s_1, ..., s_n$, are statistically independent;*

3. *The function $g(.)$ is odd, i.e., $g(y) = -g(-y)$ for all y, and at least twice differentiable everywhere;*

4. *The function $g(.)$ and the density of $\mathbf{s}$ are such that the following conditions hold for all $i = 1, ..., n$:*

$$A_i = E\{s_i^2 g'(\alpha_i s_i)\} - 2\alpha_i E\{g(\alpha_i s_i)g'(\alpha_i s_i)s_i\} - E\{g^2(\alpha_i s_i)\} < 0 \qquad (12.47)$$

where the α_i, $i = 1, ..., n$ are scalars satisfying

$$E\{s_i g(\alpha_i s_i)\} = \alpha_i E\{g^2(\alpha_i s_i)\} \tag{12.48}$$

5. Denoting

$$
\begin{aligned}
\sigma_i^2 &= E\{s_i^2\} & (12.49)\\
B_i &= E\{g^2(\alpha_i s_i)\} & (12.50)\\
C_i &= E\{g'(\alpha_i s_i)\} & (12.51)\\
F_i &= E\{u_i g(\alpha_i s_i)\} & (12.52)
\end{aligned}
$$

both eigenvalues of the 2×2 matrix

$$
\left(
\begin{array}{cc}
\sigma_i^2 C_j - \alpha_i C_j F_i - B_j & -\alpha_i C_i F_j \\
-\alpha_j C_j F_i & \sigma_j^2 C_i - \alpha_j C_i F_j - B_i
\end{array}
\right) \tag{12.53}
$$

have strictly negative real parts for all $i, j = 1, ..., n$.

Then the matrix

$$\mathbf{H} = diag(\alpha_1, ..., \alpha_n) \tag{12.54}$$

is an asymptotically stable stationary point of (12.46), where α_i satisfies Eq. (12.48).

The proof, as well as explanations of the rather technical conditions of the theorem, are given in [327]. The main point is that the algorithm indeed converges to a diagonal matrix, if the initial value $\mathbf{H}(0)$ is not too far from it. Transforming back to the original learning rule (12.41) for $\mathbf{W}$, it follows that $\mathbf{W}$ converges to a separating matrix.

Some special cases were given in [327]. For example, if the nonlinearity is chosen as the simple odd polynomial

$$g(y) = y^q, \quad q = 1, 3, 5, 7, ... \tag{12.55}$$

then all the relevant variables in the conditions of the theorem, for any probability density, will become moments of s_i. It can be shown (see exercises) that the stability condition becomes

$$E\{s^{q+1}\} - qE\{s^2\}E\{s^{q-1}\} > 0 \tag{12.56}$$

Based on this, it can be shown that the linear function $g(y) = y$ never gives asymptotic stability, while the cubic function $g(y) = y^3$ leads to asymptotic stability provided that the density of s satisfies

$$E\{s^4\} - 3(E\{s^2\})^2 > 0 \tag{12.57}$$

This expression is exactly the kurtosis or the fourth order cumulant of s [319]. If and only if the density is positively kurtotic (supergaussian), the stability condition is satisfied for the cubic polynomial $g(y) = y^3$.

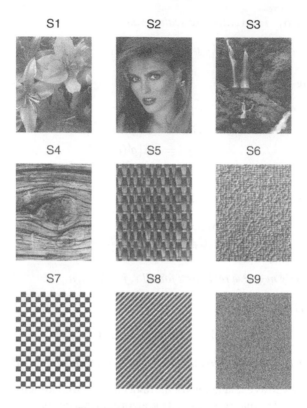

Fig. 12.3 The original images.

Example 12.3 The learning rule was applied to a signal separation problem in [235]. Consider the 9 digital images shown in Fig. 12.3. They were linearly mixed with a randomly chosen mixing matrix **A** into 9 mixture images, shown in Fig. 12.4.

Whitening, shown in Fig. 12.5, is not able to separate the images. When the learning rule (12.41) was applied to the mixtures with a tanh nonlinearity and the matrix **W** was allowed to converge, it was able to separate the images as shown in Fig. 12.6. In the figure, the images have been scaled to fit the gray levels in use; in some cases, the sign has also been reversed to avoid image negatives.

12.8.3   The nonlinear recursive least-squares learning rule

It is also possible to effectively minimize the prewhitened NLPCA criterion (12.27) using approximative recursive least-squares (RLS) techniques. Generally, RLS algorithms converge clearly faster than their stochastic gradient counterparts, and achieve a good final accuracy at the expense of a somewhat higher computational load. These

M1   M2   M3

M4   M5   M6

M7   M8   M9

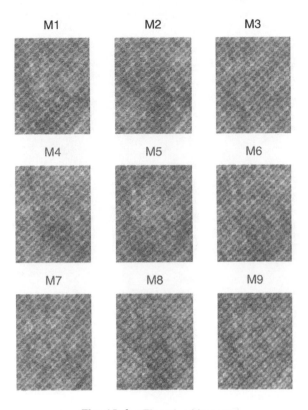

Fig. 12.4 The mixed images.

advantages are the result of the automatic determination of the learning rate parameter from the input data, so that it becomes roughly optimal.

The basic symmetric algorithm for the prewhitened problem was derived by one of the authors [347]. This is a nonlinear modification of the PAST algorithm introduced by Yang for the standard linear PCA [466, 467]; the PAST algorithm is covered in Chapter 6. Using index t to denote the iteration step, the algorithm is

$$\mathbf{q}(t) = \mathbf{g}(\mathbf{W}(t-1)\mathbf{z}(t)) = \mathbf{g}(\mathbf{y}(t)) \tag{12.58}$$

$$\mathbf{h}(t) = \mathbf{P}(t-1)\mathbf{q}(t) \tag{12.59}$$

$$\mathbf{m}(t) = \mathbf{h}(t)/[\beta + \mathbf{q}^{T}(t)\mathbf{h}(t)] \tag{12.60}$$

$$\mathbf{P}(t) = \frac{1}{\beta}\mathrm{Tri}[\mathbf{P}(t-1) - \mathbf{m}(t)\mathbf{h}^{T}(t)] \tag{12.61}$$

$$\mathbf{r}(t) = \mathbf{z}(t) - \mathbf{W}^{T}(t-1)\mathbf{q}(t) \tag{12.62}$$

$$\mathbf{W}(t) = \mathbf{W}(t-1) + \mathbf{m}(t)\mathbf{r}^{T}(t). \tag{12.63}$$

The vector variables $\mathbf{q}(t), \mathbf{h}(t), \mathbf{m}(t)$, and $\mathbf{r}(t)$ and the matrix variable $\mathbf{P}(t)$ are auxiliary variables, internal to the algorithm. As before, $\mathbf{z}(t)$ is the whitened input

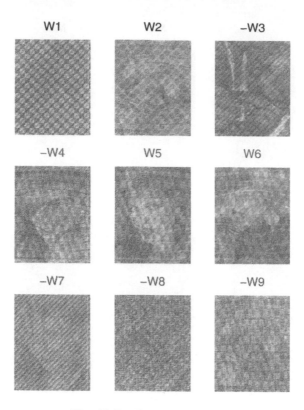

Fig. 12.5 The whitened images.

vector, $\mathbf{y}(t)$ is the output vector, $\mathbf{W}(t)$ is the weight matrix, and $\mathbf{g}$ is the nonlinearity in the NLPCA criterion. The parameter β is a kind of "forgetting constant" that should be close to unity. The notation Tri means that only the upper triangular part of the matrix is computed and its transpose is copied to the lower triangular part, making the resulting matrix symmetric. The initial values $\mathbf{W}(0)$ and $\mathbf{P}(0)$ can be chosen as identity matrices.

This algorithm updates the whole weight matrix $\mathbf{W}(t)$ simultaneously, treating all the rows of $\mathbf{W}(t)$ in a symmetric way. Alternatively, it is possible to compute the weight vectors $\mathbf{w}_i(t)$ in a sequential manner using a deflation technique. The sequential algorithm is presented in [236]. The authors show there experimentally that the recursive least-squares algorithms perform better and have faster convergence than stochastic gradient algorithms like the nonlinear subspace learning rule. Yet, the recursive algorithms are adaptive and can be used for tracking if the statistics of the data or the mixing model are slowly varying. They seem to be robust to initial values and have relatively low computational load. Also batch versions of the recursive algorithm are derived in [236].

Fig. 12.6   The separated images using the nonlinear PCA criterion and learning rule.

12.9   CONCLUDING REMARKS AND REFERENCES

The first part of this chapter reviewed some of the early research efforts in ICA, especially the technique based on nonlinear decorrelations. It was based on the work of Jutten, Hérault, and Ans [178, 179, 16]. A good overview is [227]. The exact relation between the nonlinear decorrelation criterion and independence was analyzed in the series of papers [228, 93, 408]. The Cichocki-Unbehauen algorithm was introduced in [82, 85, 84]; see also [83]. For estimating functions, the reference is [8]. The EASI algorithm was derived in [71]. The efficiency of estimating functions can in fact be extended to the notion of superefficiency [7].

Somewhat related methods specialized for discrete-valued ICs were proposed in [286, 379].

The review of nonlinear PCA is based on the authors' original works [332, 232, 233, 450, 235, 331, 327, 328, 347, 236]. A good review is also [149]. Nonlinear PCA is a versatile and useful starting point for blind signal processing. It has close connections to other well-known ICA approaches, as was shown in this chapter; see [236, 329]. It is unique in the sense that it is based on a least mean-square error formulation of the ICA problem. Due to this, recursive least mean-square algorithms can be derived; several versions like the symmetric, sequential, and batch algorithms are given in [236].

Problems

12.1   In the Hérault-Jutten algorithm (12.9,12.10), let $f(y_1) = y_1^3$ and $f(y_2) = y_2$. Write the update equations so that only x_1, x_2, m_{12}, and m_{21} appear on the right-hand side.

12.2   Consider the cost function (12.29). Assuming $\mathbf{y} = \mathbf{W}\mathbf{z}$, compute the matrix gradient of this cost function with respect to $\mathbf{W}$. Show that, except for the diagonal elements, the matrix gradient is an estimating function, i.e., its off-diagonal elements become zero when $\mathbf{W}$ is a true separating matrix for which $\mathbf{W}\mathbf{z} = \mathbf{s}$. What are the diagonal elements?

12.3   Repeat the previous problem for the maximum likelihood cost function (12.32).

12.4   Consider the stationary points of (12.46). Show that the diagonal matrix (12.54) is a stationary point if (12.48) holds.

12.5   * In Theorem 12.1, let the nonlinearity be a simple polynomial: $g(y) = y^q$ with q an odd positive integer. Assume for simplicity that all the sources s_i have the same density, so the subscript i can be dropped in the Theorem.

   12.5.1. Solve α from Eq. (12.48).

   12.5.2. Show that the stability conditions reduce to Eq. (12.56).

   12.5.3. Show that the linear function $g(y) = y$ does not fulfill the stability condition.

12.6   Consider the nonlinear subspace learning rule for whitened inputs, Eq. (12.41). Let us combine this rule with the whitening rule (12.21) in the same way as was done to derive the EASI algorithm (12.23): writing $\mathbf{B} = \mathbf{W}\mathbf{V}$ and $\Delta\mathbf{B} = \Delta\mathbf{W}\mathbf{V} + \mathbf{W}\Delta\mathbf{V}$. Like in the EASI derivation, assume that $\mathbf{W}$ is approximately orthogonal. Show that we get the new learning rule

$$\Delta\mathbf{B} = \mu[\mathbf{g}(\mathbf{y})\mathbf{y}^T - \mathbf{g}(\mathbf{y})\mathbf{g}(\mathbf{y}^T) + \mathbf{I} - \mathbf{y}\mathbf{y}^T].$$

13

Practical Considerations

In the preceding chapters, we presented several approaches for the estimation of the independent component analysis (ICA) model. In particular, several algorithms were proposed for the estimation of the basic version of the model, which has a square mixing matrix and no noise. Now we are, in principle, ready to apply those algorithms on real data sets. Many such applications will be discussed in Part IV.

However, when applying the ICA algorithms to real data, some practical considerations arise and need to be taken into account. In this chapter, we discuss different problems that may arise, in particular, overlearning and noise in the data. We also propose some preprocessing techniques (dimension reduction by principal component analysis, time filtering) that may be useful and even necessary before the application of the ICA algorithms in practice.

13.1 PREPROCESSING BY TIME FILTERING

The success of ICA for a given data set may depend crucially on performing some application-dependent preprocessing steps. In the basic methods discussed in the previous chapters, we always used centering in preprocessing, and often whitening was done as well. Here we discuss further preprocessing methods that are not necessary in theory, but are often very useful in practice.

13.1.1 Why time filtering is possible

In many cases, the observed random variables are, in fact, time signals or time series, which means that they describe the time course of some phenomenon or system. Thus the sample index t in $x_i(t)$ is a time index. In such a case, it may be very useful to filter the signals. In other words, this means taking moving averages of the time series. Of course, in the ICA model no time structure is assumed, so filtering is not always possible: If the sample points $\mathbf{x}(t)$ cannot be ordered in any meaningful way with respect to t, filtering is not meaningful, either.

For time series, any linear filtering of the signals is allowed, since it does not change the ICA model. In fact, if we filter linearly the observed signals $x_i(t)$ to obtain new signals, say $x_i^*(t)$, the ICA model still holds for $x_i^*(t)$, with the same mixing matrix. This can be seen as follows. Denote by $\mathbf{X}$ the matrix that contains the observations $\mathbf{x}(1), ..., \mathbf{x}(T)$ as its columns, and similarly for $\mathbf{S}$. Then the ICA model can be expressed as:

$$\mathbf{X} = \mathbf{AS} \tag{13.1}$$

Now, time filtering of $\mathbf{X}$ corresponds to multiplying $\mathbf{X}$ *from the right* by a matrix, let us call it $\mathbf{M}$. This gives

$$\mathbf{X}^* = \mathbf{XM} = \mathbf{ASM} = \mathbf{AS}^* \tag{13.2}$$

which shows that the ICA model still remains valid. The independent components are filtered by the same filtering that was applied on the mixtures. They are not mixed with each other in $\mathbf{S}^*$ because the matrix $\mathbf{M}$ is by definition a component-wise filtering matrix.

Since the mixing matrix remains unchanged, we can use the filtered data in the ICA estimating method only. After estimating the mixing matrix, we can apply the same mixing matrix on the original data to obtain the independent components.

The question then arises what kind of filtering could be useful. In the following, we consider three different kinds of filtering: high-pass and low-pass filtering, as well as their compromise.

13.1.2 Low-pass filtering

Basically, low-pass filtering means that every sample point is replaced by a weighted average of that point and the points immediately before it.[1] This is a form of *smoothing* the data. Then the matrix $\mathbf{M}$ in (13.2) would be something like

$$
\mathbf{M} = \frac{1}{3}
\begin{pmatrix}
& & & & \vdots & & & & \\
\ldots 1 & 1 & 1 & 0 & 0 & 0 & 0 & 0 \ldots \\
\ldots 0 & 1 & 1 & 1 & 0 & 0 & 0 & 0 \ldots \\
\ldots 0 & 0 & 1 & 1 & 1 & 0 & 0 & 0 \ldots \\
\ldots 0 & 0 & 0 & 1 & 1 & 1 & 0 & 0 \ldots \\
\ldots 0 & 0 & 0 & 0 & 1 & 1 & 1 & 0 \ldots \\
\ldots 0 & 0 & 0 & 0 & 0 & 1 & 1 & 1 \ldots \\
& & & & \vdots & & & &
\end{pmatrix}
\tag{13.3}
$$

Low-pass filtering is often used because it tends to reduce noise. This is a well-known property in signal processing that is explained in most basic signal processing textbooks.

In the basic ICA model, the effect of noise is more or less neglected; see Chapter 15 for a detailed discussion. Thus basic ICA methods work much better with data that does not have much noise, and reducing noise is thus useful and sometimes even necessary.

A possible problem with low-pass filtering is that it reduces the information in the data, since the fast-changing, high-frequency features of the data are lost. It often happens that this leads to a reduction of independence as well (see next section).

13.1.3 High-pass filtering and innovations

High-pass filtering is the opposite of low-pass filtering. The point is to remove slowly changing trends from the data. Thus a low-pass filtered version is subtracted from the signal. A classic way of doing high-pass filtering is differencing, which means replacing every sample point by the difference between the value at that point and the value at the preceding point. Thus, the matrix $\mathbf{M}$ in (13.2) would be

$$
\mathbf{M} =
\begin{pmatrix}
& & & & \vdots & & & \\
\ldots 1 & -1 & 0 & 0 & 0 & 0 & 0 \ldots \\
\ldots 0 & 1 & -1 & 0 & 0 & 0 & 0 \ldots \\
\ldots 0 & 0 & 1 & -1 & 0 & 0 & 0 \ldots \\
\ldots 0 & 0 & 0 & 1 & -1 & 0 & 0 \ldots \\
\ldots 0 & 0 & 0 & 0 & 1 & -1 & 0 \ldots \\
\ldots 0 & 0 & 0 & 0 & 0 & 1 & 1 \ldots \\
& & & & \vdots & & &
\end{pmatrix}
\tag{13.4}
$$

---

[1] To have a causal filter, points after the current point may be left out of the averaging.

High-pass filtering may be useful in ICA because in certain cases it increases the independence of the components. It often happens in practice that the components have slowly changing trends or fluctuations, in which case they are not very independent. If these slow fluctuations are removed by high-pass filtering the filtered components are often much more independent. A more principled approach to high-pass filtering is to consider it in the light of innovation processes.

Innovation processes   Given a stochastic process $s(t)$, we define its innovation process $\tilde{s}(t)$ as the error of the best prediction of $s(t)$, given its past. Such a best prediction is given by the conditional expectation of $s(t)$ given its past, because it is the expected value of the conditional distribution of $s(t)$ given its past. Thus the innovation process of $\tilde{s}(t)$ is defined by

$$\tilde{s}(t) = s(t) - E\{s(t)|s(t-1), s(t-2), ...\} \tag{13.5}$$

The expression "innovation" describes the fact that $\tilde{s}(t)$ contains all the new information about the process that can be obtained at time t by observing $s(t)$.

The concept of innovations can be utilized in the estimation of the ICA model due to the following property:

Theorem 13.1 *If* $x(t)$ *and* $s(t)$ *follow the basic ICA model, then the innovation processes* $\tilde{x}(t)$ *and* $\tilde{s}(t)$ *follow the ICA model as well. In particular, the components* $\tilde{s}_i(t)$ *are independent from each other.*

On the other hand, independence of the innovations does *not* imply the independence of the $s_i(t)$. Thus, the *innovations are more often independent* from each other than the original processes. Moreover, one could argue that the innovations are usually *more nongaussian* than the original processes. This is because the $s_i(t)$ is a kind of moving average of the innovation process, and sums tend to be more gaussian than the original variable. Together these mean that the innovation process is more susceptible to be independent and nongaussian, and thus to fulfill the basic assumptions in ICA.

Innovation processes were discussed in more detail in [194], where it was also shown that using innovations, it is possible to separate signals (images of faces) that are otherwise strongly correlated and very difficult to separate.

The connection between innovations and ordinary filtering techniques is that the computation of the innovation process is often rather similar to high-pass filtering. Thus, the arguments in favor of using innovation processes apply at least partly in favor of high-pass filtering.

A possible problem with high-pass filtering, however, is that it may increase noise for the same reasons that low-pass filtering decreases noise.

13.1.4 Optimal filtering

Both of the preceding types of filtering have their pros and cons. The optimum would be to find a filter that increases the independence of the components while reducing noise. To achieve this, some compromise between high- and low-pass filtering may be the best solution. This leads to band-pass filtering, in which the highest and the lowest frequencies are filtered out, leaving a suitable frequency band in between. What this band should be depends on the data and general answers are impossible to give.

In addition to simple low-pass/high-pass filtering, one might also use more so-phisticated techniques. For example, one might take the (1-D) wavelet transforms of the data [102, 290, 17]. Other time-frequency decompositions could be used as well.

13.2 PREPROCESSING BY PCA

A common preprocessing technique for multidimensional data is to reduce its dimen-sion by principal component analysis (PCA). PCA was explained in more detail in Chapter 6. Basically, the data is projected linearly onto a subspace

$$\tilde{\mathbf{x}} = \mathbf{E}_n \mathbf{x} \tag{13.6}$$

so that the maximum amount of information (in the least-squares sense) is preserved. Reducing dimension in this way has several benefits which we discuss in the next subsections.

13.2.1 Making the mixing matrix square

First, let us consider the case where the the number of independent components n is smaller than the number of mixtures, say m. Performing ICA on the mixtures directly can cause big problems in such a case, since the basic ICA model does not hold anymore. Using PCA we can reduce the dimension of the data to n. After such a reduction, the number of mixtures and ICs are equal, the mixing matrix is square, and the basic ICA model holds.

The question is whether PCA is able to find the subspace correctly, so that the n ICs can be estimated from the reduced mixtures. This is not true in general, but in a special case it turns out to be the case. If the data consists of n ICs only, with no noise added, the whole data is contained in an n-dimensional subspace. Using PCA for dimension reduction clearly finds this n-dimensional subspace, since the eigenvalues corresponding to that subspace, and only those eigenvalues, are nonzero. Thus reducing dimension with PCA works correctly. In practice, the data is usually not exactly contained in the subspace, due to noise and other factors, but if the noise level is low, PCA still finds approximately the right subspace; see Section 6.1.3. In the general case, some "weak" ICs may be lost in the dimension reduction process, but PCA may still be a good idea for optimal estimation of the "strong" ICs [313].

Performing first PCA and then ICA has an interesting interpretation in terms of factor analysis. In factor analysis, it is conventional that after finding the factor subspace, the actual basis vectors for that subspace are determined by some criteria that make the mixing matrix as simple as possible [166]. This is called *factor rotation*. Now, ICA can be interpreted as one method for determining this factor rotation, based on higher-order statistics instead of the structure of the mixing matrix.

13.2.2 Reducing noise and preventing overlearning

A well-known benefit of reducing the dimension of the data is that it reduces noise, as was already discussed in Chapter 6. Often, the dimensions that have been omitted consist mainly of noise. This is especially true in the case where the number of ICs is smaller than the number of mixtures.

Another benefit of reducing dimensions is that it prevents overlearning, to which the rest of this subsection is devoted. Overlearning means that if the number of parameters in a statistical model is too large when compared to the number of available data points, the estimation of the parameters becomes difficult, maybe impossible. The estimation of the parameters is then too much determined by the available sample points, instead of the actual process that generated the data, which is what we are really interested in.

Overlearning in ICA [214] typically produces estimates of the ICs that have a single spike or bump, and are practically zero everywhere else. This is because in the space of source signals of unit variance, nongaussianity is more or less maximized by such spike/bump signals. This becomes easily comprehensible if we consider the extreme case where the sample size T equals the dimension of the data m, and these are both equal to the number of independent components n. Let us collect the realizations $\mathbf{x}(t)$ of $\mathbf{x}$ as the columns of the matrix $\mathbf{X}$, and denote by $\mathbf{S}$ the corresponding matrix of the realizations of $\mathbf{s}(t)$, as in (13.1). Note that now all the matrices in (13.1) are square. This means that by changing the values of $\mathbf{A}$ (and keeping $\mathbf{X}$ fixed), we can give any values whatsoever to the elements of $\mathbf{S}$. This is a case of serious overlearning, not unlike the classic case of regression with equal numbers of data points and parameters.

Thus it is clear that in this case, the estimate of $\mathbf{S}$ that is obtained by ICA estimation depends little on the observed data. Let us assume that the densities of the source signals are known to be supergaussian (i.e., positively kurtotic). Then the ICA estimation basically consists of finding a separating matrix $\mathbf{B}$ that maximizes a measure of the supergaussianities (or sparsities) of the estimates of the source signals. Intuitively, it is easy to see that sparsity is maximized when the source signals each have only one nonzero point. Thus we see that ICA estimation with an insufficient sample size leads to a form of overlearning that gives artifactual (spurious) source signals. Such source signals are characterized by large *spikes*.

An important fact shown experimentally [214] is that a similar phenomenon is much more likely to occur if the source signals are not independently and identically distributed (i.i.d.) in time, but have strong time-dependencies. In such cases the sample size needed to get rid of overlearning is much larger, and the source signals are better characterized by *bumps*, i.e., low-pass filtered versions of spikes. An intuitive way of explaining this phenomenon is to consider such a signal as being constant on N/k blocks of k consecutive sample points. This means that the data can be considered as really having only N/k sample points; each sample point has simply been repeated k times. Thus, in the case of overlearning, the estimation procedure gives "spikes" that have a width of k time points, i.e., bumps.

Here we illustrate the phenomenon by separation of artificial source signals. Three positively kurtotic signals, with 500 sample points each, were used in these simulations, and are depicted in Fig. 13.1 *a*. Five hundred mixtures were produced, and a very small amount of gaussian noise was added to each mixture separately.

As an example of a successful ICA estimation, Fig. 13.1 *b* shows the result of applying the FastICA and maximum likelihood (ML) gradient ascent algorithms (denoted by "Bell-Sejnowski") to the mixed signals. In both approaches, the preprocessing (whitening) stage included a dimension reduction of the data into the first three principal components. It is evident that both algorithms are able to extract all the initial signals.

In contrast, when the whitening is made with very small dimension reduction (we took 400 dimensions), we see the emergence of spiky solutions (like Dirac functions), which is an extreme case of kurtosis maximization (Fig. 13.1 *c*). The algorithm used in FastICA was of a deflationary type, from which we plot the first five components extracted. As for the ML gradient ascent, which was of a symmetric type, we show five representative solutions to the 400 extracted.

Thus, we see here that without dimension reduction, we are not able to estimate the source signals.

Fig. 13.1 *d* presents an intermediate stage of dimension reduction (from the original 500 mixtures we took 50 whitened vectors). We see that the actual source signals are revealed by both methods, even though each resulting vector is more noisy than the ones shown in Fig. 13.1 *b*.

For the final example, in Fig. 13.1 *e*, we low-pass filtered the mixed signals, prior to the independent component analysis, using a 10 delay moving average filter. Taking the same amount of principal components as in *d*, we can see that we lose all the original source signals: the decompositions show a bumpy structure corresponding to the low-pass filtering of the spiky outputs presented in *c*. Through low-pass filtering, we have reduced the information contained in the data, and so the estimation is rendered impossible even with this, not very weak, dimension reduction. Thus, we see that with this low-pass filtered data, a much stronger dimension reduction by PCA is necessary to prevent overlearning.

In addition to PCA, some kind of prior information on the mixing matrix could be useful in preventing overlearning. This is considered in detail in Section 20.1.3.

13.3 HOW MANY COMPONENTS SHOULD BE ESTIMATED?

Another problem that often arises in practice is to decide the number of ICs to be estimated. This problem does not arise if one simply estimates the same number of components as the dimension of the data. This may not always be a good idea, however.

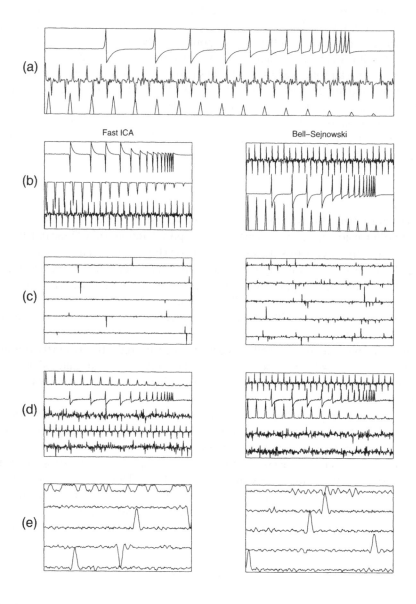

Fig. 13.1 (From [214]) Illustration of the importance of the degree of dimension reduction and filtering in artificially generated data, using FastICA and a gradient algorithm for ML estimation. (*a*) Original positively kurtotic signals. (*b*) ICA decomposition in which the preprocessing includes a dimension reduction to the first 3 principal components. (*c*) Poor, i.e., too weak dimension reduction. (*d*) Decomposition using an intermediate dimension reduction (50 components retained). (*e*) Same results as in (d) but using low-pass filtered mixtures

First, since dimension reduction by PCA is often necessary, one must choose the number of principal components to be retained. This is a classic problem; see Chapter 6. It is usually solved by choosing the minimum number of principal components that explain the data well enough, containing, for example, 90% of the variance. Often, the dimension is actually chosen by trial and error with no theoretical guidelines.

Second, for computational reasons we may prefer to estimate only a smaller number of ICs than the dimension of the data (after PCA preprocessing). This is the case when the dimension of the data is very large, and we do not want to reduce the dimension by PCA too much, since PCA always contains the risk of not including the ICs in the reduced data. Using FastICA and other algorithms that allow estimation of a smaller number of components, we can thus perform a kind of dimension reduction by ICA. In fact, this is an idea somewhat similar to projection pursuit. Here, it is even more difficult to give any guidelines as to how many components should be estimated. Trial and error may be the only method applicable.

Information-theoretic, Bayesian, and other criteria for determining the number of ICs are discussed in more detail in [231, 81, 385].

13.4 CHOICE OF ALGORITHM

Now we shall briefly discuss the choice of ICA algorithm from a practical viewpoint. As will be discussed in detail in Chapter 14, most estimation principles and objective functions for ICA are equivalent, at least in theory. So, the main choice is reduced to a couple of points:

- One choice is between estimating all the independent components in parallel, or just estimating a few of them (possibly one-by-one). This corresponds to choosing between symmetric and hierarchical decorrelation. In most cases, symmetric decorrelation is recommended. Deflation is mainly useful in the case where we want to estimate only a very limited number of ICs, and other special cases. The disadvantage with deflationary orthogonalization is that the estimation errors in the components that are estimated first accumulate and increase the errors in the later components.

- One must also choose the nonlinearity used in the algorithms. It seems that the robust, nonpolynomial nonlinearities are to be preferred in most applications. The simplest thing to do is to just use the tanh function as the nonlinearity g. This is sufficient when using FastICA. (When using gradient algorithms, especially in the ML framework, a second function needs to be used as well; see Chapter 9.)

- Finally, there is the choice between on-line and batch algorithms. In most cases, the whole data set is available before the estimation, which is called in different contexts batch, block, or off-line estimation. This is the case where FastICA can be used, and it is the algorithm that we recommend. On-line or adaptive algorithms are needed in signal-processing applications where the mixing matrix may change on-line, and fast tracking is needed. In the on-line case, the recommended algorithms are those obtained by stochastic gradient methods. It should also be noted that in some cases, the FastICA algorithm may not converge well as Newton-type algorithms sometimes exhibit oscillatory behavior. This problem can be alleviated by using gradient methods, or combinations of the two (see [197]).

13.5 CONCLUDING REMARKS AND REFERENCES

In this chapter, we considered some practical problems in ICA. When dealing with time signals, low-pass filtering of the data is useful to reduce noise. On the other hand, high-pass filtering, or computing innovation processes is useful to increase the independence and nongaussianity of the components. One of these, or their combination may be very useful in practice. Another very useful thing to do is to reduce the dimension of the data by PCA. This reduces noise and prevents overlearning. It may also solve the problems with data that has a smaller number of ICs than mixtures.

Problems

13.1 Take a Fourier transform on every observed signal $x_i(t)$. Does the ICA model still hold, and in what way?

13.2 Prove the theorem on innovations.

Computer assignments

13.1 Take a gaussian white noise sequence. Low-pass filter it by a low-pass filter with coefficients (...,0,0,1,1,1,1,1,0,0,0,...). What does the signal look like?

13.2 High-pass filter the gaussian white noise sequence. What does the signal look like?

13.3 Generate 100 samples of 100 independent components. Run FastICA on this data without any mixing. What do the estimated ICs look like? Is the estimate of the mixing matrix close to identity?

14

Overview and Comparison of Basic ICA Methods

In the preceding chapters, we introduced several different estimation principles and algorithms for independent component analysis (ICA). In this chapter, we provide an overview of these methods. First, we show that all these estimation principles are intimately connected, and the main choices are between cumulant-based vs. negentropy/likelihood-based estimation methods, and between one-unit vs. multi-unit methods. In other words, one must choose the nonlinearity and the decorrelation method. We discuss the choice of the nonlinearity from the viewpoint of statistical theory. In practice, one must also choose the optimization method. We compare the algorithms experimentally, and show that the main choice here is between on-line (adaptive) gradient algorithms vs. fast batch fixed-point algorithms.

At the end of this chapter, we provide a short summary of the whole of Part II, that is, of basic ICA estimation.

14.1 OBJECTIVE FUNCTIONS VS. ALGORITHMS

A distinction that has been used throughout this book is between the formulation of the objective function, and the algorithm used to optimize it. One might express this in the following "equation":

$$\text{ICA method} = \text{objective function} + \text{optimization algorithm.}$$

In the case of explicitly formulated objective functions, one can use any of the classic optimization methods, for example, (stochastic) gradient methods and Newton

273

methods. In some cases, however, the algorithm and the estimation principle may be difficult to separate.

The properties of the ICA method depend on both of the objective function and the optimization algorithm. In particular:

- the statistical properties (e.g., consistency, asymptotic variance, robustness) of the ICA method depend on the choice of the objective function,

- the algorithmic properties (e.g., convergence speed, memory requirements, numerical stability) depend on the optimization algorithm.

Ideally, these two classes of properties are independent in the sense that different optimization methods can be used to optimize a single objective function, and a single optimization method can be used to optimize different objective functions. In this section, we shall first treat the choice of the objective function, and then consider optimization of the objective function.

14.2 CONNECTIONS BETWEEN ICA ESTIMATION PRINCIPLES

Earlier, we introduced several different statistical criteria for estimation of the ICA model, including mutual information, likelihood, nongaussianity measures, cumulants, and nonlinear principal component analysis (PCA) criteria. Each of these criteria gave an objective function whose optimization enables ICA estimation. We have already seen that some of them are closely connected; the purpose of this section is to recapitulate these results. In fact, almost all of these estimation principles can be considered as different versions of the same general criterion. After this, we discuss the differences between the principles.

14.2.1 Similarities between estimation principles

Mutual information gives a convenient starting point for showing the similarity between different estimation principles. We have for an invertible linear transformation $\mathbf{y} = \mathbf{B}\mathbf{x}$:

$$I(y_1, y_2, ..., y_n) = \sum_i H(y_i) - H(\mathbf{x}) - \log|\det \mathbf{B}| \tag{14.1}$$

If we constrain the y_i to be uncorrelated and of unit variance, the last term on the right-hand side is constant; the second term does not depend on $\mathbf{B}$ anyway (see Chapter 10). Recall that entropy is maximized by a gaussian distribution, when variance is kept constant (Section 5.3). Thus we see that *minimization of mutual information* means *maximizing the sum of the nongaussianities* of the estimated components. If these entropies (or the corresponding negentropies) are approximated by the approximations used in Chapter 8, we obtain the same algorithms as in that chapter.

Alternatively, we could approximate mutual information by approximating the densities of the estimated ICs by some parametric family, and using the obtained log-density approximations in the definition of entropy. Thus we obtain a method that is essentially equivalent to *maximum likelihood (ML) estimation*.

The connections to other estimation principles can easily be seen using likelihood. First of all, to see the connection to *nonlinear decorrelation*, it is enough to compare the natural gradient methods for ML estimation shown in (9.17) with the nonlinear decorrelation algorithm (12.11): they are of the same form. Thus, ML estimation gives a principled method for choosing the nonlinearities in nonlinear decorrelation. The nonlinearities used are determined as certain functions of the probability density functions (pdf's) of the independent components. Mutual information does the same thing, of course, due to the equivalency discussed earlier. Likewise, the nonlinear PCA methods were shown to be essentially equivalent to ML estimation (and, therefore, most other methods) in Section 12.7.

The connection of the preceding principles to cumulant-based criteria can be seen by considering the approximation of negentropy by cumulants as in Eq. (5.35):

$$J(y) \approx \frac{1}{12}E\{y^3\}^2 + \frac{1}{48}\text{kurt}(y)^2 \qquad (14.2)$$

where the first term could be omitted, leaving just the term containing kurtosis. Likewise, cumulants could be used to approximate mutual information, since mutual information is based on entropy. More explicitly, we could consider the following approximation of mutual information:

$$I(\mathbf{y}) \approx c_1 - c_2 \sum_i \text{kurt}(y_i)^2 \qquad (14.3)$$

where c_1 and c_2 are some constants. This shows clearly the connection between cumulants and minimization of mutual information. Moreover, the tensorial methods in Chapter 11 were seen to lead to the same fixed-point algorithm as the maximization of nongaussianity as measured by kurtosis, which shows that they are doing very much the same thing as the other kurtosis-based methods.

14.2.2   Differences between estimation principles

There are, however, a couple of differences between the estimation principles as well.

1. Some principles (especially maximum nongaussianity) are able to estimate single independent components, whereas others need to estimate all the components at the same time.

2. Some objective functions use nonpolynomial functions based on the (assumed) probability density functions of the independent components, whereas others use polynomial functions related to cumulants. This leads to different non-quadratic functions in the objective functions.

3. In many estimation principles, the estimates of the ICs are constrained to be uncorrelated. This reduces somewhat the space in which the estimation is

performed. Considering, for example, mutual information, there is no reason why mutual information would be exactly minimized by a decomposition that gives uncorrelated components. Thus, this decorrelation constraint slightly reduces the theoretical performance of the estimation methods. In practice, this may be negligible.

4. One important difference in practice is that often in ML estimation, the densities of the ICs are fixed in advance, using prior knowledge on the independent components. This is possible because the pdf's of the ICs need not be known with any great precision: in fact, it is enough to estimate whether they are sub- or supergaussian. Nevertheless, if the prior information on the nature of the independent components is not correct, ML estimation will give completely wrong results, as was shown in Chapter 9. Some care must be taken with ML estimation, therefore. In contrast, using approximations of negentropy, this problem does not usually arise, since the approximations we have used in this book do not depend on reasonable approximations of the densities. Therefore, these approximations are less problematic to use.

14.3 STATISTICALLY OPTIMAL NONLINEARITIES

Thus, from a statistical viewpoint, the choice of estimation method is more or less reduced to the choice of the nonquadratic function G that gives information on the higher-order statistics in the form of the expectation $E\{G(\mathbf{b}_i^T \mathbf{x})\}$. In the algorithms, this choice corresponds to the choice of the nonlinearity g that is the derivative of G. In this section, we analyze the statistical properties of different nonlinearities. This is based on the family of approximations of negentropy given in (8.25). This family includes kurtosis as well. For simplicity, we consider here the estimation of just one IC, given by maximizing this nongaussianity measure. This is essentially equivalent to the problem

$$\max_{E\{(\mathbf{b}^T \mathbf{x})^2\}=1} E\{\pm G(\mathbf{b}^T \mathbf{x})\} \tag{14.4}$$

where the sign of G depends of the estimate on the sub- or supergaussianity of $\mathbf{b}^T \mathbf{x}$. The obtained vector is denoted by $\widehat{\mathbf{b}}$. The two fundamental statistical properties of $\widehat{\mathbf{b}}$ that we analyze are asymptotic variance and robustness.

14.3.1 Comparison of asymptotic variance *

In practice, one usually has only a finite sample of T observations of the vector $\mathbf{x}$. Therefore, the expectations in the theoretical definition of the objective function are in fact replaced by sample averages. This results in certain errors in the estimator $\widehat{\mathbf{b}}$, and it is desired to make these errors as small as possible. A classic measure of this error is asymptotic (co)variance, which means the limit of the covariance matrix of $\widehat{\mathbf{b}}\sqrt{T}$ as $T \to \infty$. This gives an approximation of the mean-square error of $\widehat{\mathbf{b}}$, as was already discussed in Chapter 4. Comparison of, say, the traces of the asymptotic variances of two estimators enables direct comparison of the accuracy of two estimators. One can solve analytically for the asymptotic variance of $\widehat{\mathbf{b}}$, obtaining the following theorem [193]:

Theorem 14.1 *The trace of the asymptotic variance of* $\widehat{\mathbf{b}}$ *as defined above for the estimation of the independent component* s_i, *equals*

$$V_G = C(\mathbf{A})\frac{E\{g^2(s_i)\} - (E\{s_i g(s_i)\})^2}{(E\{s_i g(s_i) - g'(s_i)\})^2}, \qquad (14.5)$$

where g is the derivative of G, and $C(\mathbf{A})$ is a constant that depends only on $\mathbf{A}$.

The theorem is proven at the appendix of this chapter.

Thus the comparison of the asymptotic variances of two estimators for two different nonquadratic functions G boils down to a comparison of the V_G. In particular, one can use variational calculus to find a G that minimizes V_G. Thus one obtains the following theorem [193]:

Theorem 14.2 *The trace of the asymptotic variance of* $\widehat{\mathbf{b}}$ *is minimized when G is of the form*

$$G_{opt}(y) = c_1 \log p_i(y) + c_2 y^2 + c_3 \qquad (14.6)$$

where p_i is the density function of s_i, and c_1, c_2, c_3 are arbitrary constants.

For simplicity, one can choose $G_{opt}(y) = \log p_i(y)$. Thus, we see that the optimal nonlinearity is in fact the one used in the definition of negentropy. This shows that *negentropy is the optimal measure of nongaussianity*, at least inside those measures that lead to estimators of the form considered here.[1] Also, one sees that the optimal function is the same as the one obtained for several units by the maximum likelihood approach.

14.3.2 Comparison of robustness *

Another very desirable property of an estimator is robustness against outliers. This means that single, highly erroneous observations do not have much influence on the estimator. In this section, we shall treat the question: How does the robustness of the estimator $\widehat{\mathbf{b}}$ depend on the choice of the function G? The main result is that the function $G(y)$ should not grow fast as a function of $|y|$ if we want robust estimators. In particular, this means that kurtosis gives nonrobust estimators, which may be very disadvantagous in some situations.

---

[1] One has to take into account, however, that in the definition of negentropy, the nonquadratic function is not fixed in advance, whereas in our nongaussianity measures, G *is* fixed. Thus, the statistical properties of negentropy can be only approximatively derived from our analysis.

First, note that the robustness of $\hat{\mathbf{b}}$ depends also on the method of estimation used in constraining the variance of $\hat{\mathbf{b}}^T \mathbf{x}$ to equal unity, or, equivalently, the whitening method. This is a problem independent of the choice of G. In the following, we assume that this constraint is implemented in a robust way. In particular, we assume that the data is sphered (whitened) in a robust manner, in which case the constraint reduces to $\|\hat{\mathbf{w}}\| = 1$, where $\mathbf{w}$ is the value of $\mathbf{b}$ for whitened data. Several robust estimators of the variance of $\hat{\mathbf{w}}^T \mathbf{z}$ or of the covariance matrix of $\mathbf{x}$ are presented in the literature; see reference [163].

The robustness of the estimator $\hat{\mathbf{w}}$ can be analyzed using the theory of M-estimators. Without going into technical details, the definition of an M-estimator can be formulated as follows: an estimator is called an M-estimator if it is defined as the solution $\hat{\theta}$ for θ of

$$E\{\psi(\mathbf{z}, \theta)\} = 0 \tag{14.7}$$

where $\mathbf{z}$ is a random vector and ψ is some function defining the estimator. Now, the point is that the estimator $\hat{\mathbf{w}}$ is an M-estimator. To see this, define $\theta = (\mathbf{w}, \lambda)$, where λ is the Lagrangian multiplier associated with the constraint. Using the Lagrange conditions, the estimator $\hat{\mathbf{w}}$ can then be formulated as the solution of Eq. (14.7) where ψ is defined as follows (for sphered data):

$$\psi(\mathbf{z}, \theta) = \left(\begin{array}{c} \mathbf{z}g(\mathbf{w}^T \mathbf{z}) + c\lambda \mathbf{w} \\ \|\mathbf{w}\|^2 - 1 \end{array} \right) \tag{14.8}$$

where $c = (E_{\mathbf{z}}\{G(\hat{\mathbf{w}}^T \mathbf{z})\} - E_{\nu}\{G(\nu)\})^{-1}$ is an irrelevant constant.

The analysis of robustness of an M-estimator is based on the concept of an influence function, $IF(\mathbf{z}, \hat{\theta})$. Intuitively speaking, the influence function measures the influence of single observations on the estimator. It would be desirable to have an influence function that is bounded as a function of $\mathbf{z}$, as this implies that even the influence of a far-away outlier is "bounded", and cannot change the estimate too much. This requirement leads to one definition of robustness, which is called B-robustness. An estimator is called B-robust, if its influence function is bounded as a function of $\mathbf{z}$, i.e., $\sup_{\mathbf{z}} \|IF(\mathbf{z}, \hat{\theta})\|$ is finite for every $\hat{\theta}$. Even if the influence function is not bounded, it should grow as slowly as possible when $\|\mathbf{z}\|$ grows, to reduce the distorting effect of outliers.

It can be shown that the influence function of an M-estimator equals

$$IF(\mathbf{z}, \hat{\theta}) = \mathbf{B}\psi(\mathbf{z}, \hat{\theta}) \tag{14.9}$$

where $\mathbf{B}$ is an irrelevant invertible matrix that does not depend on $\mathbf{z}$. On the other hand, using our definition of ψ, and denoting by $\gamma = \mathbf{w}^T \mathbf{z}/\|\mathbf{z}\|$ the cosine of the angle between $\mathbf{z}$ and $\mathbf{w}$, one obtains easily

$$\|\psi(\mathbf{z}, (\mathbf{w}, \lambda))\|^2 = C_1 \frac{1}{\gamma^2} h^2(\mathbf{w}^T \mathbf{z}) + C_2 h(\mathbf{w}^T \mathbf{z}) + C_3 \tag{14.10}$$

where C_1, C_2, C_3 are constants that do not depend on $\mathbf{z}$, and $h(y) = yg(y)$. Thus we see that the robustness of $\hat{\mathbf{w}}$ essentially depends on the behavior of the function $h(u)$.

The slower $h(u)$ grows, the more robust the estimator. However, the estimator really cannot be B-robust, because the γ in the denominator prevents the influence function from being bounded for all $\mathbf{z}$. In particular, outliers that are almost orthogonal to $\hat{\mathbf{w}}$, and have large norms, may still have a large influence on the estimator. These results are stated in the following theorem:

Theorem 14.3 *Assume that the data $\mathbf{z}$ is whitened (sphered) in a robust manner. Then the influence function of the estimator $\hat{\mathbf{w}}$ is never bounded for all $\mathbf{z}$. However, if $h(y) = yg(y)$ is bounded, the influence function is bounded in sets of the form $\{\mathbf{z} \mid \hat{\mathbf{w}}^T\mathbf{z}/\|\mathbf{z}\| > \epsilon\}$ for every $\epsilon > 0$, where g is the derivative of G.*

In particular, if one chooses *a function $G(y)$ that is bounded*, h is also bounded, and $\hat{\mathbf{w}}$ is quite robust against outliers. If this is not possible, one should at least choose a function $G(y)$ that does not grow very fast when $|y|$ grows. If, in contrast, $G(y)$ grows very fast when $|y|$ grows, the estimates depend mostly on a few observations far from the origin. This leads to highly nonrobust estimators, which can be completely ruined by just a couple of bad outliers. This is the case, for example, when kurtosis is used, which is equivalent to using $\hat{\mathbf{w}}$ with $G(y) = y^4$.

14.3.3 Practical choice of nonlinearity

It is useful to analyze the implications of the preceding theoretical results by considering the following family of density functions:

$$p_\alpha(s) = C_1 \exp(C_2|s|^\alpha) \tag{14.11}$$

where α is a positive constant, and C_1, C_2 are normalization constants that ensure that p_α is a probability density of unit variance. For different values of alpha, the densities in this family exhibit different shapes. For $0 < \alpha < 2$, one obtains a sparse, supergaussian density (i.e., a density of positive kurtosis). For $\alpha = 2$, one obtains the gaussian distribution, and for $\alpha > 2$, a subgaussian density (i.e., a density of negative kurtosis). Thus the densities in this family can be used as examples of different nongaussian densities.

Using Theorem 14.1, one sees that in terms of asymptotic variance, the optimal nonquadratic function is of the form:

$$G_{opt}(y) = |y|^\alpha \tag{14.12}$$

where the arbitrary constants have been dropped for simplicity. This implies roughly that for supergaussian (resp. subgaussian) densities, the optimal function is a function that grows *slower than quadratically* (resp. *faster than quadratically*). Next, recall from Section 14.3.2 that if $G(y)$ grows fast with $|y|$, the estimator becomes highly nonrobust against outliers. Also taking into account the fact that most ICs encountered in practice are supergaussian, one reaches the conclusion that as a general-purpose function, one should choose a function G that resembles rather

$$G_{opt}(u) = |y|^\alpha, \quad \text{where } \alpha < 2 \tag{14.13}$$

The problem with such functions is, however, that they are not differentiable at 0 for $\alpha \leq 1$. This can lead to problems in the numerical optimization. Thus it is better to use approximating differentiable functions that have the same kind of qualitative behavior. Considering $\alpha = 1$, in which case one has a Laplacian density, one could use instead the function $G_1(y) = \log \cosh a_1 y$ where a_1 is a constant. This is very similar to the so-called Huber function that is widely used in robust statistics as a robust alternative of the square function. Note that the derivative of G_1 is then the familiar tanh function (for $a_1 = 1$). We have found $1 \leq a_1 \leq 2$ to provide a good approximation. Note that there is a trade-off between the precision of the approximation and the smoothness of the resulting objective function.

In the case of $\alpha < 1$, i.e., highly supergaussian ICs, one could approximate the behavior of G_{opt} for large u using a gaussian function (with a minus sign): $G_2(y) = -\exp(-y^2/2)$. The derivative of this function is like a sigmoid for small values, but goes to 0 for larger values. Note that this function also fulfills the condition in Theorem 14.3, thus providing an estimator that is as robust as possible in this framework.

Thus, we reach the following general conclusions:

- A good general-purpose function is $G(y) = \log \cosh a_1 y$, where $1 \leq a_1 \leq 2$ is a constant.

- When the ICs are highly supergaussian, or when robustness is very important, $G(y) = -\exp(-y^2/2)$ may be better.

- Using kurtosis is well justified only if the ICs are subgaussian and there are no outliers.

In fact, these two nonpolynomial functions are those that we used in the nongaussianity measures in Chapter 8 as well, and illustrated in Fig. 8.20. The functions in Chapter 9 are also essentially the same, since addition of a linear function does not have much influence on the estimator. Thus, the analysis of this section justifies the use of the nonpolynomial functions that we used previously, and shows why caution should be taken when using kurtosis.

In this section, we have used purely statistical criteria for choosing the function G. One important criterion for comparing ICA methods that is completely independent of statistical considerations is the computational load. Since most of the objective functions are computationally very similar, the computational load is essentially a function of the optimization algorithm. The choice of the optimization algorithm will be considered in the next section.

14.4 EXPERIMENTAL COMPARISON OF ICA ALGORITHMS

The theoretical analysis of the preceding section gives some guidelines as to which nonlinearity (corresponding to a nonquadratic function G) should be chosen. In this section, we compare the ICA algorithms experimentally. Thus we are able to analyze the computational efficiency of the different algorithms as well. This is done by experiments, since a satisfactory theoretical analysis of convergence speed does not seem possible. We saw previously, though, that FastICA has quadratic or cubic convergence whereas gradient methods have only linear convergence, but this result is somewhat theoretical because it does not say anything about the global convergence. In the same experiments, we validate experimentally the earlier analysis of statistical performance in terms of asymptotic variance.

14.4.1 Experimental set-up and algorithms

Experimental setup In the following experimental comparisons, artificial data generated from known sources was used. This is quite necessary, because only then are the correct results known and a reliable comparison possible. The experimental setup was the same for each algorithm in order to make the comparison as fair as possible. We have also compared various ICA algorithms using real-world data in [147], where experiments with artificial data also are described in somewhat more detail. At the end of this section, conclusions from experiments with real-world data are presented.

The algorithms were compared along the two sets of criteria, statistical and computational, as was outlined in Section 14.1. The computational load was measured as flops (basic floating-point operations, such as additions or divisions) needed for convergence. The statistical performance, or accuracy, was measured using a performance index, defined as

$$
E_1 = \sum_{i=1}^{m} (\sum_{j=1}^{m} \frac{|p_{ij}|}{\max_k |p_{ik}|} - 1) + \sum_{j=1}^{m} (\sum_{i=1}^{m} \frac{|p_{ij}|}{\max_k |p_{kj}|} - 1)
\tag{14.14}
$$

where p_{ij} is the ijth element of the $m \times m$ matrix $\mathbf{P} = \mathbf{BA}$. If the ICs have been separated perfectly, $\mathbf{P}$ becomes a permutation matrix (where the elements may have different signs, though). A permutation matrix is defined so that on each of its rows and columns, only one of the elements is equal to unity while all the other elements are zero. Clearly, the index (14.14) attains its minimum value zero for an ideal permutation matrix. The larger the value E_1 is, the poorer the statistical performance of a separation algorithm. In certain experiments, another fairly similarly behaving performance index, E_2, was used. It differs slightly from E_1 in that squared values p_{ij}^2 are used instead of the absolute ones in (14.14).

ICA algorithms used The following algorithms were included in the comparison (their abbreviations are in parentheses):

- The FastICA fixed-point algorithm. This has three variations: using kurtosis with deflation (FP) or with symmetric orthogonalization (FPsym), and using the tanh nonlinearity with symmetric orthogonalization (FPsymth).

- Gradient algorithms for maximum likelihood estimation, using a fixed nonlinearity given by $-$ tanh. First, we have the ordinary gradient ascent algorithm, or the Bell-Sejnowski algorithm (BS). Second, we have the natural gradient algorithm proposed by Amari, Cichocki and Yang [12], which is abbreviated as ACY.

- Natural gradient MLE using an adaptive nonlinearity. (Abbreviated as ExtBS, since this is called the "extended Bell-Sejnowski" algorithm by some authors.) The nonlinearity was adapted using the sign of kurtosis as in reference [149], which is essentially equivalent to the density parameterization we used in Section 9.1.2.

- The EASI algorithm for nonlinear decorrelation, as discussed in Section 12.5. Again, the nonlinearity used was $\tanh$.

- The recursive least-squares algorithm for a nonlinear PCA criterion (NPCA-RLS), discussed in Section 12.8.3. In this algorithm, the plain $\tanh$ function could not be used for stability reasons, but a slightly modified nonlinearity was chosen: $y - \tanh(y)$.

Tensorial algorithms were excluded from this comparison due to the problems of scalability discussed in Chapter 11. Some tensorial algorithms have been compared rather thoroughly in [315]. However, the conclusions are of limited value, because the data used in [315] always consisted of the same three subgaussian ICs.

14.4.2 Results for simulated data

Statistical performance and computational load The basic experiment measures the computational load and statistical performance (accuracy) of the tested algorithms. We performed experiments with 10 independent components that were chosen supergaussian, because for this source type all the algorithms in the comparison worked, including ML estimation with a fixed $-\tanh$ nonlinearity. The mixing matrix **A** used in our simulations consisted of uniformly distributed random numbers. For achieving statistical reliability, the experiment was repeated over 100 different realizations of the input data. For each of the 100 realizations, the accuracy was measured using the error index E_1. The computational load was measured in floating point operations needed for convergence.

Fig. 14.1 shows a schematic diagram of the computational load vs. the statistical performance. The boxes typically contain 80% of the 100 trials, thus representing standard outcomes.

As for *statistical performance*, Fig. 14.1 shows that best results are obtained by using a $\tanh$ nonlinearity (with the right sign). This was to be expected according to the theoretical analysis of Section 14.3. $\tanh$ is a good nonlinearity especially for supergaussian ICs as in this experiment. The kurtosis-based FastICA is clearly inferior, especially in the deflationary version. Note that the statistical performance only depends on the nonlinearity, and not on the optimization method, as explained in Section 14.1. All the algorithms using $\tanh$ have pretty much the same statistical performance. Note also that no outliers were added to the data, so the robustness of the algorithms is not measured here.

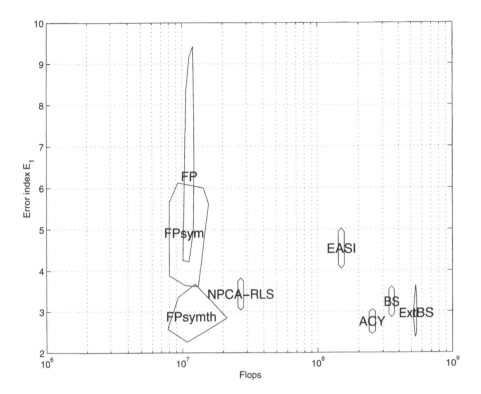

Fig. 14.1 Computational requirements in flops versus the statistical error index E_1. (Reprinted from [147], reprint permission and copyright by World Scientific, Singapore.)

Looking at the *computational load*, one sees clearly that FastICA requires the smallest amount of computation. Of the on-line algorithms, NPCA-RLS converges fastest, probably due to its roughly optimal determination of learning parameters. For the other on-line algorithms, the learning parameter was a constant, determined by making some preliminary experiments so that a value providing good convergence was found. These ordinary gradient-type algorithms have a computational load that is about 20–50 times larger than for FastICA.

To conclude, the best results from a statistical viewpoint are obtained when using the tanh nonlinearity with any algorithm. (Some algorithms, especially the tensorial ones, cannot use the tanh nonlinearity, but these were excluded from this comparison for reasons discussed earlier.) As for the computational load, the experiments show that the FastICA algorithm is much faster than the gradient algorithms.

Convergence speed of on-line algorithms   Next, we studied the convergence speeds of the *on-line* algorithms. Fixed-point algorithms do not appear in this comparison, because they are of a different type and a direct comparison is not possible.

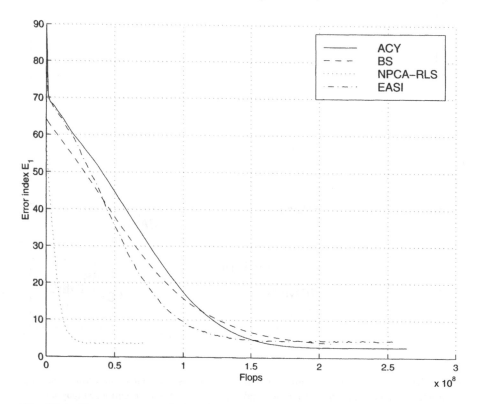

Fig. 14.2 Convergence speed of *on-line* ICA algorithms as a function of required floating-point operations for 10 supergaussian ICs. (Reprinted from [147], reprint permission and copyright by World Scientific, Singapore.)

The results (shown in Fig. 14.2) are averages of 10 trials for 10 supergaussian ICs (for which all the algorithms worked without on-line estimation of kurtosis). The main observation is that the recursive least-squares version of the nonlinear PCA algorithm (NPCA-RLS) is clearly the fastest converging of the on-line algorithms. The difference between NPCA-RLS and the other algorithms could probably be reduced by using simulated annealing or other more sophisticated technique for determining the learning parameters.

For subgaussian ICs, the results were qualitatively similar to those in Fig. 14.2, except that sometimes the EASI algorithm may converge even faster than NPCA-RLS. However, sometimes its convergence speed was the poorest among the compared algorithms. Generally, a weakness of on-line algorithms using stochastic gradients is that they are fairly sensitive to the choice of the learning parameters.

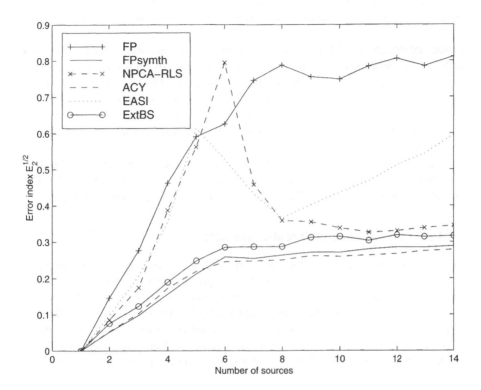

Fig. 14.3 Error as a function of the number of ICs. (Reprinted from [147], reprint permission and copyright by World Scientific, Singapore.)

Error for increasing number of components We also made a short investigation on how the statistical performances of the algorithms change with increasing number of components. In Fig. 14.3, the error (square root of the error index E_2) is plotted as the function of the number of supergaussian ICs. The results are median values over 50 different realizations of the input data. For more than five ICs, the number of data samples was increased so that it was proportional to the square of the number of ICs. The natural gradient ML algorithm (ACY), and its version with adaptive nonlinearity (ExtBS), achieve some of the best accuracies, behaving very similarly. The basic fixed-point algorithm (FP) using a cubic nonlinearity has the poorest accuracy, but its error increases only slightly after seven ICs. On the other hand, the version of the fixed-point algorithm which uses symmetric orthogonalization and tanh nonlinearity (FPsymth) performs as well as the natural gradient ML algorithm. Again, we see that it is the nonlinearity that is the most important in determining statistical performance. For an unknown reason, the errors of the EASI and NPCA-RLS algorithms have a peak around 5–6 ICs. For a larger number of ICs, the accuracy of the NPCA-RLS algorithm is close to the best algorithms, while the error of EASI increases linearly with the number of independent components. However, the error of all the algorithms is tolerable for most practical purposes.

Effect of noise In [147], the effect of additive gaussian noise on the performance of ICA algorithms has been studied, too. The first conclusion is that the estimation accuracy degrades fairly smoothly until the noise power increases up to -20 dB of the signal power. If the amount of noise is increased even more, it may happen that the studied ICA algorithms are not able to separate all the sources. In practice, noise smears the separated ICs or sources, making the separation results almost useless if there is a lot of noise present.

Another observation is that once there is even a little noise present in the data, the error strongly depends on the condition number of the mixing matrix **A**. The condition number of a matrix [320, 169] describes how close to singularity it is.

14.4.3 Comparisons with real-world data

We have compared in [147] the preceding ICA algorithms using three different real-world data sets. The applications were projection pursuit for well-known crab and satellite data sets, and finding interesting source signals from the biomedical magnetoencephalographic data (see Chapter 22). For the real-world data, the true independent components are unknown, and the assumptions made in the standard ICA model may not hold, or hold only approximately. Hence it is only possible to compare the performances of the ICA algorithms with each other, in the application at hand.

The following general conclusions can be made from these experiments [147]:

1. ICA is a robust technique. Even though the assumption of statistical independence is not strictly fulfilled, the algorithms converge towards a clear set of components (MEG data), or a subspace of components whose dimension is much smaller than the dimension of the problem (satellite data). This is a good characteristic encouraging the use of ICA as a general data analysis tool.

2. The FastICA algorithm and the natural gradient ML algorithm with adaptive nonlinearity (ExtBS) yielded usually similar results with real-world data. This is not surprising, because there exists a close theoretical connection between these algorithms, as discussed in Chapter 9. Another pair of similarly behaving algorithms consisted of the EASI algorithm and the nonlinear PCA algorithm using recursive least-squares (NPCA-RLS).

3. In difficult real-world problems, it is useful to apply several different ICA algorithms, because they may reveal different ICs from the data. For the MEG data, none of the compared algorithms was best in separating all types of source signals.

14.5 REFERENCES

The fundamental connection between cumulants and negentropy and mutual information was introduced in [89]. A similar approximation of likelihood by cumulants
was introduced in [140]. Approximation of negentropy by cumulants was originally
considered in [222]. The connection between infomax and likelihood was shown
in [363, 64], and the connection between mutual information and likelihood has
been explicitly discussed in [69]. The interpretation of nonlinear PCA criteria as
maximum likelihood estimation was presented in [236]. The connections between
different methods were discussed in such review papers as [201, 65, 269].

The theoretical analysis of the performance of the estimators is taken from [193].
See [69] for more information, especially on the effect of the decorrelation constraint
on the estimator. On robustness and influence functions, see such classic texts as
[163, 188].

More details on the experimental comparison can be found in [147].

14.6 SUMMARY OF BASIC ICA

Now we summarize Part II. This part treated the estimation of the basic ICA model,
i.e., the simplified model with no noise or time-structure, and a square mixing
matrix. The observed data $\mathbf{x} = (x_1, ..., x_n)^T$ is modeled as a linear transformation
of components $\mathbf{s} = (s_1, ..., s_n)^T$ that are statistically independent:

$$\mathbf{x} = \mathbf{As} \tag{14.15}$$

This is a rather well-understood problem for which several approaches have been
proposed. What distinguished ICA from PCA and classic factor analysis is that the
nongaussian structure of the data is taken into account. This *higher-order statistical
information* (i.e., information not contained in the mean and the covariance matrix)
can be utilized, and therefore, the independent components can be actually separated,
which is not possible by PCA and classic factor analysis.

Often, the data is preprocessed by *whitening* (sphering), which exhausts the second
order information that is contained in the covariance matrix, and makes it easier to
use the higher-order information:

$$\mathbf{z} = \mathbf{Vx} = (\mathbf{VA})\mathbf{s} \tag{14.16}$$

The linear transformation $\mathbf{VA}$ in the model is then reduced to an *orthogonal* one,
i.e., a rotation. Thus, we are searching for an orthogonal matrix $\mathbf{W}$ so that $\mathbf{y} = \mathbf{Wz}$
should be good estimates of the independent components.

Several approaches can then be taken to utilize the higher-order information. A
principled, yet intuitive approach is given by finding linear combinations of *maximum
nongaussianity*, as motivated by the central limit theorem. Sums of nongaussian
random variables tend to be closer to gaussian that the original ones. Therefore if
we take a linear combination $y = \sum_i w_i z_i$ of the observed (whitened) variables,

this will be maximally nongaussian if it equals one of the independent components. Nongaussianity can be measured by kurtosis or by (approximations of) negentropy. This principle shows the very close connection between ICA and *projection pursuit*, in which the most nongaussian projections are considered as the interesting ones.

Classic estimation theory directly gives another method for ICA estimation: *maximum likelihood estimation*. An information-theoretic alternative is to *minimize the mutual information* of the components. All these principles are essentially equivalent or at least closely related. The principle of maximum nongaussianity has the additional advantage of showing how to estimate the independent components one-by-one. This is possible by a deflationary orthogonalization of the estimates of the individual independent components.

With every estimation method, we are optimizing functions of expectations of *nonquadratic functions*, which is necessary to gain access to higher-order information. Nonquadratic functions usually cannot be maximized simply by solving the equations: Sophisticated numerical algorithms are necessary.

The *choice of the ICA algorithm* is basically a choice between on-line and batch-mode algorithms. In the on-line case, the algorithms are obtained by stochastic gradient methods. If all the independent components are estimated in parallel, the most popular algorithm in this category is *natural gradient ascent of likelihood*. The fundamental equation in this method is

$$\mathbf{W} \leftarrow \mathbf{W} + \mu[\mathbf{I} + \mathbf{g}(\mathbf{y})\mathbf{y}^T]\mathbf{W} \tag{14.17}$$

where the component-wise nonlinear function $\mathbf{g}$ is determined from the log-densities of the independent components; see Table 9.1 for details.

In the more usual case, where the computations are made in batch-mode (off-line), much more efficient algorithms are available. The *FastICA* algorithm is a very efficient batch algorithm that can be derived either from a fixed-point iteration or as an approximate Newton method. The fundamental iteration in FastICA is, for one row $\mathbf{w}$ of $\mathbf{W}$:

$$\mathbf{w} \leftarrow E\{\mathbf{z}g(\mathbf{w}^T\mathbf{z})\} - E\{g'(\mathbf{w}^T\mathbf{z})\}\mathbf{w} \tag{14.18}$$

where the nonlinearity g can be almost any smooth function, and $\mathbf{w}$ should be normalized to unit norm at every iteration. FastICA can be used to estimate the components either one-by-one by finding maximally nongaussian directions (see Tables 8.3), or in parallel by maximizing nongaussianity or likelihood (see Table 8.4 or Table 9.2).

In practice, before application of these algorithms, suitable *preprocessing* is often necessary (Chapter 13). In addition to the compulsory centering and whitening, it is often advisable to perform principal component analysis to reduce the dimension of the data, or some time filtering by taking moving averages.

Appendix Proofs

Here we prove Theorem 14.1. Making the change of variable $\mathbf{q} = \mathbf{A}^T \mathbf{b}$, the equation defining the optimal solutions $\hat{\mathbf{q}}$ becomes

$$\sum_t \mathbf{s}_t g(\hat{\mathbf{q}}^T \mathbf{s}_t) = \lambda \sum_t \mathbf{s}_t \mathbf{s}_t^T \hat{\mathbf{q}} \qquad (\text{A.1})$$

where $t = 1, ..., T$ is the sample index, T is the sample size, and λ is a Lagrangian multiplier.. Without loss of generality, let us assume that $\hat{\mathbf{q}}$ is near the ideal solution $\mathbf{q} = (1, 0, 0, ...)$. Note that due to the constraint $E\{(\mathbf{b}^T \mathbf{x})^2\} = \|\mathbf{q}\|^2 = 1$, the variance of the first component of $\hat{\mathbf{q}}$, denoted by $\hat{\mathbf{q}}_1$, is of a smaller order than the variance of the vector of other components, denoted by $\hat{\mathbf{q}}_{-1}$. Excluding the first component in (A.1), and making the first-order approximation $g(\hat{\mathbf{q}}^T \mathbf{s}) = g(s_1) + g'(s_1)\hat{\mathbf{q}}_{-1}^T \mathbf{s}_{-1}$, where also $\mathbf{s}_{-1}$ denotes $\mathbf{s}$ without its first component, one obtains after some simple manipulations

$$\frac{1}{\sqrt{T}} \sum_t \mathbf{s}_{-1}[g(s_1) - \lambda s_1] = \frac{1}{T} \sum_t \mathbf{s}_{-1}[-\mathbf{s}_{-1}^T g'(s_1) + \lambda \mathbf{s}_{-1}^T]\hat{\mathbf{q}}_{-1}\sqrt{T}$$
$$(\text{A.2})$$

where the sample index t has been dropped for simplicity. Making the first-order approximation $\lambda = E\{s_1 g(s_1)\}$, one can write (A.2) in the form $u = v\hat{\mathbf{q}}_{-1}\sqrt{T}$ where v converges to the identity matrix multiplied by $E\{s_1 g(s_1)\} - E\{g'(s_1)\}$, and u converges to a variable that has a normal distribution of zero mean whose covariance matrix equals the identity matrix multiplied by $E\{g^2(s_1)\} - (E\{s_1 g(s_1)\})^2$. This implies the theorem, since $\hat{\mathbf{q}}_{-1} = \mathbf{B}_{-1}\hat{\mathbf{b}}$, where $\mathbf{B}_{-1}$ is the inverse of $\mathbf{A}^T$ without its first row.

Part III

EXTENSIONS AND RELATED METHODS

15

Noisy ICA

In real life, there is always some kind of noise present in the observations. Noise can correspond to actual physical noise in the measuring devices, or to inaccuracies of the model used. Therefore, it has been proposed that the independent component analysis (ICA) model should include a noise term as well. In this section, we consider different methods for estimating the ICA model when noise is present.

However, estimation of the mixing matrix seems to be quite difficult when noise is present. It could be argued that in practice, a better approach could often be to reduce noise in the data before performing ICA. For example, simple filtering of time-signals is often very useful in this respect, and so is dimension reduction by principal component analysis (PCA); see Sections 13.1.2 and 13.2.2.

In noisy ICA, we also encounter a new problem: estimation of the noise-free realizations of the independent components (ICs). The noisy model is not invertible, and therefore estimation of the noise-free components requires new methods. This problem leads to some interesting forms of denoising.

15.1 DEFINITION

Here we extend the basic ICA model to the situation where noise is present. The noise is assumed to be additive. This is a rather realistic assumption, standard in factor analysis and signal processing, and allows for a simple formulation of the noisy model. Thus, the noisy ICA model can be expressed as

$$\mathbf{x} = \mathbf{As} + \mathbf{n} \qquad (15.1)$$

where $\mathbf{n} = (n_1, ..., n_n)$ is the noise vector. Some further assumptions on the noise are usually made. In particular, it is assumed that

1. The noise is independent from the independent components.

2. The noise is gaussian.

The covariance matrix of the noise, say $\boldsymbol{\Sigma}$, is often assumed to of the form $\sigma^2\mathbf{I}$, but this may be too restrictive in some cases. In any case, the noise covariance is assumed to be known. Little work on estimation of an unknown noise covariance has been conducted; see [310, 215, 19].

The identifiability of the mixing matrix in the noisy ICA model is guaranteed under the same restrictions that are sufficient in the basic case, [1] basically meaning independence and nongaussianity. In contrast, the realizations of the independent components s_i can no longer be identified, because they cannot be completely separated from noise.

15.2 SENSOR NOISE VS. SOURCE NOISE

In the typical case where the noise covariance is assumed to be of the form $\sigma^2\mathbf{I}$, the noise in Eq. (15.1) could be considered as "sensor" noise. This is because the noise variables are separately added on each sensor, i.e., observed variable x_i. This is in contrast to "source" noise, in which the noise is added to the independent components (sources). Source noise can be modeled with an equation slightly different from the preceding, given by

$$\mathbf{x} = \mathbf{A}(\mathbf{s} + \mathbf{n}) \tag{15.2}$$

where again the covariance of the noise is diagonal. In fact, we could consider the noisy independent components, given by $\tilde{s}_i = s_i + n_i$, and rewrite the model as

$$\mathbf{x} = \mathbf{A}\tilde{\mathbf{s}} \tag{15.3}$$

We see that this is just the basic ICA model, with modified independent components. What is important is that the assumptions of the basic ICA model are still valid: the components of $\tilde{\mathbf{s}}$ are nongaussian and independent. Thus we can estimate the model in (15.3) by any method for basic ICA. This gives us a perfectly suitable estimator for the noisy ICA model. This way we can estimate the mixing matrix and the noisy independent components. The estimation of the original independent components from the noisy ones is an additional problem, though; see below.

This idea is, in fact, more general. Assume that the noise covariance has the form

$$\boldsymbol{\Sigma} = \mathbf{A}\mathbf{A}^T\sigma^2 \tag{15.4}$$

---

[1]This seems to be admitted by the vast majority of ICA researchers. We are not aware of any rigorous proofs of this property, though.

Then the noise vector can be transformed into another one $\tilde{\mathbf{n}} = \mathbf{A}^{-1}\mathbf{n}$, which can be called equivalent source noise. Then the equation (15.1) becomes

$$\mathbf{x} = \mathbf{As} + \mathbf{A\tilde{n}} = \mathbf{A}(\mathbf{s} + \tilde{\mathbf{n}}) \qquad (15.5)$$

The point is that the covariance of $\tilde{\mathbf{n}}$ is $\sigma^2\mathbf{I}$, and thus the transformed components in $\mathbf{s} + \tilde{\mathbf{n}}$ are independent. Thus, we see again that the mixing matrix $\mathbf{A}$ can be estimated by basic ICA methods.

To recapitulate: if the noise is added to the independent components and not to the observed mixtures, or has a particular covariance structure, the mixing matrix can be estimated by ordinary ICA methods. The denoising of the independent components is another problem, though; it will be treated in Section 15.5 below.

15.3   FEW NOISE SOURCES

Another special case that reduces to the basic ICA model can be found, when the number of noise components and independent components is not very large. In particular, if their total number is not larger than the number of mixtures, we again have an ordinary ICA model, in which some of the components are gaussian noise and others are the real independent components. Such a model could still be estimated by the basic ICA model, using one-unit algorithms with less units than the dimension of the data.

In other words, we could define the vector of the independent components as $\tilde{\mathbf{s}} = (s_1, ..., s_k, n_1, ..., n_l)^T$ where the $s_i, i = 1, ..., k$ are the "real" independent components and the $n_i, i = 1, ..., l$ are the noise variables. Assume that the number of mixtures equals $k + l$, that is the number of real ICs plus the number of noise variables. In this case, the ordinary ICA model holds with $\mathbf{x} = \mathbf{A\tilde{s}}$, where $\mathbf{A}$ is a matrix that incorporates the mixing of the real ICs and the covariance structure of the noise, and the number of the independent components in $\tilde{\mathbf{s}}$ is equal to the number of observed mixtures. Therefore, finding the k most nongaussian directions, we can estimate the real independent components. We cannot estimate the remaining dummy independent components that are actually noise variables, but we did not want to estimate them in the first place.

The applicability of this idea is quite limited, though, since in most cases we want to assume that the noise is added on each mixture, in which case $k + l$, the number of real ICs plus the number of noise variables, is necessarily larger than the number of mixtures, and the basic ICA model does not hold for $\tilde{\mathbf{s}}$.

15.4   ESTIMATION OF THE MIXING MATRIX

Not many methods for noisy ICA estimation exist in the general case. The estimation of the noiseless model seems to be a challenging task in itself, and thus the noise is usually neglected in order to obtain tractable and simple results. Moreover, it may be unrealistic in many cases to assume that the data could be divided into signals and noise in any meaningful way.

Here we treat first the problem of estimating the mixing matrix. Estimation of the independent components will be treated below.

15.4.1 Bias removal techniques

Perhaps the most promising approach to noisy ICA is given by bias removal techniques. This means that ordinary (noise-free) ICA methods are modified so that the bias due to noise is removed, or at least reduced.

Let us denote the noise-free data in the following by

$$\mathbf{v} = \mathbf{A}\mathbf{s} \tag{15.6}$$

We can now use the basic idea of finding projections, say $\mathbf{w}^T\mathbf{v}$, in which nongaussianity, is locally maximized for whitened data, with constraint $\|\mathbf{w}\| = 1$. As shown in Chapter 8, projections in such directions give consistent estimates of the independent components, if the measure of nongaussianity is well chosen. This approach could be used for noisy ICA as well, if only we had measures of nongaussianity which are immune to gaussian noise, or at least, whose values for the original data can be easily estimated from noisy observations. We have $\mathbf{w}^T\mathbf{x} = \mathbf{w}^T\mathbf{v} + \mathbf{w}^T\mathbf{n}$, and thus the point is to measure the nongaussianity of $\mathbf{w}^T\mathbf{v}$ from the observed $\mathbf{w}^T\mathbf{x}$ so that the measure is not affected by the noise $\mathbf{w}^T\mathbf{n}$.

Bias removal for kurtosis If the measure of nongaussianity is kurtosis (the fourth-order cumulant), it is almost trivial to construct one-unit methods for noisy ICA, because kurtosis is immune to gaussian noise. This is because the kurtosis of $\mathbf{w}^T\mathbf{x}$ equals the kurtosis of $\mathbf{w}^T\mathbf{v}$, as can be easily proven by the basic properties of kurtosis.

It must be noted, however, that in the preliminary whitening, the effect of noise must be taken into account; this is quite simple if the noise covariance matrix is known. Denoting by $\mathbf{C} = E\{\mathbf{x}\mathbf{x}^T\}$ the covariance matrix of the observed noisy data, the ordinary whitening should be replaced by the operation

$$\tilde{\mathbf{x}} = (\mathbf{C} - \mathbf{\Sigma})^{-1/2}\mathbf{x} \tag{15.7}$$

In other words, the covariance matrix $\mathbf{C} - \mathbf{\Sigma}$ of the noise-free data should be used in whitening instead of the covariance matrix $\mathbf{C}$ of the noisy data. In the following, we call this operation "quasiwhitening". After this operation, the quasiwhitened data $\tilde{\mathbf{x}}$ follows a noisy ICA model as well:

$$\tilde{\mathbf{x}} = \mathbf{B}\mathbf{s} + \tilde{\mathbf{n}} \tag{15.8}$$

where $\mathbf{B}$ is *orthogonal*, and $\tilde{\mathbf{n}}$ is a linear transform of the original noise in (15.1). Thus, the theorem in Chapter 8 is valid for $\tilde{\mathbf{x}}$, and finding local maxima of the absolute value of kurtosis is a valid method for estimating the independent components.

Bias removal for general nongaussianity measures As was argued in Chapter 8, it is important in many applications to use measures of nongaussianity that have better statistical properties than kurtosis. We introduced the following measure:

$$J_G(\mathbf{w}^T\mathbf{v}) = [E\{G(\mathbf{w}^T\mathbf{v})\} - E\{G(\nu)\}]^2 \tag{15.9}$$

where the function G is a sufficiently regular nonquadratic function, and ν is a standardized gaussian variable.

Such a measure could be used for noisy data as well, if only we were able to estimate $J_G(\mathbf{w}^T\mathbf{v})$ of the noise-free data from the noisy observations $\mathbf{x}$. Denoting by z a nongaussian random variable, and by n a gaussian noise variable of variance σ^2, we should be able to express the relation between $E\{G(z)\}$ and $E\{G(z+n)\}$ in simple algebraic terms. In general, this relation seems quite complicated, and can be computed only using numerical integration.

However, it was shown in [199] that for certain choices of G, a similar relation becomes very simple. The basic idea is to choose G to be the density function of a zero-mean gaussian random variable, or a related function. These nonpolynomial moments are called *gaussian moments*.

Denote by

$$\varphi_c(x) = \frac{1}{c}\varphi(\frac{x}{c}) = \frac{1}{\sqrt{2\pi}c}\exp(-\frac{x^2}{2c^2}) \tag{15.10}$$

the gaussian density function with variance c^2, and by $\varphi_c^{(k)}(x)$ the kth ($k > 0$) derivative of $\varphi_c(x)$. Denote further by $\varphi_c^{(-k)}$ the kth integral function of $\varphi_c(x)$, obtained by $\varphi_c^{(-k)}(x) = \int_0^x \varphi_c^{(-k+1)}(\xi)d\xi$, where we define $\varphi_c^{(0)}(x) = \varphi_c(x)$. (The lower integration limit 0 is here quite arbitrary, but has to be fixed.) Then we have the following theorem [199]:

Theorem 15.1 *Let z be any nongaussian random variable, and n an independent gaussian noise variable of variance σ^2. Define the gaussian function φ as in (15.10). Then for any constant $c > \sigma^2$, we have*

$$E\{\varphi_c(z)\} = E\{\varphi_d(z+n)\} \tag{15.11}$$

with $d = \sqrt{c^2 - \sigma^2}$. Moreover, (15.11) still holds when φ is replaced by $\varphi^{(k)}$ for any integer index k.

The theorem means that we can estimate the independent components from noisy observations by maximizing a general contrast function of the form (15.9), where the direct estimation of the statistics $E\{G(\mathbf{w}^T\mathbf{v})\}$ of the noise-free data is made possible by using $G(u) = \varphi_c^{(k)}(u)$. We call the statistics of the form $E\{\varphi_c^{(k)}(\mathbf{w}^T\mathbf{v})\}$ the gaussian moments of the data. Thus, for quasiwhitened data $\tilde{\mathbf{x}}$, we maximize the following contrast function:

$$\max_{\|\mathbf{w}\|=1} [E\{\varphi_{d(\mathbf{w})}^{(k)}(\mathbf{w}^T\tilde{\mathbf{x}})\} - E\{\varphi_c^{(k)}(\nu)\}]^2 \tag{15.12}$$

with $d(\mathbf{w}) = \sqrt{c^2 - \mathbf{w}^T \tilde{\boldsymbol{\Sigma}} \mathbf{w}}$. This gives a consistent (i.e., convergent) method of estimating the noisy ICA model, as was shown in Chapter 8.

To use these results in practice, we need to choose some values for k. In fact, c disappears from the final algorithm, so value for this parameter need not be chosen. Two indices k for the gaussian moments seem to be of particular interest: $k = 0$ and $k = -2$. The first corresponds to the gaussian density function; its use was proposed in Chapter 8. The case $k = -2$ is interesting because the contrast function is then of the form of a (negative) log-density of a supergaussian variable. In fact, $\varphi^{(-2)}(u)$ can be very accurately approximated by $G(u) = 1/2 \log \cosh u$, which was also used in Chapter 8.

FastICA for noisy data   Using the unbiased measures of nongaussianity given in this section, we can derive a variant of the FastICA algorithm [198]. Using kurtosis or gaussian moments give algorithms of a similar form, just like in the noise-free case.

The algorithm takes the form [199, 198]:

$$\boxed{\mathbf{w}^* = E\{\tilde{\mathbf{x}} g(\mathbf{w}^T \tilde{\mathbf{x}})\} - (\mathbf{I} + \tilde{\boldsymbol{\Sigma}}) \mathbf{w} E\{g'(\mathbf{w}^T \tilde{\mathbf{x}})\}} \tag{15.13}$$

where $\mathbf{w}^*$, the new value of $\mathbf{w}$, is normalized to unit norm after every iteration, and $\tilde{\boldsymbol{\Sigma}}$ is given by

$$\tilde{\boldsymbol{\Sigma}} = E\{\tilde{\mathbf{n}} \tilde{\mathbf{n}}^T\} = (\mathbf{C} - \boldsymbol{\Sigma})^{-1/2} \boldsymbol{\Sigma} (\mathbf{C} - \boldsymbol{\Sigma})^{-1/2} \tag{15.14}$$

The function g is here the derivative of G, and can thus be chosen among the following:

$$g_1(u) = \tanh(u), \quad g_2(u) = u \exp(-u^2/2), \quad g_3(u) = u^3 \tag{15.15}$$

where g_1 is an approximation of $\varphi^{(-1)}$, which is the gaussian cumulative distribution function (these relations hold up to some irrelevant constants). These functions cover essentially the nonlinearities ordinarily used in the FastICA algorithm.

15.4.2   Higher-order cumulant methods

A different approach to estimation of the mixing matrix is given by methods using higher-order cumulants only. Higher-order cumulants are unaffected by gaussian noise (see Section 2.7), and therefore any such estimation method would be immune to gaussian noise. Such methods can be found in [63, 263, 471]. The problem is, however, that such methods often use cumulants of order 6. Higher-order cumulants are sensitive to outliers, and therefore methods using cumulants of orders higher than 4 are unlikely to be very useful in practice. A nice feature of this approach is, however, that we do not need to know the noise covariance matrix.

Note that the cumulant-based methods in Part II used both second- and fourth-order cumulants. Second-order cumulants are *not* immune to gaussian noise, and therefore the cumulant-based method introduced in the previous chapters would not be immune either. Most of the cumulant-based methods could probably be modified to work in the noisy case, as we did in this chapter for methods maximizing the absolute value of kurtosis.

15.4.3 Maximum likelihood methods

Another approach for estimation of the mixing matrix with noisy data is given by maximum likelihood (ML) estimation. First, one could maximize the joint likelihood of the mixing matrix and the realizations of the independent components, as in [335, 195, 80]. This is given by

$$
\log L(\mathbf{A}, \mathbf{s}(1), ..., \mathbf{s}(T)) =
$$
$$
- \sum_{t=1}^{T} \left[\frac{1}{2} \|\mathbf{A}\mathbf{s}(t) - \mathbf{x}(t)\|_{\Sigma^{-1}}^2 + \sum_{i=1}^{n} f_i(s_i(t)) \right] + C \quad (15.16)
$$

where $\|\mathbf{m}\|_{\Sigma^{-1}}^2$ is defined as $\mathbf{m}^T \Sigma^{-1} \mathbf{m}$, the $\mathbf{s}(t)$ are the realizations of the independent components, and C is an irrelevant constant. The f_i are the logarithms of the probability density functions (pdf's) of the independent components. Maximization of this joint likelihood is, however, computationally very expensive.

A more principled method would be to maximize the (marginal) likelihood of the mixing matrix, and possibly that of the noise covariance, which was done in [310]. This was based on the idea of approximating the densities of the independent components as gaussian mixture densities; the application of the EM algorithm then becomes feasible. In [42], the simpler case of discrete-valued independent components was treated. A problem with the EM algorithm is, however, that the computational complexity grows exponentially with the dimension of the data.

A more promising approach might be to use bias removal techniques so as to modify existing ML algorithms to be consistent with noisy data. Actually, the bias removal techniques given here can be interpreted as such methods; a related method was given in [119].

Finally, let us mention a method based on the geometric interpretation of the maximum likelihood estimator, introduced in [33], and a rather different approach for narrow-band sources, introduced in [76].

15.5 ESTIMATION OF THE NOISE-FREE INDEPENDENT COMPONENTS

15.5.1 Maximum a posteriori estimation

In noisy ICA, it is not enough to estimate the mixing matrix. Inverting the mixing matrix in (15.1), we obtain

$$
\mathbf{W}\mathbf{x} = \mathbf{s} + \mathbf{W}\mathbf{n} \quad (15.17)
$$

In other words, we only get noisy estimates of the independent components. Therefore, we would like to obtain estimates of the original independent components $\hat{s}_i$ that are somehow optimal, i.e., contain minimum noise.

A simple approach to this problem would be to use the maximum a posteriori (MAP) estimates. See Section 4.6.3 for the definition. Basically, this means that we take the values that have maximum probability, given the $\mathbf{x}$. Equivalently, we take as $\hat{s}_i$ those values that maximize the joint likelihood in (15.16), so this could also be called a maximum likelihood (ML) estimator.

To compute the MAP estimator, let us take the gradient of the log-likelihood (15.16) with respect to the $\mathbf{s}(t), t = 1, ..., T$ and equate this to 0. Thus we obtain the equation

$$\hat{\mathbf{A}}^T \mathbf{\Sigma}^{-1} \hat{\mathbf{A}} \hat{\mathbf{s}}(t) - \hat{\mathbf{A}}^T \mathbf{\Sigma}^{-1} \mathbf{x}(t) + f'(\hat{\mathbf{s}}(t)) = 0 \tag{15.18}$$

where the derivative of the log-density, denoted by f', is applied separately on each component of the vector $\hat{\mathbf{s}}(t)$.

In fact, this method gives a nonlinear generalization of classic Wiener filtering presented in Section 4.6.2. An alternative approach would be to use the time-structure of the ICs (see Chapter 18) for denoising. This results in a method resembling the Kalman filter; see [250, 249].

15.5.2 Special case of shrinkage estimation

Solving for the $\hat{s}$ is not easy, however. In general, we must use numerical optimization. A simple special case is obtained if the noise covariance is assumed to be of the same form as in (15.4) [200, 207]. This corresponds to the case of (equivalent) source noise. Then (15.18) gives

$$\hat{\mathbf{s}} = g(\hat{\mathbf{A}}^{-1} \mathbf{x}) \tag{15.19}$$

where the scalar component-wise function g is obtained by inverting the relation

$$g^{-1}(u) = u + \sigma^2 f'(u) \tag{15.20}$$

Thus, the MAP estimator is obtained by inverting a certain function involving f', or the score function [395] of the density of s. For nongaussian variables, the score function is nonlinear, and so is g.

In general, the inversion required in (15.20) may be impossible analytically. Here we show three examples, which will be shown to have great practical value in Chapter 21, where the inversion can be done easily.

Example 15.1 Assume that s has a Laplacian (or double exponential) distribution of unit variance. Then $p(s) = \exp(-\sqrt{2}|s|)/\sqrt{2}$, $f'(s) = \sqrt{2}\,\text{sign}(s)$, and g takes the form

$$g(u) = \text{sign}(u) \max(0, |u| - \sqrt{2}\sigma^2) \tag{15.21}$$

(Rigorously speaking, the function in (15.20) is not invertible in this case, but approximating it by a sequence of invertible functions, (15.21) is obtained as the limit.) The function in (15.21) is a *shrinkage* function that reduces the absolute value of its argument by a fixed amount, as depicted in Fig 15.1. Intuitively, the utility of such a function can be seen as follows. Since the density of a supergaussian random variable (e.g., a Laplacian random variable) has a sharp peak at zero, it can be assumed that small values of the noisy variable correspond to pure noise, i.e., to $s = 0$. Thresholding such values to zero should thus reduce noise, and the shrinkage function can indeed be considered a soft thresholding operator.

Example 15.2 More generally, assume that the score function is approximated as a linear combination of the score functions of the gaussian and the Laplacian distributions:

$$f'(s) = as + b \, \text{sign}(s) \tag{15.22}$$

with $a, b > 0$. This corresponds to assuming the following density model for s:

$$p(s) = C \exp(-as^2/2 - b|s|) \tag{15.23}$$

where C is an irrelevant scaling constant. This is depicted in Fig. 15.2. Then we obtain

$$g(u) = \frac{1}{1 + \sigma^2 a} \text{sign}(u) \max(0, |u| - b\sigma^2) \tag{15.24}$$

This function is a shrinkage with additional scaling, as depicted in Fig 15.1.

Example 15.3 Yet another possibility is to use the following strongly supergaussian probability density:

$$p(s) = \frac{1}{2d} \frac{(\alpha + 2) \left[\alpha (\alpha + 1)/2 \right]^{(\alpha/2+1)}}{\left[\sqrt{\alpha (\alpha + 1)/2} + |s/d| \right]^{(\alpha+3)}} \tag{15.25}$$

with parameters $\alpha, d > 0$, see Fig. 15.2. When $\alpha \to \infty$, the Laplacian density is obtained as the limit. The strong sparsity of the densities given by this model can be seen e.g., from the fact that the kurtosis [131, 210] of these densities is always larger than the kurtosis of the Laplacian density, and reaches infinity for $\alpha \leq 2$. Similarly, $p(0)$ reaches infinity as α goes to zero. The resulting shrinkage function given by (15.20) can be obtained after some straightforward algebraic manipulations as:

$$g(u) = \text{sign}(u) \max(0, \frac{|u| - ad}{2} + \frac{1}{2}\sqrt{(|u| + ad)^2 - 4\sigma^2(\alpha + 3)} \,) \tag{15.26}$$

where $a = \sqrt{\alpha(\alpha + 1)/2}$, and $g(u)$ is set to zero in case the square root in (15.26) is imaginary. This is a shrinkage function that has a stronger thresholding flavor, as depicted in Fig. 15.1.

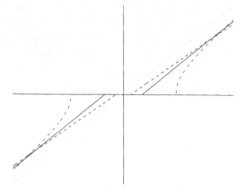

Fig. 15.1 Plots of the shrinkage functions. The effect of the functions is to reduce the absolute value of its argument by a certain amount which depends on the noise level. Small arguments are set to zero. This reduces gaussian noise for sparse random variables. Solid line: shrinkage corresponding to Laplacian density as in (15.21). Dashed line: typical shrinkage function obtained from (15.24). Dash-dotted line: typical shrinkage function obtained from (15.26). For comparison, the line $x = y$ is given by dotted line. All the densities were normalized to unit variance, and noise variance was fixed to .3.

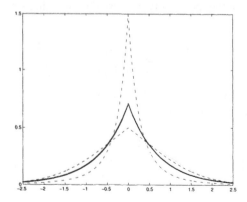

Fig. 15.2 Plots of densities corresponding to models (15.23) and (15.25) of the sparse components. Solid line: Laplacian density. Dashed line: a typical moderately supergaussian density given by (15.23). Dash-dotted line: a typical strongly supergaussian density given by (15.25). For comparison, gaussian density is given by dotted line.

15.6 DENOISING BY SPARSE CODE SHRINKAGE

Although the basic purpose of noisy ICA estimation is to estimate the ICs, the model can be used to develop an interesting denoising method as well.

Assume that we observe a noisy version,

$$\mathbf{x} = \mathbf{v} + \mathbf{n} \qquad (15.27)$$

of the data $\mathbf{x}$, which has previously been modeled by ICA

$$\mathbf{v} = \mathbf{As} \qquad (15.28)$$

To denoise $\mathbf{x}$, we can compute estimates $\hat{\mathbf{s}}$ of the independent components by the above MAP estimation procedure. Then we can reconstruct the data as

$$\hat{\mathbf{v}} = \mathbf{A}\hat{\mathbf{s}} \qquad (15.29)$$

The point is that if the mixing matrix is orthogonal and the noise covariance is of the form $\sigma^2 \mathbf{I}$, the condition in (15.4) is fulfilled. This condition of the noise is a common one. Thus we could approximate the mixing matrix by an orthogonal one, for example the one obtained by orthogonalization of the mixing matrix as in (8.48).

This method is called *sparse code shrinkage* [200, 207], since it means that we transform the data into a sparse, i.e., supergaussian code, and then apply shrinkage on that code. To summarize, the method is as follows.

1. First, using a noise-free training set of $\mathbf{v}$, estimate ICA and orthogonalize the mixing matrix. Denote the orthogonal mixing matrix by $\mathbf{W}^T$. Estimate a density model $p_i(s_i)$ for each sparse component, using the models in (15.23) and (15.25).

2. Compute for each noisy observation $\mathbf{x}(t)$ the corresponding noisy sparse components $\mathbf{u}(t) = \mathbf{Wx}(t)$. Apply the shrinkage nonlinearity $g_i(.)$ as defined in (15.24), or in (15.26), on each component $u_i(t)$, for every observation index t. Denote the obtained components by $\hat{s}_i(t) = g_i(u_i(t))$.

3. Invert the transform to obtain estimates of the noise-free data, given by $\hat{\mathbf{v}}(t) = \mathbf{W}^T\hat{\mathbf{s}}(t)$.

For experiments using sparse code shrinkage on image denoising, see Chapter 21. In that case, the method is closely related to wavelet shrinkage and "coring" methods [116, 403].

15.7 CONCLUDING REMARKS

In this chapter, we treated the estimation of the ICA model when additive sensor noise is present. First of all, it was shown that in some cases, the mixing matrix can be estimated with basic ICA methods without any further complications. In cases where this is not possible, we discussed bias removal techniques for estimation of the mixing matrix, and introduced a bias-free version of the FastICA algorithm.

Next, we considered how to estimate the noise-free independent components, i.e., how to denoise the initial estimates of the independent components. In the case of supergaussian data, it was shown that this led to so-called shrinkage estimation. In fact, we found an interesting denoising procedure called sparse code shrinkage.

Note that in contrast to Part II where we considered the estimation of the basic ICA model, the material in this chapter is somewhat speculative in character. The utility of many of the methods in this chapter has not been demonstrated in practice. We would like to warn the reader not to use the noisy ICA methods lightheartedly: It is always advisable to first attempt to denoise the data so that basic ICA methods can be used, as discussed in Chapter 13.

16

ICA with Overcomplete Bases

A difficult problem in independent component analysis (ICA) is encountered if the number of mixtures x_i is smaller than the number of independent components s_i. This means that the mixing system is not invertible: We cannot obtain the independent components (ICs) by simply inverting the mixing matrix $\mathbf{A}$. Therefore, even if we knew the mixing matrix exactly, we could not recover the exact values of the independent components. This is because information is lost in the mixing process.

This situation is often called ICA with overcomplete bases. This is because we have in the ICA model

$$\mathbf{x} = \mathbf{As} = \sum_i \mathbf{a}_i s_i \tag{16.1}$$

where the number of "basis vectors", $\mathbf{a}_i$, is larger than the dimension of the space of $\mathbf{x}$: thus this basis is "too large", or overcomplete. Such a situation sometimes occurs in feature extraction of images, for example.

As with noisy ICA, we actually have two different problems. First, how to estimate the mixing matrix, and second, how to estimate the realizations of the independent components. This is in stark contrast to ordinary ICA, where these two problems are solved at the same time. This problem is similar to the noisy ICA in another respect as well: It is much more difficult than the basic ICA problem, and the estimation methods are less developed.

16.1   ESTIMATION OF THE INDEPENDENT COMPONENTS

16.1.1   Maximum likelihood estimation

Many methods for estimating the mixing matrix use as subroutines methods that estimate the independent components for a known mixing matrix. Therefore, we shall first treat methods for reconstructing the independent components, assuming that we know the mixing matrix. Let us denote by m the number of mixtures and by n the number of independent components. Thus, the mixing matrix has size $m \times n$ with $n > m$, and therefore it is not invertible.

The simplest method of estimating the independent components would be to use the pseudoinverse of the mixing matrix. This yields

$$\hat{\mathbf{s}} = \mathbf{A}^T (\mathbf{A}\mathbf{A}^T)^{-1} \mathbf{x} \tag{16.2}$$

In some situations, such a simple pseudoinverse gives a satisfactory solution, but in many cases we need a more sophisticated estimate.

A more sophisticated estimator of $\mathbf{s}$ can be obtained by maximum likelihood (ML) estimation [337, 275, 195], in a manner similar to the derivation of the ML or maximum a posteriori (MAP) estimator of the noise-free independent components in Chapter 15. We can write the posterior probability of $\mathbf{s}$ as follows:

$$p(\mathbf{s}|\mathbf{x}, \mathbf{A}) = 1_{\mathbf{x}=\mathbf{A}\mathbf{s}} \prod_i p_i(s_i) \tag{16.3}$$

where $1_{\mathbf{x}=\mathbf{A}\mathbf{s}}$ is an indicator function that is 1 if $\mathbf{x} = \mathbf{A}\mathbf{s}$ and 0 otherwise. The (prior) probability densities of the independent components are given by $p_i(s_i)$. Thus, we obtain the maximum likelihood estimator of $\mathbf{s}$ as

$$\hat{\mathbf{s}} = \arg \max_{\mathbf{x}=\mathbf{A}\mathbf{s}} \sum_i \log p_i(s_i) \tag{16.4}$$

Alternatively, we could assume that there is noise present as well. In this case, we get a likelihood that is formally the same as with ordinary noisy mixtures in (15.16). The only difference is in the number of components in the formula.

The problem with the maximum likelihood estimator is that it is not easy to compute. This optimization cannot be expressed as a simple function in analytic form in any interesting case. It can be obtained in closed form if the s_i have gaussian distribution: In this case the optimum is given by the pseudoinverse in (16.2). However, since ICA with gaussian variables is of little interest, the pseudoinverse is not a very satisfactory solution in many cases.

In general, therefore, the estimator given by (16.4) can only be obtained by numerical optimization. A gradient ascent method can be easily derived. One case where the optimization is easier than usual is when the s_i have a Laplacian distribution:

$$p_i(s_i) = \frac{1}{\sqrt{2}} \exp(\sqrt{2}|s_i|) \tag{16.5}$$

Ignoring uninteresting constants, we have

$$\hat{\mathbf{s}} = \arg \max_{\mathbf{x} = \mathbf{A}\mathbf{s}} \sum_i |s_i| \tag{16.6}$$

which can be formulated as a linear program and solved by classic methods for linear programming [275].

16.1.2 The case of supergaussian components

Using a supergaussian distribution, such as the Laplacian distribution, is well justified in feature extraction, where the components are supergaussian. Using the Laplacian density also leads to an interesting phenomenon: The ML estimator gives coefficients $\hat{s}_i$ of which only m are nonzero. Thus, only the minimum number of the components are activated. Thus we obtain a sparse decomposition in the sense that the components are quite often equal to zero.

It may seem at first glance that it is useless to try to estimate the ICs by ML estimation, because they cannot be estimated exactly in any case. This is not so, however; due to this phenomenon of sparsity, the ML estimation is very useful. In the case where the independent components are very supergaussian, most of them are very close to zero because of the large peak of the pdf at zero. (This is related to the principle of sparse coding that will be treated in more detail in Section 21.2.)

Thus, those components that are not zero may not be very many, and the system may be invertible for those components. If we first determine which components are likely to be clearly nonzero, and then invert that part of the linear system, we may be able to get quite accurate reconstructions of the ICs. This is done implicitly in the ML estimation method. For example, assume that there are three speech signals mixed into two mixtures. Since speech signals are practically zero most of the time (which is reflected in their strong supergaussianity), we could assume that only two of the signals are nonzero at the same time, and successfully reconstruct those two signals [272].

16.2 ESTIMATION OF THE MIXING MATRIX

16.2.1 Maximizing joint likelihood

To estimate the mixing matrix, one can use maximum likelihood estimation. In the simplest case of ML estimation, we formulate the joint likelihood of $\mathbf{A}$ and the realization of the s_i, and maximize it with respect to all these variables. It is slightly simpler to use a noisy version of the joint likelihood. This is of the same form as the one in Eq. (15.16):

$$\log L(\mathbf{A}, \mathbf{s}(1), ..., \mathbf{s}(T)) = -\sum_{t=1}^{T} \left[\frac{1}{2\sigma^2} \|\mathbf{A}\mathbf{s}(t) - \mathbf{x}(t)\|^2 + \sum_{i=1}^{n} f_i(s_i(t)) \right] + C \tag{16.7}$$

where σ^2 is the noise variance, here assumed to be infinitely small, the $s(t)$ are the realizations of the independent components, and C is an irrelevant constant. The functions f_i are the log-densities of the independent components.

Maximization of (16.7) with respect to $\mathbf{A}$ and s_i could be accomplished by a global gradient descent with respect to all the variables [337]. Another approach to maximization of the likelihood is to use an alternating variables technique [195], in which we first compute the ML estimates of the $s_i(t)$ for a fixed $\mathbf{A}$ and then, using this new $\mathbf{A}$, we compute the ML estimates of the $s_i(t)$, and so on. The ML estimate of the $s_i(t)$ for a given $\mathbf{A}$ is given by the methods of the preceding section, considering the noise to be infinitely small. The ML estimate of $\mathbf{A}$ for given $s_i(t)$ can be computed as

$$\mathbf{A} = \left(\sum_t \mathbf{x}(t)\mathbf{x}(t)^T\right)^{-1} \sum_t \mathbf{x}(t)\mathbf{s}(t)^T \tag{16.8}$$

This algorithm needs some extra stabilization, however. For example, normalizing the estimates of the s_i to unit norm is necessary. Further stabilization can be obtained by first whitening the data. Then we have (considering infinitely small noise)

$$E\{\mathbf{x}\mathbf{x}^T\} = \mathbf{A}\mathbf{A}^T = \mathbf{I} \tag{16.9}$$

which means that the rows of $\mathbf{A}$ form an orthonormal system. This orthonormality could be enforced after every step of (16.8), for further stabilization.

16.2.2 Maximizing likelihood approximations

Maximization of the joint likelihood is a rather crude method of estimation. From a Bayesian viewpoint, what we really want to maximize is the *marginal* posterior probability of the mixing matrix. (For basic concepts of Bayesian estimation, see Section 4.6.)

A more sophisticated form of maximum likelihood estimation can be obtained by using a Laplace approximation of the posterior distribution of $\mathbf{A}$. This improves the stability of the algorithm, and has been successfully used for estimation of overcomplete bases from image data [274], as well as for separation of audio signals [272]. For details on the Laplace approximation, see [275]. An alternative for the Laplace approximation is provided by ensemble learning; see Section 17.5.1.

A promising direction of research is given by Monte Carlo methods. These are a class of methods often used in Bayesian estimation, and are based on numerical integration using stochastic algorithms. One method in this class, Gibbs sampling, has been used in [338] for overcomplete basis estimation. Monte Carlo methods typically give estimators with good statistical properties; the drawback is that they are computationally very demanding.

Also, one could use an expectation-maximization (EM) algorithm [310, 19]. Using gaussian mixtures as models for the distributions of the independent components, the algorithm can be derived in analytical form. The problem is, however, that its complexity grows exponentially with the dimension of s, and thus it can only be used in small dimensions. Suitable approximations of the algorithm might alleviate this limitation [19].

A very different approximation of the likelihood method was derived in [195], in which a form of competitive neural learning was used to estimate overcomplete bases with supergaussian data. This is a computationally powerful approximation that seems to work for certain data sets. The idea is that the extreme case of sparsity or supergaussianity is encountered when at most one of the ICs is nonzero at any one time. Thus we could simply assume that only one of the components is nonzero for a given data point, for example, the one with the highest value in the pseudoinverse reconstruction. This is not a realistic assumption in itself, but it may give an interesting approximation of the real situation in some cases.

16.2.3 Approximate estimation using quasiorthogonality

The maximum likelihood methods discussed in the preceding sections give a well-justified approach to ICA estimation with overcomplete bases. The problem with most of the methods in the preceding section is that they are computationally quite expensive. A typical application of ICA with overcomplete bases is, however, feature extraction. In feature extraction, we usually have spaces of very high dimensions. Therefore, we show here a method [203] that is more heuristically justified, but has the advantage of being not more expensive computationally than methods for basic ICA estimation. This method is based on the FastICA algorithm, combined with the concept of quasiorthogonality.

Sparse approximately uncorrelated decompositions Our heuristic approach is justified by the fact that in feature extraction for many kinds of natural data, the ICA model is only a rather coarse approximation. In particular, the number of potential "independent components" seems to be infinite: The set of such components is closer to a continuous manifold than a discrete set. One evidence for this is that classic ICA estimation methods give different basis vectors when started with different initial values, and the number of components thus produced does not seem to be limited. Any classic ICA estimation method gives a rather arbitrary collection of components which are somewhat independent, and have sparse marginal distributions.

We can also assume, for simplicity, that the data is prewhitened as a preprocessing step, as in most ICA method in Part II. Then the independent components are simply given by the dot-products of the whitened data vector $\mathbf{z}$ with the basis vectors $\mathbf{a}_i$.

Due to the preceding considerations, we assume in our approach that what is usually needed, is a collection of basis vectors that has the following two properties:

1. The dot-products $\mathbf{a}_i^T \mathbf{z}$ of the observed data with the basis vectors have sparse (supergaussian) marginal distributions.

2. The $\mathbf{a}_i^T \mathbf{z}$ should be approximately uncorrelated ("quasiuncorrelated"). Equivalently, the vectors $\mathbf{a}_i$ should be approximately orthogonal ("quasiorthogonal").

A decomposition with these two properties seems to capture the essential properties of the decomposition obtained by estimation of the ICA model. Such decompositions could be called sparse approximately uncorrelated decompositions.

It is clear that it is possible to find highly overcomplete basis sets that have the first property of these two. Classic ICA estimation is usually based on maximizing the sparseness (or, in general, nongaussianity) of the dot-products, so the existence of several different classic ICA decompositions for a given image data set shows the existence of decompositions with the first property.

What is not obvious, however, is that it is possible to find strongly overcomplete decompositions such that the dot-products are approximately uncorrelated. The main point here is that this is possible because of the phenomenon of quasiorthogonality.

Quasiorthogonality in high-dimensional spaces   Quasiorthogonality [247] is a somewhat counterintuitive phenomenon encountered in very high-dimensional spaces. In a certain sense, there is much more room for vectors in high-dimensional spaces. The point is that in an n-dimensional space, where n is large, it is possible to have (say) $2n$ vectors that are practically orthogonal, i.e., their angles are close to 90 degrees. In fact, when n grows, the angles can be made arbitrarily close to 90 degrees. This must be contrasted with small-dimensional spaces: If, for example, $n = 2$, even the maximally separated $2n = 4$ vectors exhibit angles of 45 degrees.

For example, in image decomposition, we are usually dealing with spaces whose dimensions are of the order of 100. Therefore, we can easily find decompositions of, say, 400 basis vectors, such that the vectors are quite orthogonal, with practically all the angles between basis vectors staying above 80 degrees.

FastICA with quasiorthogonalization   To obtain a quasiuncorrelated sparse decomposition as defined above, we need two things. First, a method for finding vectors $\mathbf{a}_i$ that have maximally sparse dot-products, and second, a method of quasiorthogonalization of such vectors. Actually, most classic ICA algorithms can be considered as maximizing the nongaussianity of the dot-products with the basis vectors, provided that the data is prewhitened. (This was shown in Chapter 8.) Thus the main problem here is constructing a proper method for quasidecorrelation.

We have developed two methods for quasidecorrelation: one of them is symmetric and the other one is deflationary. This dichotomy is the same as in ordinary decorrelation methods used in ICA. As above, it is here assumed that the data is whitened.

A simple way of achieving quasiorthogonalization is to modify the ordinary deflation scheme based on a Gram-Schmidt-like orthogonalization. This means that we estimate the basis vectors one by one. When we have estimated p basis vectors $\mathbf{a}_1, ..., \mathbf{a}_p$, we run the one-unit fixed-point algorithm for $\mathbf{a}_{p+1}$, and after every iteration step subtract from $\mathbf{a}_{p+1}$ a certain proportion of the 'projections' $\mathbf{a}_{p+1}^T \mathbf{a}_j \mathbf{a}_j, j = 1, ..., p$ of the previously estimated p vectors, and then renormalize

$\mathbf{a}_{p+1}$:

$$1.\mathbf{a}_{p+1} \leftarrow \mathbf{a}_{p+1} - \alpha \sum_{j=1}^{p} \mathbf{a}_{p+1}^{T} \mathbf{a}_{j} \mathbf{a}_{j}$$
$$2.\mathbf{a}_{p+1} \leftarrow \mathbf{a}_{p+1}/\|\mathbf{a}_{p+1}\| \tag{16.10}$$

where α is a constant determining the force of quasiorthogonalization. If $\alpha = 1$, we have ordinary, perfect orthogonalization. We have found in our experiments that an α in the range $[0.1 \ldots 0.3]$ is sufficient in spaces where the dimension is 64.

In certain applications it may be desirable to use a symmetric version of quasiorthogonalization, in which no vectors are "privileged" over others [210, 197]. This can be accomplished, for example, by the following algorithm:

1. $\mathbf{A} \leftarrow \frac{3}{2}\mathbf{A} - \frac{1}{2}\mathbf{A}\mathbf{A}^{T}\mathbf{A}$.
2. Normalize each column of $\mathbf{A}$ to unit norm $\tag{16.11}$

which is closely related to the iterative symmetric orthogonalization method used for basic ICA in Section 8.4.3. The present algorithm is simply doing one iteration of the iterative algorithm. In some cases, it may be necessary to do two or more iterations, although in the experiments below, just one iteration was sufficient.

Thus, the algorithm that we propose is similar to the FastICA algorithm as described, e.g. in Section 8.3.5 in all other respects than the orthogonalization, which is replaced by one of the preceding quasiorthogonalization methods.

Experiments with overcomplete image bases We applied our algorithm on image windows (patches) of 8×8 pixels taken from natural images. Thus, we used ICA for feature extraction as explained in detail in Chapter 21.

The mean of the image window (DC component) was removed as a preprocessing step, so the dimension of the data was 63. Both deflationary and symmetric quasiorthogonalization were used. The nonlinearity used in the FastICA algorithm was the hyperbolic tangent. Fig. 16.1 shows an estimated approximately 4 times overcomplete basis (with 240 components). The sample size was 14000. The results shown here were obtained using the symmetric approach; the deflationary approach yielded similar results, with the parameter α fixed at 0.1.

The results show that the estimated basis vectors are qualitatively quite similar to those obtained by other, computationally more expensive methods [274]; they are also similar to those obtained by basic ICA (see Chapter 21). Moreover, by computing the dot-products between different basis vectors, we see that the basis is, indeed, quasiorthogonal. This validates our heuristic approach.

16.2.4 Other approaches

We mention here some other algorithms for estimation of overcomplete bases. First, in [341], independent components with binary values were considered, and a geometrically motivated method was proposed. Second, a tensorial algorithm for the overcomplete estimation problem was proposed in [63]. Related theoretical results were derived in [58]. Third, a natural gradient approach was developed in [5]. Fur-

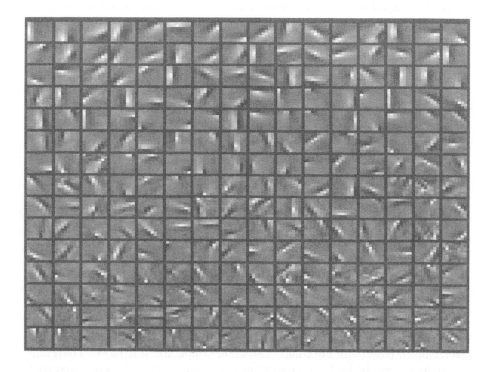

Fig. 16.1 The basis vectors of a 4 times overcomplete basis. The dimension of the data is 63 (excluding the DC component, i.e., the local mean) and the number of basis vectors is 240. The results are shown in the original space, i.e., the inverse of the preprocessing (whitening) was performed. The symmetric approach was used. The basis vectors are very similar to Gabor functions or wavelets, as is typical with image data (see Chapter 21).

ther developments on estimation of overcomplete bases using methods similar to the preceding quasiorthogonalization algorithm can be found in [208].

16.3 CONCLUDING REMARKS

The ICA problem becomes much more complicated if there are more independent components than observed mixtures. Basic ICA methods cannot be used as such. In most practical applications, it may be more useful to use the basic ICA model as an approximation of the overcomplete basis model, because the estimation of the basic model can be performed with reliable and efficient algorithms.

When the basis is overcomplete, the formulation of the likelihood is difficult, since the problem belongs to the class of missing data problems. Methods based on maximum likelihood estimation are therefore computationally rather inefficient. To obtain computationally efficient algorithms, strong approximations are necessary. For example, one can use a modification of the FastICA algorithm that is based on finding a quasidecorrelating sparse decomposition. This algorithm is computationally very efficient, reducing the complexity of overcomplete basis estimation to that of classic ICA estimation.

17

Nonlinear ICA

This chapter deals with independent component analysis (ICA) for nonlinear mixing models. A fundamental difficulty in the nonlinear ICA problem is that it is highly nonunique without some extra constraints, which are often realized by using a suitable regularization. We also address the nonlinear blind source separation (BSS) problem. Contrary to the linear case, we consider it different from the respective nonlinear ICA problem. After considering these matters, some methods introduced for solving the nonlinear ICA or BSS problems are discussed in more detail. Special emphasis is given to a Bayesian approach that applies ensemble learning to a flexible multilayer perceptron model for finding the sources and nonlinear mixing mapping that have most probably given rise to the observed mixed data. The efficiency of this method is demonstrated using both artificial and real-world data. At the end of the chapter, other techniques proposed for solving the nonlinear ICA and BSS problems are reviewed.

17.1 NONLINEAR ICA AND BSS

17.1.1 The nonlinear ICA and BSS problems

In many situations, the basic linear ICA or BSS model

$$x = As = \sum_{j=1}^{n} s_j a_j \qquad (17.1)$$

is too simple for describing the observed data x adequately. Hence, it is natural to consider extension of the linear model to nonlinear mixing models. For instantaneous mixtures, the nonlinear mixing model has the general form

$$\mathbf{x} = \mathbf{f}(\mathbf{s}) \tag{17.2}$$

where $\mathbf{x}$ is the observed m-dimensional data (mixture) vector, $\mathbf{f}$ is an unknown real-valued m-component mixing function, and $\mathbf{s}$ is an n-vector whose elements are the n unknown independent components.

Assume now for simplicity that the number of independent components n equals the number of mixtures m. The general nonlinear ICA problem then consists of finding a mapping $\mathbf{h} : \mathbb{R}^n \to \mathbb{R}^n$ that gives components

$$\mathbf{y} = \mathbf{h}(\mathbf{x}) \tag{17.3}$$

that are statistically independent. A fundamental characteristic of the nonlinear ICA problem is that in the general case, solutions always exist, and they are highly nonunique. One reason for this is that if x and y are two independent random variables, any of their functions $f(x)$ and $g(y)$ are also independent. An even more serious problem is that in the nonlinear case, x and y can be mixed and still statistically independent, as will be shown below. This is not unlike in the case of gaussian ICs in a linear mixing.

In this chapter, we define BSS in a special way to clarify the distinction between finding independent components, and finding the original sources. Thus, in the respective nonlinear BSS problem, one should find the original source signals $\mathbf{s}$ that have generated the observed data. This is usually a clearly more meaningful and unique problem than nonlinear ICA defined above, provided that suitable prior information is available on the sources and/or the mixing mapping. It is worth emphasizing that if some arbitrary independent components are found for the data generated by (17.2), they may be quite different from the true source signals. Hence the situation differs greatly from the basic linear data model (17.1), for which the ICA or BSS problems have the same solution. Generally, solving the nonlinear BSS problem is not easy, and requires additional prior information or suitable regularizing constraints.

An important special case of the general nonlinear mixing model (17.2) consists of so-called *post-nonlinear mixtures*. There each mixture has the form

$$x_i = f_i \left(\sum_{j=1}^{n} a_{ij} s_j \right), \qquad i = 1, \dots, n \tag{17.4}$$

Thus the sources s_j, $j = 1, \dots, n$ are first mixed linearly according to the basic ICA/BSS model (17.1), but after that a nonlinear function f_i is applied to them to get the final observations x_i. It can be shown [418] that for the post-nonlinear mixtures, the indeterminacies are usually the same as for the basic linear instantaneous mixing model (17.1). That is, the sources can be separated or the independent components estimated up to the scaling, permutation, and sign indeterminacies under weak conditions on the mixing matrix $\mathbf{A}$ and source distributions. The post-nonlinearity assumption is useful and reasonable in many signal processing applications, because it can be thought of as a model for a nonlinear sensor distortion. In more general situations, it is a restrictive and somewhat arbitrary constraint. This model will be treated in more detail below.

Another difficulty in the general nonlinear BSS (or ICA) methods proposed thus far is that they tend to be computationally rather demanding. Moreover, the computational load usually increases very rapidly with the dimensionality of the problem, preventing in practice the application of nonlinear BSS methods to high-dimensional data sets.

The nonlinear BSS and ICA methods presented in the literature could be divided into two broad classes: *generative approaches* and *signal transformation approaches* [438]. In the generative approaches, the goal is to find a specific model that explains how the observations were generated. In our case, this amounts to estimating both the source signals s and the unknown mixing mapping $\mathbf{f}(\cdot)$ that have generated the observed data x through the general mapping (17.2). In the signal transformation methods, one tries to estimate the sources directly using the inverse transformation (17.3). In these methods, the number of estimated sources is the same as the number of observed mixtures [438].

17.1.2 Existence and uniqueness of nonlinear ICA

The question of existence and uniqueness of solutions for nonlinear independent component analysis has been addressed in [213]. The authors show that there always exists an infinity of solutions if the space of the nonlinear mixing functions $\mathbf{f}$ is not limited. They also present a method for constructing parameterized families of nonlinear ICA solutions. A unique solution (up to a rotation) can be obtained in the two-dimensional special case if the mixing mapping $\mathbf{f}$ is constrained to be a conformal mapping together with some other assumptions; see [213] for details.

In the following, we present in more detail the constructive method introduced in [213] that always yields at least one solution to the nonlinear ICA problem. This procedure might be considered as a generalization of the well-known Gram-Schmidt orthogonalization method. Given m independent variables $\mathbf{y} = (y_1, \dots, y_m)$ and a variable x, a new variable $y_{m+1} = g(\mathbf{y}, x)$ is constructed so that the set $y_1, \dots, y_{m+1}$ is mutually independent.

The construction is defined recursively as follows. Assume that we have already m independent random variables $y_1, \dots, y_m$ which are jointly uniformly distributed in $[0, 1]^m$. Here it is not a restriction to assume that the distributions of the y_i are uniform, since this follows directly from the recursion, as will be seen below; for a single variable, uniformity can be attained by the probability integral transformation; see (2.85). Denote by x any random variable, and by $a_1, \dots, a_m, b$ some nonrandom scalars. Define

$$g(a_1, \ldots, a_m, b; p_{\mathbf{y},x}) = P(x \leq b | y_1 = a_1, \ldots, y_m = a_m)$$
(17.5)

$$= \frac{\int_{-\infty}^{b} p_{\mathbf{y},x}(a_1, \ldots, a_m, \xi) d\xi}{p_{\mathbf{y}}(a_1, \ldots, a_m)}$$

where $p_{\mathbf{y}}(\cdot)$ and $p_{\mathbf{y},x}(\cdot)$ are the marginal probability densities of $\mathbf{y}$ and $(\mathbf{y}, x)$, respectively (it is assumed here implicitly that such densities exist), and $P(\cdot|\cdot)$ denotes the conditional probability. The $p_{\mathbf{y},x}$ in the argument of g is to remind that g depends on the joint probability distribution of $\mathbf{y}$ and x. For $m = 0$, g is simply the cumulative distribution function of x. Now, g as defined above gives a nonlinear decomposition, as stated in the following theorem.

Theorem 17.1 *Assume that $y_1, \ldots, y_m$ are independent scalar random variables that have a joint uniform distribution in the unit cube $[0, 1]^m$. Let x be any scalar random variable. Define g as in* (17.5), *and set*

$$y_{m+1} = g(y_1, \ldots, y_m, x; p_{\mathbf{y},x})$$
(17.6)

Then y_{m+1} is independent from the $y_1, \ldots, y_m$, and the variables $y_1, \ldots, y_{m+1}$ are jointly uniformly distributed in the unit cube $[0, 1]^{m+1}$.

The theorem is proved in [213]. The constructive method given above can be used to decompose n variables $x_1, \ldots, x_n$ into n independent components $y_1, \ldots, y_n$, giving a solution for the nonlinear ICA problem.

This construction also clearly shows that the decomposition in independent components is by no means unique. For example, we could first apply a linear transformation on the $\mathbf{x}$ to obtain another random vector $\mathbf{x}' = \mathbf{Lx}$, and then compute $\mathbf{y}' = \mathbf{g}'(\mathbf{x}')$ with $\mathbf{g}'$ being defined using the above procedure, where $\mathbf{x}$ is replaced by $\mathbf{x}'$. Thus we obtain another decomposition of $\mathbf{x}$ into independent components. The resulting decomposition $\mathbf{y}' = \mathbf{g}'(\mathbf{Lx})$ is in general different from $\mathbf{y}$, and cannot be reduced to $\mathbf{y}$ by any simple transformations. A more rigorous justification of the nonuniqueness property has been given in [213].

Lin [278] has recently derived some interesting theoretical results on ICA that are useful in describing the nonuniqueness of the general nonlinear ICA problem. Let the matrices $\mathbf{H}_s$ and $\mathbf{H}_x$ denote the Hessians of the logarithmic probability densities $\log p_s(\mathbf{s})$ and $\log p_x(\mathbf{x})$ of the source vector $\mathbf{s}$ and mixture (data) vector $\mathbf{x}$, respectively. Then for the basic linear ICA model (17.1) it holds that

$$\mathbf{H}_s = \mathbf{A}^T \mathbf{H}_x \mathbf{A}$$
(17.7)

where $\mathbf{A}$ is the mixing matrix. If the components of $\mathbf{s}$ are truly independent, $\mathbf{H}_s$ should be a diagonal matrix. Due to the symmetry of the Hessian matrices $\mathbf{H}_s$ and $\mathbf{H}_x$, Eq. (17.7) imposes $n(n-1)/2$ constraints for the elements of the $n \times n$ matrix $\mathbf{A}$. Thus a constant mixing matrix $\mathbf{A}$ can be solved by estimating $\mathbf{H}_x$ at two different points, and assuming some values for the diagonal elements of $\mathbf{H}_s$.

If the nonlinear mapping (17.2) is twice differentiable, we can approximate it locally at any point by the linear mixing model (17.1). There $\mathbf{A}$ is defined by the first order term $\partial \mathbf{f}(\mathbf{s})/\partial \mathbf{s}$ of the Taylor series expansion of $\mathbf{f}(\mathbf{s})$ at the desired point. But now $\mathbf{A}$ generally changes from point to point, so that the constraint conditions (17.7) still leave $n(n-1)/2$ degrees of freedom for determining the mixing matrix $\mathbf{A}$ (omitting the diagonal elements). This also shows that the nonlinear ICA problem is highly nonunique.

Taleb and Jutten have considered separability of nonlinear mixtures in [418, 227]. Their general conclusion is the same as earlier: Separation is impossible without additional prior knowledge on the model, since the independence assumption alone is not strong enough in the general nonlinear case.

17.2 SEPARATION OF POST-NONLINEAR MIXTURES

Before discussing approaches applicable to general nonlinear mixtures, let us briefly consider blind separation methods proposed for the simpler case of post-nonlinear mixtures (17.4). Especially Taleb and Jutten have developed BSS methods for this case. Their main results have been represented in [418], and a short overview of their studies on this problem can be found in [227]. In the following, we present the the main points of their method.

A separation method for the post-nonlinear mixtures (17.4) should generally consist of two subsequent parts or stages:

1. A *nonlinear stage*, which should cancel the nonlinear distortions f_i, $i = 1, \ldots, n$. This part consists of nonlinear functions $g_i(\boldsymbol{\theta}_i, u)$. The parameters $\boldsymbol{\theta}_i$ of each nonlinearity g_i are adjusted so that cancellation is achieved (at least roughly).

2. A *linear stage* that separates the approximately linear mixtures $\mathbf{v}$ obtained after the nonlinear stage. This is done as usual by learning a $n \times n$ separating matrix $\mathbf{B}$ for which the components of the output vector $\mathbf{y} = \mathbf{B}\mathbf{v}$ of the separating system are statistically independent (or as independent as possible).

Taleb and Jutten [418] use the mutual information $I(\mathbf{y})$ between the components $y_1, \ldots, y_n$ of the output vector (see Chapter 10) as the cost function and independence criterion in both stages. For the linear part, minimization of the mutual information leads to the familiar Bell-Sejnowski algorithm (see Chapters 10 and 9)

$$\frac{\partial I(\mathbf{y})}{\partial \mathbf{B}} = -\mathrm{E}\{\boldsymbol{\psi}\mathbf{x}^T\} - (\mathbf{B}^T)^{-1} \tag{17.8}$$

where components ψ_i of the vector $\boldsymbol{\psi}$ are score functions of the components y_i of the output vector $\mathbf{y}$:

$$\psi_i(u) = \frac{d}{du} \log p_i(u) = \frac{p_i'(u)}{p_i(u)} \tag{17.9}$$

Here $p_i(u)$ is the probability density function of y_i and $p_i'(u)$ its derivative. In practice, the natural gradient algorithm is used instead of the Bell-Sejnowski algorithm (17.8); see Chapter 9.

For the nonlinear stage, one can derive the gradient learning rule [418]

$$
\frac{\partial I(\mathbf{y})}{\partial \boldsymbol{\theta}_k} = -\mathrm{E}\left\{ \frac{\partial \log | g_k'(\boldsymbol{\theta}_k, x_k) |}{\partial \boldsymbol{\theta}_k} \right\} - \mathrm{E}\left\{ \sum_{i=1}^{n} \psi_i(y_i) b_{ik} \frac{\partial g_k(\boldsymbol{\theta}_k, x_k) |}{\partial \boldsymbol{\theta}_k} \right\}
$$

Here x_k is the kth component of the input vector, b_{ik} is the element ik of the matrix $\mathbf{B}$, and g_k' is the derivative of the kth nonlinear function g_k. The exact computation algorithm depends naturally on the specific parametric form of the chosen nonlinear mapping $g_k(\boldsymbol{\theta}_k, x_k)$. In [418], a multilayer perceptron network is used for modeling the functions $g_k(\boldsymbol{\theta}_k, x_k)$, $k = 1, \ldots, n$.

In linear BSS, it suffices that the score functions (17.9) are of the right type for achieving separation. However, their appropriate estimation is critical for the good performance of the proposed nonlinear separation method. The score functions (17.9) must be estimated adaptively from the output vector $\mathbf{y}$. Several alternative ways to do this are considered in [418]. An estimation method based on the Gram-Charlier expansion performs appropriately only for mild post-nonlinear distortions. However, another method, which estimates the score functions directly, also provides very good results for hard nonlinearities. Experimental results are presented in [418]. A well performing batch type method for estimating the score functions has been introduced in a later paper [417].

Before proceeding, we mention that separation of post-nonlinear mixtures also has been studied in [271, 267, 469] using mainly extensions of the natural gradient algorithm.

17.3 NONLINEAR BSS USING SELF-ORGANIZING MAPS

One of the earliest ideas for achieving nonlinear BSS (or ICA) is to use Kohonen's self-organizing map (SOM) to that end. This method was originally introduced by Pajunen et al. [345]. The SOM [247, 172] is a well-known mapping and visualization method that in an unsupervised manner learns a nonlinear mapping from the data to a usually two-dimensional grid. The learned mapping from often high-dimensional data space to the grid is such that it tries to preserve the structure of the data as well as possible. Another goal in the SOM method is to map the data so that it would be uniformly distributed on the rectangular (or hexagonal) grid. This can be roughly achieved with suitable choices [345].

If the joint probability density of two random variables is uniformly distributed inside a rectangle, then clearly the marginal densities along the sides of the rectangle are statistically independent. This observation gives the justification for applying self-organizing map to nonlinear BSS or ICA. The SOM mapping provides the regularization needed in nonlinear BSS, because it tries to preserve the structure of the data. This implies that the mapping should be as simple as possible while achieving the desired goals.

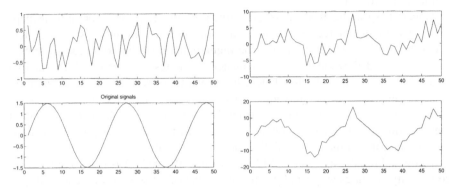

Fig. 17.1 Original source signals. **Fig. 17.2** Nonlinear mixtures.

The following experiment [345] illustrates the use of the self-organizing map in nonlinear blind source separation. There were two subgaussian source signals s_i shown in Fig. 17.1, consisting of a sinusoid and uniformly distributed white noise. Each source vector s was first mixed linearly using the mixing matrix

$$\mathbf{A} = \left[\begin{array}{cc} 0.7 & 0.3 \\ 0.3 & 0.7 \end{array} \right] \qquad (17.10)$$

After this, the data vectors x were obtained as post-nonlinear mixtures of the sources by applying the formula (17.4), where the nonlinearity $f_i(t) = t^3 + t, i = 1, 2$. These mixtures x_i are depicted in Fig. 17.2.

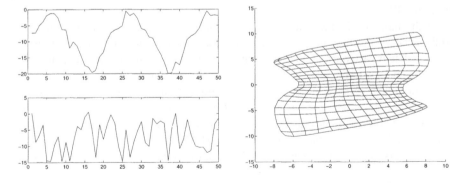

Fig. 17.3 Signals separated by SOM. **Fig. 17.4** Converged SOM map.

The sources separated by the SOM method are shown in Fig. 17.3, and the converged SOM map is illustrated in Fig. 17.4. The estimates of the source signals in Fig. 17.3 are obtained by mapping each data vector x onto the map of Fig. 17.4, and reading the coordinates of the mapped data vector. Even though the preceding experiment was carried out with post-nonlinear mixtures, the use of the SOM method is not limited to them.

Generally speaking, there are several difficulties in applying self-organizing maps to nonlinear blind source separation. If the sources are uniformly distributed, then it can be heuristically justified that the regularization of the nonlinear separating mapping provided by the SOM approximately separates the sources. But if the true sources are not uniformly distributed, the separating mapping providing uniform densities inevitably causes distortions, which are in general the more serious the farther the true source densities are from the uniform ones. Of course, the SOM method still provides an approximate solution to the nonlinear ICA problem, but this solution may have little to do with the true source signals.

Another difficulty in using SOM for nonlinear BSS or ICA is that computational complexity increases very rapidly with the number of the sources (dimensionality of the map), limiting the potential application of this method to small-scale problems. Furthermore, the mapping provided by the SOM is discrete, where the discretization is determined by the number of grid points.

17.4 A GENERATIVE TOPOGRAPHIC MAPPING APPROACH TO NONLINEAR BSS *

17.4.1 Background

The self-organizing map discussed briefly in the previous section is a nonlinear mapping method that is inspired by neurobiological modeling arguments. Bishop, Svensen and Williams introduced the generative topographic mapping (GTM) method as a statistically more principled alternative to SOM. Their method is presented in detail in [49].

In the basic GTM method, mutually similar impulse (delta) functions that are equispaced on a rectangular grid are used to model the discrete uniform density in the space of latent variables, or the joint density of the sources in our case. The mapping from the sources to the observed data, corresponding in our nonlinear BSS problem to the nonlinear mixing mapping (17.2), is modeled using a mixture-of-gaussians model. The parameters of the mixture-of-gaussians model, defining the mixing mapping, are then estimated using a maximum likelihood (ML) method (see Section 4.5) realized by the expectation-maximization (EM) algorithm [48, 172]. After this, the inverse (separating) mapping from the data to the latent variables (sources) can be determined.

It is well-known that any continuous smooth enough mapping can be approximated with arbitrary accuracy using a mixture-of-gaussians model with sufficiently many gaussian basis functions [172, 48]. Roughly stated, this provides the theoretical basis of the GTM method. A fundamental difference of the GTM method compared with SOM is that GTM is based on a *generative* approach that starts by assuming a model for the latent variables, in our case the sources. On the other hand, SOM tries to separate the sources directly by starting from the data and constructing a suitable separating signal transformation. A key benefit of GTM is its firm theoretical foundation which helps to overcome some of the limitations of SOM. This also provides the basis of generalizing the GTM approach to arbitrary source densities.

Using the basic GTM method instead of SOM for nonlinear blind source separation does not yet bring out any notable improvement, because the densities of the sources are still assumed to be uniform. However, it is straightforward to generalize the GTM method to *arbitrary known* source densities. The advantage of this approach is that one can directly regularize the inverse of the mixing mapping by using the known source densities. This modified GTM method is then used for finding a noncomplex mixing mapping. This approach is described in the following.

17.4.2   The modified GTM method

The modified GTM method introduced in [346] differs from the standard GTM [49] only in that the required joint density of the latent variables (sources) is defined as a *weighted* sum of delta functions instead of plain delta functions. The weighting coefficients are determined by discretizing the known source densities. Only the main points of the GTM method are presented here, with emphasis on the modifications made for applying it to nonlinear blind source separation. Readers wishing to gain a deeper understanding of the GTM method should look at the original paper [49].

The GTM method closely resembles SOM in that it uses a discrete grid of points forming a regular array in the m-dimensional latent space. As in SOM, the dimension of the latent space is usually $m = 2$. Vectors lying in the latent space are denoted by $s(t)$; in our application they will be source vectors. The GTM method uses a set of L fixed nonlinear basis functions $\{\phi_j(s)\}$, $j = 1, \ldots, L$, which form a nonorthogonal basis set. These basis functions typically consist of a regular array of spherical gaussian functions, but the basis functions can at least in principle be of other types.

The mapping from the m-dimensional latent space to the n-dimensional data space, which is in our case the mixing mapping of Eq. (17.2), is in GTM modeled as a linear combination of basis functions φ_j:

$$x = f(s) = M\varphi(s), \quad \varphi = [\varphi_1, \varphi_2, \ldots, \varphi_L]^T \qquad (17.11)$$

Here M is an $n \times L$ matrix of weight parameters.

Denote the node locations in the latent space by α_i. Eq. (17.11) then defines a corresponding set of reference vectors

$$m_i = M\varphi(\alpha_i) \qquad (17.12)$$

in data space. Each of these reference vectors then forms the center of an isotropic gaussian distribution in data space. Denoting the common variance of these gaussians by β^{-1}, we get

$$p_x(x \mid i) = \left(\frac{\beta}{2\pi}\right)^{n/2} \exp\left(-\frac{\beta}{2} \parallel m_i - x \parallel^2\right) \qquad (17.13)$$

The probability density function for the GTM model is obtained by summing over all of the gaussian components, yielding

$$
\begin{aligned}
p_{\mathbf{x}}(\mathbf{x}(t) \mid \mathbf{M}, \beta) &= \sum_{i=1}^{K} P(i) p_{\mathbf{x}}(\mathbf{x} \mid i) \\
&= \sum_{i=1}^{K} \frac{1}{K} \left(\frac{\beta}{2\pi} \right)^{n/2} \exp \left(-\frac{\beta}{2} \parallel \mathbf{m}_i - \mathbf{x} \parallel^2 \right) \quad (17.14)
\end{aligned}
$$

Here K is the total number of gaussian components, which is equal to the number of grid points in latent space, and the prior probabilities $P(i)$ of the gaussian components are all equal to $1/K$.

GTM tries to represent the distribution of the observed data $\mathbf{x}$ in the n-dimensional data space in terms of a smaller m-dimensional nonlinear manifold [49]. The gaussian distribution in (17.13) represents a noise or error model which is needed because the data usually does not lie exactly in such a lower dimensional manifold. It is important to realize that the K gaussian distributions defined in (17.13) have nothing to do with the basis function φ_i, $i = 1, \dots, L$. Usually it is advisable that the number L of the basis functions is clearly smaller than the number K of node locations and their respective noise distributions (17.13). In this way, one can avoid overfitting and prevent the mixing mapping (17.11) to become overly complicated.

The unknown parameters in this model are the weight matrix $\mathbf{M}$ and the inverse variance β. These parameters are estimated by fitting the model (17.14) to the observed data vectors $\mathbf{x}(1), \mathbf{x}(2), \dots, \mathbf{x}(T)$ using the maximum likelihood method discussed earlier in Section 4.5. The log likelihood function of the observed data is given by

$$
L(\mathbf{M}, \beta) = \sum_{t=1}^{T} \log p_{\mathbf{x}}(\mathbf{x}(t) | \mathbf{M}, \beta) = \sum_{t=1}^{T} \log \int p_{\mathbf{x}}(\mathbf{x}(t) | \mathbf{s}, \mathbf{M}, \beta) p_{\mathbf{s}}(\mathbf{s}) d\mathbf{s} \quad (17.15)
$$

where β^{-1} is the variance of $\mathbf{x}$ given $\mathbf{s}$ and $\mathbf{M}$, and T is the total number of data vectors $\mathbf{x}(t)$.

For applying the modified GTM method, the probability density function $p_{\mathbf{s}}(\mathbf{s})$ of the source vectors $\mathbf{s}$ should be *known*. Assuming that the sources $s_1, s_2, \dots, s_m$ are statistically independent, this joint density can be evaluated as the product of the marginal densities of the individual sources:

$$
p_{\mathbf{s}}(\mathbf{s}) = \prod_{i=1}^{m} p_i(s_i) \quad (17.16)
$$

Each marginal density is here a discrete density defined at the sampling points corresponding to the locations of the node vectors.

The latent space in the GTM method usually has a small dimension, typically $m = 2$. The method can be applied in principle for $m > 2$, but its computational load then increases quite rapidly just like in the SOM method. For this reason, only two sources s_1 and s_2 are considered in the following. The dimension of the latent space is chosen to be $m = 2$, and we use a rectangular $K_1 \times K_2$ grid with equispaced nodes, so that the total number of nodes is $K = K_1 \times K_2$. The locations of the node points in the grid are denoted by α_{ij}, $i = 1, \ldots, K_1$, $j = 1, \ldots, K_2$. One can then write

$$p_{\mathbf{s}}(\mathbf{s}) = \sum_{i=1}^{K_1} \sum_{j=1}^{K_2} a_{ij}\delta(\mathbf{s} - \alpha_{ij}) = \sum_{q=1}^{K} a_q\delta(\mathbf{s} - \alpha_q) \qquad (17.17)$$

The coefficients $a_{ij} = p_1(i)p_2(j)$, where $p_1(i)$ and $p_2(j)$ are the values of the marginal densities $p_1(s_1)$ and $p_2(s_2)$ corresponding to the location of the node point α_{ij}. In (17.17), $\delta(\cdot)$ is the Dirac delta function or impulse function which has the special property that $\int g(\mathbf{s})\delta(\mathbf{s} - \mathbf{s}_0)ds = g(\mathbf{s}_0)$ if the integration extends over the point $\mathbf{s}_0$, otherwise the integral is zero. In the last phase, the node points and their respective probabilities have been reindexed again using a single index q for easier notation. This can be done easily by going through all the node points in some prescribed order, for example rowwise.

Inserting (17.17) into (17.15) yields

$$L(\mathbf{M}, \beta) = \sum_{t=1}^{T} \log \left(\sum_{q=1}^{K} a_q p_{\mathbf{x}}(\mathbf{x}(t)|\alpha_q, \mathbf{M}, \beta) \right) \qquad (17.18)$$

Computing the gradient of this expression with respect to the weight matrix $\mathbf{M}$ and setting it to zero, after some manipulations yields the following equation for updating the weight matrix $\mathbf{M}$:

$$(\mathbf{\Phi}^T \mathbf{G}_{\text{old}} \mathbf{\Phi})\mathbf{M}_{\text{new}}^T = \mathbf{\Phi}^T \mathbf{R}_{\text{old}} \mathbf{X} \qquad (17.19)$$

In this formula, $\mathbf{X} = [\mathbf{x}(1), \ldots, \mathbf{x}(T)]^T$ is the data matrix, and the (q, j)-th element $f_{qj} = \varphi_j(\alpha_q)$ of the $K \times L$ matrix $\mathbf{\Phi}$ is the value of the jth basis function $\varphi_j(\cdot)$ at the qth node point α_q. Furthermore, $\mathbf{G}$ is a diagonal matrix with elements

$$G_{qq} = \sum_{t=1}^{T} R_{qt}(\mathbf{M}, \beta) \qquad (17.20)$$

and the elements of the responsibility matrix $\mathbf{R}$ are

$$R_{qt} = \frac{a_q p_{\mathbf{x}}(\mathbf{x}(t)|\alpha_q, \mathbf{M}, \beta)}{\sum_{k=1}^{K} a_k p_{\mathbf{x}}(\mathbf{x}(t)|\alpha_k, \mathbf{M}, \beta)} \qquad (17.21)$$

Then the variance parameter β can be updated using the formula

$$\frac{1}{\beta_{\text{new}}} = \frac{1}{Tn} \sum_{q=1}^{K} \sum_{t=1}^{T} R_{qt} \| \mathbf{M}_{\text{new}}\varphi(\alpha_q) - \mathbf{x}(t) \|^2 \qquad (17.22)$$

where n is the dimension of the data space.

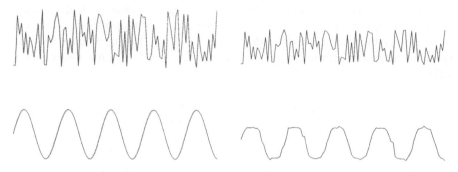

Fig. 17.5 Source signals.   **Fig. 17.6** Separated signals.

In GTM, the EM algorithm is used for maximizing the likelihood. Here the E-step (17.21) consists of computing the responsibilities R_{qt}, and the M-steps (17.19), (17.22) of updating the parameters $\mathbf{M}$ and β. The preceding derivation is quite similar to the one as in the original GTM method [49], only the prior density coefficients $a_{ij} = a_q$ have been added to the model.

After a few iterations, the EM algorithm converges to the parameter values $\mathbf{M}^*$ and β^* that maximize the log likelihood (17.15), at least locally. The optimum values $\mathbf{M}^*$ and β^* then specify the estimated probability density (17.14) that GTM provides for the data vectors $\mathbf{x}$. Because the prior density $p_{\mathbf{s}}(\mathbf{s})$ of the sources s is assumed to be known, it is then straightforward to compute the posterior density $p(\mathbf{s}(t) \mid \mathbf{x}(t), \mathbf{M}^*, \beta^*)$ of the sources given the observed data using the Bayes' rule. As mentioned in Chapter 4, this posterior density contains all the relevant information about the sources.

However, it is often convenient to choose a specific source estimate $\mathbf{s}(t)$ corresponding to each data vector $\mathbf{x}(t)$ for visualizing purposes. An often used estimate is the mean $\mathrm{E}\{\mathbf{s}(t) \mid \mathbf{x}(t), \mathbf{M}^*, \beta^*\}$ of the posterior density. It can be computed in the GTM method from the simple formula [49]

$$\hat{\mathbf{s}}(t) = \mathrm{E}\{\mathbf{s}(t) \mid \mathbf{x}(t), \mathbf{M}^*, \beta^*\} = \sum_{q=1}^{K} R_{qt}\alpha_q \qquad (17.23)$$

If the posterior density of the sources is multimodal, the posterior mean (17.23) can give misleading results. Then it is better to use for example the maximum a posteriori (MAP) estimate, which is simply the source value corresponding to the maximum responsibility $q_{\max} = \mathrm{argmax}(R_{qt}), q = 1, \ldots, K$, for each sample index t.

17.4.3 An experiment

In the following, a simple experiment involving two sources shown in Fig. 17.5 and three noisy nonlinear mixtures is described. The mixed data was generated by transforming linear mixtures of the original sources using a multilayer perceptron

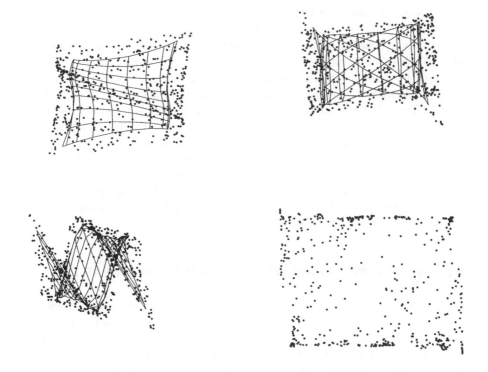

Fig. 17.7 Joint mixture densities with superimposed maps. Top left: Joint density $p(x_1, x_2)$ of mixtures x_1 and x_2. Top right: Joint density $p(x_1, x_3)$ of mixtures x_1 and x_3. Bottom left: Joint density $p(x_2, x_3)$ of mixtures x_2 and x_3. Bottom right: Joint density of the estimated two source signals.

network with a volume conserving architecture (see [104]). Such an architecture was chosen for ensuring that the total mixing mapping is bijective and therefore reversible, and for avoiding highly complex distortions of the source densities. However, this choice has the advantage that it makes the total mixing mapping more complex than the post-nonlinear model (17.4). Finally, gaussian noise was added to the mixtures.

The mixtures were generated using the model

$$\mathbf{x} = \mathbf{A}\mathbf{s} + \tanh(\mathbf{U}\mathbf{A}\mathbf{s}) + \mathbf{n} \tag{17.24}$$

where $\mathbf{U}$ is an upper-diagonal matrix with zero diagonal elements. The nonzero elements of $\mathbf{U}$ were drawn from a standard gaussian distribution. The matrix $\mathbf{U}$ ensures volume conservation of the nonlinearity applied to $\mathbf{A}\mathbf{s}$.

The modified GTM algorithm presented above was used to learn a separating mapping. For reducing scaling effects, the mixtures were first whitened. After whitening the mixtures are uncorrelated and have unit variance. Then the modified GTM algorithm was run for eight iterations using a 5×5 grid. The number of basis functions was $L = 7$.

The separated sources depicted in Fig. 17.6 can be compared with the original sources in Fig. 17.5. The waveforms of the original sources are approximately recovered, even though there is some inevitable distortion due to the noise, discretization, and the difficulty of the problem.

The mixtures are not shown here directly because they are in three dimensions. The two-dimensional marginal densities of the mixtures are shown in Fig. 17.7, however. They clearly reveal the nonlinearity of the mixing mapping; in particular, the joint density $p(x_2, x_3)$ of the second and third mixtures (components of the data vector x) is highly nonlinear. Also the joint density of the separated sources is shown in Fig. 17.7 in the subfigure at bottom right. It indicates that a factorizable density has been approximately obtained. The superimposed maps in the first 3 subfigures of Figure 17.7 were obtained by mapping a 10×10 grid of source vectors s to the mixture (data) space using the mapping (17.11) learned by the modified GTM algorithm.

17.5 AN ENSEMBLE LEARNING APPROACH TO NONLINEAR BSS

In this section, we present a new generative approach for nonlinear blind source separation or independent component analysis. Here the nonlinear mapping (17.2) from the unknown sources s to the known observations x is modeled using the familiar multilayer perceptron (MLP) network structure [172, 48]. MLP networks have the universal approximation property [172] for smooth continuous mappings, and they suit well for modeling both strongly and mildly nonlinear mappings. However, the learning procedure is based on unsupervised Bayesian ensemble learning. It is quite different from standard back-propagation learning which minimizes the mean-square representation error in MLP networks in a supervised manner [172, 48].

17.5.1 Ensemble learning

A flexible model family, such as MLP networks, provides infinitely many possible explanations of different complexity for the observed data. Choosing too complex a model results in overfitting, where the model tries to make up meaningless explanations for the noise in addition to the true sources or independent components. Choosing too simple a model results in underfitting, leaving hidden some of the true sources that have generated the data.

An appropriate solution to this problem is that no single model should actually be chosen. Instead, all the possible explanations should be taken into account and weighted according to their posterior probabilities. This approach, known as Bayesian learning [48], optimally solves the trade-off between under- and overfitting. All the relevant information needed in choosing an appropriate model is contained in the posterior probability density functions (pdf's) of different model structures. The posterior pdf's of too simple models are low, because they leave a lot of the data unexplained. On the other hand, too complex models occupy little probability mass, even though they often show a high but very narrow peak in their posterior pdf's corresponding to the overfitted parameters.

In practice, exact computation of the posterior pdf's of the models is impossible. Therefore, some suitable approximation method must be used. *Ensemble learning* [180, 25, 260], also known as variational learning, is a method for parametric approximation of posterior pdf's where the search takes into account the probability *mass* of the models. Therefore, it does not suffer from overfitting. The basic idea in ensemble learning is to minimize the misfit between the posterior pdf and its parametric approximation.

Let us denote by $X = \{\mathbf{x}(t)|t\}$ the set of available mixture (data) vectors, and by $S = \{\mathbf{s}(t)|t\}$ the respective source vectors. Denote by $\boldsymbol{\theta}$ all the unknown parameters of the mixture (data) model, which will be described in more detail in the next subsection. Furthermore, let $p(S, \boldsymbol{\theta}|X)$ denote the exact posterior pdf and $q(S, \boldsymbol{\theta}|X)$ its parametric approximation. The misfit is measured with the Kullback-Leibler (KL) divergence $\mathcal{J}_{KL}$ between the densities p and q, which is defined by the cost function

$$\mathcal{J}_{KL} = \mathrm{E}_q \left\{ \log \frac{q}{p} \right\} = \int q(S, \boldsymbol{\theta}|X) \log \frac{q(S, \boldsymbol{\theta}|X)}{p(S, \boldsymbol{\theta}|X)} d\boldsymbol{\theta} dS \tag{17.25}$$

The Kullback-Leibler divergence or distance measures the difference in the probability mass between the densities p and q. Its minimum value 0 is achieved when the two densities are the same; see Chapter 5.

17.5.2 Model structure

As mentioned before, MLP networks are now used for modeling the nonlinear mixing mapping $\mathbf{f}(\cdot)$ in (17.2). In the following, we represent the mixing model [259, 436] in more detail. Let $\mathbf{x}(t)$ denote the observed data vector at time t, and $\mathbf{s}(t)$ the vector of independent components (source signals) at time t. The matrices $\mathbf{Q}$ and $\mathbf{A}$ contain the weights of the output and the hidden layers of the network, respectively, and $\mathbf{b}$ and $\mathbf{a}$ are the respective bias vectors of the output and hidden layers. The vector of nonlinear activation functions, applied componentwise, is denoted by $\mathbf{g}(\cdot)$, and $\mathbf{n}(t)$ is a zero mean gaussian noise vector corrupting the observations. Using these notations, the mixing (data) model can be written

$$\mathbf{x}(t) = \mathbf{Q}\mathbf{g}(\mathbf{A}\mathbf{s}(t) + \mathbf{a}) + \mathbf{b} + \mathbf{n}(t) \tag{17.26}$$

The hyperbolic tangent $g(y) = \tanh(y)$ is used as the nonlinear activation function, which is a typical choice in MLP networks. Other continuous activation functions could be used, too. The sources are assumed to be independent, and they are modeled by mixtures of gaussians. The independence assumption is natural, because the goal of the model is to find the underlying independent components of the observations. By using mixtures of gaussians, one can model sufficiently well any nongaussian distribution of the sources [48]. This type of representation has earlier been successfully applied to the standard linear ICA model in [258].

The parameters of the model are (1) the weight matrices $\mathbf{A}$ and $\mathbf{Q}$, and the bias vectors $\mathbf{a}$ and $\mathbf{b}$; (2) the parameters of the distributions of the noise, source signals and column vectors of the weight matrices; (3) the hyperparameters used for defining the distributions of the biases and the parameters in the group (2). For a more detailed description, see [259, 436]. All the parameterized distributions are assumed to be gaussian, except for the sources, which are modeled as mixtures of gaussians. This does not limit the generality of the approach severely, but makes computational implementation simpler and much more efficient. The hierarchical description of the distributions of the parameters of the model used here is a standard procedure in probabilistic Bayesian modeling. Its strength lies in that knowledge about equivalent status of different parameters can be easily incorporated. For example, all the variances of the noise components have a similar status in the model. This is reflected by the fact that their distributions are assumed to be governed by common hyperparameters.

17.5.3   Computing Kullback-Leibler cost function *

In this subsection, the Kullback-Leibler cost function $\mathcal{J}_{KL}$ defined earlier in Eq. (17.25) in considered in more detail. For approximating and then minimizing it, we need two things: the exact formulation of the posterior density $p(S, \boldsymbol{\theta}|X)$ and its parametric approximation $q(S, \boldsymbol{\theta}|X)$.

According to the Bayes' rule, the posterior pdf of the unknown variables S and $\boldsymbol{\theta}$ is

$$p(S, \boldsymbol{\theta}|X) = \frac{p(X|S, \boldsymbol{\theta})p(S|\boldsymbol{\theta})p(\boldsymbol{\theta})}{p(X)} \tag{17.27}$$

The probability density $p(X|S, \boldsymbol{\theta})$ of the data X given the sources S and the parameters $\boldsymbol{\theta}$ is obtained from Eq. (17.26). Let us denote the variance of the ith component of the noise vector $\mathbf{n}(t)$ by by σ_i^2. The distribution $p(x_i(t)|\mathbf{s}(t), \boldsymbol{\theta})$ of the ith component $x_i(t)$ of the data vector $\mathbf{x}(t)$ is thus gaussian with the mean $\mathbf{q}_i^T \mathbf{g}(\mathbf{A}\mathbf{s}(t) + \mathbf{a}) + b_i$ and variance σ_i^2. Here $\mathbf{q}_i^T$ denotes the ith row vector of the weight matrix $\mathbf{Q}$, and b_i is the ith component of the bias vector $\mathbf{b}$. As usually, the noise components $n_i(t)$ are assumed to be independent, and therefore

$$p(X|S, \boldsymbol{\theta}) = \prod_{t=1}^{T} \prod_{i=1}^{n} p(x_i(t)|\mathbf{s}(t), \boldsymbol{\theta}) \tag{17.28}$$

The terms $p(S|\boldsymbol{\theta})$ and $p(\boldsymbol{\theta})$ in (17.27) are also products of simple gaussian distributions, and they are obtained directly from the definition of the model structure [259, 436]. The term $p(X)$ does not depend on the model parameters and can be neglected.

The approximation $q(S, \boldsymbol{\theta} \mid X)$ must be simple enough for mathematical tractability and computational efficiency. First, we assume that the source signals S are independent of the other parameters $\boldsymbol{\theta}$, so that $q(S, \boldsymbol{\theta} \mid X)$ decouples into

$$q(S, \boldsymbol{\theta} \mid X) = q(S \mid X)q(\boldsymbol{\theta} \mid X) \tag{17.29}$$

For the parameters $\boldsymbol{\theta}$, a gaussian density with a diagonal covariance matrix is used. This implies that the approximation $q(\boldsymbol{\theta} \mid X)$ is a product of independent gaussian distributions:

$$q(\boldsymbol{\theta} \mid X) = \prod_j q_j(\theta_j \mid X) \tag{17.30}$$

The parameters of each gaussian component density $q_j(\theta_j \mid X)$ are its mean $\bar{\theta}_i$ and variance $\tilde{\theta}_i$.

The source signals $s_i(t)$, $i = 1, \ldots, n$ are assumed to be mutually statistically independent, and also at different time instants (sample values) $t = 1, \ldots, T$, so that

$$q(S \mid X) = \prod_{t=1}^{T} \prod_{i=1}^{n} q_{ti}(s_i(t) \mid X) \tag{17.31}$$

Here the component densities $q_{ti}(s_i(t) \mid X)$ are modeled as mixtures of gaussian densities. Inserting Eqs. (17.31) and (17.30) into (17.29) provides the desired approximation $q(S, \boldsymbol{\theta} \mid X)$ of the posterior density.

Both the posterior density $p(S, \boldsymbol{\theta}|X)$ and its approximation $q(S, \boldsymbol{\theta}|X)$ are products of simple gaussian or mixture-of-gaussians terms, which simplifies the cost function (17.25) considerably: it splits into expectations of many simple terms. The terms of the form $E_q\{\log q_j(S, \theta_j \mid X)\}$ are negative entropies of gaussians, having the exact values $-(1 + \log 2\pi\tilde{\theta}_j)/2$. The terms of the form $-E_q\{\log p(x_i(t)|\mathbf{s}(t), \boldsymbol{\theta})\}$ are most difficult to handle. They are approximated by applying second order Taylor series expansions of the nonlinear activation functions as explained in [259, 436]. The remaining terms are expectations of simple gaussian terms, which can be computed as in [258].

The cost function $\mathcal{J}_{KL}$ depends on the posterior means $\bar{\theta}_i$ and variances $\tilde{\theta}_i$ of the parameters of the network and the source signals. This is because instead of finding a point estimate, the joint posterior pdf of the sources and parameters is estimated in ensemble learning. The variances give information about the reliability of the estimates.

Let us denote the two parts of the cost function (17.25) arising from the denominator and numerator of the logarithm by, respectively, $\mathcal{J}_p = -E_q\{\log p\}$ and $\mathcal{J}_q = E_q\{\log q\}$. The variances $\tilde{\theta}_i$ are obtained by differentiating (17.25) with respect to $\tilde{\theta}_i$ [259, 436]:

$$\frac{\partial \mathcal{J}_{KL}}{\partial \tilde{\theta}} = \frac{\partial \mathcal{J}_p}{\partial \tilde{\theta}} + \frac{\partial \mathcal{J}_q}{\partial \tilde{\theta}} = \frac{\partial \mathcal{J}_p}{\partial \tilde{\theta}} - \frac{1}{2\tilde{\theta}} \tag{17.32}$$

Equating this to zero yields a fixed-point iteration for updating the variances:

$$\tilde{\theta} = \left[2\frac{\partial \mathcal{J}_p}{\partial \tilde{\theta}} \right]^{-1} \tag{17.33}$$

The means $\bar{\theta}_i$ can be estimated from the approximate Newton iteration [259, 436]

$$\bar{\theta} \leftarrow \bar{\theta} - \frac{\partial \mathcal{J}_p}{\partial \bar{\theta}} \left[\frac{\partial^2 \mathcal{J}_{KL}}{\partial \bar{\theta}^2} \right]^{-1} \approx \bar{\theta} - \frac{\partial \mathcal{J}_p}{\partial \bar{\theta}} \tilde{\theta} \tag{17.34}$$

The formulas (17.33) and (17.34) have a central role in learning.

17.5.4 Learning procedure *

Usually MLP networks learn the nonlinear input–output mapping in a supervised manner using known input–output training pairs, for which the mean-square mapping error is minimized using the back-propagation algorithm or some alternative method [172, 48]. In our case, the inputs are the *unknown* source signals $s(t)$, and only the outputs of the MLP network, namely the observed data vectors $x(t)$, are known. Hence, unsupervised learning must be applied. Detailed account of the learning method and discussion of potential problems can be found in [259, 436]. In the following, we give an overall description of the learning method.

The practical learning procedure used in all the experiments was the same. First, linear principal component analysis (PCA) (see Chapter 6) was applied to find sensible initial values for the posterior means of the sources. Even though PCA is a linear method, it yields much better initial values than a random choice. The posterior variances of the sources are initialized to small values. Good initial values are important for the method because it can effectively prune away unused parts of the MLP network[1].

Initially the weights of the MLP network have random values, and the network has quite a bad representation for the data. If the sources were adapted from random values, too, the network would consider many of the sources useless for the representation and prune them away. This would lead to a local minimum from which the network might not recover. Therefore the sources were fixed at the values given by linear PCA for the first 50 sweeps through the entire data set. This allows the MLP network to find a meaningful mapping from the sources to the observations, thereby justifying using the sources for the representation. For the same reason, the parameters controlling the distributions of the sources, weights, noise, and the hyperparameters are not adapted during the first 100 sweeps. They are adapted only after the MLP network has found sensible values for the variables whose distributions these parameters control.

Furthermore, we first used a simpler nonlinear model where the sources had standard gaussian distributions instead of mixtures of gaussians. This is called *nonlinear factor analysis* model in the following. After this phase, the sources were rotated using the FastICA algorithm. The rotation of the sources was compensated by an inverse rotation of the weight matrix A of the hidden layer. The final representation of the data was then found by continuing learning, but using now the mixture-of-gaussians model for the sources. In [259], this representation is called nonlinear independent factor analysis.

---

[1]Pruning techniques for neural networks are discussed, for example, in [172, 48]

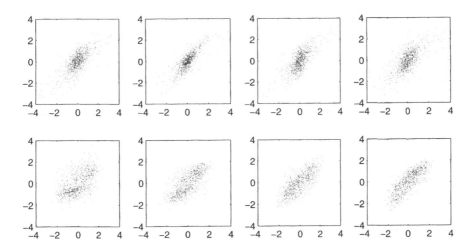

Fig. 17.8 Original sources are on the x-axis of each scatter plot and the sources estimated by a linear ICA are on the y-axis. Signal-to-noise ratio is 0.7 dB.

17.5.5 Experimental results

In all the simulations, the total number of sweeps was 7500, where one sweep means going through all the observations $\mathbf{x}(t)$ once. As explained before, a nonlinear factor analysis (or nonlinear PCA subspace) representation using plain gaussians as model distributions for the sources was estimated first. In the experiments, 2000 first sweeps were used for finding this intermediate representation. After a linear ICA rotation, the final mixture-of-gaussians representation of the sources was then estimated during the remaining 5500 sweeps. In the following, experiments with artificially generated nonlinear data are first described, followed by separation results on real-world process data.

Simulated data   In the first experiment, there were eight sources, four subgaussian and four supergaussian ones. The data were generated from these sources through a nonlinear mapping, which was obtained by using a randomly initialized MLP network having 30 hidden neurons and 20 output neurons. Gaussian noise having a standard deviation of 0.1 was added to the data. The nonlinearity used in the hidden neurons was chosen to be the inverse hyperbolic sine $\sinh^{-1}(x)$. This guarantees that the nonlinear source separation algorithm using the MLP network with tanh nonlinearities cannot use exactly the same weights.

Several different numbers of hidden neurons were tested in order to optimize the structure of the MLP network, but the number of sources was assumed to be known. This assumption is reasonable because it seems to be possible to optimize the number of sources simply by minimizing the cost function, as experiments with purely gaussian sources have shown [259, 438]. The MLP network which minimized

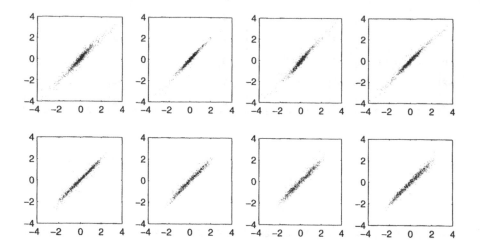

Fig. 17.9 Scatter plots of the sources after 2000 sweeps of nonlinear factor analysis followed by a rotation with a linear ICA. Signal-to-noise ratio is 13.2 dB.

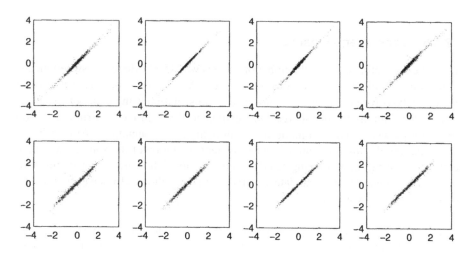

Fig. 17.10 The final separation results after using a mixture-of-gaussians model for the sources for the last 5500 sweeps. Signal to noise ratio is 17.3 dB. The original sources have clearly been found.

the Kullback-Leibler cost function turned out to have 50 hidden neurons. The number of gaussians in the mixtures modeling the distribution of each source was chosen to be three, and no attempt was made to optimize this number.

The results are depicted in Figs. 17.8, 17.9, and 17.10. Each figure shows eight scatter plots, corresponding to each of the eight original sources. The original source

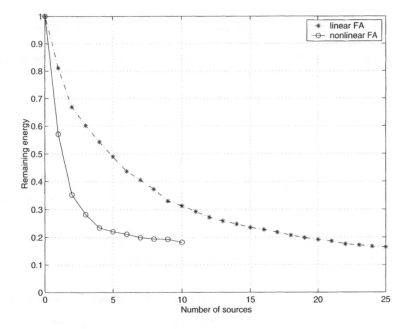

Fig. 17.11 The remaining energy of the process data as a function of the number of extracted components using linear and nonlinear factor analysis

which was used for generating the data appears on the x-axis, and the respective estimated source in on the y-axis of each plot. Each point corresponds to one sample point $\mathbf{x}(t)$. The upper plots of each figure correspond to the supergaussian and the lower plots to the subgaussian sources. The optimal result is a straight line implying that the estimated values of the sources coincide with the true values.

Figure 17.8 shows the result given by the linear FastICA algorithm alone. Linear ICA is able to retrieve the original sources with only 0.7 dB signal-to-noise ratio (SNR). In practice a linear method could not deduce the number of sources, and the result would be even worse. The poor signal-to-noise ratio shows that the data really lies in a nonlinear subspace. Figure 17.9 depicts the results after 2000 sweeps with gaussian sources (nonlinear factor analysis) followed by a rotation with linear FastICA. Now the SNR is considerably better, 13.2 dB, and the sources have clearly been retrieved. Figure 17.10 shows the final results after another 5500 sweeps when the mixture-of-gaussians model has been used for the sources. The SNR has further improved to 17.3 dB.

Industrial process data Another data set consisted of 30 time series of length 2480 measured using different sensors from an industrial pulp process. A human expert preprocessed the measured signals by roughly compensating for the time lags of the process originating from the finite speed of pulp flow through the process.

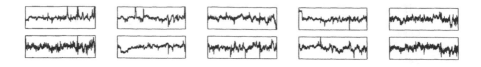

Fig. 17.12 The ten estimated sources of the industrial pulp process. Time increases from left to right.

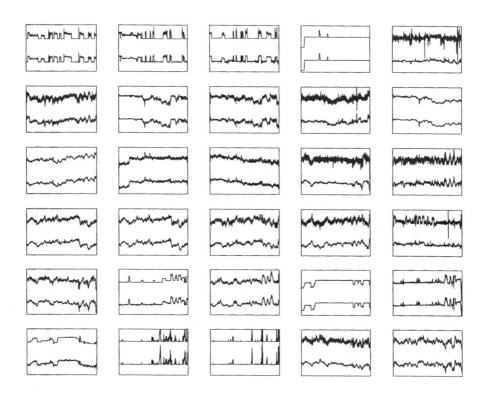

Fig. 17.13 The 30 original time series are shown on each plot on top of the reconstruction made from the sources shown in Fig. 17.12

For studying the intrinsic dimensionality of the data, linear factor analysis was applied to the data. The result is shown in Fig. 17.11. The figure also shows the results with nonlinear factor analysis. It is obvious that the data are quite nonlinear, because nonlinear factor analysis is able to explain the data with 10 components equally well as linear factor analysis (PCA) with 21 components.

Different numbers of hidden neurons and sources were tested with random initializations using gaussian model for sources (nonlinear factor analysis). It turned out that the Kullback-Leibler cost function was minimized for an MLP network having

10 sources (outputs) and 30 hidden neurons. The same network size was chosen for nonlinear blind source separation based on the mixture-of-gaussians model for the sources. After 2000 sweeps using the nonlinear factor analysis model, the sources were rotated with FastICA, and each source was modeled with a mixture of three gaussian distributions. The learning process was then continued using this refined model for 5500 more sweeps. The resulting sources are shown in Fig. 17.12.

Figure 17.13 shows 30 subimages, each corresponding to a specific measurement made from the process. The original measurement appears as the upper time series in each subimage, and the lower time series is the reconstruction of this measurement given by the network. These reconstructions are the posterior means of the outputs of the network when the inputs were the estimated sources shown in Fig. 17.12. The original measurements show a great variability, but the reconstructions are strikingly accurate. In some cases it seems that the reconstruction is less noisy than the original signal. This is not so surprising, because generally an estimate of data reconstructed from a smaller number of most significant components is often able to remove some noise.

The experiments suggest that the estimated source signals can have meaningful physical interpretations. The results are encouraging, but further studies are needed to verify the interpretations of the signals.

The proposed ensemble learning method for nonlinear blind source separation can be extended in several ways. An obvious extension is inclusion of time delays into the data model, making use of the temporal information often present in the sources. This would probably help in describing the process data even better. Using the Bayesian framework, it is also easy to treat missing observations or only partly known inputs.

17.6 OTHER APPROACHES

In this section, other methods proposed for nonlinear independent component analysis or blind source separation are briefly reviewed. The interested reader can find more information on them in the given references.

Already in 1987, Jutten [226] used soft nonlinear mixtures for assessing the robustness and performance of the seminal Hérault-Jutten algorithm introduced for the linear BSS problem (see Chapter 12). However, Burel [57] was probably the first to introduce an algorithm specifically for nonlinear ICA. His method is based on back-propagation type neural learning for parametric nonlinearities, and suffers from high computational complexity and problems with local minima. In a series of papers [104, 357, 355, 105, 358], Deco and Parra with their co-authors developed methods based on volume conserving symplectic transformations for nonlinear ICA. The constraint of volume conservation is, however, somewhat arbitrary, and so these methods are usually not able to recover the original sources. Also their computational load tends to be high.

The idea of using the self-organizing map [247] for the general nonlinear ICA problem, as discussed in Section 17.3, was introduced in [345]. However, this approach is limited mainly to separation of sources having distributions not too far from the uniform distribution. The nonlinear mixing process is considered to be regular: it is required that the inverse of the mixing mapping should be the least complex mapping which yields independent components [345]. Lin, Grier, and Cowan [279] have independently proposed using SOM for nonlinear ICA in a different manner by treating ICA as a computational geometry problem.

The ensemble learning approach to nonlinear ICA, discussed in more detail earlier in this chapter, is based on using multilayer perceptron networks as a flexible model for the nonlinear mixing mapping (17.2). Several authors have used autoassociative MLP networks [172] discussed briefly in Chapter 6 for learning a similar type of mappings. Both the generative model and its inversion are learned simultaneously, but separately without utilizing the fact that the models are connected. Autoassociative MLPs have shown some success in nonlinear data representation [172], but generally they suffer from slow learning prone to local minima.

Most works on autoassociative MLPs use point estimates for weights and sources obtained by minimizing the mean-square representation error for the data. It is then impossible to reliably choose the structure of the model, and problems with over- or underfitting can be severe. Hecht-Nielsen [176, 177] proposed so-called replicator networks for universal optimal nonlinear coding of input data. These networks are autoassociate MLP networks, where the data vectors are mapped onto a unit hypercube so that the mapped data is uniformly distributed inside the hypercube. The coordinates of the mapped data on the axes of the hypercube, called natural coordinates, then in fact form a nonlinear ICA solution, even though this has not been noted in the original papers [176, 177].

Hochreiter and Schmidhuber [181] have used in context with MLP networks a method based on minimum description length, called LOCOCODE. This method does estimate the distribution of the weights, but has no model for the sources. It is then impossible to measure the description length of the sources. Anyway, their method shows interesting connections with ICA; sometimes it provides a nonlinear ICA solution, sometimes it does not [181]. Another well-known information theoretic criterion, mutual information, is applied to measuring independence in [2, 469]. In these papers, methods based on various MLP network structures are also introduced for nonlinear blind separation. In particular, Yang, Amari, and Cichocki [469] deal with extensions of the basic natural gradient method (see Chapter 9), for nonlinear BSS, presenting also an extension based on entropy maximization and experiments with post-nonlinear mixtures.

Xu has developed the general Bayesian Ying-Yang framework which can be applied on ICA as well; see e.g. [462, 463].

Other general approaches proposed for solving the nonlinear ICA or BSS problems include a pattern repulsion based method [295], a state-space modeling approach [86], and an entropy-based method [134]. Various separation methods for the considerably simpler case of post-nonlinear mixtures (17.4) have been introduced at least in [271, 267, 365, 418, 417].

In the ensemble learning method discussed before, the necessary regularization for nonlinear ICA is achieved by choosing the model and sources that have most probably generated the observed data. Attias has applied a similar generative model to the linear ICA problem in [19]. The ensemble learning based method described in this chapter differs from the method introduced in [19] in that it uses a more general nonlinear data model, and applies a fully Bayesian treatment to the hyperparameters of the network or graphical model, too. A related extension was proposed in [20]. Connections of the ensemble learning method with other Bayesian approaches are discussed in more detail in [259, 438].

Nonlinear independent component analysis or blind source separation are generally difficult problems both computationally and conceptually. Therefore, local linear ICA/BSS methods have received some attention recently as a practical compromise between linear ICA and completely nonlinear ICA or BSS. These methods are more general than standard linear ICA in that several different linear ICA models are used to describe the observed data. The local linear ICA models can be either overlapping, as in the mixture-of-ICA methods introduced in [273], or nonoverlapping, as in the clustering-based methods proposed in [234, 349].

17.7   CONCLUDING REMARKS

In this chapter, generalizations of standard linear independent component analysis (ICA) or blind source separation (BSS) problems to nonlinear data models have been considered. We made a distinction between the ICA and BSS problems, because in the nonlinear case their relation is more complicated than in the linear case. In particular, the nonlinear ICA problem is ill-posed without some suitable extra constraints or regularization, having in general infinitely many qualitatively different solutions.

Solving the nonlinear BSS problem appropriately using only the independence assumption of the source signals is possible only in simple special cases, for example, when the mixtures are post-nonlinear. Otherwise, suitable additional information on the problem is required. This extra information is often provided in the form of regularizing constraints. Various methods proposed for regularizing the nonlinear ICA or BSS problems have been briefly reviewed in the previous section. Another possibility is to have more information about the sources or mixtures themselves. An example of such an approach is the method based on the generative topographic mapping (GTM), which requires knowledge of the probability distributions of the sources.

A large part of this chapter has been devoted to a recently introduced fully Bayesian approach based on ensemble learning for solving the nonlinear BSS problem. This method applies the well-known MLP network, which is well-suited to modeling both mildly and strongly nonlinear mappings. The proposed unsupervised ensemble-learning method tries to find the sources and the mapping that together have most probably generated the observed data. This regularization principle has a firm theoretical foundation, and it is intuitively satisfying for the nonlinear source separation problem. The results are encouraging for both artificial and real-world data. The ensemble-learning method allows nonlinear source separation for larger scale problems than some previously proposed computationally quite demanding methods, and it can be easily extended in various directions.

A lot of work remains to be done in developing suitable methods for the nonlinear ICA and BSS problems, and understanding better which constraints are most suitable in each situation. A number of different approaches have been proposed, but no comparisons are yet available for assessing their strengths and weaknesses.

18

Methods using Time Structure

The model of independent component analysis (ICA) that we have considered so far consists of mixing independent random variables, usually linearly. In many applications, however, what is mixed is not random variables but time signals, or time series. This is in contrast to the basic ICA model in which the samples of x have no particular order: We could shuffle them in any way we like, and this would have no effect on the validity of the model, nor on the estimation methods we have discussed. If the independent components (ICs) are time signals, the situation is quite different.

In fact, if the ICs are time signals, they may contain much more structure than simple random variables. For example, the autocovariances (covariances over different time lags) of the ICs are then well-defined statistics. One can then use such additional statistics to improve the estimation of the model. This additional information can actually make the estimation of the model possible in cases where the basic ICA methods cannot estimate it, for example, if the ICs are gaussian but correlated over time.

In this chapter, we consider the estimation of the ICA model when the ICs are time signals, $s_i(t), t = 1, ..., T$, where t is the time index. In the previous chapters, we denoted by t the sample index, but here t has a more precise meaning, since it defines an order between the ICs. The model is then expressed by

$$\mathbf{x}(t) = \mathbf{As}(t) \tag{18.1}$$

where $\mathbf{A}$ is assumed to be square as usual, and the ICs are of course independent. In contrast, *the ICs need not be nongaussian.*

In the following, we shall make some assumptions on the time structure of the ICs that allow for the estimation of the model. These assumptions are alternatives to the assumption of nongaussianity made in other chapters of this book. First, we shall assume that the ICs have different autocovariances (in particular, they are all different from zero). Second, we shall consider the case where the variances of the ICs are nonstationary. Finally, we discuss Kolmogoroff complexity as a general framework for ICA with time-correlated mixtures.

We do not here consider the case where it is the mixing matrix that changes in time; see [354].

18.1 SEPARATION BY AUTOCOVARIANCES

18.1.1 Autocovariances as an alternative to nongaussianity

The simplest form of time structure is given by (linear) autocovariances. This means covariances between the values of the signal at different time points: $\text{cov}(x_i(t)x_i(t - \tau))$ where τ is some lag constant, $\tau = 1, 2, 3, \ldots$. If the data has time-dependencies, the autocovariances are often different from zero.

In addition to the autocovariances of one signal, we also need covariances between two signals: $\text{cov}(x_i(t)x_j(t - \tau))$ where $i \neq j$. All these statistics for a given time lag can be grouped together in the time-lagged covariance matrix

$$\mathbf{C}_\tau^\mathbf{x} = E\{\mathbf{x}(t)\mathbf{x}(t - \tau)^T\} \tag{18.2}$$

The theory of time-dependent signals was briefly discussed in Section 2.8.

As we saw in Chapter 7, the problem in ICA is that the simple zero-lagged covariance (or correlation) matrix $\mathbf{C}$ does not contain enough parameters to allow the estimation of $\mathbf{A}$. This means that simply finding a matrix $\mathbf{V}$ so that the components of the vector

$$\mathbf{z}(t) = \mathbf{V}\mathbf{x}(t) \tag{18.3}$$

are white, is not enough to estimate the independent components. This is because there is an infinity of different matrices $\mathbf{V}$ that give decorrelated components. This is why in basic ICA, we have to use the nongaussian structure of the independent components, for example, by minimizing the higher-order dependencies as measured by mutual information.

The key point here is that the information in a time-lagged covariance matrix $\mathbf{C}_\tau^\mathbf{x}$ could be used instead of the higher-order information [424, 303]. What we do is to find a matrix $\mathbf{B}$ so that in addition to making the instantaneous covariances of $\mathbf{y}(t) = \mathbf{B}\mathbf{x}(t)$ go to zero, the *lagged* covariances are made zero as well:

$$E\{y_i(t)y_j(t - \tau)\} = 0, \quad \text{for all } i, j, \tau \tag{18.4}$$

The motivation for this is that for the ICs $s_i(t)$, the lagged covariances are all zero due to independence. Using these lagged covariances, we get enough extra information to estimate the model, under certain conditions specified below. No higher-order information is then needed.

18.1.2 Using one time lag

In the simplest case, we can use just one time lag. Denote by τ such a time lag, which is very often taken equal to 1. A very simple algorithm can now be formulated to find a matrix that cancels both the instantaneous covariances and the ones corresponding to lag τ.

Consider whitened data (see Chapter 6), denoted by $\mathbf{z}$. Then we have for the orthogonal separating matrix $\mathbf{W}$:

$$\mathbf{W}\mathbf{z}(t) = \mathbf{s}(t) \tag{18.5}$$

$$\mathbf{W}\mathbf{z}(t - \tau) = \mathbf{s}(t - \tau) \tag{18.6}$$

Let us consider a slightly modified version of the lagged covariance matrix as defined in (18.2), given by

$$\bar{\mathbf{C}}_\tau^{\mathbf{z}} = \frac{1}{2}[\mathbf{C}_\tau^{\mathbf{z}} + (\mathbf{C}_\tau^{\mathbf{z}})^T] \tag{18.7}$$

We have by linearity and orthogonality the relation

$$\bar{\mathbf{C}}_\tau^{\mathbf{z}} = \frac{1}{2}\mathbf{W}^T[E\{\mathbf{s}(t)\mathbf{s}(t - \tau)^T\} + E\{\mathbf{s}(t - \tau)\mathbf{s}(t)^T\}]\mathbf{W} = \mathbf{W}^T\bar{\mathbf{C}}_\tau^{\mathbf{s}}\mathbf{W} \tag{18.8}$$

Due to the independence of the $s_i(t)$, the time-lagged covariance matrix $\mathbf{C}_\tau^{\mathbf{s}} = E\{\mathbf{s}(t)\mathbf{s}(t - \tau)\}$ is diagonal; let us denote it by $\mathbf{D}$. Clearly, the matrix $\bar{\mathbf{C}}_\tau^{\mathbf{s}}$ equals this same matrix. Thus we have

$$\bar{\mathbf{C}}_\tau^{\mathbf{z}} = \mathbf{W}^T\mathbf{D}\mathbf{W} \tag{18.9}$$

What this equation shows is that the matrix $\mathbf{W}$ is part of the eigenvalue decomposition of $\bar{\mathbf{C}}_\tau^{\mathbf{z}}$. The eigenvalue decomposition of this symmetric matrix is simple to compute. In fact, the reason why we considered this matrix instead of the simple time-lagged covariance matrix (as in [303]) was precisely that we wanted to have a symmetric matrix, because then the eigenvalue decomposition is well defined and simple to compute. (It is actually true that the lagged covariance matrix is symmetric if the data exactly follows the ICA model, but estimates of such matrices are not symmetric.)

The AMUSE algorithm Thus we have a simple algorithm, called AMUSE [424], for estimating the separating matrix $\mathbf{W}$ for whitened data:

1. Whiten the (zero-mean) data $\mathbf{x}$ to obtain $\mathbf{z}(t)$.

2. Compute the eigenvalue decomposition of $\bar{\mathbf{C}}_\tau^{\mathbf{z}} = \frac{1}{2}[\mathbf{C}_\tau + \mathbf{C}_\tau^T]$, where $\mathbf{C}_\tau = E\{\mathbf{z}(t)\mathbf{z}(t - \tau)\}$ is the time-lagged covariance matrix, for some lag τ.

3. The rows of the separating matrix $\mathbf{W}$ are given by the eigenvectors.

An essentially similar algorithm was proposed in [303].

This algorithm is very simple and fast to compute. The problem is, however, that it only works when the eigenvectors of the matrix $\bar{\mathbf{C}}_\tau$ are uniquely defined. This is the case if the eigenvalues are all distinct (not equal to each other). If some of the eigenvalues are equal, then the corresponding eigenvectors are not uniquely defined, and the corresponding ICs cannot be estimated. This restricts the applicability of this method considerably. These eigenvalues are given by $\mathrm{cov}(s_i(t)s_i(t - \tau))$, and thus the eigenvalues are distinct if and only if the lagged covariances are different for all the ICs.

As a remedy to this restriction, one can search for a suitable time lag τ so that the eigenvalues are distinct, but this is not always possible: If the signals $s_i(t)$ have identical power spectra, that is, identical autocovariances, then no value of τ makes estimation possible.

18.1.3   Extension to several time lags

An extension of the AMUSE method that improves its performance is to consider several time lags τ instead of a single one. Then, it is enough that the covariances for *one* of these time lags are different. Thus the choice of τ is a somewhat less serious problem.

In principle, using several time lags, we want to *simultaneously diagonalize* all the corresponding lagged covariance matrices. It must be noted that the diagonalization is not possible exactly, since the eigenvectors of the different covariance matrices are unlikely to be identical, except in the theoretical case where the data is exactly generated by the ICA model. So here we formulate functions that express the degree of diagonalization obtained and find its maximum.

One simple way of measuring the diagonality of a matrix $\mathbf{M}$ is to use the operator

$$\mathrm{off}(\mathbf{M}) = \sum_{i \neq j} m_{ij}^2 \tag{18.10}$$

which gives the sum of squares of the off-diagonal elements $\mathbf{M}$. What we now want to do is to minimize the sum of the off-diagonal elements of several lagged covariances of $\mathbf{y} = \mathbf{Wz}$. As before, we use the symmetric version $\bar{\mathbf{C}}_\tau^{\mathbf{y}}$ of the lagged covariance matrix. Denote by S the set of the chosen lags τ. Then we can write this as an objective function $\mathcal{J}(\mathbf{w})$:

$$\mathcal{J}_1(\mathbf{W}) = \sum_{\tau \in S} \mathrm{off}(\mathbf{W}\bar{\mathbf{C}}_\tau^{\mathbf{z}}\mathbf{W}^T) \tag{18.11}$$

Minimizing $\mathcal{J}_1$ under the constraint that $\mathbf{W}$ is orthogonal gives us the estimation method. This minimization could be performed by (projected) gradient descent. Another alternative is to adapt the existing methods for eigenvalue decomposition to this simultaneous approximate diagonalization of several matrices. The algorithm called SOBI (second-order blind identification) [43] is based on these principles, and so is TDSEP [481].

The criterion $\mathcal{J}_1$ can be simplified. For an orthogonal transformation, $\mathbf{W}$, the sum of the squares of the elements of $\mathbf{WMW}^T$ is constant.[1] Thus, the "off" criterion could be expressed as the difference of the total sum of squares minus the sum of the squares on the diagonal. Thus we can formulate

$$\mathcal{J}_2(\mathbf{W}) = - \sum_{\tau \in S} \sum_i (\mathbf{w}_i^T \bar{\mathbf{C}}_\tau^{\mathbf{z}} \mathbf{w}_i)^2 \tag{18.12}$$

where the $\mathbf{w}_i^T$ are the rows of $\mathbf{W}$. Thus, minimizing $\mathcal{J}_2$ is equivalent to minimizing $\mathcal{J}_1$.

An alternative method for measuring the diagonality can be obtained using the approach in [240]. For any positive-definite matrix $\mathbf{M}$, we have

$$\sum_i \log m_{ii} \geq \log |\det \mathbf{M}| \tag{18.13}$$

and the equality holds only for diagonal $\mathbf{M}$. Thus, we could measure the nondiagonality of $\mathbf{M}$ by

$$F(\mathbf{M}) = \sum_i \log m_{ii} - \log |\det \mathbf{M}| \tag{18.14}$$

Again, the total nondiagonality of the $\mathbf{C}_\tau$ at different time lags can be measured by the sum of these measures for different time lags. This gives us the following objective function to minimize:

$$\mathcal{J}_3(\mathbf{W}) = \frac{1}{2} \sum_{\tau \in S} F(\bar{\mathbf{C}}_\tau^{\mathbf{y}}) = \frac{1}{2} \sum_{\tau \in S} F(\mathbf{W}\bar{\mathbf{C}}_\tau^{\mathbf{z}}\mathbf{W}^T) \tag{18.15}$$

Just as in maximum likelihood (ML) estimation, $\mathbf{W}$ decouples from the term involving the logarithm of the determinant. We obtain

$$\mathcal{J}_3(\mathbf{W}) = \sum_{\tau \in S} \sum_i \frac{1}{2} \log(\mathbf{w}_i^T \bar{\mathbf{C}}_\tau^{\mathbf{z}} \mathbf{w}_i) - \log |\det \mathbf{W}| - \frac{1}{2} \log |\det \bar{\mathbf{C}}_\tau^{\mathbf{z}}| \tag{18.16}$$

Considering whitened data, in which case $\mathbf{W}$ can be constrained orthogonal, we see that the term involving the determinant is constant, and we finally have

$$\mathcal{J}_3(\mathbf{W}) = \sum_{\tau \in S} \sum_i \frac{1}{2} \log(\mathbf{w}_i^T \mathbf{C}_\tau^{\mathbf{z}} \mathbf{w}_i) + \text{const.} \tag{18.17}$$

This is in fact rather similar to the function $\mathcal{J}_2$ in (18.12). The only difference is that the function $-u^2$ has been replaced by $1/2 \log(u)$. What these functions have

---

[1]This is because it equals $\text{trace}(\mathbf{WMW}^T(\mathbf{WMW}^T)^T) = \text{trace}(\mathbf{WMM}^T\mathbf{W}^T) = \text{trace}(\mathbf{W}^T\mathbf{WMM}^T) = \text{trace}(\mathbf{MM}^T)$.

in common is concavity, so one might speculate that many other concave functions could be used as well.

The gradient of $\mathcal{J}_3$ can be evaluated as

$$\frac{\partial \mathcal{J}_3}{\partial \mathbf{W}} = \sum_{\tau \in S} \mathbf{Q}_\tau \mathbf{W} \bar{\mathbf{C}}_\tau^{\mathbf{z}} \tag{18.18}$$

with

$$\mathbf{Q}_\tau = \operatorname{diag}(\mathbf{W} \bar{\mathbf{C}}_\tau^{\mathbf{z}} \mathbf{W}^T)^{-1} \tag{18.19}$$

Thus we obtain the gradient descent algorithm

$$\boxed{\Delta \mathbf{W} \propto \sum_{\tau \in S} \mathbf{Q}_\tau \mathbf{W} \bar{\mathbf{C}}_\tau^{\mathbf{z}}} \tag{18.20}$$

Here, $\mathbf{W}$ should be orthogonalized after every iteration. Moreover, care must be taken so that in the inverse in (18.19), very small entries do not cause numerical problems. A very similar gradient descent can be obtained for (18.12), the main difference being the scalar function in the definition of $\mathbf{Q}$.

Thus we obtain an algorithm that estimates $\mathbf{W}$ based on autocorrelations with several time lags. This gives a simpler alternative to methods based on joint approximative diagonalization. Such an extension allows estimation of the model in some cases where the simple method using a single time lag fails. The basic limitation cannot be avoided, however: *if the ICs have identical autocovariances (i.e., identical power spectra), they cannot be estimated* by the methods using time-lagged covariances only. This is in contrast to ICA using higher-order information, where the independent components are allowed to have identical distributions.

Further work on using autocovariances for source separation can be found in [11, 6, 106]. In particular, the optimal weighting of different lags has be considered in [472, 483].

18.2 SEPARATION BY NONSTATIONARITY OF VARIANCES

An alternative approach to using the time structure of the signals was introduced in [296], where it was shown that ICA can be performed by using the nonstationarity of the signals. The nonstationarity we are using here is the nonstationarity of the variances of the ICs. Thus the variances of the ICs are assumed to change smoothly in time. Note that this nonstationarity of the signals is independent from nongaussianity or the linear autocovariances in the sense that none of them implies or presupposes any of the other assumptions.

To illustrate the variance nonstationarity in its purest form, let us look at the signal in Fig. 18.1. This signal was created so that it has a gaussian marginal density, and no linear time correlations, i.e., $E\{\mathbf{x}(t)\mathbf{x}(t - \tau)^T\} = 0$ for any lag τ. Thus,

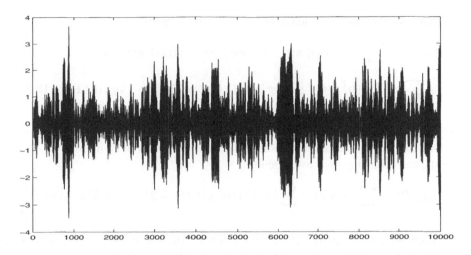

Fig. 18.1 A signal with nonstationary variance.

ICs of this kind could not be separated by basic ICA methods, or using linear time-correlations. On the other hand, the nonstationarity of the signal is clearly visible. It is characterized by bursts of activity.

Below, we review some basic approaches to this problem. Further work can be found in [40, 370, 126, 239, 366].

18.2.1 Using local autocorrelations

Separation of nonstationary signals could be achieved by using a variant of autocorrelations, somewhat similar to the case of Section 18.1. It was shown in [296] that if we find a matrix $\mathbf{B}$ so that the components of $\mathbf{y}(t) = \mathbf{B}\mathbf{x}(t)$ are uncorrelated *at every time point t*, we have estimated the ICs. Note that due to nonstationarity, the covariance of $\mathbf{y}(t)$ depends on t, and thus if we force the components to be uncorrelated for every t, we obtain a much stronger condition than simple whitening.

The (local) uncorrelatedness of $\mathbf{y}(t)$ could be measured using the same measures of diagonality as used in Section 18.1.3. We use here a measure based on (18.14):

$$Q(\mathbf{B}, t) = \sum_i \log E_t\{y_i(t)^2\} - \log|\det E_t\{\mathbf{y}(t)\mathbf{y}(t)^T\}|$$
(18.21)

The subscript t in the expectation emphasizes that the signal is nonstationary, and the expectation is the expectation around the time point t. This function is minimized by the separating matrix $\mathbf{B}$.

Expressing this as a function of $\mathbf{B} = (\mathbf{b}_1, ..., \mathbf{b}_n)^T$ we obtain

$$Q(\mathbf{B}, t) = \sum_i \log E_t\{(\mathbf{b}_i^T \mathbf{x}(t))^2\} - \log |\det E_t\{\mathbf{B}\mathbf{x}(t)\mathbf{x}(t)^T \mathbf{B}^T\}|$$

$$= \sum_i \log E_t\{(\mathbf{b}_i^T \mathbf{x}(t))^2\} - \log |\det E_t\{\mathbf{x}(t)\mathbf{x}(t)^T\}| - 2\log |\det \mathbf{B}| \quad (18.22)$$

Note that the term $\log |\det E_t\{\mathbf{x}(t)\mathbf{x}(t)^T\}|$ does not depend on $\mathbf{B}$ at all. Furthermore, to take into account all the time points, we sum the values of Q in different time points, and obtain the objective function

$$\mathcal{J}_4(\mathbf{B}) = \sum_t Q(\mathbf{B}, t) = \sum_{i,t} \log E_t\{(\mathbf{b}_i^T \mathbf{x}(t))^2\} - 2\log |\det \mathbf{B}| + \text{const.}$$
$$(18.23)$$

As usual, we can whiten the data to obtain whitened data $\mathbf{z}$, and force the separating matrix $\mathbf{W}$ to be orthogonal. Then the objective function simplifies to

$$\mathcal{J}_4(\mathbf{W}) = \sum_t Q(\mathbf{W}, t) = \sum_{i,t} \log E_t\{(\mathbf{w}_i^T \mathbf{z}(t))^2\} + \text{const.}$$
$$(18.24)$$

Thus we can compute the gradient of $\mathcal{J}_4$ as

$$\frac{\partial \mathcal{J}_4}{\partial \mathbf{W}} = 2 \sum_t \text{diag}(E_t\{(\mathbf{w}_i^T \mathbf{z}(t))^2\}^{-1}) \mathbf{W} E_t\{\mathbf{z}(t)\mathbf{z}(t)^T\}.$$
$$(18.25)$$

The question is now: How to estimate the local variances $E_t\{(\mathbf{w}_i^T \mathbf{z}(t))^2\}$? We cannot simply use the sample variances, due to nonstationarity, which leads to dependence between these variances and the $\mathbf{z}(t)$. Instead, we have to use some local estimates at time point t. A natural thing to do is to assume that the *variance changes slowly*. Then we can estimate the local variance by local sample variances. In other words:

$$\hat{E}_t\{(\mathbf{w}_i^T \mathbf{z}(t))^2\} = \sum_\tau h(\tau)(\mathbf{w}_i^T \mathbf{z}(t - \tau))^2 \quad (18.26)$$

where h is a moving average operator (low-pass filter), normalized so that the sum of its components is one.

Thus we obtain the following algorithm:

$$\boxed{\Delta \mathbf{W} \propto \sum_t \text{diag}(\hat{E}_t\{(\mathbf{w}_i^T \mathbf{z}(t))^2\}^{-1}) \mathbf{W} \mathbf{z}(t)\mathbf{z}(t)^T} \quad (18.27)$$

where after every iteration, $\mathbf{W}$ is symmetrically orthogonalized (see Chapter 6), and $\hat{E}_t$ is computed as in (18.26). Again, care must be taken that taking the inverse of very small local variances does not cause numerical problems. This is the basic method for estimating signals with nonstationary variances. It is a simplified form of the algorithm in [296].

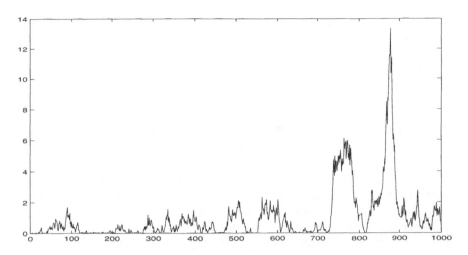

Fig. 18.2 The energy (i.e., squares) of the initial part of the signal in Fig. 18.1. This is clearly time-correlated.

The algorithm in (18.27) enables one to estimate the ICs using the information on the nonstationarity of their variances. This principle is different from the ones considered in preceding chapters and the preceding section. It was implemented by considering simultaneously different *local* autocorrelations. An alternative method for using nonstationarity will be considered next.

18.2.2   Using cross-cumulants

Nonlinear autocorrelations   A second method of using nonstationarity is based on interpreting variance nonstationarity in terms of higher-order cross-cumulants. Thus we obtain a very simple criterion that expresses nonstationarity of variance. To see how this works, consider the energy (i.e., squared amplitude) of the signal in Fig. 18.1. The energies of the initial 1000 time points are shown in Fig. 18.2. What is clearly visible is that the energies are correlated in time. This is of course a consequence of the assumption that the variance changes smoothly in time.

Before proceeding, note that the nonstationarity of a signal depends on the time-scale and the level of the detail in the model of the signal. If the nonstationarity of the variance is incorporated in the model (by hidden Markov models, for example), the signal no longer needs to be considered nonstationary [370]. This is the approach that we choose in the following. In particular, the energies are *not* considered nonstationary, but rather they are considered as stationary signals that are time-correlated. This is simply a question of changing the viewpoint.

So, we could measure the variance nonstationarity of a signal $y(t), t = 1, ...t$ using a measure based on the time-correlation of energies: $E\{y(t)^2 y(t-\tau)^2\}$ where τ is some lag constant, often equal to one. For the sake of mathematical simplicity, it is often useful to use cumulants instead of such basic higher-order correlations. The cumulant corresponding to the correlation of energies is given by the fourth-order cross cumulant

$$\text{cum}(y(t), y(t), y(t - \tau), y(t - \tau))$$
$$= E\{y(t)^2 y(t - \tau)^2\} - E\{y(t)^2\} E\{y(t - \tau)^2\} - 2(E\{y(t)y(t - \tau)\})^2$$
$$(18.28)$$

This could be considered as a normalized version of the cross-correlation of energies. In our case, where the variances are changing smoothly, this cumulant is positive because the first term dominates the two normalizing terms.

Note that although cross-cumulants are zero for random variables with *jointly* gaussian distributions, they need not be zero for variables with gaussian marginal distributions. Thus positive cross-cumulants do not imply nongaussian marginal distributions for the ICs, which shows that the property measured by this cross-cumulant is indeed completely different from the property of nongaussianity.

The validity of this criterion can be easily proven. Consider a linear combination of the observed signals $x_i(t)$ that are mixtures of original ICs, as in (18.1). This linear combination, say $\mathbf{b}^T\mathbf{x}(t)$, is a linear combination of the ICs $\mathbf{b}^T\mathbf{x}(t) = \mathbf{b}^T\mathbf{A}\mathbf{s}(t)$, say $\mathbf{q}^T\mathbf{s}(t) = \sum_i q_i s_i(t)$. Using the basic properties of cumulants, the nonstationarity of such a linear combination can be evaluated as

$$\text{cum}(\mathbf{b}^T\mathbf{x}(t), \mathbf{b}^T\mathbf{x}(t), \mathbf{b}^T\mathbf{x}(t - \tau), \mathbf{b}^T\mathbf{x}(t - \tau))$$
$$= \sum_i q_i^4 \text{cum}(s_i(t), s_i(t), s_i(t - \tau), s_i(t - \tau)) \quad (18.29)$$

Now, we can constrain the variance of $\mathbf{b}^T\mathbf{x}$ to be equal to unity to normalize the scale (cumulants are not scale-invariant). This implies var $\sum_i q_i s_i = \|\mathbf{q}\|^2 = 1$. Let us consider what happens if we maximize nonstationarity with respect to $\mathbf{b}$. This is equivalent to the optimization problem

$$\max_{\|\mathbf{q}\|^2 = 1} \sum_i q_i^4 \text{cum}(s_i(t), s_i(t), s_i(t - \tau), s_i(t - \tau)) \quad (18.30)$$

This optimization problem is formally identical to the one encountered when kurtosis (or in general, its absolute value) is maximized to find the most nongaussian directions, as in Chapter 8. It was proven that solutions to this optimization problem give the ICs. In other words, the maxima of (18.30) are obtained when only one of the q_i is nonzero. This proof applies directly in our case as well, and thus we see that *the maximally nonstationary linear combinations give the ICs*.[2] Since the cross-cumulants are assumed to be all positive, the problem we have here is in fact slightly simpler since we can then simply maximize the cross-cumulant of the linear combinations, and need not consider its absolute value as is done with kurtosis in Chapter 8.

---

[2]Note that this statement requires that we *identify* nonstationarity with the energy correlations, which may or may not be meaningful depending on the context.

Thus we see that maximization of the nonstationarity, as measured by the cross-cumulant, of a linear combination of the observed mixtures allows for the estimation of one IC. This also gives a one-unit approach to source separation by nonstationarity.

A fixed-point algorithm   To maximize the variance nonstationarity as measured by the cross-cumulant, one can use a fixed-point algorithm derived along the same lines as the FastICA algorithm for maximizing nongaussianity.

To begin with, let us whiten the data to obtain $z(t)$. Now, using the principle of fixed-point iteration as in Chapter 8, let us equate $\mathbf{w}$ to the gradient of the cross-cumulant of $\mathbf{w}^T z(t)$. This gives, after straightforward calculations, the following update for $\mathbf{w}$:

$$\mathbf{w} \leftarrow E\{\mathbf{z}(t)\mathbf{w}^T \mathbf{z}(t)(\mathbf{w}^T \mathbf{z}(t - \tau))^2\} + E\{\mathbf{z}(t - \tau)\mathbf{w}^T \mathbf{z}(t - \tau)(\mathbf{w}^T \mathbf{z}(t))^2\}$$
$$- 2\mathbf{w} - 4\bar{\mathbf{C}}_\tau^{\mathbf{z}}\mathbf{w}(\mathbf{w}^T \bar{\mathbf{C}}_\tau^{\mathbf{z}}\mathbf{w}) \quad (18.31)$$

where we have multiplied the gradient by $1/2$ for notational simplicity, and the matrix $\bar{\mathbf{C}}_\tau^{\mathbf{z}}$ is equal to $\frac{1}{2}[E\{\mathbf{z}(t)\mathbf{z}(t - \tau)\} + E\{\mathbf{z}(t)\mathbf{z}(t - \tau)\}]$, as above. The algorithm thus consists of iteratively computing the new value of $\mathbf{w}$ according to (18.31), and normalizing $\mathbf{w}$ to unit norm after every step.

The convergence of the algorithm can be proven to be cubic, i.e., very fast. A detailed proof can be constructed as with kurtosis. Let us only note here that if the algorithm is expressed with respect to the transformed variable $\mathbf{q}$, which can be simply obtained by computing the gradient of (18.29), we have

$$q_i \leftarrow q_i^3[4\text{cum}(s_i(t), s_i(t), s_i(t - \tau), s_i(t - \tau))] \quad (18.32)$$

followed by normalization of the norm of $\mathbf{q}$. This can be easily seen to lead to convergence of $\mathbf{q}$ to a vector where only one of the q_i is nonzero. The index i for which q_i will be nonzero depends on the initial value of $\mathbf{q}$.

Thus we have obtained a fast fixed-point algorithm for separating ICs by non-stationarity, using cross-cumulants. This gives an alternative for the algorithm in the preceding subsection. This algorithm is similar to the FastICA algorithm. The convergence of the algorithm is cubic, like the convergence of the cumulant-based FastICA. This was based on the interpretation of a particular cross-cumulant as a measure of nonstationarity.

18.3   SEPARATION PRINCIPLES UNIFIED

18.3.1   Comparison of separation principles

In this chapter we have discussed the separation of ICs (sources) using their time-dependencies. In particular, we showed how to use autocorrelations and variance nonstationarities. These two principles complement the principle of nongaussianity that was the basis of estimation in the basic ICA model in Part II.

This raises the question: In what situations should one use each of these methods. The answer is basically simple: The different criteria make different assumptions on the data, and the choice of the criterion should be made depending on the data that we want to analyze.

First of all, in many cases the data has no time structure at all, i.e., x is a random variable, and not a time signal. This means that the order of the samples is arbitrary and has no significance. In that case, only the basic ICA method based on nongaussianity can be used. So the alternatives are only meaningful if the data comes from a source that has a time structure.

If the data does have clear time-dependencies, these usually imply nonzero autocorrelation functions, and the methods based on autocorrelations can be used. However, such methods only work if the autocorrelations are different for each IC. In the case where some of the autocorrelations are identical, one could then try to use the methods based on nonstationarity, since these methods utilize the time structure but do not require the time structure of each IC to be different. The nonstationarity methods work best, of course, if the time structure does consist of a changing variance, as with the signal depicted in Fig. 18.1.

On the other hand, the basic ICA methods often work well even when the ICs have time-dependencies. The basic methods do not use the time structure, but this does not mean that they would be disturbed by such structure. It must be noted, though, that the basic ICA methods may then be very far from optimal, since they do not use the whole structure of the data.

18.3.2 Kolmogoroff complexity as a unifying framework

It is also possible to combine different types of information. For example, methods that combine nongaussianity with autocorrelations were proposed in [202, 312]. What is interesting is that one can formulate a general framework that encompasses all these different principles. This framework that includes basic ICA and methods using time structure was proposed by Pajunen [342, 343], based on the information-theoretic concept of Kolmogoroff complexity.

As has been argued in Chapters 8 and 10, ICA can be seen as a method of finding a transformation into components that are somehow *structured*. It was argued that nongaussianity is a measure of structure. Nongaussianity can be measured by the information-theoretic concept of entropy.

Entropy of a random variable measures the structure of its marginal distribution only. On the other hand, in this section we have been dealing with time signals that have different kinds of time structure, like autocorrelations and nonstationarity. How could one measure such a more general type of structure using information-theoretic criteria? The answer lies in Kolmogoroff complexity. One can define a very general form of linear ICA by finding projections of the data that have low complexity in this sense. First, let us look at the definition of Kolmogoroff complexity.

Definition of Kolmogoroff complexity The information-theoretic measures of structure are based on the interpretation of coding length as structure. Suppose that we want to code a signal $s(t), t = 1, ..., T$. For simplicity, let us assume that the signal is binary, so that every value $s(t)$ is 0 or 1. We can code such a signal so that every bit in the code gives the value of $s(t)$ for one t; this is the obvious and trivial code. In general, it is not possible to code this signal with less that T bits. However, most natural signals have *redundancy*, i.e., parts of the signal can be efficiently predicted from other parts. Such a signal can be coded, or compressed, so that the code length is shorter than the original signal length T. It is well known that audio or image signals, for example, can be coded so that the code length is decreased considerably. This is because such natural signals are highly structured. For example, image signals do not consist of random pixels, but of such higher-order regularities as edges, contours, and areas of constant color [154].

We could thus measure the amount of structure of the signal $s(t)$ by the amount of compression that is possible in coding the signal. For signals of fixed length T, the *structure could be measured by the length of the shortest possible code for the signal.* Note that the signal could be compressed by many different kinds of coding schemes developed in the coding theory literature, but we are here considering the shortest possible code, thus maximizing the compression over all possible coding schemes. For a more rigorous definition of the concept, see [342, 343].

ICA and Kolmogoroff complexity   We can now define a generalization of ICA as follows: Find the transformation of the data where the sum of the coding lengths of the components is as short as possible. However, an additional operation is needed here. We also need to consider the code length of the transformation itself [342, 343]. This leads to a framework that is closely related to the minimum description length (MLD) principle [380, 381]. Thus the objective function we are using can be written as follows:

$$J(\mathbf{W}) = \frac{1}{T} \sum_i K(\mathbf{b}_i^T \mathbf{x}) - \log |\det \mathbf{W}| \qquad (18.33)$$

where $K(.)$ denotes the complexity. The latter term is related to the coding length of the transformation, using the MDL principle.

This objective function can be considered as a generalization of mutual information. If the signals have no time structure, their Kolmogoroff complexities are given by their entropies, and thus we have in (18.33) the definition of mutual information. Furthermore, in [344] it was shown how to approximate $K(.)$ by criteria using the time structure of the signals, which proves that Kolmogoroff complexity gives a generalization of methods using time correlations as well. (More on this connection can be found in Section 23.3.)

Kolmogoroff complexity is a rather theoretical measure, since its computation involves finding the best coding scheme for the signal. The number of possible coding schemes is infinite, so this optimization cannot be performed in practice with much accuracy. In some special cases of Kolmogoroff complexity the optimization can be performed rather accurately however. These include the above-mentioned cases of mutual information and time-correlations.

18.4 CONCLUDING REMARKS

In addition to the fundamental assumption of independence, another assumption that assures that the signals have enough structure is needed for successful separation of signals. This is because the ICA model cannot be estimated for gaussian random variables. In basic ICA, which was treated in Part II, the assumption was the nongaussianity of the ICs. In this chapter, we used the alternative assumption that the ICs are time signals that have some time dependencies. Here we have at least two cases in which separation is possible. First, the case where the signals have different power spectra, i.e., different autocovariance functions. Second, the case where they have nonstationary variances. All these assumptions can be considered in the unifying framework of Kolmogoroff complexity: They can all be derived as special cases of complexity minimization.

19

Convolutive Mixtures and Blind Deconvolution

This chapter deals with blind deconvolution and blind separation of convolutive mixtures.

Blind deconvolution is a signal processing problem that is closely related to basic independent component analysis (ICA) and blind source separation (BSS). In communications and related areas, blind deconvolution is often called *blind equalization*. In blind deconvolution, we have only one observed signal (output) and one source signal (input). The observed signal consists of an unknown source signal mixed with itself at different time delays. The task is to estimate the source signal from the observed signal only, without knowing the convolving system, the time delays, and mixing coefficients.

Blind separation of convolutive mixtures considers the combined blind deconvolution and instantaneous blind source separation problem. This estimation task appears under many different names in the literature: ICA with convolutive mixtures, multichannel blind deconvolution or identification, convolutive signal separation, and blind identification of multiple-input-multiple-output (MIMO) systems. In blind separation of convolutive mixtures, there are several source (input) signals and several observed (output) signals just like in the instantaneous ICA problem. However, the source signals have different time delays in each observed signal due to the finite propagation speed in the medium. Each observed signal may also contain time-delayed versions of the same source due to multipath propagation caused typically by reverberations from some obstacles. Figure 23.3 in Chapter 23 shows an example of multipath propagation in mobile communications.

In the following, we first consider the simpler blind deconvolution problem, and after that separation of convolutive mixtures. Many techniques for convolutive mixtures have in fact been developed by extending methods designed originally for either the blind deconvolution or standard ICA/BSS problems. In the appendix, certain basic concepts of discrete-time filters needed in this chapter are briefly reviewed.

19.1 BLIND DECONVOLUTION

19.1.1 Problem definition

In blind deconvolution [170, 171, 174, 315], it is assumed that the observed discrete-time signal $x(t)$ is generated from an unknown source signal $s(t)$ by the convolution model

$$x(t) = \sum_{k=-\infty}^{\infty} a_k s(t-k) \tag{19.1}$$

Thus, delayed versions of the source signal are mixed together. This situation appears in many practical applications, for example, in communications and geophysics.

In blind deconvolution, both the source signal $s(t)$ and the convolution coefficients a_k are unknown. Observing $x(t)$ only, we want to estimate the source signal $s(t)$. In other words, we want to find a deconvolution filter

$$y(t) = \sum_{k=-\infty}^{\infty} h_k x(t-k) \tag{19.2}$$

which provides a good estimate of the source signal $s(t)$ at each time instant. This is achieved by choosing the coefficients h_k of the deconvolution filter suitably. In practice the deconvolving finite impulse response (FIR) filter (see the Appendix for definition) in Eq. (19.2) is assumed to be of sufficient but finite length. Other structures are possible, but this one is the standard choice.

To estimate the deconvolving filter, certain assumptions on the source signal $s(t)$ must be made. Usually it is assumed that the source signal values $s(t)$ at different times t are nongaussian, statistically independent and identically distributed (i.i.d.). The probability distribution of the source signal $s(t)$ may be known or unknown. The indeterminacies remaining in the blind deconvolution problem are that the estimated source signal may have an arbitrary scaling (and sign) and time shift compared with the true one. This situation is similar to the permutation and sign indeterminacy encountered in ICA; the two models are, in fact, intimately related as will be explained in Section 19.1.4.

Of course, the preceding ideal model usually does not exactly hold in practice. There is often additive noise present, though we have omitted noise from the model (19.1) for simplicity. The source signal sequence may not satisfy the i.i.d condition, and its distribution is often unknown, or we may only know that the source signal is subgaussian or supergaussian. Hence blind deconvolution often is a difficult signal processing task that can be solved only approximately, in practice.

If the linear time-invariant system (19.1) is *minimum phase* (see the Appendix), then the blind deconvolution problem *can* be solved in a straightforward way. On the above assumptions, the deconvolving filter is simply a *whitening filter* that temporally whitens the observed signal sequence $\{x(t)\}$ [171, 174]. However, in many applications, for example, in telecommunications, the system is typically nonminimum phase [174] and this simple solution cannot be used.

We shall next discuss some popular approaches to blind deconvolution. Blind deconvolution is frequently needed in communications applications where it is convenient to use complex-valued data. Therefore we present most methods for this general case. The respective algorithms for real data are obtained as special cases. Methods for estimating the ICA model with complex-valued data are discussed later in Section 20.3.

19.1.2 Bussgang methods

Bussgang methods [39, 171, 174, 315] include some of the earliest algorithms [152, 392] proposed for blind deconvolution, but they are still widely used. In Bussgang methods, a noncausal FIR filter structure

$$y(t) = \sum_{k=-L}^{L} w_k^*(t) x(t-k) \tag{19.3}$$

of length $2L+1$ is used. Here $*$ denotes the complex conjugate. The weights $w_k(t)$ of the FIR filter depend on the time t, and they are adapted using the least-mean-square (LMS) type algorithm [171]

$$w_k(t+1) = w_k(t) + \mu x(t-k) e^*(t), \qquad k = -L, \dots, L \tag{19.4}$$

where the error signal is defined by

$$e(t) = g(y(t)) - y(t) \tag{19.5}$$

In these equations, μ is a positive learning parameter, $y(t)$ is given by (19.3), and $g(.)$ is a suitably chosen nonlinearity. It is applied separately to the real and imaginary components of $y(t)$. The algorithm is initialized by setting $w_0(0) = 1$, $w_k(0) = 0$, $k \neq 0$.

Assume that the filter length $2L + 1$ is large enough and the learning algorithm has converged. It can be shown that then the following condition holds for the output $y(t)$ of the FIR filter (19.3):

$$E\{y(t)y(t-k)\} \approx E\{y(t)g(y(t-k))\} \tag{19.6}$$

A stochastic process that satisfies the condition (19.6) is called a Bussgang process.

The nonlinearity $g(t)$ can be chosen in several ways, leading to different Bussgang type algorithms [39, 171]. The *Godard algorithm* [152] is the best performing Bussgang algorithm in the sense that it is robust and has the smallest mean-square error after convergence; see [171] for details. The Godard algorithm minimizes the nonconvex cost function

$$\mathcal{J}_p(t) = E\{[|y(t)|^p - \gamma_p]^2\} \tag{19.7}$$

where p is a positive integer and γ_p is a positive real constant defined by the statistics of the source signal:

$$\gamma_p = \frac{E\{|s(t)|^{2p}\}}{E\{|s(t)|^p\}} \tag{19.8}$$

The constant γ_p is chosen in such a way that the gradient of the cost function $\mathcal{J}_p(t)$ is zero when perfect deconvolution is attained, that is, when $y(t) = s(t)$. The error signal (19.5) in the gradient algorithm (19.4) for minimizing the cost function (19.7) with respect to the weight $w_k(t)$ has the form

$$e(t) = y(t)|y(t)|^{p-2}[\gamma_p - |y(t)|^p] \tag{19.9}$$

In computing $e(t)$, the expectation in (19.7) has been omitted for getting a simpler stochastic gradient type algorithm. The respective nonlinearity $g(y(t))$ is given by [171]

$$g(y(t)) = y(t) + y(t)|y(t)|^{p-2}[\gamma_p - |y(t)|^p] \tag{19.10}$$

Among the family of Godard algorithms, the so-called *constant modulus algorithm* (CMA) is widely used. It is obtained by setting $p = 2$ in the above formulas. The cost function (19.7) is then related to the minimization of the kurtosis. The CMA and more generally Godard algorithms perform appropriately for subgaussian sources only, but in communications applications the source signals are subgaussian.[1]. The CMA algorithm is the most successful blind equalization (deconvolution) algorithm used in communications due to its low complexity, good performance, and robustness [315].

Properties of the CMA cost function and algorithm have been studied thoroughly in [224]. The constant modulus property possessed by many types of communications signals has been exploited also in developing efficient algebraic blind equalization and source separation algorithms [441]. A good general review of Bussgang type blind deconvolution methods is [39].

19.1.3   Cumulant-based methods

Another popular group of blind deconvolution methods consists of cumulant-based approaches [315, 170, 174, 171]. They apply explicitly higher-order statistics of the observed signal $x(t)$, while in the Bussgang methods higher-order statistics

---

[1]The CMA algorithm can be applied to blind deconvolution of supergaussian sources by using a negative learning parameter μ in (19.4); see [11]

are involved into the estimation process implicitly via the nonlinear function $g(\cdot)$. Cumulants have been defined and discussed briefly in Chapter 2.

Shalvi and Weinstein [398] have derived necessary and sufficient conditions and a set of cumulant-based criteria for blind deconvolution. In particular, they introduced a stochastic gradient algorithm for maximizing a constrained kurtosis based criterion. We shall next describe this algorithm briefly, because it is computationally simple, converges globally, and can be applied equally well to both subgaussian and supergaussian source signals $s(t)$.

Assume that the source (input) signal $s(t)$ is complex-valued and symmetric, satisfying the condition $E\{s(t)^2\} = 0$. Assume that the length of the causal FIR deconvolution filter is M. The output $z(t)$ of this filter at discrete time t can then be expressed compactly as the inner product

$$z(t) = \mathbf{w}^T(t)\mathbf{y}(t) \tag{19.11}$$

where the M-dimensional filter weight vector $\mathbf{w}(t)$ and output vector $\mathbf{y}(t)$ at time t are respectively defined by

$$\mathbf{y}(t) = [y(t), y(t-1), \dots, y(t-M+1)]^T \tag{19.12}$$

$$\mathbf{w}(t) = [w(t), w(t-1), \dots, w(t-M+1)]^T \tag{19.13}$$

Shalvi and Weinstein's algorithm is then given by [398, 351]

$$\mathbf{u}(t+1) = \mathbf{u}(t) + \mu\,\mathrm{sign}(\kappa_s)[|z(t)|^2 z(t)]\mathbf{y}^*(t)$$
$$\mathbf{w}(t+1) = \mathbf{u}(t+1)/\parallel \mathbf{u}(t+1) \parallel \tag{19.14}$$

Here κ_s is the kurtosis of $s(t)$, $\parallel \cdot \parallel$ is the usual Euclidean norm, and the unnormalized filter weight vector $\mathbf{u}(t)$ is defined quite similarly as $\mathbf{w}(t)$ in (19.13).

It is important to notice that Shalvi and Weinstein's algorithm (19.14) requires whitening of the output signal $y(t)$ for performing appropriately (assuming that $s(t)$ is white, too). For a single complex-valued signal sequence (time series) $\{y(t)\}$, the temporal whiteness condition is

$$E\{y(t)y^*(t-k)\} = \sigma_y^2\delta_{tk} = \begin{cases} \sigma_y^2, & t = k \\ 0, & t \neq k \end{cases} \tag{19.15}$$

where the variance of $y(t)$ is often normalized to unity: $\sigma_y^2 = 1$. Temporal whitening can be achieved by spectral prewhitening in the Fourier domain, or by using time-domain techniques such as linear prediction [351]. Linear prediction techniques have been discussed for example in the books [169, 171, 419].

Shalvi and Weinstein have presented a somewhat more complicated algorithm for the case $E\{s(t)^2\} \neq 0$ in [398]. Furthermore, they showed that there exists a close relationship between their algorithm and the CMA algorithm discussed in the previous subsection; see also [351]. Later, they derived fast converging but more involved super-exponential algorithms for blind deconvolution in [399]. Shalvi and

Weinstein have reviewed their blind deconvolution methods in [170]. Closely related algorithms were proposed earlier in [114, 457].

It is interesting to note that Shalvi and Weinstein's algorithm (19.14) can be derived by maximizing the absolute value of the kurtosis of the filtered (deconvolved) signal $z(t)$ under the constraint that the output signal $y(t)$ is temporally white [398, 351]. The temporal whiteness condition leads to the normalization constraint of the weight vector $\mathbf{w}(t)$ in (19.14). The corresponding criterion for standard ICA is familiar already from Chapter 8, where gradient algorithms similar to (19.14) have been discussed. Also Shalvi and Weinstein's super-exponential algorithm [399] is very similar to the cumulant-based FastICA as introduced in Section 8.2.3. The connection between blind deconvolution and ICA is discussed in more detail in the next subsection.

Instead of cumulants, one can resort to *higher-order spectra* or *polyspectra* [319, 318]. They are defined as Fourier transforms of the cumulants quite similarly as the power spectrum is defined as a Fourier transform of the autocorrelation function (see Section 2.8.5). Polyspectra provide a basis for blind deconvolution and more generally identification of nonminimum-phase systems, because they preserve phase information of the observed signal. However, blind deconvolution methods based on higher-order spectra tend to be computationally more complex than Bussgang methods, and converge slowly [171]. Therefore, we shall not discuss them here. The interested reader can find more information on those methods in [170, 171, 315].

19.1.4 Blind deconvolution using linear ICA

In defining the blind deconvolution problem, the values of the original signal $s(t)$ were assumed to be independent for different t and nongaussian. Therefore, the blind deconvolution problem is formally closely related to the standard ICA problem. In fact, one can define a vector

$$\tilde{\mathbf{s}}(t) = [s(t), s(t-1), ..., s(t-n+1)]^T \tag{19.16}$$

by collecting n last values of the source signal, and similarly define

$$\tilde{\mathbf{x}}(t) = [x(t), x(t-1), ..., x(t-n+1)]^T \tag{19.17}$$

Then $\tilde{\mathbf{x}}$ and $\tilde{\mathbf{s}}$ are n-dimensional vectors, and the convolution (19.1) can be expressed for a finite number of values of the summation index k as

$$\tilde{\mathbf{x}} = \mathbf{A}\tilde{\mathbf{s}} \tag{19.18}$$

where $\mathbf{A}$ is a matrix that contains the coefficients a_k of the convolution filter as its rows, at different positions for each row. This is the classic matrix representation of a filter. This representation is not exact near the top and bottom rows, but for a sufficiently large n, it is good enough in practice.

From (19.18) we see that the blind deconvolution problem is actually (approximately) a special case of ICA. The components of s are independent, and the mixing is linear, so we get the standard linear ICA model.

In fact, the one-unit (deflationary) ICA algorithms in Chapter 8 can be directly used to perform blind deconvolution. As defined above, the inputs $\mathbf{x}(t)$ should then consist of sample sequences $x(t), x(t-1), ..., x(t-n+1)$ of the signal $x(t)$ to be deconvolved. Estimating just one "independent component", we obtain the original deconvolved signal $s(t)$. If several components are estimated, they correspond to translated versions of the original signal, so it is enough to estimate just one component.

19.2 BLIND SEPARATION OF CONVOLUTIVE MIXTURES

19.2.1 The convolutive BSS problem

In several practical applications of ICA, some kind of convolution takes place simultaneously with the linear mixing. For example, in the classic cocktail-party problem, or separation of speech signals recorded by a set of microphones, the speech signals do not arrive in the microphones at the same time. This is because the sound travels in the atmosphere with a very limited speed. Moreover, the microphones usually record echos of the speakers' voices caused by reverberations from the walls of the room or other obstacles. These two phenomena can be modeled in terms of convolutive mixtures. Here we have not considered noise and other complications that often appear in practice; see Section 24.2 and [429, 430].

Blind source separation of convolutive mixtures is basically a combination of standard instantaneous linear blind source separation and blind deconvolution. In the convolutive mixture model, each element of the mixing matrix $\mathbf{A}$ in the model $\mathbf{x}(t)$ = $\mathbf{As}(t)$ is a *filter* instead of a scalar. Written out for each mixture, the data model for convolutive mixtures is given by

$$x_i(t) = \sum_{j=1}^{n} \sum_{k} a_{ikj} s_j(t - k), \text{ for } i = 1, ..., n \qquad (19.19)$$

This is a FIR filter model, where each FIR filter (for fixed indices i and j) is defined by the coefficients a_{ikj}. Usually these coefficients are assumed to be time-independent constants, and the number of terms over which the convolution index k runs is finite. Again, we observe only the mixtures $x_i(t)$, and both the independent source signals $s_i(t)$ and all the coefficients a_{ikj} must be estimated.

To invert the convolutive mixtures (19.19), a set of similar FIR filters is typically used:

$$y_i(t) = \sum_{j=1}^{n} \sum_{k} w_{ikj} x_j(t - k), \text{ for } i = 1, ..., n \qquad (19.20)$$

The output signals $y_1(t), \dots, y_n(t)$ of the separating system are estimates of the source signals $s_1(t), \dots, s_n(t)$ at discrete time t. The w_{ikj} give the coefficients of the FIR filters of the separating system. The FIR filters used in separation can be either causal or noncausal depending on the method. The number of coefficient in each separating filter must sometimes be very large (hundreds or even thousands) for achieving sufficient inversion accuracy. Instead of the feedforward FIR structure, feedback (IIR type) filters have sometimes been used for separating convolutive mixtures, an example is presented in Section 23.4. See [430] for a discussion of mutual advantages and drawbacks of these filter structures in convolutive BSS.

At this point, it is useful to discuss relationships between the convolutive BSS problem and the standard ICA problem on a general level [430]. Recall first than in standard linear ICA and BSS, the indeterminacies are the scaling and the order of the estimated independent components or sources (and their sign, which can be included in scaling). With convolutive mixtures the indeterminacies are more severe: the order of the estimated sources $y_i(t)$ is still arbitrary, but scaling is replaced by (arbitrary) filtering. In practice, many of the methods proposed for convolutive mixtures filter the estimated sources $y_i(t)$ so that they are temporally uncorrelated (white). This follows from the strong independence condition that most of the blind separation methods introduced for convolutive mixtures try to realize as well as possible. The temporal whitening effect causes some inevitable distortion if the original source signals themselves are not temporally white. Sometimes it is possible to get rid of this by using a feedback filter structure; see [430].

Denote by

$$\mathbf{y}(t) = [y_1(t), y_2(t), \ldots, y_n(t)]^T \tag{19.21}$$

the vector of estimated source signals. They are both temporally and spatially white if

$$E\{\mathbf{y}(t)\mathbf{y}^H(t-k)\} = \delta_{tk}\mathbf{I} = \begin{cases} \mathbf{I}, & t = k \\ \mathbf{0}, & t \neq k \end{cases} \tag{19.22}$$

where H denotes complex conjugate transpose (Hermitian operator). The standard spatial whitening condition $E\{\mathbf{y}(t)\mathbf{y}^H(t)\} = \mathbf{I}$ is obtained as a special case when $t = k$. The condition (19.22) is required to hold for all the lag values k for which the separating filters (19.20) are defined. Douglas and Cichocki have introduced a simple adaptive algorithm for whitening convolutive mixtures in [120]. Lambert and Nikias have given an efficient temporal whitening method based on FIR matrix algebra and Fourier transforms in [257].

Standard ICA makes use of *spatial* statistics of the mixtures to learn a *spatial* blind separation system. In general, higher-order spatial statistics are needed for achieving this goal. However, if the source signals are temporally correlated, second-order spatiotemporal statistics are sufficient for blind separation under some conditions, as shown in [424] and discussed in Chapter 18. In contrast, blind separation of convolutive mixtures must utilize *spatiotemporal* statistics of the mixtures to learn a *spatiotemporal* separation system.

Stationarity of the sources has a decisive role in separating convolutive mixtures, too. If the sources have nonstationary variances, second-order spatiotemporal statistics are enough as briefly discussed in [359, 456].

For convolutive mixtures, stationary sources require higher than second-order statistics, just as basic ICA, but the following simplification is possible [430]. Spatiotemporal second-order statistics can be used to decorrelate the mixtures. This step returns the problem to that of conventional ICA, which again requires higher-order spatial statistics. Examples of such approaches are can be found in [78, 108, 156]. This simplification is not very widely used, however.

Alternatively, one can resort to higher-order spatiotemporal statistics from the beginning for sources that cannot be assumed nonstationary. This approach has been adopted in many papers, and it will be discussed briefly later in this chapter.

19.2.2 Reformulation as ordinary ICA

The simplest approach to blind separation of convolutive mixtures is to reformulate the problem using the standard linear ICA model. This is possible because blind deconvolution can be formulated as a special case of ICA, as we saw in (19.18). Define now a vector $\tilde{s}$ by concatenating M time-delayed versions of every source signal:

$$\tilde{s}(t) = [s_1(t), s_1(t-1), ..., s_1(t-M+1), s_2(t), s_2(t-1), ..., s_2(t-M+1),$$
$$... , s_n(t), s_n(t-1), ..., s_n(t-M+1)]^T \quad (19.23)$$

and define similarly a vector

$$\tilde{x}(t) = [x_1(t), x_1(t-1), ..., x_1(t-M+1), x_2(t), x_2(t-1), ..., x_2(t-M+1),$$
$$... , x_n(t), x_n(t-1), ..., x_n(t-M+1)]^T \quad (19.24)$$

Using these definitions, the convolutive mixing model (19.19) can be written

$$\tilde{x} = \tilde{A}\tilde{s} \quad (19.25)$$

where $\tilde{A}$ is a matrix containing the coefficients a_{ikj} of the FIR filters in a suitable order. Now one can estimate the convolutive BSS model by applying ordinary ICA methods to the standard linear ICA model (19.25).

Deflationary estimation is treated in [108, 401, 432]. These methods are based on finding maxima of the absolute value of kurtosis, thus generalizing the kurtosis-based methods of Chapter 8. Other examples of approaches in which the convolutive BSS problem has been solved using conventional ICA can be found in [156, 292].

A problem with the formulation (19.25) is that when the original data vector x is expanded to $\tilde{x}$, its dimension grows very much. The number M of time delays that needs to be taken into account depends on the application, but it is often tens or hundreds, and the dimension of model (19.25) grows with the same factor, to nM. This may lead to prohibitively high dimensions. Therefore, depending on the application and the dimensions n and M, this reformulation can solve the convolutive BSS problem satisfactorily, or not.

In blind deconvolution, this is not such a big problem because we have just one signal to begin with, and we only need to estimate one independent component, which is easier than estimating all of them. In convolutive BSS, however, we often need to estimate all the independent components, and their number is nM in the model (19.25). Thus the computations may be very burdensome, and the number of data points needed to estimate such a large number of parameters can be prohibitive in practical applications. This is especially true if we want to estimate the separating system adaptively, trying to track changes in the mixing system. Estimation should then be fast both in terms of computations and data collection time.

Regrettably, these remarks hold largely for other approaches proposed for blind separation of convolutive mixtures, too. A fundamental reason of the computational difficulties encountered with convolutive mixtures is the fact that the number of the unknown parameters in the model (19.19) is so large. If the filters have length M, it is M-fold compared with the respective instantaneous ICA model. This basic problem cannot be avoided in any way.

19.2.3 Natural gradient methods

In Chapter 9, the well-known Bell-Sejnowski and natural gradient algorithms were derived from the maximum likelihood principle. This principle was shown to be quite closely related to the maximization of the output entropy, which is often called the information maximization (infomax) principle; see Chapter 9. These ICA estimation criteria and algorithms can be extended to convolutive mixtures in a straightforward way. Early results and derivations of algorithms can be found in [13, 79, 121, 268, 363, 426, 427]. An application to CDMA communication signals will be described later in Chapter 23.

Amari, Cichocki, and Douglas presented an elegant and systematic approach for deriving natural gradient type algorithms for blind separation of convolutive mixtures and related tasks. It is based on algebraic equivalences and their nice properties. Their work has been summarized in [11], where rather general natural gradient learning rules have been given for complex-valued data both in the time domain and z-transform domain. The derived natural gradient rules can be implemented in either batch, on-line, or block on-line forms [11]. In the batch form, one can use a noncausal FIR filter structure, while the on-line algorithms require the filters to be causal.

In the following, we represent an efficient natural gradient type algorithm [10, 13] described also in [430] for blind separation of convolutive mixtures. It can be implemented on-line using a feedforward (FIR) filter structure in the time domain. The algorithm is given for complex-valued data.

The separating filters are represented as a sequence of coefficient matrices $\mathbf{W}_k(t)$ at discrete time t and lag (delay) k. The separated output with this notation and causal FIR filters is

$$\mathbf{y}(t) = \sum_{k=0}^{L} \mathbf{W}_k(t)\mathbf{x}(t - k) \tag{19.26}$$

Here $\mathbf{x}(t - k)$ is n-dimensional data vector containing the values of the n mixtures (19.19) at the time instant $t - k$, and $\mathbf{y}(t)$ is the output vector whose components are estimates of the source signals $s_i(t), i = 1, \ldots, m$. Hence $\mathbf{y}(t)$ has m components, with $m \leq n$.

This matrix notation allows the derivation of a separation algorithm using the natural gradient approach. The resulting weight matrix update algorithm, which takes into account the causal approximation of a doubly infinite filter by delaying the output by L samples, is as follows [13, 430]:

$$\boxed{\Delta\mathbf{W}_k(t) \propto \mathbf{W}_k(t) - \mathbf{g}(\mathbf{y}(t - L))\mathbf{v}^H(t - k), \quad k = 0, \ldots, L}$$

(19.27)

Quite similarly as in Chapter 9, each component of the vector $\mathbf{g}$ applies the nonlinearity $g_i(.)$ to the respective component of the argument vector. The optimal nonlinearity $g_i(.)$ is the negative score function $g_i = p_i'/p_i$ of the distribution p_i of the source s_i. In (19.27), $\mathbf{v}(t)$ is reverse-filtered output computed using the L latest samples backwards from the current sample:

$$\mathbf{v}(t) = \sum_{q=0}^{L} \mathbf{W}_{L-q}^H(t)\mathbf{y}(t - q)$$

(19.28)

The vector $\mathbf{v}$ needs to be stored for the latest L samples to compute the update $\Delta\mathbf{W}_k(t)$ of the weight matrix $\mathbf{W}_k(t)$ for all lags $l = 0, \ldots, L$. The algorithm has rather modest computational and memory requirements.

Note that if $L = 0$, the formulas (19.27) and (19.28) reduce to the standard natural gradient algorithm. In [13], the authors present a speech separation experiment where about 50 seconds of mixed data were needed to achieve about 10-15 dB enhancement in the quality of separated signals.

19.2.4 Fourier transform methods

Fourier transform techniques are useful in dealing with convolutive mixtures, because convolutions become products between Fourier transforms in the frequency domain.

It was shown in Chapter 13 that filtering the data is allowed before performing ICA, since filtering does not change the mixing matrix. Using the same proof, one can see that applying Fourier transform to the data does not change the mixing matrix either. Thus we can apply Fourier transform to both sides of Eq. (19.19). Denoting by $X_i(\omega)$, $S_i(\omega)$, and $A_{ij}(\omega)$ the Fourier transforms of $x_i(t)$, $s_i(t)$, and $a_{ij}(t)$, respectively, we obtain

$$X_i(\omega) = \sum_{j=1}^{n} A_{ij}(\omega)S_i(\omega), \quad \text{for } i = 1, ..., n$$

(19.29)

This shows that the convolutive mixture model (19.19) is transformed into a instantaneous linear ICA model in the frequency domain. The price that we have to pay for this is that the mixing matrix is now a function of the angular frequency ω while in the standard ICA/BSS problem it is constant.

To utilize standard ICA in practice in the Fourier domain, one can take *short-time* Fourier transforms of the data, instead of the global transform. This means that the data is windowed, usually by a smooth windowing function such as a gaussian envelope, and the Fourier transform is applied separately to each data window. The dependency of $X_i(\omega)$ on ω can be simplified by dividing the values of ω into a certain number of frequency bins (intervals). For every frequency bin, we have then a number of observations of $X_i(\omega)$, and we can estimate the ICA model separately for each frequency bin. Note that the ICs and the mixing matrix are now complex-valued. See Section 20.3 on how to estimate the ICA model with complex-valued data.

The problem with this Fourier approach is the indeterminacy of permutation and sign that is ubiquitous in ICA. The permutation and signs of the sources are usually different in each frequency interval. For reconstructing a source signal $s_i(t)$ in the time domain, we need all its frequency components. Hence we a need some method for choosing which source signals in different frequency intervals belong together. To this end, various continuity criteria have been introduced by many authors; see [15, 59, 216, 356, 397, 405, 406, 430].

Another major group of Fourier methods developed for convolutive mixtures avoids the preceding problem by performing the actual separation in the time domain. Only selected parts of the separation procedure are carried out in the frequency domain. Separating filters may be easier to learn in the frequency domain because components are now orthogonal and do not depend on each other like the time domain coefficients [21, 430]. Examples of methods that apply their separation criterion in the time domain but do the rest in the frequency domain are reported in [21, 257]. A frequency domain representation of the filters is learned, and they are also applied in the frequency domain. The final time-domain result is reconstructed using for example the overlap-save technique of digital signal processing (see [339]). Thus, the permutation and scaling problem does not exist.

The work by Lambert and Nikias deserves special attention, see the review in [257]. They have introduced methods that utilize the Bussgang family of cost functions and standard adaptive filtering algorithms in blind separation of convolutive mixtures. FIR matrix algebra introduced in [256] is employed as an efficient tool for systematic development of methods. Lambert and Nikias [257] have considered three general classes of Bussgang type cost functions, namely blind least mean-squares (LMS), Infomax, and direct Bussgang costs. Most of these costs can be implemented in either the time or frequency domain, or in the batch or continuously adaptive modes. Lambert and Nikias have introduced several efficient and practical algorithms for blind separation of convolutive mixtures having different computational complexities and convergence speeds. For example, block-oriented frequency domain implementations can be used to perform robust blind separation on convolutive mixtures which have hundreds or thousands of time delays [257].

19.2.5   Spatiotemporal decorrelation methods

Consider first the noisy instantaneous linear ICA model

$$\mathbf{x}(t) = \mathbf{A}\mathbf{s}(t) + \mathbf{n}(t) \tag{19.30}$$

which has been discussed in more detail in Chapter 15. Making the standard realistic assumption that the additive noise $\mathbf{n}(t)$ is independent of the source signals $\mathbf{s}(t)$, the spatial covariance matrix $\mathbf{C_x}(t)$ of $\mathbf{x}(t)$ at time t is

$$\mathbf{C_x}(t) = \mathbf{A}\mathbf{C_s}(t)\mathbf{A}^T + \mathbf{C_n}(t) \tag{19.31}$$

where $\mathbf{C_s}(t)$ and $\mathbf{C_n}(t)$ are respectively the covariance matrices of the sources and the noise at time t. If the sources $\mathbf{s}(t)$ are nonstationary with respect to their covariances, then in general $\mathbf{C_s}(t) \neq \mathbf{C_s}(t + \tau)$ for $\tau \neq 0$. This allows to write multiple conditions for different choices of τ to solve for $\mathbf{A}$, $\mathbf{C_s}(t)$, and $\mathbf{C_n}(t)$. Note that the covariances matrices $\mathbf{C_s}(t)$ and $\mathbf{C_n}(t)$ are diagonal. The diagonality of $\mathbf{C_s}(t)$ follows from the independence of the sources, and $\mathbf{C_n}(t)$ can be taken diagonal because the components of the noise vector $\mathbf{n}(t)$ are assumed to be uncorrelated.

We can also look at cross-covariance matrices $\mathbf{C_x}(t, t + \tau) = \mathrm{E}\{\mathbf{x}(t)\mathbf{x}(t + \tau)^T\}$ over time. This approach has been mentioned in the context of convolutive mixtures in [456], and it can be used with instantaneous mixtures as described in Chapter 18. For convolutive mixtures, we can write in frequency domain for sample averages [359, 356]

$$\bar{\mathbf{C}}_\mathbf{x}(\omega, t) = \mathbf{A}(\omega)\mathbf{C_s}(\omega, t)\mathbf{A}^H(\omega) + \mathbf{C_n}(\omega, t) \tag{19.32}$$

where $\bar{\mathbf{C}}_\mathbf{x}$ is the averaged spatial covariance matrix. If $\mathbf{s}$ is nonstationary, one can again write multiple linearly independent equations for different time lags and solve for unknowns or find LMS estimates of them by diagonalizing a number of matrices in the frequency domain [123, 359, 356].

If the mixing system is minimum phase, decorrelation alone can provide a unique solution, and the nonstationarity of the signals is not needed [55, 280, 402]. Many methods have been proposed for this case, for example, in [113, 120, 149, 281, 280, 296, 389, 390, 456]. More references are given in [430]. However, such decorrelating methods cannot necessarily be applied to practical communications and audio separation problems, because the mixtures encountered there are often not minimum-phase. For example in the cocktail-party problem the system is minimum phase if each speaker is closest to his or her "own" microphone, otherwise not [430].

19.2.6   Other methods for convolutive mixtures

Many methods proposed for blind separation of convolutive mixtures are extensions of earlier methods originally designed for either the standard linear instantaneous BSS (ICA) problem or for the blind deconvolution problem. We have already discussed some extensions of the natural gradient method in Section 19.2.3 and Bussgang methods in Section 19.2.4. Bussgang methods have been generalized for convolutive mixtures also in [351]. Matsuoka's method [296] for BSS of nonstationary sources is modified for convolutive mixtures in [239] using natural gradient learning. Nguyen Thi and Jutten [420] have generalized the seminal Hérault-Jutten algorithm described in Chapter 12 to BSS of convolutive mixtures. Their approach has also been studied in [74, 101]. A state-space approach for blind separation of convolutive mixtures has been studied in [479].

There exist numerous approaches to convolutive BSS which are based on criteria utilizing directly spatiotemporal higher-order statistics. Methods based on the maximization of the sum of the squares of the kurtoses to estimate the whole separating system were introduced in [90], and further developed in [307]. Other methods based on spatiotemporal higher-order statistics have been presented in [1, 124, 145, 155, 218, 217, 400, 416, 422, 434, 433, 470, 471, 474]. More references can be found in [91, 430].

19.3   CONCLUDING REMARKS

Historically, many ideas used in ICA were originally developed in the context of blind deconvolution, which is an older topic of research than ICA. Later, it was found that many methods developed for blind deconvolution can be directly applied for ICA, and vice versa. Blind deconvolution can thus be considered an intellectual ancestor of ICA. For example, Donoho proposed in [114] that the deconvolution filter (19.2) could be found by finding the filter whose output is maximally nongaussian. This is the same principle as used for ICA in Chapter 8. Douglas and Haykin have explored relationships between blind deconvolution and blind source separation in [122]. Elsewhere, it has been pointed out that Bussgang criteria are closely related to nonlinear PCA criteria [236] and several other ICA methods [11].

In this chapter, we have briefly discussed Bussgang, cumulant, and ICA based methods for blind deconvolution. Still one prominent class of blind deconvolution and separation methods for convolutive mixtures consists of subspace approaches [143, 171, 311, 315, 425]. They can be used only if the number of output signals (observed mixtures) strictly exceeds the number of sources. Subspace methods resort to second-order statistics and fractional sampling, and they are applicable to cyclostationary source signals which are commonplace in communications [91].

General references on blind deconvolution are [170, 171, 174, 315]. Blind deconvolution and separation methods for convolutive mixtures have often been developed in context with blind channel estimation and identification problems in communications. These topics are beyond the scope of our book, but the interested reader can find useful review chapters on blind methods in communications in [143, 144].

In the second half of this chapter, we have considered separation of convolutive mixtures. The mixing process then takes place both temporally and spatially, which complicates the blind separation problem considerably. Numerous methods for handling this problem have been proposed, but it is somewhat difficult to assess their usefulness, because comparison studies are still lacking. The large number of parameters is a problem, making it difficult to apply convolutive BSS methods to large scale problems. Other practical problems in audio and communications applications have been discussed in Torkkola's tutorial review [430]. More information can be found in the given references and recent reviews [257, 425, 429, 430] on convolutive BSS.

Appendix Discrete-time filters and the z-transform

In this appendix, we briefly discuss certain basic concepts and results of discrete-time signal processing which are needed in this chapter.

Linear causal discrete-time filters [169, 339] can generally be described by the difference equation

$$y(n) + \sum_{i=1}^{M} \alpha_i y(n-i) = x(n) + \sum_{i=1}^{N} \beta_i x(n-i) \tag{A.1}$$

which is mathematically equivalent to the ARMA model (2.127) in Section 2.8.6. In (A.1), n is discrete time, $x(n)$ is the input signal of the filter, and $y(n)$ its output at time instant n. Causality means that in (A.1) there are no quantities that depend on future time instants $n+j, j > 0$, making it possible to compute the filter output $y(n)$ in real time. The constant coefficients $\beta_i, i = 1, \ldots, N$ define the *FIR (Finite Impulse Response)* part of the filter (A.1), having the order M. Respectively, the coefficients $\alpha_i, i = 1, \ldots, M$ define the *IIR (Infinite Impulse Response)* part of the filter (A.1) with the order M.

If $M = 0$, (A.1) defines a pure FIR filter, and if $N = 0$, a pure IIR filter results. Either of these filter structures is typically used in separating convolutive mixtures. The FIR filter is more popular, because it is always stable, which means that its output $y(n)$ is bounded for bounded input values $x(n-i)$ and coefficients β_i. On the other hand, IIR filter can be unstable because of its feedback (recurrent) structure.

The stability and other properties of the discrete-time filter (A.1) can be analyzed conveniently in terms of the *z-transform* [169, 339]. For a discrete-time real sequence $\{x(k)\}$, the z-transform is defined as the series

$$X(z) = \sum_{k=-\infty}^{\infty} x(k) z^{-k} \tag{A.2}$$

where z is a complex variable with real and imaginary part. For specifying the z-transform of a sequence uniquely, one must also know its region of convergence.

The z-transform has several useful properties that follow from its definition. Of particular importance in dealing with convolutive mixtures is the property that the z-transform of the *convolution sum*

$$y(n) = \sum_{k} h_k x(n-k) \tag{A.3}$$

is the product of the z-transforms of the sequences $\{h_k\}$ and $\{x(n)\}$:

$$Y(z) = H(z) X(z) \tag{A.4}$$

The weights h_k in (A.3) are called *impulse response* values, and the quantity $H(z) = Y(z)/X(z)$ is called *transfer function*. The transfer function of the convolution sum (A.3) is the z-transform of its impulse response sequence.

The *Fourier transform* of a sequence is obtained from its z-transform as a special case by constraining the variable z to lie on the unit circle in the complex plane. This can be done by setting

$$z = \exp(\jmath\omega) = \cos(\omega) + \jmath\sin(\omega) \tag{A.5}$$

where $\jmath$ is the imaginary unit and ω the angular frequency. The Fourier transform has similar convolution and other properties as the z-transform [339].

Applying the z-transform to both sides of Eq. (A.1) yields

$$A(z)Y(z) = B(z)X(z) \tag{A.6}$$

where

$$A(z) = 1 + \sum_{k=1}^{M} \alpha_k z^{-k}, \qquad Y(z) = \sum_{k=0}^{M} y(n-k)z^{-k} \tag{A.7}$$

$A(z)$ is the z-transform of the coefficients $1, \alpha_1, \dots, \alpha_M$ where the coefficient $\alpha_0 = 1$ corresponds to $y(n)$, and $Y(z)$ is the z-transform of the output sequence $y(n), \dots, y(n-M)$. $B(z)$ and $X(z)$ are defined quite similarly as z-transform of the coefficients $1, \beta_1, \dots, \beta_N$, and the respective input signal sequence $x(n), \dots, x(n-N)$.

From (A.6), we get for the transfer function of the linear filter (A.1)

$$H(z) = \frac{Y(z)}{X(z)} = \frac{B(z)}{A(z)} \tag{A.8}$$

Note that for a pure FIR filter, $A(z) = 1$, and for pure IIR filter $B(z) = 1$. The zeros of denominator polynomial $A(z)$ are called the *poles* of the transfer function (A.8), and the zeros of numerator $B(z)$ are called the zeros of (A.8). It can be shown (see for example [339]) that the linear causal discrete-time filter (A.1) is stable if all the poles of the transfer function lie inside the unit circle in the complex plane. This is also the stability condition for a pure IIR filter.

From (A.8), $X(z) = G(z)Y(z)$, where the *inverse filter* $G(z)$ has the transfer function $1/H(z) = A(z)/B(z)$. Hence, the inverse filter of a pure FIR filter is a pure IIR filter and vice versa. Clearly, the general stability condition for the inverse filter $G(z)$ is that the zeros of $B(z)$ (and hence the zeros of the filter $H(z)$) in (A.8) are inside the unit circle in the complex plane. This is also the stability condition for the inverse of a pure FIR filter.

Generally, it is desirable that both the poles and the zeros of the transfer function (A.8) lie inside the unit circle. Then both the filter and its inverse filter exist and are stable. Such filters are called *minimum phase* filters. The minimum phase property is a necessity in many methods developed for convolutive mixtures. It should be noted that a filter that has no stable causal inverse may have a stable noncausal inverse, realized by a nonminimum-phase filter.

These matters are discussed much more thoroughly in many textbooks of digital signal processing and related areas; see for example [339, 302, 169, 171].

20

Other Extensions

In this chapter, we present some additional extensions of the basic independent component analysis (ICA) model. First, we discuss the use of prior information on the mixing matrix, especially on its sparseness. Second, we present models that somewhat relax the assumption of the independence of the components. In the model called independent subspace analysis, the components are divided into subspaces that are independent, but the components inside the subspaces are *not* independent. In the model of topographic ICA, higher-order dependencies are modeled by a topographic organization. Finally, we show how to adapt some of the basic ICA algorithms to the case where the data is complex-valued instead of real-valued.

20.1 PRIORS ON THE MIXING MATRIX

20.1.1 Motivation for prior information

No prior knowledge on the mixing matrix is used in the basic ICA model. This has the advantage of giving the model great generality. In many application areas, however, information on the form of the mixing matrix is available. Using prior information on the mixing matrix is likely to give better estimates of the matrix for a given number of data points. This is of great importance in situations where the computational costs of ICA estimation are so high that they severely restrict the amount of data that can be used, as well as in situations where the amount of data is restricted due to the nature of the application.

This situation can be compared to that found in nonlinear regression, where overlearning or overfitting is a very general phenomenon [48]. The classic way of avoiding overlearning in regression is to use regularizing priors, which typically penalize regression functions that have large curvatures, i.e., lots of "wiggles". This makes it possible to use regression methods even when the number of parameters in the model is very large compared to the number of observed data points. In the extreme theoretical case, the number of parameters is infinite, but the model can still be estimated from finite amounts of data by using prior information. Thus suitable priors can reduce overlearning that was discussed in Section 13.2.2.

One example of using prior knowledge that predates modern ICA methods is the literature on beamforming (see the discussion in [72]), where a very specific form of the mixing matrix is represented by a small number of parameters. Another example is in the application of ICA to magnetoencephalography (see Chapter 22), where it has been found that the independent components (ICs) can be modeled by the classic dipole model, which shows how to constrain the form of the mixing coefficients [246]. The problem with these methods, however, is that they may be applicable to a few data sets only, and lose the generality that is one of the main factors in the current flood of interest in ICA.

Prior information can be taken into account in ICA estimation by using Bayesian prior distributions for the parameters. This means that the parameters, which in this case are the elements of the mixing matrix, are treated as random variables. They have a certain distribution and are thus more likely to assume certain values than others. A short introduction to Bayesian estimation was given in Section 4.6.

In this section, we present a form of prior information on the mixing matrix that is both general enough to be used in many applications and strong enough to increase the performance of ICA estimation. To give some background, we first investigate the possibility of using two simple classes of priors for the mixing matrix $\mathbf{A}$: Jeffreys' prior and quadratic priors. We come to the conclusion that these two classes are not very useful in ICA. Then we introduce the concept of sparse priors. These are priors that enforce a sparse structure on the mixing matrix. In other words, the prior penalizes mixing matrices with a larger number of significantly nonzero entries. Thus this form of prior is analogous to the widely-used prior knowledge on the supergaussianity or sparseness of the independent components. In fact, due to this similarity, sparse priors are so-called conjugate priors, which implies that estimation using this kind of priors is particularly easy: Ordinary ICA methods can be simply adapted to using such priors.

20.1.2   Classic priors

In the following, we assume that the estimator $\mathbf{B}$ of the inverse of the mixing matrix $\mathbf{A}$ is constrained so that the estimates of the independent components $\mathbf{y} = \mathbf{B}\mathbf{x}$ are *white*, i.e., decorrelated and of unit variance: $\mathbf{E}\{\mathbf{y}\mathbf{y}^T\} = \mathbf{I}$. This restriction greatly facilitates the analysis. It is basically equivalent to first whitening the data and then restricting $\mathbf{B}$ to be orthogonal, but here we do not want to restrict the generality of these results by whitening. We concentrate here on formulating priors for $\mathbf{B} = \mathbf{A}^{-1}$. Completely analogue results hold for prior on $\mathbf{A}$.

Jeffreys' prior The classic prior in Bayesian inference is Jeffreys' prior. It is considered a maximally uninformative prior, which already indicates that it is probably not useful for our purpose.

Indeed, it was shown in [342] that Jeffreys' prior for the basic ICA model has the form:

$$p(\mathbf{B}) \propto |\det \mathbf{B}^{-1}| \tag{20.1}$$

Now, the constraint of whiteness of the $\mathbf{y} = \mathbf{B}\mathbf{x}$ means that $\mathbf{B}$ can be expressed as $\mathbf{B} = \mathbf{W}\mathbf{V}$, where $\mathbf{V}$ is a constant whitening matrix, and $\mathbf{W}$ is restricted to be orthogonal. But we have $\det \mathbf{B} = \det \mathbf{W} \det \mathbf{V} = \det \mathbf{V}$, which implies that Jeffreys's prior is constant in the space of allowed estimators (i.e., decorrelating $\mathbf{B}$). Thus we see that Jeffreys' prior has no effect on the estimator, and therefore cannot reduce overlearning.

Quadratic priors In regression, the use of quadratic regularizing priors is very common [48]. It would be tempting to try to use the same idea in the context of ICA. Especially in feature extraction, we could require the columns of $\mathbf{A}$, i.e. the features, to be smooth in the same sense as smoothness is required from regression functions. In other words, we could consider every column of $\mathbf{A}$ as a discrete approximation of a smooth function, and choose a prior that imposes smoothness for the underlying continuous function. Similar arguments hold for priors defined on the rows of $\mathbf{B}$, i.e., the filters corresponding to the features.

The simplest class of regularizing priors is given by quadratic priors. We will show here, however, that such quadratic regularizers, at least the simple class that we define below, do not change the estimator. Consider priors that are of the form

$$\log p(\mathbf{B}) = \sum_{i=1}^{n} \mathbf{b}_i^T \mathbf{M} \mathbf{b}_i + \text{const.} \tag{20.2}$$

where the $\mathbf{b}_i^T$ are the rows of $\mathbf{B} = \mathbf{A}^{-1}$, and $\mathbf{M}$ is a matrix that defines the quadratic prior. For example, for $\mathbf{M} = \mathbf{I}$ we have a "weight decay" prior $\log p(\mathbf{B}) = \sum_i \|\mathbf{b}_i\|^2$ that is often used to penalize large elements in $\mathbf{B}$. Alternatively, we could include in $\mathbf{M}$ some differential operators so that the prior would measure the "smoothnesses" of the $\mathbf{b}_i$, in the sense explained above. The prior can be manipulated algebraically to yield

$$\sum_{i=1}^{n} \mathbf{b}_i^T \mathbf{M} \mathbf{b}_i = \sum_{i=1}^{n} \text{tr}(\mathbf{M}\mathbf{b}_i\mathbf{b}_i^T) = \text{tr}(\mathbf{M}\mathbf{B}^T\mathbf{B}) \tag{20.3}$$

Quadratic priors have little significance in ICA estimation, however. To see this, let us constrain the estimates of the independent components to be white as previously.

This means that we have

$$E\{\mathbf{yy}^T\} = E\{\mathbf{Bxx}^T\mathbf{B}^T\} = \mathbf{BCB}^T = \mathbf{I} \qquad (20.4)$$

in the space of allowed estimates, which gives after some algebraic manipulations $\mathbf{B}^T\mathbf{B} = \mathbf{C}^{-1}$. Now we see that

$$\sum_{i=1}^{n} \mathbf{b}_i^T \mathbf{M} \mathbf{b}_i = \text{tr}(\mathbf{M}\mathbf{C}^{-1}) = \text{const.} \qquad (20.5)$$

In other words, the quadratic prior is constant. The same result can be proven for a quadratic prior on $\mathbf{A}$. Thus, quadratic priors are of little interest in ICA.

20.1.3 Sparse priors

Motivation A much more satisfactory class of priors is given by what we call *sparse priors*. This means that the prior information says that most of the elements of each row of $\mathbf{B}$ are zero; thus their distribution is supergaussian or sparse. The motivation for considering sparse priors is both empirical and algorithmic.

Empirically, it has been observed in feature extraction of images (see Chapter 21) that the obtained filters tend to be localized in space. This implies that the distribution of the elements b_{ij} of the filter $\mathbf{b}_i$ tends to be sparse, i.e., most elements are practically zero. A similar phenomenon can be seen in analysis of magnetoencephalography, where each source signal is usually captured by a limited number of sensors. This is due to the spatial localization of the sources and the sensors.

The algorithmic appeal of sparsifying priors, on the other hand, is based on the fact that sparse priors can be made to be conjugate priors (see below for definition). This is a special class of priors, and means that estimation of the model using this prior requires only very simple modifications in ordinary ICA algorithms.

Another motivation for sparse priors is their neural interpretation. Biological neural networks are known to be sparsely connected, i.e., only a small proportion of all possible connections between neurons are actually used. This is exactly what sparse priors model. This interpretation is especially interesting when ICA is used in modeling of the visual cortex (Chapter 21).

Measuring sparsity The sparsity of a random variable, say s, can be measured by expectations of the form $E\{G(s)\}$, where G is a nonquadratic function, for example, the following

$$G(s) = -|s|. \qquad (20.6)$$

The use of such measures requires that the variance of s is normalized to a fixed value, and its mean is zero. These kinds of measures were widely used in Chapter 8 to probe the higher-order structure of the estimates of the ICs. Basically, this is a robust nonpolynomial moment that typically is a monotonic function of kurtosis. Maximizing this function is maximizing kurtosis, thus supergaussianity and sparsity.

In feature extraction and probably several other applications as well, the distributions of the elements of of the mixing matrix and its inverse are zero-mean due to symmetry. Let us assume that the data $\mathbf{x}$ is whitened as a preprocessing step. Denote by $\mathbf{z}$ the whitened data vector whose components are thus uncorrelated and have unit variance. Constraining the estimates $\mathbf{y} = \mathbf{Wz}$ of the independent components to be white implies that $\mathbf{W}$, the inverse of the whitened mixing matrix, is orthogonal. This implies that the sum of the squares of the elements $\sum_j w_{ij}$ is equal to one for every i. The elements of each row $\mathbf{w}_i^T$ of $\mathbf{W}$ can be then considered a realization of a random variable of zero mean and unit variance. This means we could measure the sparsities of the rows of $\mathbf{W}$ using a sparsity measure of the form (20.6).

Thus, we can define a sparse prior of the form

$$\log p(\mathbf{W}) = \sum_{i=1}^{n} \sum_{j=1}^{n} G(w_{ij}) + \text{const.} \tag{20.7}$$

where G is the logarithm of some supergaussian density function. The function G in (20.6) is such log-density, corresponding to the Laplacian density, so we see that we have here a measure of sparsity of the $\mathbf{w}_i$.

The prior in (20.7) has the nice property of being a conjugate prior. Let us assume that the independent components are supergaussian, and for simplicity, let us further assume that they have identical distributions, with log-density G. Now we can take that same log-density as the log-prior density G in (20.7). Then we can write the prior in the form

$$\log p(\mathbf{W}) = \sum_{i=1}^{n} \sum_{j=1}^{n} G(\mathbf{w}_i^T \mathbf{e}_j) + \text{const.} \tag{20.8}$$

where we denote by $\mathbf{e}_i$ the canonical basis vectors, i.e., the ith element of $\mathbf{e}_i$ is equal to one, and all the others are zero. Thus the posterior distribution has the form:

$$\log p(\mathbf{W}|\mathbf{z}(1), ..., z(T)) = \sum_{i=1}^{n} [\sum_{t=1}^{T} G(\mathbf{w}_i^T \mathbf{z}(t)) + \sum_{j=1}^{n} G(\mathbf{w}_i^T \mathbf{e}_j)] + \text{const.} \tag{20.9}$$

This form shows that the posterior distribution has the same form as the prior distribution (and, in fact, the original likelihood). Priors with this property are called conjugate priors in Bayesian theory. The usefulness of conjugate priors resides in the property that the prior can be considered to correspond to a "virtual" sample. The posterior distribution in (20.9) has the same form as the likelihood of a sample of size $T + n$, which consists of both the observed $\mathbf{z}(t)$ and the canonical basis vectors $\mathbf{e}_i$. In other words, the posterior in (20.9) is the likelihood of the augmented (whitened) data sample

$$\mathbf{z}^*(t) = \begin{cases} \mathbf{z}(t), & \text{if } 1 \leq t \leq T \\ \mathbf{e}_{t-T}, & \text{if } T < t \leq T + n \end{cases} \tag{20.10}$$

Thus, using conjugate priors has the additional benefit that we can use exactly the same algorithm for maximization of the posterior as in ordinary maximum likelihood estimation of ICA. All we need to do is to add this virtual sample to the data; the virtual sample is of the same size n as the dimension of the data.

For experiments using sparse priors in image feature extraction, see [209].

Modifying prior strength The conjugate priors given above can be generalized by considering a family of supergaussian priors given by

$$\log p(\mathbf{W}) = \sum_{i=1}^{n} \sum_{j=1}^{n} \alpha G(\mathbf{w}_i^T \mathbf{e}_j) + \text{const.} \tag{20.11}$$

Using this kind of prior means that the virtual sample points are weighted by some parameter α. This parameter expresses the degree of belief that we have in the prior. A large α means that the belief in the prior is strong. Also, the parameter α could be different for different i, but this seems less useful here. The posterior distribution then has the form:

$$\log p(\mathbf{W}|\mathbf{z}(1), ..., \mathbf{z}(T)) = \sum_{i=1}^{n} [\sum_{t=1}^{T} G(\mathbf{w}_i^T \mathbf{z}(t)) + \sum_{j=1}^{n} \alpha G(\mathbf{w}_i^T \mathbf{e}_j)] + \text{const.} \tag{20.12}$$

The preceding expression can be further simplified in the case where the assumed density of the independent components is Laplacian, i.e., $G(y) = -|y|$. In this case, the α can multiply the $\mathbf{e}_j$ themselves:

$$\log p(\mathbf{W}|\mathbf{z}(1), ..., \mathbf{z}(T)) = \sum_{i=1}^{n} [-\sum_{t=1}^{T} |\mathbf{w}_i^T \mathbf{z}(t)| - \sum_{j=1}^{n} |\mathbf{w}_i^T (\alpha \mathbf{e}_j)|] + \text{const.} \tag{20.13}$$

which is simpler than (20.12) from the algorithmic viewpoint: It amounts to the addition of just n virtual data vectors of the form $\alpha \mathbf{e}_j$ to the data. This avoids all the complications due to the differential weighting of sample points in (20.12), and ensures that any conventional ICA algorithm can be used by simply adding the virtual sample to the data. In fact, the Laplacian prior is most often used in ordinary ICA algorithms, sometimes in the form of the log cosh function that can be considered as a smoother approximation of the absolute value function.

Whitening and priors In the preceding derivation, we assumed that the data is preprocessed by whitening. It should be noted that the effect of the sparse prior is dependent on the whitening matrix. This is because sparseness is imposed on the separating matrix of the whitened data, and the value of this matrix depends on the whitening matrix. There is an infinity of whitening matrices, so imposing sparseness on the whitened separating matrix may have different meanings.

On the other hand, it is not necessary to whiten the data. The preceding framework can be used for non-white data as well. If the data is not whitened, the meaning of the sparse prior is somewhat different, though. This is because every row of $\mathbf{b}_i$ is not constrained to have unit norm for general data. Thus our measure of sparsity does not anymore measure the sparsities of each $\mathbf{b}_i$. On the other hand, the developments of the preceding section show that the sum of squares of the whole matrix $\sum_{ij} b_{ij}$ does stay constant. This means that the sparsity measure is now measuring rather the global sparsity of $\mathbf{B}$, instead of the sparsities of individual rows.

In practice, one usually wants to whiten the data for technical reasons. Then the problems arises: How to impose the sparseness on the original separating matrix even when the data used in the estimation algorithm needs to be whitened? The preceding framework can be easily modified so that the sparseness is imposed on the original separating matrix. Denote by $\mathbf{V}$ the whitening matrix and by $\mathbf{B}$ the separating matrix for original data. Thus, we have $\mathbf{WV} = \mathbf{B}$ and $\mathbf{z} = \mathbf{Vx}$ by definition. Now, we can express the prior in (20.8) as

$$\log p(\mathbf{B}) = \sum_{i=1}^{n}\sum_{j=1}^{n} G(\mathbf{b}_i^T \mathbf{e}_j) + \text{const.} = \sum_{i=1}^{n}\sum_{j=1}^{n} G(\mathbf{w}_i^T(\mathbf{Ve}_j)) + \text{const.} \tag{20.14}$$

Thus, we see that the virtual sample added to $\mathbf{z}(t)$ now consists of the columns of the whitening matrix, instead of the identity matrix.

Incidentally, a similar manipulation of (20.8) shows how to put the prior on the original mixing matrix instead of the separating matrix. We always have $\mathbf{VA} = (\mathbf{W})^{-1} = \mathbf{W}^T$. Thus, we obtain $\mathbf{a}_i^T \mathbf{e}_j = \mathbf{a}_i^T \mathbf{V}^T(\mathbf{V}^{-1})^T \mathbf{e}_j = \mathbf{w}_i^T(\mathbf{V}^{-1})^T \mathbf{e}_j$. This shows that imposing a sparse prior on $\mathbf{A}$ is done by using the virtual sample given by the rows of the inverse of the whitening matrix. (Note that for whitened data, the mixing matrix is the transpose of the separating matrix, so the fourth logical possibility of formulating prior for the whitened mixing matrix is not different from using a prior on the whitened separating matrix.)

In practice, the problems implied by whitening can often be solved by using a whitening matrix that is sparse in itself. Then imposing sparseness on the whitened separating matrix is meaningful. In the context of image feature extraction, a sparse whitening matrix is obtained by the zero-phase whitening matrix (see [38] for discussion), for example. Then it is natural to impose the sparseness for the whitend separating matrix, and the complications discussed in this subsection can be ignored.

20.1.4 Spatiotemporal ICA

When using sparse priors, we typically make rather similar assumptions on both the ICs and the mixing matrix. Both are assumed to be generated so that the values are taken from independent, typically sparse, distributions. At the limit, we might develop a model where the very same assumptions are made on the mixing matrix and the ICs. Such a model [412] is called *spatiotemporal* ICA since it does ICA both in the temporal domain (assuming that the ICs are time signals), and in the spatial domain, which corresponds to the spatial mixing defined by the mixing matrix.

In spatiotemporal ICA, the distinction between ICs and the mixing matrix is completely abolished. To see why this is possible, consider the data as a single matrix of the observed vectors as its columns: $\mathbf{X} = (\mathbf{x}(1), ..., \mathbf{x}(T))$, and likewise for the ICs. Then the ICA model can be expressed as

$$\mathbf{X} = \mathbf{AS} \tag{20.15}$$

Now, taking a transpose of this equation, we obtain

$$\mathbf{X}^T = \mathbf{S}^T \mathbf{A}^T \tag{20.16}$$

Now we see that the matrix $\mathbf{S}$ is like a mixing matrix, with $\mathbf{A}^T$ giving the realizations of the "independent components". Thus, by taking the transpose, we flip the roles of the mixing matrix and the ICs.

In the basic ICA model, the difference between s and $\mathbf{A}$ is due to the statistical assumptions made on s, which are the independent random variables, and on $\mathbf{A}$, which is a constant matrix of parameters. But with sparse priors, we made assumptions on $\mathbf{A}$ that are very similar to those usually made on s. So, we can simply consider both $\mathbf{A}$ and $\mathbf{S}$ as being generated by independent random variables, in which case either one of the mixing equations (with or without transpose) are equally valid. This is the basic idea in spatiotemporal ICA.

There is another important difference between $\mathbf{S}$ and $\mathbf{A}$, though. The dimensions of $\mathbf{A}$ and $\mathbf{S}$ are typically very different: $\mathbf{A}$ is square whereas $\mathbf{S}$ has many more columns than rows. This difference can be abolished by considering that there $\mathbf{A}$ has many fewer columns than rows, that is, there is some redundancy in the signal.

The estimation of the spatiotemporal ICA model can be performed in a manner rather similar to using sparse priors. The basic idea is to form a virtual sample where the data consists of two parts, the original data and the data obtained by transposing the data matrix. The dimensions of these data sets must be strongly reduced and made equal to each other, using PCA-like methods. This is possible because it was assumed that both $\mathbf{A}$ and $\mathbf{S}^T$ have the same kind of redundancy: many more rows than columns. For details, see [412], where the infomax criterion was applied on this estimation task.

20.2 RELAXING THE INDEPENDENCE ASSUMPTION

In the ICA data model, it is assumed that the components s_i are independent. However, ICA is often applied on data sets, for example, on image data, in which the obtained estimates of the independent components are not very independent, even approximately. In fact, it is not possible, in general, to decompose a random vector x linearly into components that are independent. This raises questions on the utility and interpretation of the components given by ICA. Is it useful to perform ICA on real data that does not give independent components, and if it is, how should the results be interpreted?

One approach to this problem is to reinterpret the estimation results. A straightforward reinterpretation was offered in Chapter 10: ICA gives components that are as independent as possible. Even in cases where this is not enough, we can still justify the utility by other arguments. This is because ICA simultaneously serves certain other useful purposes than dependence reduction. For example, it can be interpreted as projection pursuit (see Section 8.5) or sparse coding (see Section 21.2). Both of these methods are based on the maximal nongaussianity property of the independent components, and they give important insight into what ICA algorithms are really doing.

A different approach to the problem of not finding independent components is to relax the very assumption of independence, thus explicitly formulating new data models. In this section, we consider this approach, and present three recently developed methods in this category. In multidimensional ICA, it is assumed that only certain sets (subspaces) of the components are mutually independent. A closely related method is independent subspace analysis, where a particular distribution structure inside such subspaces is defined. Topographic ICA, on the other hand, attempts to utilize the dependence of the estimated "independent" components to define a topographic order.

20.2.1   Multidimensional ICA

In multidimensional independent component analysis [66, 277], a linear generative model as in basic ICA is assumed. In contrast to basic ICA, however, the components (responses) s_i are not assumed to be all mutually independent. Instead, it is assumed that the s_i can be divided into couples, triplets or in general k-tuples, such that the s_i inside a given k-tuple may be dependent on each other, but dependencies between different k-tuples are not allowed.

Every k-tuple of s_i corresponds to k basis vectors a_i. In general, the dimensionality of each independent subspace need not be equal, but we assume so for simplicity. The model can be simplified by two additional assumptions. First, even though the components s_i are not all independent, we can always define them so that they are uncorrelated, and of unit variance. In fact, linear correlations inside a given k-tuple of dependent components could always be removed by a linear transformation. Second, we can assume that the data is whitened (sphered), just as in basic ICA.

These two assumptions imply that the a_i are orthonormal. In particular, the independent subspaces become orthogonal after whitening. These facts follow directly from the proof in Section 7.4.2, which applies here as well, due to our present assumptions.

Let us denote by J the number of independent feature subspaces, and by $S_j, j = 1, ..., J$ the set of the indices of the s_i belonging to the subspace of index j. Assume that the data consists of T observed data points $x(t), t = 1, ..., T$. Then we can express the likelihood L of the data, given the model as follows

$$L(\mathbf{x}(t), t = 1, ..., T; \mathbf{b}_i, i = 1, ..., n)$$
$$= \prod_{t=1}^{T} [|\det \mathbf{B}| \prod_{j=1}^{J} p_j(\mathbf{b}_i^T \mathbf{x}(t), i \in S_j)] \qquad (20.17)$$

where $p_j(.)$, which is a function of the k arguments $\mathbf{b}_i^T \mathbf{x}(t), i \in S_j$, gives the probability density inside the jth k-tuple of s_i. The term $|\det \mathbf{B}|$ appears here as in any expression of the probability density of a transformation, giving the change in volume produced by the linear transformation, as in Chapter 9.

The k-dimensional probability density $p_j(.)$ is not specified in advance in the general definition of multidimensional ICA [66]. Thus, the question arises how to estimate the model of multidimensional ICA. One approach is to estimate the basic ICA model, and then group the components into k-tuples according to their dependence structure [66]. This is meaningful only if the independent components are well defined and can be accurately estimated; in general we would like to utilize the subspace structure in the estimation process. Another approach is to model the distributions inside the subspaces by a suitable model. This is potentially very difficult, since we then encounter the classic problem of estimating k-dimensional distributions. One solution for this problem is given by independent subspaces analysis, to be explained next.

20.2.2 Independent subspace analysis

Independent subspace analysis [204] is a simple model that models some dependencies between the components. It is based on combining multidimensional ICA with the principle of invariant-feature subspaces.

Invariant-feature subspaces   To motivate independent subspace analysis, let us consider the problem of feature extraction, treated in more detail in Chapter 21. In the most basic case, features are given by linear transformations, or filters. The presence of a given feature is detected by computing the dot-product of input data with a given feature vector. For example, wavelet, Gabor, and Fourier transforms, as well as most models of V1 simple cells, use such linear features (see Chapter 21). The problem with linear features, however, is that they necessarily lack any invariance with respect to such transformations as spatial shift or change in (local) Fourier phase [373, 248].

Kohonen [248] developed the principle of invariant-feature subspaces as an abstract approach to representing features with some invariances. The principle of invariant-feature subspaces states that one can consider an invariant feature as a linear subspace in a feature space. The value of the invariant, higher-order feature is given by (the square of) the norm of the projection of the given data point on that subspace, which is typically spanned by lower-order features.

A feature subspace, as any linear subspace, can always be represented by a set of orthogonal basis vectors, say $\mathbf{b}_i, i = 1, ..., k$, where k is the dimension of the subspace. Then the value $F(\mathbf{x})$ of the feature F with input vector $\mathbf{x}$ is given by

$$F(\mathbf{x}) = \sum_{i=1}^{k} (\mathbf{b}_i^T \mathbf{x})^2 \qquad (20.18)$$

In fact, this is equivalent to computing the distance between the input vector $\mathbf{x}$ and a general linear combination of the vectors (possibly filters) $\mathbf{b}_i$ of the feature subspace [248].

Spherical symmetry Invariant-feature subspaces can be embedded in multidimensional independent component analysis by considering probability distributions for the k-tuples of s_i that are *spherically symmetric*, i.e., depend only on the norm. In other words, the probability density $p_j(.)$ of a k-tuple can be expressed as a function of the sum of the squares of the $s_i, i \in S_j$ only. For simplicity, we assume further that the $p_j(.)$ are equal for all j, i.e., for all subspaces.

This means that the logarithm of the likelihood L of the data $\mathbf{x}(t), t = 1, ..., T$, can be expressed as

$$\log L(\mathbf{x}(t), t = 1, ..., T; \mathbf{b}_i, i = 1, ..., n)$$

$$= \sum_{t=1}^{T} \sum_{j=1}^{J} \log p(\sum_{i \in S_j} (\mathbf{b}_i^T \mathbf{x}(t))^2) + T \log | \det \mathbf{B}| \qquad (20.19)$$

where $p(\sum_{i \in S_j} s_i^2) = p_j(s_i, i \in S_j)$ gives the probability density inside the jth k-tuple of s_i.

Recall that prewhitening allows us to consider the $\mathbf{b}_i$ to be orthonormal, which implies that $\log | \det \mathbf{B}|$ is zero. This shows that the likelihood in Eq. (20.19) is a function of the norms of the projections of $\mathbf{x}$ on the subspaces indexed by j, which are spanned by the orthonormal basis sets given by $\mathbf{b}_i, i \in S_j$.

In the case of clearly supergaussian components, we can use the following probability distribution:

$$\log p(\sum_{i \in S_j} s_i^2) = -\alpha[\sum_{i \in S_j} s_i^2]^{1/2} + \beta \qquad (20.20)$$

which could be considered a multi-dimensional version of the exponential distribution. The scaling constant α and the normalization constant β are determined so as to give a probability density that is compatible with the constraint of unit variance of the s_i, but they are irrelevant in the following. Thus we see that the estimation of the model consists of finding subspaces such that the *norms of the projections of the (whitened) data on those subspaces have maximally sparse distributions*.

Independent subspace analysis is a natural generalization of ordinary ICA. In fact, if the projections on the subspaces are reduced to dot-products, i.e., projections on one-dimensional (1-D) subspaces, the model reduces to ordinary ICA, provided that, in addition, the independent components are assumed to have symmetric distributions. It is to be expected that the norms of the projections on the subspaces represent some higher-order, invariant features. The exact nature of the invariances has not been specified in the model but will emerge from the input data, using only the prior information on their independence.

If the subspaces have supergaussian (sparse) distributions, the dependency implied by the model is such that components in the same subspace tend to be nonzero at the same time. In other words, the subspaces are somehow "activated" as a whole, and then the values of the individual components are generated according to how strongly the subspaces are activated. This is the particular kind of dependency that is modeled by independent subspaces in most applications, for example, with image data.

For more details on independent subspace analysis, the reader is referred to [204]. Some experiments on image data are reported in Section 21.5 as well.

20.2.3 Topographic ICA

Another way of approaching the problem of nonexistence of independent components is to try to somehow make the dependency structure of the estimated components visible. This dependency structure is often very informative and could be utilized in further processing.

Estimation of the "residual" dependency structure of estimates of independent components could be based, for example, on computing the cross-cumulants. Typically these would be higher-order cumulants, since second-order cross-cumulants, i.e., covariance, are typically very small, and can in fact be forced to be zero, as we did by orthogonalization after whitening in Part II. However, using such measures raises the question as to how such numerical estimates of the dependence structure should be visualized or otherwise utilized. Moreover, there is another serious problem associated with simple estimation of some dependency measures from the estimates of the independent components. This is due to the fact that often the independent components do not form a well-defined set. Especially in image decomposition (Chapter 21), the set of potential independent components seems to be larger than what can be estimated at one time, in fact the set might be infinite. A classic ICA method gives an arbitrarily chosen subset of such independent components. Thus, it is important in many applications that the dependency information is utilized during the estimation of the independent components, so that the estimated set of independent components is one whose residual dependencies can be represented in a meaningful way. (This is something we already argued in connection with independent subspace analysis.)

Topographic ICA, introduced in [206], is a modification of the classic ICA model in which the dependencies of the components are explicitly represented. In particular, we propose that the residual dependency structure of the independent components, i.e., dependencies that cannot be canceled by ICA, could be used to define a *topographical order* between the components. The topographical order is easy to represent by visualization, and is important in image feature extraction due to its connections to brain modeling [206].

Our model gives a topographic map where the distance of the components in the topographic representation is a function of the dependencies of the components. Components that are near to each other in the topographic representation are strongly dependent in the sense of higher-order correlations.

To obtain topographic ICA, we generalize the model defined by (20.19) so that it models a dependence not only inside the k-tuples, but among all neighboring components. A neighborhood relation defines a topographical order. We define the likelihood of the model as follows:

$$\log L(\mathbf{B}) = \sum_{t=1}^{T} \sum_{j=1}^{n} G(\sum_{i=1}^{n} h(i,j)(\mathbf{b}_i^T \mathbf{x}(t))^2) + T \log|\det \mathbf{B}| + \text{const.} \quad (20.21)$$

Here, the $h(i, j)$ is a neighborhood function, which expresses the strength of the connection between the ith and jth units. It can be defined in the same way as in other topographic maps, like the self-organizing map (SOM) [247]. The function G is similar to the one in independent subspace analysis. The additive constant depends only on $h(i, j)$.

This model thus can be considered a generalization of the model of independent subspace analysis. In independent subspace analysis, the latent variables s_i are clearly divided into k-tuples or subspaces, whereas in topographic ICA, such subspaces are completely overlapping: Every neighborhood corresponds to one subspace.

Just as independent subspace analysis, topographic ICA usually models a situation where nearby components tend to be active (nonzero) at the same time. This seems to be a common dependency structure for natural sparse data [404]. In fact, the likelihood given earlier can also be derived as an approximation of the likelihood of a model where the variance of the ICs is controlled by some higher-order variables, so that the variances of near-by components are strongly dependent.

For more details on topographic ICA, the reader is referred to [206]. Some experiments on image data are reported in Chapter 21 as well.

20.3   COMPLEX-VALUED DATA

Sometimes in ICA, the ICs and/or the mixing matrix are complex-valued. For example, in signal processing in some cases frequency (Fourier) domain representations of signals have advantages over time-domain representations. Especially in the separation of convolutive mixtures (see Chapter 19) it is quite common to Fourier transform the signals, which results in complex-valued signals.

In this section we show how the FastICA algorithm can be extended to complex-valued signals. Both the ICs s and the observed mixtures x assume complex values. For simplicity, we assume that the number of independent component variables is the same as the number of observed linear mixtures. The mixing matrix $\mathbf{A}$ is of full rank and it may be complex as well, but this need not be the case.

In addition to the assumption of the independence of the components s_i, an assumption on the dependence of the real and complex parts of a single IC is made here. We assume that every s_i is white in the sense that the real and imaginary parts of s_j are uncorrelated and their variances are equal; this is quite realistic in practical problems.

Related work on complex ICA can be found in [21, 132, 305, 405].

20.3.1   Basic concepts of complex random variables

First, we review some basic concepts of complex random variables; see [419] for more details.

A complex random variable y can be represented as $y = u + iv$ where u and v are real-valued random variables. The density of y is $f(y) = f(u, v) \in \mathbb{R}$. The expectation of y is $E\{y\} = E\{u\} + iE\{v\}$. Two complex random variables y_1 and y_2 are uncorrelated if $E\{y_1 y_2^*\} = E\{y_1\}E\{y_2^*\}$, where $y^* = u - iv$ designates the complex conjugate of y. The covariance matrix of a zero-mean complex random vector $\mathbf{y} = (y_1, \ldots, y_n)$ is

$$E\{\mathbf{yy}^H\} = \begin{bmatrix} C_{11} & \cdots & C_{1n} \\ \vdots & \ddots & \vdots \\ C_{n1} & \cdots & C_{nn} \end{bmatrix} \tag{20.22}$$

where $C_{jk} = E\{y_j y_k^*\}$ and $\mathbf{y}^H$ stands for the Hermitian of $\mathbf{y}$, that is, $\mathbf{y}$ transposed and conjugated. The data can be whitened in the usual way.

In our complex ICA model, all ICs s_i have zero mean and unit variance. Moreover, we require that they have uncorrelated real and imaginary parts of equal variances. This can be equivalently expressed as $E\{\mathbf{ss}^H\} = \mathbf{I}$ and $E\{\mathbf{ss}^T\} = \mathbf{O}$. In the latter, the expectation of the outer product of a complex random vector without the conjugate is a null matrix. These assumptions imply that s_i must be strictly complex; that is, the imaginary part of s_i may not in general vanish.

The definition of kurtosis can be easily generalized. For a zero-mean complex random variable it could be defined, for example, as [305, 319]

$$\text{kurt}(y) = E\{|y|^4\} - E\{yy^*\}E\{yy^*\} - E\{yy\}E\{y^*y^*\} - E\{yy^*\}E\{y^*y\} \tag{20.23}$$

but the definitions vary with respect to the placement of conjugates (*) — actually, there are 2^4 ways to define the kurtosis [319]. We choose the definition in [419], where

$$\text{kurt}(y) = E\{|y|^4\} - 2(E\{|y|^2\})^2 - |E\{y^2\}|^2 = E\{|y|^4\} - 2 \tag{20.24}$$

where the last equality holds if y is white, i.e., the real and imaginary parts of y are uncorrelated and their variances are equal to $1/2$. This definition of kurtosis is intuitive since it vanishes if y is gaussian.

20.3.2   Indeterminacy of the independent components

The independent components $\mathbf{s}$ in the ICA model are found by searching for a matrix $\mathbf{B}$ such that $\mathbf{s} = \mathbf{Bx}$. However, as in basic ICA, there are some indeterminacies. In the real case, a scalar factor α_i can be exchanged between s_i and a column $\mathbf{a}_i$ of $\mathbf{A}$ without changing the distribution of $\mathbf{x}$: $\mathbf{a}_i s_i = (\alpha_i \mathbf{a}_i)(\alpha_i^{-1} s_i)$. In other words, the order, the signs and the scaling of the independent components cannot be determined. Usually one defines the absolute scaling by defining $E\{s_i^2\} = 1$; thus only the signs of the independent components are indetermined.

Similarly in the complex case there is an unknown phase v_j for each s_j. Let us write the decomposition

$$\mathbf{a}_i s_i = (v_i \mathbf{a}_i)(v_i^{-1} s_j) \tag{20.25}$$

where the modulus of v_i is equal to one. If s_i has a spherically symmetric distribution, i.e., the distribution depends on the modulus of s_i only, the multiplication by a variable v_i does not change the distribution of s_i. Thus the distribution of $\mathbf{x}$ remains unchanged as well. From this indeterminacy it follows that it is impossible to retain the phases of s_i, and $\mathbf{BA}$ is a matrix where in each row and each column there is one nonzero element that is of unit modulus.

20.3.3 Choice of the nongaussianity measure

Now we generalize the framework in Chapter 8 for complex-valued signals. In the complex case, the distributions for the complex variables are often spherically symmetric, so only the modulus is interesting. Thus we could use a nongaussianity measure that is based on the modulus only. Based on the measure of nongaussianity as in (8.25), we use the following:

$$J_G(\mathbf{w}) = E\{G(|\mathbf{w}^H \mathbf{z}|^2)\} \tag{20.26}$$

where G is a smooth even function, $\mathbf{w}$ is an n-dimensional complex vector and $E\{|\mathbf{w}^H \mathbf{z}|^2\} = \|\mathbf{w}\|^2 = 1$. The data is whitened, as the notation $\mathbf{z}$ already indicates.

This can be compared with (20.24), which gives the kurtosis of complex variables: if we choose $G(y) = y^2$, then $J_G(\mathbf{w}) = E\{|\mathbf{w}^H \mathbf{z}|^4\}$. Thus J essentially measures the kurtosis of $\mathbf{w}^H \mathbf{z}$, which is a classic measure in higher-order statistics.

Maximizing J_G we estimate one IC. Estimating n independent components is possible, just as in the real case, by using a sum of n measures of nongaussianity, and a constraint of orthogonality. Thus one obtains the following optimization problem:

$$\text{maximize } \sum_{j=1}^{n} J_G(\mathbf{w}_i) \text{ with respect to } \mathbf{w}_i, \, j = 1, \dots, n$$

$$\text{under constraint } E\{\mathbf{w}_k^H \mathbf{w}_i\} = \delta_{jk} \tag{20.27}$$

where $\delta_{jk} = 1$ for $j = k$ and $\delta_{jk} = 0$ otherwise.

It is highly preferable that the estimator given by the contrast function is robust against outliers. The more slowly G grows as its argument increases, the more robust is the estimator. For the choice of G we propose now three different functions, the derivatives g of which are also given:

$$G_1(y) = \sqrt{a_1 + y}, \quad g_1(y) = \frac{1}{2\sqrt{a_1 + y}} \tag{20.28}$$

$$G_2(y) = \log(a_2 + y), \quad g_2(y) = \frac{1}{a_2 + y} \tag{20.29}$$

$$G_3(y) = \frac{1}{2}y^2, \quad g_3(y) = y \tag{20.30}$$

where a_1 and a_2 are some arbitrary constants (for example, $a_1 \approx 0.1$ and $a_2 \approx 0.1$ seem to be suitable). Of the preceding functions, G_1 and G_2 grow more slowly than G_3 and thus they give more robust estimators. G_3 is motivated by kurtosis (20.24).

20.3.4   Consistency of estimator

In Chapter 8 it was stated that any nonlinear learning function G divides the space of probability distributions into two half-spaces. Independent components can be estimated by either maximizing or minimizing a function similar to (20.26), depending on which half-space their distribution lies in. A theorem for real valued signals was presented that distinguished between maximization and minimization and gave the exact conditions for convergence. Now we show how this idea can be generalized to complex-valued random variables. We have the following theorem on the local consistency of the estimators [47]:

Theorem 20.1 *Assume that the input data follows the complex ICA model. The observed mixtures are prewhitened so that $E\{\mathbf{zz}^H\} = \mathbf{I}$. The independent components have zero mean, unit variance, and uncorrelated real and imaginary parts of equal variances. Also, $G : \mathbb{R}^+ \cup \{0\} \to \mathbb{R}$ is a sufficiently smooth even function. Then the local maxima (resp. minima) of $E\{G(|\mathbf{w}^H\mathbf{z}|^2)\}$ under the constraint $E\{|\mathbf{w}^H\mathbf{z}|^2\} = \|\mathbf{w}\|^2 = 1$ include those rows of the inverse of the whitened mixing matrix $\mathbf{VA}$ such that the corresponding independent components s_k satisfy*

$$E\{g(|s_k|^2) + |s_k|^2\, g'(|s_k|^2) - |s_k|^2\, g(|s_k|^2)\} < 0 \quad (> 0, \text{ resp.})$$
(20.31)

where $g()$ is the derivative of $G()$ and $g'()$ is the derivative of $g()$.

A special case of the theorem is when $g(y) = y$, $g'(y) = 1$. Condition (20.31) now reads

$$E\{|s_k|^2 + |s_k|^2 - |s_k|^2|s_k|^2\} = -E\{|s_k|^4\} + 2 < 0 \quad (> 0, \text{ resp.}).$$
(20.32)

Thus the local maxima of $E\{G(|\mathbf{w}^H\mathbf{z}|^2)\}$ should be found when $E\{|s_k|^4\} - 2 > 0$, i.e., the kurtosis (20.24) of s_k is positive. This implies that we are actually maximizing the absolute values of kurtosis, just like in the basic case in Chapter 8.

20.3.5   Fixed-point algorithm

We now give the fixed-point algorithm for complex signals under the complex ICA model. The algorithm searches for the extrema of $E\{G(|\mathbf{w}^H\mathbf{z}|^2)\}$. The derivation is presented in [47].

The FastICA algorithm for one unit, using whitened data, is

$$\mathbf{w} \leftarrow E\{\mathbf{x}(\mathbf{w}^H\mathbf{x})^* g(|\mathbf{w}^H\mathbf{x}|^2)\} - E\{g(|\mathbf{w}^H\mathbf{x}|^2) + |\mathbf{w}^H\mathbf{x}|^2 g'(|\mathbf{w}^H\mathbf{x}|^2)\}\mathbf{w}$$

$$\mathbf{w} \leftarrow \frac{\mathbf{w}}{\|\mathbf{w}\|}$$
(20.33)

The one-unit algorithm can be extended to the estimation of the whole ICA transformation in exactly the same way as in the real case. The orthogonalization methods in Section 8.4 can be used by simply replacing every transpose operation in the formulas by the Hermitian operation [47].

20.3.6 Relation to independent subspaces

Our approach to complex ICA closely resembles independent subspace methods, discussed in Section 20.2.2, and multidimensional ICA, discussed in Section 20.2.1.

In our complex ICA, the nongaussianity measure operates on $|\mathbf{w}^H\mathbf{z}|^2$ which can be interpreted as the norm of a projection onto a subspace. The subspace is two-dimensional, corresponding to the real and imaginary parts of a complex number. In contrast to the subspace method, one of the basis vectors is determined straightforward from the other basis vector. In independent subspace analysis, the independent subspace is determined only up to an orthogonal $k \times k$ matrix factor. In complex ICA however, the indeterminacy is less severe: the sources are determined up to a complex factor v, $|v| = 1$.

It can be concluded that complex ICA is a restricted form of independent subspace methods.

20.4 CONCLUDING REMARKS

The methods presented in the first two sections of the chapter were all related to the case where we know more about the data than just the blind assumption of independence. Using sparse priors, we incorporate some extra knowledge on the sparsity of the mixing matrix in the estimation procedure. This was made very easy by the algorithmic trick of conjugate priors.

In the methods of independent subspaces or topographic ICA, on the other hand, we assume that we cannot really find independent components; instead we can find groups of independent components, or components whose dependency structure can be visualized. A special case of the subspace formalism is encountered if the independent components are complex-valued.

Another class of extensions that we did not treat in this chapter are the so-called *semiblind* methods, that is, methods in which much prior information on the mixing is available. In the extreme case, the mixing could be almost completely known, in which case the "blind" aspect of the method disappears. Such semiblind methods are quite application-dependent. Some methods related to telecommunications are treated in Chapter 23. A closely related theoretical framework is the "principal" ICA proposed in [285]. See also [415] for a semiblind method in a brain imaging application.

Part IV

APPLICATIONS OF ICA

21

Feature Extraction by ICA

A fundamental approach in signal processing is to design a statistical generative model of the observed signals. The components in the generative model then give a representation of the data. Such a representation can then be used in such tasks as compression, denoising, and pattern recognition. This approach is also useful from a neuroscientific viewpoint, for modeling the properties of neurons in primary sensory areas.

In this chapter, we consider a certain class of widely used signals, which we call natural images. This means images that we encounter in our lives all the time; images that depict wild-life scenes, human living environments, etc. The working hypothesis here is that this class is sufficiently homogeneous so that we can build a statistical model using observations of those signals, and then later use this model for processing the signals, for example, to compress or denoise them.

Naturally, we shall use independent component analysis (ICA) as the principal model for natural images. We shall also consider the extensions of ICA introduced in Chapter 20. We will see that ICA does provide a model that is very similar to the most sophisticated low-level image representations used in image processing and vision research. ICA gives a statistical justification for using those methods that have often been more heuristically justified.

21.1 LINEAR REPRESENTATIONS

21.1.1 Definition

Image representations are often based on discrete linear transformations of the observed data. Consider a black-and-white image whose gray-scale value at the pixel indexed by x and y is denoted by $I(x, y)$. Many basic models in image processing express the image $I(x, y)$ as a linear superposition of some features or basis functions $a_i(x, y)$:

$$I(x, y) = \sum_{i=1}^{n} a_i(x, y) s_i \qquad (21.1)$$

where the s_i are stochastic coefficients, different for each image $I(x, y)$. Alternatively, we can just collect all the pixel values in a single vector $\mathbf{x} = (x_1, x_2, ..., x_m)^T$, in which case we can express the representation as

$$\mathbf{x} = \mathbf{As} \qquad (21.2)$$

just like in basic ICA. We assume here that the number of transformed components equals the number of observed variables, although this need not be the case in general. This kind of a linear superposition model gives a useful description on a low level where we can ignore such higher-level nonlinear phenomena as occlusion.

In practice, we may not model a whole image using the model in (21.1). Rather, we apply it on image patches or windows. Thus we partition the image into patches of, for example, 8×8 pixels and model the patches with the model in (21.1). Care must then be taken to avoid border effects.

Standard linear transformations widely used in image processing are, for example, the Fourier, Haar, Gabor, and cosine transforms. Each of them has its own favorable properties [154]. Recently, a lot of interest has been aroused by methods that attempt to combine the good qualities of frequency-based methods (Fourier and cosine transforms) with the basic pixel-by-pixel representation. Here we succinctly explain some of these methods; for more details see textbooks on the subject, e.g., [102], or see [290].

21.1.2 Gabor analysis

Gabor functions or Gabor filters [103, 128] are functions that are extensively used in image processing. These functions are localized with respect to three parameters: spatial location, orientation, and frequency. This is in contrast to Fourier basis function that are not localized in space, and the basic pixel-by-pixel representation that is not localized in frequency or orientation.

Let us first consider, for simplicity, one-dimensional (1-D) Gabor functions instead of the two-dimensional (2-D) functions used for images. The Gabor functions are

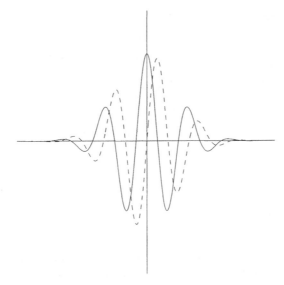

Fig. 21.1 A pair of 1-D Gabor functions. These functions are localized in space as well as in frequency. The real part is given by the solid line and the imaginary part by the dashed line.

then of the form

$$g_{1d}(x) = \exp(-\alpha^2(x - x_0)^2)[\cos(2\pi\beta(x - x_0) + \gamma) + i\sin(2\pi\beta(x - x_0) + \gamma)]$$
(21.3)

where

- α is the constant in the gaussian modulation function, which determines the width of the function in space.

- x_0 defines the center of the gaussian function, i.e., the location of the function.

- β is the frequency of oscillation, i.e., the location of the function in Fourier space.

- γ is the phase of the harmonic oscillation.

Actually, one Gabor function as in (21.3) defines two scalar functions: One as its real part and the other one as its imaginary part. Both of these are equally important, and the representation as a complex function is done mainly for algebraic convenience. A typical pair of 1-D Gabor functions is plotted in Fig. 21.1.

Two-dimensional Gabor functions are created by first taking a 1-D Gabor function along one of the dimensions and multiplying it by a gaussian envelope in the other dimension:

$$g_{2d}(x, y) = \exp(-\alpha^2(y - y_0)^2)g_{1d}(x)$$
(21.4)

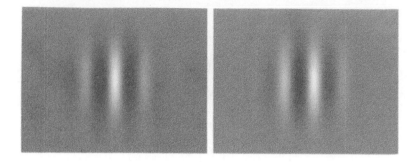

Fig. 21.2 A pair of 2-D Gabor functions. These functions are localized in space, frequency, and orientation. The real part is on the left, and the imaginary part on the right. These functions have not been rotated.

where the parameter α in the gaussian envelope need not be the same in both directions. Second, this function is rotated by an orthogonal transformation of (x, y) to a given angle. A typical pair of the real and imaginary parts of a Gabor functions are shown in Fig. 21.2.

Gabor analysis is an example of multi-resolution analysis, which means that the image is analyzed separately at different resolutions, or frequencies. This is because Gabor functions can be generated at different sizes by varying the parameter α, and at different frequencies by varying β.

An open question is what set of values should one choose for the parameters to obtain a useful representation of the data. Many different solutions exist; see, e.g., [103, 266]. The wavelet bases, discussed next, give one solution.

21.1.3 Wavelets

Another closely related method of multiresolution analysis is given by wavelets [102, 290]. Wavelet analysis is based on a single prototype function called the mother wavelet $\phi(x)$. The basis functions (in one dimension) are obtained by translations $\phi(x + l)$ and dilations or rescalings $\phi(2^{-s}x)$ of this basic function. Thus we use the family of functions

$$\phi_{s,l}(x) = 2^{-s/2}\phi(2^{-s}x - l) \tag{21.5}$$

The variables s and l are integers that represent scale and dilation, respectively. The scale parameter, s, indicates the width of the wavelet, while the location index, l, gives the position of the mother wavelet. The fundamental property of wavelets is thus the self-similarity at different scales. Note that ϕ is real-valued.

The mother wavelet is typically localized in space as well as in frequency. Two typical choices are shown in Fig. 21.3.

A 2-D wavelet transform is obtained in the same way as a 2-D Fourier transform: by first taking the 1-D wavelet transforms of all rows (or all columns), and then

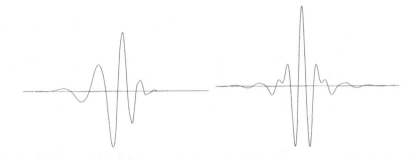

Fig. 21.3 Two typical mother wavelets. On the left, a Daubechies mother wavelet, and on the right, a Meyer mother wavelet.

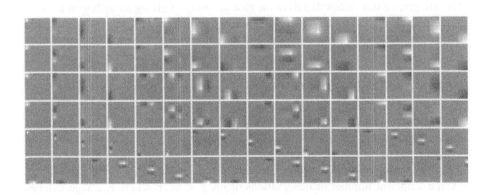

Fig. 21.4 Part of a 2-D wavelet basis.

the 1-D wavelet transform of the results of this transform. Some 2-D wavelet basis vectors are shown in Fig. 21.4.

The wavelet representation also has the important property of being localized both in space and in frequency, just like the Gabor transform. Important differences are the following:

- There is no phase parameter, and the wavelets all have the same phase. Thus, all the basis functions look the same, whereas in Gabor analysis, we have the couples given by the real and imaginary parts. Thus we have basis vectors of two different phases, and moreover the phase parameter can be modified. In Gabor analysis, some functions are similar to bars, and others are similar to edges, whereas in wavelet analysis, the basis functions are usually something in between.

- The change in size and frequency (parameters α and β in Gabor functions) are not independent. Instead, the change in size implies a strictly corresponding change in frequency.

- Usually in wavelets, there is no orientation parameter either. The only orientations encountered are horizontal and vertical, which come about when the horizontal and vertical wavelets have different scales.

- The wavelet transform gives an orthogonal basis of the 1-D space. This is in contrast to Gabor functions, which do not give an orthogonal basis.

One could say that wavelet analysis gives a basis where the size and frequency parameters are given fixed values that have the nice property of giving an orthogonal basis. On the other hand, the wavelet representation is poorer than the Gabor representation in the sense that the basis functions are not oriented, and all have the same phase.

21.2   ICA AND SPARSE CODING

The transforms just considered are fixed transforms, meaning that the basis vectors are fixed once and for all, independent of any data. In many cases, however, it would be interesting to estimate the transform from data. Estimation of the representation in Eq. (21.1) consists of determining the values of s_i and $a_i(x, y)$ for all i and (x, y), given a sufficient number of observations of images, or in practice, image patches $I(x, y)$.

For simplicity, let us restrict ourselves here to the basic case where the $a_i(x, y)$ form an invertible linear system, that is, the matrix $\mathbf{A}$ is square. Then we can invert the system as

$$s_i = \sum_{x,y} w_i(x, y) I(x, y) \tag{21.6}$$

where the w_i denote the inverse filters. Note that we have (using the standard ICA notation)

$$\mathbf{a}_i = \mathbf{A}\mathbf{A}^T\mathbf{w}_i = \mathbf{C}\mathbf{w}_i \tag{21.7}$$

which shows a simple relation between the filters w_i and the corresponding basis vectors a_i. The basis vectors are obtained by filtering the coefficients in w_i by the filtering matrix given by the autocorrelation matrix. For natural image data, the autocorrelation matrix is typically a symmetric low-pass filtering matrix, so the basis vectors a_i are basically smoothed versions of the filters w_i.

The question is then: What principles should be used to estimate a transform from the data? Our starting point here is a representation principle called *sparse coding* that has recently attracted interest both in signal processing and in theories on the visual system [29, 336]. In sparse coding, the data vector is represented using a set of basis vectors so that *only a small number of basis vectors are activated at the same time*. In a neural network interpretation, each basis vector corresponds to one neuron, and the coefficients s_i are given by their activations. Thus, only a small number of neurons is activated for a given image patch.

Equivalently, the principle of sparse coding could be expressed by the property that *a given neuron is activated only rarely*. This means that the coefficients s_i have sparse distributions. The distribution of s_i is called sparse when s_i has a probability density with a peak at zero, and heavy tails, which is the case, for example, with the Laplacian (or double exponential) density. In general, sparseness can be equated with supergaussianity.

In the simplest case, we can assume that the sparse coding is linear, in which case sparse coding fits into the framework used in this chapter. One could then estimate a linear sparse coding transformation of the data by formulating a measure of sparseness of the components, and maximizing the measure in the set of linear transformations. In fact, since sparsity is closely related to supergaussianity, ordinary measures of nongaussianity, such as kurtosis and the approximations of negentropy, could be interpreted as measures of nongaussianity as well. Maximizing sparsity is thus one method of maximizing nongaussianity, and we saw in Chapter 8 that maximizing nongaussianity of the components is one method of estimating the ICs. Thus, sparse coding can be considered as one method for ICA. At the same time, sparse coding gives a different interpretation of the goal of the transform.

The utility of sparse coding can be seen, for example, in such applications as compression and denoising. In compression, since only a small subset of the components are nonzero for a given data point, one could code the data point efficiently by coding only those nonzero components. In denoising, one could use some testing (thresholding) procedures to find out those components that are really active, and set to zero the other components, since their observations are probably almost purely noise. This is an intuitive interpretation of the denoising method given in Section 15.6.

21.3 ESTIMATING ICA BASES FROM IMAGES

Thus, ICA and sparse coding give essentially equivalent methods for estimating features from natural images, or other kinds of data sets. Here we show the results of such an estimation. The set of images that we used consisted of natural scenes previously used in [191]. An example can be found in Fig. 21.7 in Section 21.4.3, upper left-hand corner.

First, we must note that ICA applied to image data usually gives one component representing the local mean image intensity, or the DC component. This component normally has a distribution that is not sparse; often it is even subgaussian. Thus, it must be treated separately from the other, supergaussian components, at least if the sparse coding interpretation is to be used. Therefore, in all experiments we first subtract the local mean, and then estimate a suitable sparse coding basis for the rest of the components. Because the data then has lost one linear dimension, the dimension of the data must be reduced, for example, using principal component analysis (PCA).

Each image was first linearly normalized so that the pixels had zero mean and unit variance. A set of 10000 image patches (windows) of 16×16 pixels were taken at random locations from the images. From each patch the local mean was subtracted as just explained. To remove noise, the dimension of the data was reduced to 160. The preprocessed dataset was used as the input to the FastICA algorithm, using the tanh nonlinearity.

Figure 21.5 shows the obtained basis vectors. The basis vectors are clearly localized in space, as well as in frequency and orientation. Thus the features are *closely related to Gabor functions*. In fact, one can approximate these basis functions by Gabor functions, so that for each basis vector one minimizes the squared error between the basis vector and a Gabor function; see Section 4.4. This gives very good fits, and shows that Gabor functions are a good approximation. Alternatively, one could characterize the ICA basis functions by noting that many of them could be interpreted as edges or bars.

The basis vectors are also *related to wavelets* in the sense that they represent more or less the same features in different scales. This means that the frequency and the size of the envelope (i.e. the area covered by the basis vectors) are dependent. However, the ICA basis vectors have many more degrees of freedom than wavelets. In particular, wavelets have only two orientations, whereas ICA vectors have many more, and wavelets have no phase difference, whereas ICA vectors have very different phases. Some recent extensions of wavelets, such as curvelets, are much closer to ICA basis vectors, see [115] for a review.

21.4 IMAGE DENOISING BY SPARSE CODE SHRINKAGE

In Section 15.6 we discussed a denoising method based on the estimation of the noisy ICA model [200, 207]. Here we show how to apply this method to image denoising. We used as data the same images as in the preceding section. To reduce computational load, here we used image windows of 8×8 pixels. As explained in

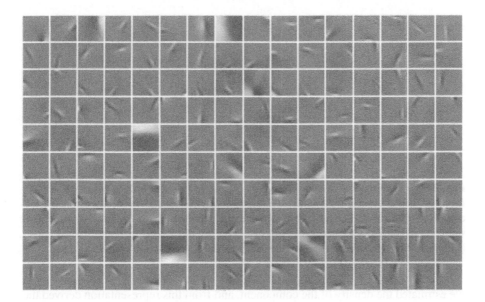

Fig. 21.5 The ICA basis vectors of natural image patches (windows). The basis vectors give features that are localized in space, frequency, and orientation, thus resembling Gabor functions.

Section 15.6, the basis vectors were further orthogonalized; thus the basis vectors could be considered as orthogonal sparse coding rather than ICA.

21.4.1 Component statistics

Since sparse code shrinkage is based on the property that individual components in the transform domain have sparse distributions, we first investigate how well this requirement holds. At the same time we can see which of the parameterizations in Section 15.5.2 can be used to approximate the underlying densities.

Measuring the sparseness of the distributions can be done by almost any nongaussianity measure. We have chosen the most widely used measure, the normalized kurtosis. Normalized kurtosis is defined as

$$\kappa(s) = \frac{E\{s^4\}}{(E\{s^2\})^2} - 3 \tag{21.8}$$

The kurtoses of components in our data set were about 5, on the average. Orthogonalization did not very significantly change the kurtosis. All the components were supergaussian.

Next, we compared various parametrizations in the task of fitting the observed densities. We picked one component at random from the orthogonal 8×8 sparse coding transform for natural scenes. First, using a nonparametric histogram technique,

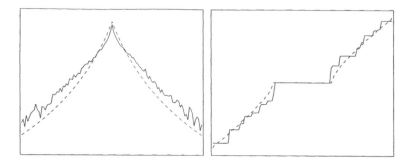

Fig. 21.6 Analysis of a randomly selected component from the orthogonalized ICA transforms of natural scenes, with window size 8 × 8. Left: Nonparametrically estimated log-densities (solid curve) vs. the best parametrization (dashed curve). Right: Nonparametric shrinkage nonlinearity (solid curve) vs. that given by our parametrization (dashed curve). (Reprinted from [207], reprint permission from the IEEE Press ©2001 IEEE.)

we estimated the density of the component, and from this representation derived the log density and the shrinkage nonlinearity shown in Fig. 21.6. Next, we fitted the parametrized densities discussed in Section 15.5.2 to the observed density. Note that in each case, the densities were sparser than the Laplacian density, and thus the very sparse parametrization in (15.25) was used. In can be seen that the density and the shrinkage nonlinearity derived from the density model match quite well those given by nonparametric estimation.

Thus we see that the components of the sparse coding bases found are highly supergaussian for natural image data; the sparsity assumption is valid.

21.4.2 Remarks on windowing

The theory of sparse code shrinkage was developed for general random vectors. When applying this framework to images, certain problems arise. The simplest way to apply the method to images would be to simply divide the image into windows, and denoise each such window separately. This approach, however, has a couple of drawbacks: statistical dependencies across the synthetic edges are ignored, resulting in a blocking artifact, and the resulting procedure is not translation invariant: The algorithm is sensitive to the precise position of the image with respect to the windowing grid.

We have solved this problem by taking a *sliding window* approach. This means that we do not divide the image into distinct windows; rather we denoise every possible $N \times N$ window of the image. We then effectively have N^2 different estimated values for each pixel, and select the final result as the mean of these values. Although originally chosen on rather heuristic grounds, the sliding window method can be justified by two interpretations.

The first interpretation is *spin-cycling*. The basic version of the recently introduced wavelet shrinkage method is not translation-invariant, because this is not a property of the wavelet decomposition in general. Thus, Coifman and Donoho [87] suggested performing wavelet shrinkage on all translated wavelet decompositions of the data, and taking the mean of these results as the final denoised signal, calling the obtained method spin-cycling. It is easily seen that our sliding window method is then precisely spin-cycling of the distinct window algorithm.

The second interpretation of sliding windows is due to the *method of frames*. Consider the case of decomposing a data vector into a linear combination of a set of given vectors, where the number of given vectors is larger than the dimensionality of the data, i.e. given x and A, where A is an m-by-n matrix ($m < n$), find the vector s, in $x = As$. This has an infinite number of solutions. The classic way is to select the solution s with minimum norm, given by $\hat{s} = A^+x$, where A^+ is the pseudoinverse of A. This solution is often referred to as the method of frames solution [102]. (More on this case of "overcomplete bases" can be found in Chapter 16.)

Now consider each basis window in each possible window position of the image as an overcomplete basis for the whole image. Then, if the transform we use is orthogonal, the sliding window algorithm is equivalent to calculating the method of frames decomposition, shrinking each component, and reconstructing the image.

21.4.3   Denoising results

To begin the actual denoising experiments, a random image from the natural scene collection was chosen for denoising, and gaussian noise of standard deviation 0.3 was added (compared to a standard deviation of 1.0 for the original image). This noisy version was subsequently denoised using the Wiener filter (Section 4.6.2) to give a baseline comparison. Then, the sparse code shrinkage method was applied using the estimated orthogonalized ICA transform (8×8 windows), with the component nonlinearities as given by the appropriate estimated parametrization. Figure 21.7 shows a typical result [207]. Visually, it seems that sparse code shrinkage gives the best noise reduction while retaining the features in the image. The Wiener filter does not really eliminate the noise. It seems as though our method is performing like a feature detector, in that it retains those features that are clearly visible in the noisy data but cuts out anything that is probably a result of the noise. Thus, it reduces noise effectively due to the nonlinear nature of the shrinkage operation.

Thus, we see that sparse code shrinkage is a promising method of image denoising. The denoising result is qualitatively quite different from those given by traditional filtering methods, and more along the lines of wavelet shrinkage and coring results [116, 403, 476].

21.5   INDEPENDENT SUBSPACES AND TOPOGRAPHIC ICA

The basic feature extraction schemes use linear features and linear filters as we have so far used in this chapter. More sophisticated methods can be obtained by introducing nonlinearities in the system. In the context of ICA, this could mean nonlinearities that take into account the dependencies of the linear features.

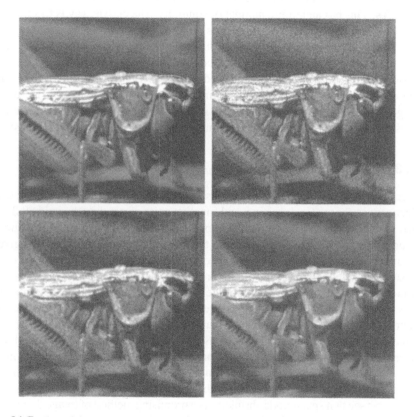

Fig. 21.7   Denoising a natural scene (noise level 0.3). Top left: The original image. Top right: Noise added. Bottom left: After wiener filtering. Bottom right: Results after sparse code shrinkage. (Reprinted from [207], reprint permission from the IEEE Press©2001 IEEE.)

In fact, it is not possible in general to decompose a random vector into independent components. One *can* always obtain uncorrelated components, and this is what we obtain with FastICA. In image feature extraction, however, one can clearly see that the ICA components are not independent by using any measure of higher-order correlations. Such higher-order correlations were discussed in Section 20.2, in which extensions of the ICA model were proposed to take into account some of the remaining dependencies.

Here we apply two of the extensions discussed in Section 20.2, independent subspace analysis and topographic ICA, to image feature extraction [205, 204, 206]. These give interesting extensions of the linear feature framework. The data and preprocessing were as in Section 21.3.

Figure 21.8 shows the basis vectors of the 40 feature subspaces, when subspace dimension was chosen to be 4. It can be seen that the basis vectors associated with a single subspace all have approximately the same orientation and frequency. Their locations are not identical, but close to each other. The phases differ considerably. Thus, the norm of the projection onto the subspace is relatively independent of the phase of the input. This is in fact what the principle of invariant-feature subspaces, one of the inspirations for independent subspace analysis, is all about. Every feature subspace can thus be considered a generalization of a quadrature-phase filter pair [373], giving a nonlinear feature that is localized in orientation and frequency, but invariant to phase and somewhat invariant to location shifts. For more details, see [204].

In topographic ICA, the neighborhood function was defined so that every neighborhood consisted of a 3×3 square of 9 units on a 2-D torus lattice [247]. The obtained basis vectors are shown in Fig. 21.9. The basis vectors are rather similar to those obtained by ordinary ICA of image data. In addition, they have a clear topographic organization. In fact, in Section 20.2.3 we discussed the connection between independent subspace analysis and topographic ICA; this connection can be found in Fig. 21.9. Two neighboring basis vectors in Fig. 21.9 tend to be of the same orientation and frequency. Their locations are near to each other as well. In contrast, their phases are very different. This means that a neighborhood of such basis vectors is similar to an independent subspace. For more details, see [206].

21.6 NEUROPHYSIOLOGICAL CONNECTIONS

In addition to the signal processing applications, we must not forget that sparse coding was originally developed as a model of the representation of images in the primary visual cortex of mammals (V1).

The filters $w_i(x, y)$ can then be identified as the receptive fields of the simple cells in the cortex, and the s_i are their activities when presented with a given image patch $I(x, y)$. It has been shown [336] that when this model is estimated with input data consisting of patches of natural scenes, the obtained filters $w_i(x, y)$ have the three principal properties of simple cells in V1: they are localized, oriented, and bandpass. The obtained filters w_i have been compared quantitatively with those measured by

Fig. 21.8 Independent subspaces of natural image data. The model gives Gabor-like basis vectors for image windows. Every group of four basis vectors corresponds to one independent feature subspace, or complex cell. Basis vectors in a subspace are similar in orientation, location and frequency. In contrast, their phases are very different. (Adapted from [204].)

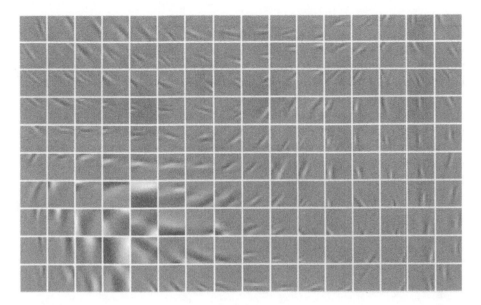

Fig. 21.9 Topographic ICA of natural image data. This gives Gabor-like basis vectors as well. Basis vectors that are similar in orientation, location and/or frequency are close to each other. The phases of nearby basis vectors are very different, giving each neighborhood a phase invariance. (Adapted from [206].)

single-cell recordings of the macaque cortex, and a good match for most of the parameters was found [443, 442].

Independent subspaces, on the other hand, can be considered as a model of complex cells [204], which is the other principal class of cells analyzing visual input in the cortex. In fact, the invariances that emerge are quite similar to those found in complex cells. Finally, topographic ICA can be seen as a model of the topographic organization of the cells in the visual cortex [206].

21.7 CONCLUDING REMARKS

ICA can be used to extract independent features from different kinds of data. This is possible by taking patches (windows) from the signals and considering these as the multi-dimensional signals on which ICA is applied. This is closely related to sparse coding, which means extracting features that have the property that only a small number of features is simultaneously active. Here we considered feature extraction from natural images, in which case we obtain a decomposition that is closely related to the one given by wavelets or Gabor analysis. The features are localized in space as well as in frequency, and are oriented. Such features can be used in image processing in the same way as wavelet and Gabor methods, and they can also be used as models of the receptive fields of simple cells in the visual cortex.

In addition to the linear feature extraction by ICA, one can use extensions of ICA to obtain nonlinear features. Independent subspace analysis gives features with invariance with respect to location and phase, and topographic ICA gives a topographic organization for the features, together with the same invariances. These models are useful for investigating the higher-order correlations between the basic "independent" components. Higher-order correlations between wavelet or Gabor coefficients have been investigated in [404, 478] as well. See also [273] for a mixture model based on ICA, where the different mixtures can be interpreted as contexts.

The ICA framework can also be used for feature extraction from other kinds of data, for example, color and stereo images [186, 187], video data [442], audio data [37], and hyperspectral data [360].

22

Brain Imaging Applications

With the advent of new anatomical and functional brain imaging methods, it is now possible to collect vast amounts of data from the living human brain. It has thus become very important to extract the essential features from the data to allow an easier representation or interpretation of their properties. This is a very promising area of application for independent component analysis (ICA). Not only is this an area of rapid growth and great importance; some kinds of brain imaging data also seem to be quite well described by the ICA model. This is especially the case with electroencephalograms (EEG) and magnetoencephalograms (MEG), which are recordings of electric and magnetic fields of signals emerging from neural currents within the brain. In this chapter, we review some of these brain imaging applications, concentrating on EEG and MEG.

22.1 ELECTRO- AND MAGNETOENCEPHALOGRAPHY

22.1.1 Classes of brain imaging techniques

Several anatomical and functional imaging methods have been developed to study the living human brain noninvasively, that is, without any surgical procedures. One class of methods gives anatomical (structural) images of the brain with a high spatial resolution, and include computerized X-ray tomography (CT) and magnetic resonance imaging (MRI). Another class of methods gives *functional* information on which parts of the brain are activated at a given time. Such brain imaging methods can help in answering the question: What parts of the brain are needed for a given task?

Well-known functional brain mapping methods include positron emission tomography (PET) and functional MRI (fMRI), which are based on probing the changes in metabolic activity. The time resolution of PET and fMRI is limited, due to the slowness of the metabolic response in the brain, which is in the range of a few seconds.

Here we concentrate on another type of functional brain imaging methods that are characterized by a high time resolution. This is possible by measuring the *electrical* activity within the brain. Electrical activity is the fundamental means by which information is transmitted and processed in the nervous system. These methods are the only noninvasive ones that provide direct information about the neural dynamics on a millisecond scale. As a trade-off, the spatial resolution is worse than in fMRI, being about 5 mm, under favorable conditions. The basic methods in this class are electroencephalography (EEG) and magnetoencephalography (MEG) [317, 165], which we describe next. Our exposition is based on the one in [447]; see also [162].

22.1.2 Measuring electric activity in the brain

Neurons and potentials The human brain consists of approximately 10^{10} to 10^{11} neurons [230]. These cells are the basic information-processing units. Signals between neurons are transmitted by means of action potentials, which are very short bursts of electrical activity. The action potential is transformed in the receiving neuron to what is called a postsynaptic potential that is longer in duration, though also weaker. Single action potentials and postsynaptic potentials are very weak and cannot be detected as such by present noninvasive measurement devices.

Fortunately, however, neurons that have relatively strong postsynaptic potentials at any given time tend to be clustered in the brain. Thus, the total electric current produced in such a cluster may be large enough to be detected. This can be done by measuring the potential distribution on the scalp by placing electrodes on it, which is the method used in EEG. A more sophisticated method is to measure the magnetic fields associated with the current, as is done in MEG.

EEG and MEG The total electric current in an activated region is often modeled as a dipole. It can be assumed that in many situations, the electric activity of the brain at any given point of time can be modeled by only a very small number of dipoles. These dipoles produce an electric potential as well as a magnetic field distribution that can be measured outside the head. The magnetic field is more local, as it does not suffer from the smearing caused by the different electric conductivities of the several layers between the brain and the measuring devices seen in EEG. This is one of the main advantages of MEG, as it leads to a *much higher spatial resolution*.

EEG is used extensively for monitoring the electrical activity within the human brain, both for research and clinical purposes. It is in fact one of the most widespread brain mapping techniques to date. EEG is used both for the measurement of spontaneous activity and for the study of evoked potentials. Evoked potentials are activity triggered by a particular stimulus that may be, for example, auditory or somatosensory. Typical clinical EEG systems use around 20 electrodes, evenly distributed over the head. State-of-the-art EEGs may consist of a couple hundred sensors. The signal-to noise ratio is typically quite low: the background potential distribution is of the order of 100 microvolts, whereas the evoked potentials may be two orders of magnitude weaker.

MEG measurements give basically very similar information to EEG, but with a higher spatial resolution. MEG is mainly used for basic cognitive brain research. To measure the weak magnetic fields of the brain, superconducting quantum interference devices (SQUIDs) are needed. The measurements are carried out inside a magnetically shielded room. The superconducting characteristics of the device are guaranteed through its immersion in liquid helium, at a temperature of $-269°$C. The experiments is this chapter were conducted using a Neuromag-122^{TM} device, manufactured by Neuromag Ltd., and located at the Low Temperature Laboratory of the Helsinki University of Technology. The whole-scalp sensor array in this device is composed of 122 sensors (planar gradiometers), organized in pairs at 61 locations around the head, measuring simultaneously the tangential derivatives of the magnetic field component normal to the helmet-shaped bottom of the dewar. The sensors are mainly sensitive to currents that are directly below them, and tangential to the scalp.

22.1.3   Validity of the basic ICA model

The application of ICA to the study of EEG and MEG signals assumes that several conditions are verified, at least approximately: the existence of statistically independent source signals, their instantaneous linear mixing at the sensors, and the stationarity of the mixing and the independent components (ICs).

The independence criterion considers solely the statistical relations between the amplitude distributions of the signals involved, and not the morphology or physiology of neural structures. Thus, its validity depends on the experimental situation, and cannot be considered in general.

Because most of the energy in EEG and MEG signals lies below 1 kHz, the so-called quasistatic approximation of Maxwell equations holds, and each time instance can be considered separately [162]. Therefore, the propagation of the signals is immediate, there is no need for introducing any time-delays, and the instantaneous mixing is valid.

The nonstationarity of EEG and MEG signals is well documented [51]. When considering the underlying source signals as stochastic processes, the requirement of stationarity is in theory necessary to guarantee the existence of a representative distribution of the ICs. Yet, in the implementation of batch ICA algorithms, the data are considered as random variables, and their distributions are estimated from the whole data set. Thus, the nonstationarity of the signals is not really a violation of the assumptions of the model. On the other hand, the stationarity of the mixing matrix **A** is crucial. Fortunately, this assumption agrees with widely accepted neuronal source models [394, 309].

22.2 ARTIFACT IDENTIFICATION FROM EEG AND MEG

As a first application of ICA on EEG and MEG signals, we consider separation of artifacts. Artifacts mean signals not generated by brain activity, but by some external disturbances, such as muscle activity. A typical example is ocular artifacts, generated by eye muscle activity.

A review on artifact identification and removal, with special emphasis on the ocular ones, can be found in [56, 445]. The simplest, and most widely used method consists of discarding the portions of the recordings containing attributes (e.g., amplitude peak, frequency contents, variance and slope) that are typical of artifacts and exceed a determined threshold. This may lead to significant loss of data, and to complete inability of studying interesting brain activity occuring near or during strong eye activity, such as in visual tracking experiments.

Other methods include the subtraction of a regression portion of one or more additional inputs (e.g., from electrooculograms, electrocardiograms, or electromyograms) from the measured signals. This technique is more likely to be used in EEG recordings, but may, in some situations, be applied to MEG. It should be noted that this technique may lead to the insertion of undesirable new artifacts into the brain recordings [221]. Further methods include the signal-space projection [190], and subtracting the contributions of modeled dipoles accounting for the artifact [45]. In both of these latter methods we need either a good model of the artifactual source or a considerable amount of data where the amplitude of the artifact is much higher than that of the EEG or MEG.

ICA gives a method for artifact removal where we do not need an accurate model of the process that generated the artifacts; this is the *blind* aspect of the method. Neither do we need specified observation intervals that contain mainly the artifact, nor additional inputs; this is the unsupervised aspect of the method. Thus ICA gives a promising method for artifact identification and removal. It was shown in [445, 446] and [225] that artifacts can indeed be estimated by ICA alone. It turns out that the artifacts are quite independent from the rest of the signal, and thus even this requirement of the model is reasonably well fulfilled.

In the experiments on MEG artifact removal [446], the MEG signals were recorded in a magnetically shielded room with the 122-channel whole-scalp magnetometer described above. The test person was asked to blink and make horizontal saccades, in order to produce typical ocular (eye) artifacts. Moreover, to produce myographic (muscle) artifacts, the subject was asked to bite his teeth for as long as 20 seconds. Yet another artifact was created by placing a digital watch one meter away from the helmet into the shielded room. Figure 22.1 presents a subset of 12 artifact-contaminated MEG signals, from a total of 122 used in the experiment. Several artifact structures are evident from this figure, such as eye and muscle activity.

The results of artifact extraction using ICA are shown in Fig. 22.2. Components IC1 and IC2 are clearly the activations of two different muscle sets, whereas IC3 and IC5 are, respectively, horizontal eye movements and blinks. Furthermore, other disturbances with weaker signal-to-noise ratio, such as the heart beat and a digital watch, are extracted as well (IC4 and IC8, respectively). IC9 is probably a faulty

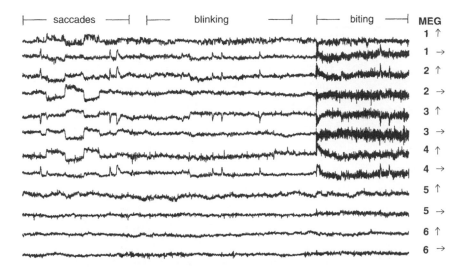

| saccades | blinking | biting | MEG |

Fig. 22.1 A subset of 12 spontaneous MEG signals from the frontal, temporal and occipital areas. The data contains several types of artifacts, including ocular and muscle activity, the cardiac cycle, and environmental magnetic disturbances. (Adapted from [446].)

sensor. ICs 6 and 7 may be breathing artifacts, or alternatively artificial bumps caused by overlearning (Section 13.2.2). For each component the left, back and right views of the field patterns are shown. These field patterns can be computed from the columns of the mixing matrix.

22.3 ANALYSIS OF EVOKED MAGNETIC FIELDS

Evoked magnetic fields, i.e., the magnetic fields triggered by an external stimulus, are one of the fundamental research methods in cognitive brain research. State-of-the-art approaches for processing magnetic evoked fields are often based on a careful expert scrutiny of the complete data, which can be either in raw format or averaged over several responses to repeating stimuli. At each time instance, one or several neural sources are modeled, often as dipoles, so as to produce as good a fit to the data as possible [238]. The choice of the time instances where this fitting should be made, as well as the type of source models employed, are therefore crucial. Using ICA, we can again obtain a blind decomposition without imposing any a priori structure on the measurements.

The application of ICA in event related studies was first introduced in the blind separation of auditory evoked potentials in [288]. This method has been further developed using magnetic auditory and somatosensory evoked fields in [449, 448]. Interestingly, the most significant independent components that were found in these studies seem to be of dipolar nature. Using a dipole model to calculate the source

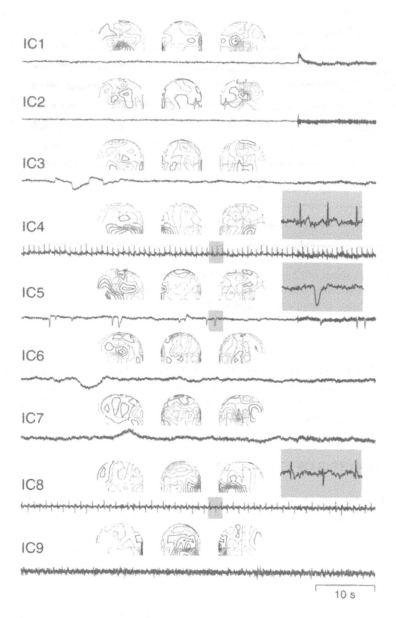

Fig. 22.2 Artifacts found from MEG data, using the FastICA algorithm. Three views of the field patterns generated by each independent component are plotted on top of the respective signal. Full lines correspond to magnetic flux exiting the head, whereas the dashed lines correspond to the flux inwards. Zoomed portions of some of the signals are shown as well. (Reprinted from [446], reprint permission and copyright by the MIT Press.)

locations, we have found them to fall on very plausible brain areas. Thus, ICA validates the conventional dipole modeling assumption in these studies. Future studies, though, will probably find cases where the dipole model is too restrictive.

In [448], ICA was shown to be able to differentiate between somatosensory and auditory brain responses in the case of vibrotactile stimulation, which, in addition to tactile stimulation, also produced a concomitant sound. Principal component analysis (PCA) has often been used to decompose signals of this kind, but as we have seen in Chapter 7, it cannot really separate independent signals. In fact, computing the principal components of these signals, we see that most of the principal components still represent combined somatosensory and auditory responses [448]. In contrast, computing the ICs, the locations of the equivalent current sources fall on the expected brain regions for the particular stimulus, showing separation by modality.

Another study was conducted in [449], using only averaged auditory evoked fields. The stimuli consisted of 200 tone bursts that were presented to the subject's right ear, using 1 s interstimulus intervals. These bursts had a duration of 100 ms, and a frequency of 1 kHz. Figure 22.3 shows the 122 averages of the auditory evoked responses over the head. The insert, on the left, shows a sample enlargement of such averages, for an easier comparison with the results depicted in the next figures.

Again, we see from Figs 22.4 *a* and 22.4 *b* that PCA is unable to resolve the complex brain response, whereas the ICA technique produces cleaner and sparser response components. For each component presented in Fig. 22.4 *a* and Fig. 22.4 *b*, left, top and, right side views of the corresponding field pattern are shown. Note that the first principal component exhibits clear dipole-like pattern both over the left and the right hemispheres, corroborating the idea of an unsuccessful segmentation of the evoked response. Subsequent principal components, however, tend to have less and less structured patterns.

From the field patterns associated with the independent components we see that the evoked responses of the left hemisphere are isolated in IC1 and IC4. IC2 has stronger presence over the right hemisphere, and IC3 fails to show any clear field pattern structure. Furthermore, we can see that IC1 and IC2 correspond to responses typically labeled as N1m, with the characteristic latency of around 100 ms after the onset of the stimulation. The shorter latency of IC1, mainly reflecting activity contralateral to the stimulated ear, agrees with the known information available for such studies.

22.4 ICA APPLIED ON OTHER MEASUREMENT TECHNIQUES

In addition to the EEG/MEG results reported here, ICA has been applied to other brain imaging and biomedical signals as well:

- Functional magnetic resonance images (fMRI). One can use ICA in two different ways, separating either independent spatial activity patterns [297], or independent temporal activation patterns [50]. A comparison of the two modes

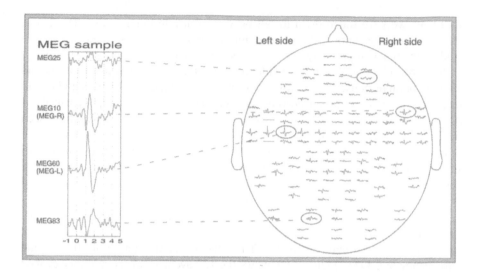

Fig. 22.3 Averaged auditory evoked responses to 200 tones, using MEG. Channels MEG10 and MEG60 are used in Fig. 22.4 as representatives of one left-hemisphere and one right-hemisphere MEG signal. Each tick in the MEG sample corresponds to 100 ms, going from 100 ms before stimulation onset to 500 ms after. (Adapted from [449].)

can be found in [367]. A combination of the two modes can be achieved by spatiotemporal ICA, see Section 20.1.4.

- Optical imaging means directly "photographing" the surface of the brain after making a hole in the skull. Application of ICA can be found in [374, 396]. As in the case of fMRI signals, this is a case of separating image mixtures as in the example in Fig. 12.4. Some theory for this particular case is further developed in [164, 301]; also the innovations processes may be useful (see Section 13.1.3 and [194]).

- Outside the area of brain imaging, let us mention applications to the removal of artifacts from cardiographic (heart) signals [31, 459] and magnetoneurographic signals [482]. The idea is very similar to the one used in MEG artifact removal. Further related work is in [32, 460]. Another neuroscientific application is in intracellular calcium spike analysis [375].

22.5   CONCLUDING REMARKS

In this chapter we have shown examples of ICA in the analysis of brain signals.

First, ICA was shown to be suitable for extracting different types of artifacts from EEG and MEG data, even in situations where these disturbances are smaller than the background brain activity.

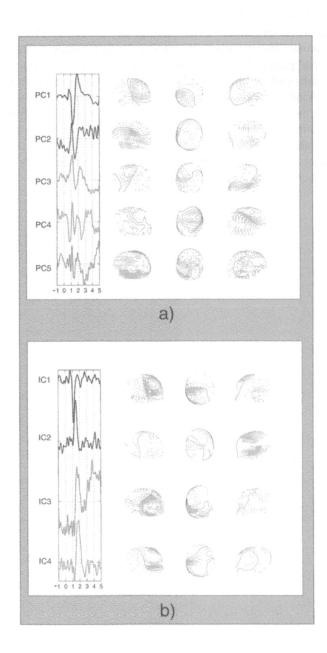

Fig. 22.4 Principal (*a*) and independent (*b*) components found from the auditory evoked field study. For each component, both the activation signal and three views of the corresponding field pattern are plotted. (Adapted from [449].)

Second, ICA can be used to decompose evoked fields or potentials. For example, ICA was able to differentiate between somatosensory and auditory brain responses in the case of vibrotactile stimulation. Also, it was able to discriminate between the ipsi- and contralateral principal responses in the case of auditory evoked potentials. In addition, the independent components, found with no other modeling assumption than their statistical independence, exhibit field patterns that agree with the conventional current dipole models. The equivalent current dipole sources corresponding to the independent components fell on the brain regions expected to be activated by the particular stimulus.

Applications of ICA have been proposed for analysis of other kinds of biomedical data as well, including fMRI, optical imaging, and ECG.

23

Telecommunications

This chapter deals with applications of independent component analysis (ICA) and blind source separation (BSS) methods to telecommunications. In the following, we concentrate on code division multiple access (CDMA) techniques, because this specific branch of telecommunications provides several possibilities for applying ICA and BSS in a meaningful way. After an introduction to multiuser detection and CDMA communications, we present mathematically the CDMA signal model and show that it can be cast in the form of a noisy matrix ICA model. Then we discuss in more detail three particular applications of ICA or BSS techniques to CDMA data. These are a simplified complexity minimization approach for estimating fading channels, blind separation of convolutive mixtures using an extension of the natural gradient algorithm, and improvement of the performance of conventional CDMA receivers using complex-valued ICA. The ultimate goal in these applications is to detect the desired user's symbols, but for achieving this intermediate quantities such as fading channel or delays must usually be estimated first. At the end of the chapter, we give references to other communications applications of ICA and related blind techniques used in communications.

23.1 MULTIUSER DETECTION AND CDMA COMMUNICATIONS

In wireless communication systems, like mobile phones, an essential issue is division of the common transmission medium among several users. This calls for a *multiple access* communication scheme. A primary goal in designing multiple access systems is to enable each user of the system to communicate despite the fact that the other

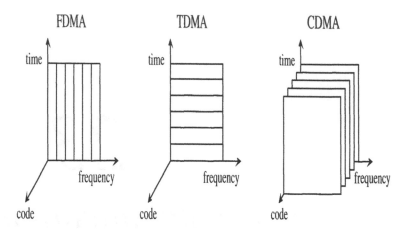

Fig. 23.1   A schematic diagram of the multiple access schemes FDMA, TDMA, and CDMA [410, 382].

users occupy the same resources, possibly simultaneously. As the number of users in the system grows, it becomes necessary to use the common resources as efficiently as possible. These two requirements have given rise to a number of multiple access schemes.

Figure 23.1 illustrates the most common multiple access schemes [378, 410, 444]. In frequency division multiple access (FDMA), each user is given a nonoverlapping frequency slot in which one and only one user is allowed to operate. This prevents interference of other users. In time division multiple access (TDMA) a similar idea is realized in the time domain, where each user is given a unique time period (or periods). One user can thus transmit and receive data only during his or her predetermined time interval while the others are silent at the same time.

In CDMA [287, 378, 410, 444], there is no disjoint division in frequency and time spaces, but each user occupies the same frequency band simultaneously. The users are now identified by their codes, which are unique to each user. Roughly speaking, each user applies his unique code to his information signal (data symbols) before transmitting it through a common medium. In transmission different users' signals become mixed, because the same frequencies are used at the same time. Each user's transmitted signal can be identified from the mixture by applying his unique code at the receiver.

In its simplest form, the code is a pseudorandom sequence of ±1s, also called a *chip sequence* or *spreading code*. In this case we speak about direct sequence (DS) modulation [378], and call the multiple access method DS-CDMA. In DS-CDMA, each user's narrow-band data symbols (information bits) are spread in frequency before actual transmission via a common medium. The spreading is carried out by multiplying each user's data symbols (information bits) by his unique wide-band chip sequence (spreading code). The chip sequence varies much faster than the

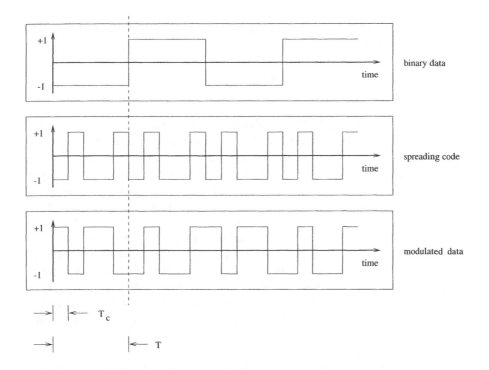

Fig. 23.2 Construction of a CDMA signal [382]. Top: Binary user's symbols to be transmitted. Middle: User's specific spreading code (chip sequence). Bottom: Modulated CDMA signal, obtained by multiplying user's symbols by the spreading code.

information bit sequence. In the frequency domain, this leads to spreading of the power spectrum of the transmitted signal. Such *spread spectrum techniques* are useful because they make the transmission more robust against disturbances caused by other signals transmitted simultaneously [444].

Example 23.1 Figure 23.2 shows an example of the formation of a CDMA signal. On the topmost subfigure, there are 4 user's symbols (information bits) $-1, +1, -1, +1$ to be transmitted. The middle subfigure shows the chip sequence (spreading code). It is now $-1, +1, -1, -1, +1$. Each symbol is multiplied by the chip sequence in a similar manner. This yields the modulated CDMA signal on the bottom row of Fig. 23.2, which is then transmitted. The bits in the spreading code change in this case 5 times faster that the symbols.

Let us denote the mth data symbol (information bit) by b_m, and the chip sequence by $s(t)$. The time period of the chip sequence is T (see Fig. 23.2), so that $s(t) \in \{-1, +1\}$ when $t \in [0, T)$, and $s(t) = 0$ when $t \notin [0, T)$. The length of the chip sequence is C chips, and the time duration of each chip is $T_c = T/C$. The number of bits in the observation interval is denoted by N. In Fig. 23.2, the observation interval contains $N = 4$ symbols, and the length of the chip sequence is $C = 5$.

Using these notations, the CDMA signal $r(t)$ at time t arising in this simple example can be written

$$r(t) = \sum_{m=1}^{N} b_m s(t - mT) \tag{23.1}$$

In the reception of the DS-CDMA signal, the final objective is to estimate the transmitted symbols. However, both code timing and channel estimation are often prerequisite tasks. Detection of the desired user's symbols is in CDMA systems more complicated than in the simpler TDMA and FDMA systems used previously in mobile communications. This is because the spreading code sequences of different users are typically nonorthogonal, and because several users are transmitting their symbols at the same time using the same frequency band. However, CDMA systems offer several advantages over more traditional techniques [444, 382]. Their capacity is larger, and it degrades gradually with increasing number of simultaneous users who can be asynchronous [444]. CDMA technology is therefore a strong candidate for future global wireless communications systems. For example, it has already been chosen as the transmission technique for the European third generation mobile system UMTS [334, 182], which will provide useful new services, especially multimedia and high-bit-rate packet data.

In mobile communications systems, the required signal processing differs in the base station (uplink) from that in the mobile phone (downlink). In the base station, all the signals sent by different users must be detected, but there is also much more signal processing capacity available. The codes of all the users are known but their time delays are unknown. For delay estimation, one can use for example the simple matched filter [378, 444], subspace approaches [44, 413], or the optimal but computationally highly demanding maximum likelihood method [378, 444]. When the delays have been estimated, one can estimate the other parameters such as the fading process and symbols [444].

In downlink (mobile phone) signal processing, each user knows only its own code, while the codes of the other users are unknown. There is less processing power than in the base station. Also the mathematical model of the signals differs slightly, since users share the same channel in the downlink communications. Especially the first two features of downlink processing call for new, efficient and simple solutions. ICA and BSS techniques provide a promising new approach to the downlink signal processing using short spreading codes and DS-CDMA systems.

Figure 23.3 shows a typical CDMA transmission situation in an urban environment. Signal 1 arrives directly from the base station to the mobile phone in the car. It has the smallest time delay and is the strongest signal, because it is not attenuated by the reflection coefficients of the obstacles in the path. Due to *multipath propagation*, the user in the car in Fig. 23.3 receives also weaker signals 2 and 3, which have longer time delays. The existence of multipath propagation allows the signal to interfere with itself. This phenomenon is known as *intersymbol interference (ISI)*. Using spreading codes and suitable processing methods, multipath interference can be mitigated [444].

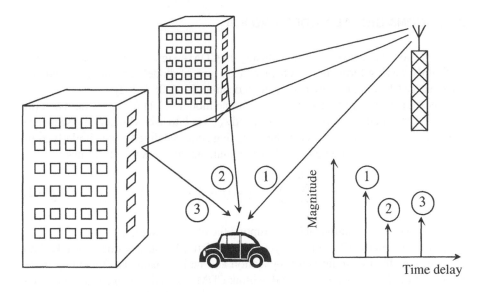

Fig. 23.3 An example of multipath propagation in urban environment.

There are several other problems that complicate CDMA reception. One of the most serious ones is *multiple access interference (MAI)*, which arises from the fact that the same frequency band is occupied simultaneously. MAI can be alleviated by increasing the length of the spreading code, but at a fixed chip rate, this decreases the data rate. In addition, the *near–far problem* arises when signals from near and far are received at the same time. If the received powers from different users become too different, a stronger user will seriously interfere with the weaker ones, even if there is a small correlation between the users' spreading codes. In the FDMA and TDMA systems, the near–far problem does not arise because different users have nonoverlapping frequency or time slots.

The near–far problem in the base station can be mitigated by power control, or by *multiuser detection*. Efficient multiuser detection requires knowledge or estimation of many system parameters such as propagation delay, carrier frequency, and received power level. This is usually not possible in the downlink. However, then *blind* multiuser detection techniques can be applied, provided that the spreading codes are short enough [382].

Still other problems appearing in CDMA systems are power control, synchronization, and fading of channels, which is present in all mobile communications systems. Fading means variation of the signal power in mobile transmission caused for example by buildings and changing terrain. See [378, 444, 382] for more information on these topics.

23.2   CDMA SIGNAL MODEL AND ICA

In this section, we represent mathematically the CDMA signal model which is studied in slightly varying forms in this chapter. This type of models and the formation of the observed data in them are discussed in detail in [444, 287, 382].

It is straightforward to generalize the simple model (23.1) for K users. The mth symbol of the kth user is denoted by b_{km}, and $s_k(\cdot)$ is k:th user's binary chip sequence (spreading code). For each user k, the spreading code is defined quite similarly as in Example 23.1. The combined signal of K simultaneous users then becomes

$$r(t) = \sum_{m=1}^{N} \sum_{k=1}^{K} b_{km} s_k(t - mT) + n(t) \tag{23.2}$$

where $n(t)$ denotes additive noise corrupting the observed signal.

The signal model (23.2) is not yet quite realistic, because it does not take into account the effect of multipath propagation and fading channels. Including these factors in (23.2) yields our desired downlink CDMA signal model for the observed data $r(t)$ at time t:

$$r(t) = \sum_{m=1}^{N} \sum_{k=1}^{K} b_{km} \sum_{l=1}^{L} a_{lm} s_k(t - mT - d_l) + n(t) \tag{23.3}$$

Here the index m refers to the symbol, k to the user, and l to the path. The term d_l denotes the delay of the lth path, which is assumed to be constant during the observation interval of N symbol bits. Each of the K simultaneous users has L independent transmission paths. The term a_{lm} is the fading factor of the lth path corresponding to the mth symbol.

In general, the fading coefficients a_{lm} are complex-valued. However, we can apply standard real-valued ICA methods to the data (23.3) by using only the real part of it. This is the case in the first two approaches to be discussed in the next two sections, while the last method in Section 23.5 directly uses complex data.

The continuous time data (23.3) is first sampled using the chip rate, so that C equispaced samples per symbol are taken. From subsequent discretized equispaced data samples $r[n]$, C-length data vectors are then collected:

$$\mathbf{r}_m = (r[mC], r[mC + 1], \ldots, r[(m + 1)C - 1])^T \tag{23.4}$$

Then the model (23.3) can be written in vector form as [44]

$$\mathbf{r}_m = \sum_{k=1}^{K} \sum_{l=1}^{L} [a_{l,m-1} b_{k,m-1} \underline{\mathbf{g}}_{kl} + a_{l,m} b_{k,m} \overline{\mathbf{g}}_{kl}] + \mathbf{n}_m \tag{23.5}$$

where $\mathbf{n}_m$ denotes the noise vector consisting of subsequent C last samples of noise $n(t)$. The vector $\underline{\mathbf{g}}_{kl}$ denotes the "early" part of the code vector, and $\overline{\mathbf{g}}_{kl}$ the "late" part, respectively. These vectors are given by

$$\underline{\mathbf{g}}_{kl} = [\, s_k[C - d_l + 1], \ldots, s_k[C], \mathbf{0}_{d_l}^T\,]^T \tag{23.6}$$

$$\overline{\mathbf{g}}_{kl} = [\mathbf{0}_{d_l}^T, s_k[1], \dots, s_k[C - d_l]]^T \tag{23.7}$$

Here d_l is the discretized index representing the time delay, $d_l \in \{0, \dots, (C - 1)/2\}$, and $\mathbf{0}_{d_l}^T$ is a row vector having d_l zeros as its elements. The early and late parts of the code vector arise because of the time delay d_l, which means that the chip sequence generally does not coincide with the time interval of a single user's symbol, but extends over two subsequent bits $b_{k,m-1}$ and $b_{k,m}$. This effect of the time delay can be easily observed by shifting the spreading code to the right in Fig. 23.2.

The vector model (23.5) can be expressed in compact form as a matrix model. Define the data matrix

$$\mathbf{R} = [\mathbf{r}_1, \mathbf{r}_2, \dots, \mathbf{r}_N] \tag{23.8}$$

consisting of N subsequent data vectors $\mathbf{r}_i$. Then $\mathbf{R}$ can be represented as

$$\mathbf{R} = \mathbf{GF} + \mathbf{N} \tag{23.9}$$

where the $C \times 2KL$ matrix $\mathbf{G}$ contains all the KL early and late code vectors

$$\mathbf{G} = [\underline{\mathbf{g}}_{11}, \overline{\mathbf{g}}_{11}, \dots, \underline{\mathbf{g}}_{KL}, \overline{\mathbf{g}}_{KL}] \tag{23.10}$$

and the $2KL \times N$ matrix $\mathbf{F} = [\mathbf{f}_1 \dots \mathbf{f}_N]$ contains the symbols and fading terms

$$\begin{aligned} \mathbf{f}_m &= [a_{1,m-1}b_{1,m-1}, a_{1m}b_{1m}, \\ &\quad \dots, a_{L,m-1}b_{K,m-1}, a_{Lm}b_{Km}]^T \end{aligned} \tag{23.11}$$

The vector $\mathbf{f}_m$ represents the $2KL$ symbols and fading terms of all the users and paths corresponding to the mth pair of early and late code vectors.

From the physical situation, it follows that each path and user are at least approximately independent of each other [382]. Hence every product $a_{i,m-1}b_{i,m-1}$ or $a_{im}b_{im}$ of a symbol and the respective fading term can be regarded as an independent source signal. Because each user's subsequent transmitted symbols are assumed to be independent, these products are also independent for a given user i. Denote the independent sources $a_{1,m-1}b_{1,m-1}, \dots, a_{Lm}b_{Km}$ by $y_i(m), i = 1, \dots, 2KL$. Here every $2L$ sources correspond to each user, where the coefficient 2 follows from the presence of the early and late parts.

To see the correspondence of (23.9) to ICA, let us write the noisy linear ICA model $\mathbf{x} = \mathbf{As} + \mathbf{n}$ in the matrix form as

$$\mathbf{X} = \mathbf{AS} + \mathbf{N} \tag{23.12}$$

The data matrix $\mathbf{X}$ has as its columns the data vectors $\mathbf{x}(1), \mathbf{x}(2), \dots$, and $\mathbf{S}$ and $\mathbf{N}$ are similarly compiled source and noise matrices whose columns consist of the source and noise vectors $\mathbf{s}(t)$ and $\mathbf{n}(t)$, respectively. Comparing the matrix CDMA signal model (23.9) with (23.12) shows that it has the same form as the noisy linear ICA model. Clearly, in the CDMA model (23.9) $\mathbf{F}$ is the matrix of source signals, $\mathbf{R}$ is the observed data matrix, and $\mathbf{G}$ is the unknown mixing matrix.

For estimating the desired user's parameters and symbols, several techniques are available [287, 444]. Matched filter (correlator) [378, 444] is the simplest estimator, but it performs well only if different users' chip sequences are orthogonal or the users have equal powers. The matched filter suffers greatly from the near–far problem, rendering it unsuitable for CDMA reception without a strict power control. The so-called RAKE detector [378] is a somewhat improved version of the basic matched filter which takes advantage of multiple propagation paths. The maximum likelihood (ML) method [378, 444] would be optimal, but it has a very high computational load, and requires knowledge of all the users' codes. However, in downlink reception, only the desired user's code is known. To remedy this problem while preserving acceptable performance, subspace approaches have been proposed for example in [44]. But they are sensitive to noise, and fail when the signal subspace dimension exceeds the processing gain. This easily occurs even with moderate system load due to the multipath propagation. Some other semiblind methods proposed for the CDMA problem such as the minimum mean-square estimator (MMSE) are discussed later in this chapter and in [287, 382, 444].

It should be noted that the CDMA estimation problem is not completely blind, because there is some prior information available. In particular, the transmitted symbols are binary (more generally from a finite alphabet), and the spreading code (chip sequence) is known. On the other hand, multipath propagation, possibly fading channels, and time delays make separation of the desired user's symbols a very challenging estimation problem which is more complicated than the standard ICA problem.

23.3 ESTIMATING FADING CHANNELS

23.3.1 Minimization of complexity

Pajunen [342] has recently introduced a complexity minimization approach as a true generalization of standard ICA. In his method, temporal information contained in the source signals is also taken into account in addition to the spatial independence utilized by standard ICA. The goal is to optimally exploit all the available information in blind source separation. In the special case where the sources are temporally white (uncorrelated), complexity minimization reduces to standard ICA [342]. Complexity minimization has been discussed in more detail in Section 18.3.

Regrettably, the original method for minimizing the Kolmogoroff complexity measure is computationally highly demanding except for small scale problems. But if the source signals are assumed to be gaussian and nonwhite with significant time correlations, the minimization task becomes much simpler [344]. Complexity minimization then reduces to principal component analysis of temporal correlation matrices. This method is actually just another example of blind source separation approaches based on second-order temporal statistics; for example [424, 43], which were discussed earlier in Chapter 18.

In the following, we apply this simplified method to the estimation of the fading channel coefficients of the desired user in a CDMA systems. Simulations with downlink data, propagated through a Rayleigh fading channel, show noticeable performance gains compared with blind minimum mean-square error channel estimation, which is currently a standard method for solving this problem. The material in this section is based on the original paper [98].

We thus assume that the fading process is gaussian and complex-valued. Then the amplitude of the fading process is Rayleigh distributed; this case is called Rayleigh fading (see [444, 378]). We also assume that a training sequence or a preamble is available for the desired user, although this may not always be the case in practice. Under these conditions, only the desired user's contribution in the sampled data is time correlated, which is then utilized. The proposed method has the advantage that it estimates code timing only implicitly, and hence it does not degrade the accuracy of channel estimation.

A standard method for separating the unknown source signals is based on minimization of the mutual information (see Chapter 10 and [197, 344]) of the separated signals $\mathbf{f}_m = [y_1(m) \ldots y_{2KL}(m)]^T = \mathbf{y}$:

$$\mathcal{J}(\mathbf{y}) = \sum_i H(y_i) + \log | \det \mathbf{G} | \qquad (23.13)$$

where $H(y_i)$ is the entropy of y_i (see Chapter 5). But entropy has the interpretation that it represents the optimum averaged code length of a random variable. Hence mutual information can be expressed by using algorithmic complexity as [344]

$$\mathcal{J}(\mathbf{y}) = \sum_i K(y_i) + \log | \det \mathbf{G} | \qquad (23.14)$$

where $K(\cdot)$ is the per-symbol Kolmogoroff complexity, given by the number of bits needed to describe y_i. By using prior information about the signals, the coding costs can be explicitly approximated. For instance, if the signals are gaussian, independence becomes equivalent to uncorrelatedness. Then the Kolmogoroff complexity can be replaced by the per-symbol differential entropy, which in this case depends on second-order statistics only.

For Rayleigh type fading transmission channels, the prior information can be formulated by considering that the probability distributions of the mutually independent source signals $y_i(m)$ have zero-mean gaussian distributions. Suppose we want to estimate the channel coefficients of the transmission paths, by sending a given length constant $b_{1m} = 1$ symbol sequence to the desired user. We consider the signals $y_i(m)$, $i = 1, \ldots, 2L$, with i representing the indexes of the $2L$ sources corresponding to the first user. Then $y_i(m)$ will actually represent the channel coefficients of all the first user's paths. Since we assume that the channel is Rayleigh fading, then these signals are gaussian and time correlated. In this case, blind separation of the sources can be achieved by using only second-order statistics. In fact, we can express the Kolmogoroff complexity by coding these signals using principal component analysis [344].

23.3.2   Channel estimation *

Let $\mathbf{y}_i(m) = [y_i(m), \ldots, y_i(m - D + 1)]$ denote the vector consisting of D last samples of every such source signal $y_i(m)$, $i = 1, \ldots, 2L$. Here D is the number of delayed terms, showing what is the range of time correlations taken into account when estimating the current symbol. The information contained in any of these sources can be approximated by the code length needed for representing the D principal components, which have variances given by the eigenvalues of the temporal correlation matrix $\mathbf{C}_i = E[\mathbf{y}_i(m)\mathbf{y}_i^T(m)]$ [344]. Since we assume that the transmission paths are mutually independent, the overall entropy of the source is given by summing up the entropies of the principal components. Using the result that the entropy of a gaussian random variable is given by the logarithm of the variance, we get for the entropy of each source signal

$$H(y_i) = \frac{1}{2L} \sum_k \log \sigma_k^2 = \frac{1}{2L} \log \det \mathbf{C}_i \tag{23.15}$$

Inserting this into the cost function (23.13) yields

$$\mathcal{J}(\mathbf{y})) = \sum_i \frac{1}{2L} \log \det \mathbf{C}_i - \log |\det \mathbf{W}| \tag{23.16}$$

where $\mathbf{W} = \mathbf{G}^{-1}$ is the separating matrix.

The separating matrix $\mathbf{W}$ can be estimated by using a gradient descent approach for minimizing the cost function (23.16), leading to the update rule [344]

$$\Delta \mathbf{W} = -\mu \frac{\partial \log \mathcal{J}(\mathbf{y})}{\partial \mathbf{W}} + \alpha \Delta \mathbf{W} \tag{23.17}$$

where μ is the learning rate and α is the momentum term [172] that can be introduced to avoid getting trapped into a local minimum corresponding to a secondary path.

Let $\mathbf{w}_i^T$ denote the ith row vector of the separating matrix $\mathbf{W}$. Since only the correlation matrix $\mathbf{C}_i$ of the ith source depends on $\mathbf{w}_i$, we can express the gradient of the cost function by computing the partial derivatives

$$\frac{\partial \log \det \mathbf{C}_i}{\partial w_{ik}}$$

with respect to the scalar elements of the vector $\mathbf{w}_i^T = [w_{i1}, \ldots, w_{iC}]$. For these partial derivatives, one can derive the formula [344]

$$\frac{\partial \log \det \mathbf{C}_i}{\partial w_{ik}} = 2 \operatorname{trace} \left(\mathbf{C}_i^{-1} E \left[\mathbf{y}_i^T \frac{\partial \mathbf{y}_i}{\partial w_{ik}} \right] \right) \tag{23.18}$$

Since $y_i(m) = \mathbf{w}_i^T \mathbf{r}_m$, we get

$$\frac{\partial \mathbf{y}_i}{\partial w_{ik}} = [r_{k,m}, \ldots, r_{k,m-L+1}] \tag{23.19}$$

where $r_{k,i}$ is the element (k, j) of the observation matrix $\mathbf{R}$ defined earlier using formulas (23.4) and (23.9).

What is left to do now is to compute the gradient update part due to the mapping information. It can be written [344]

$$\log |\det \mathbf{W}| = \sum_{i=1}^{C} \log \|(\mathbf{I} - \mathbf{P}_i)\mathbf{w}_i\| \tag{23.20}$$

where $\mathbf{P}_i = \mathbf{W}_i(\mathbf{W}_i^T\mathbf{W}_i)^{-1}\mathbf{W}_i^T$ is a projection matrix onto the subspace spanned by the column vectors of the matrix $\mathbf{W}_i = [\mathbf{w}_1, \ldots, \mathbf{w}_{i-1}]$. Now the cost function can be separated, and the different independent components can be found one by one, by taking into account the previously estimated components, contained in the subspace spanned by the columns of the matrix $\mathbf{W}_i$.

Since our principal interest lies in the transmission path having the largest power, corresponding usually to the desired user, it is sufficient to estimate the first such independent component. In this case, the projection matrix $\mathbf{P}_1$ becomes a zero matrix. Then the overall gradient (23.17) for the first row $\mathbf{w}_1^T$ of the separating matrix can be written

$$\frac{\partial \log \mathcal{J}(\mathbf{y})}{\partial \mathbf{w}_1^T} = \frac{1}{D}\text{trace}\left(\mathbf{C}_1^{-1}\text{E}\left[\mathbf{y}_1^T\frac{\partial \mathbf{y}_1}{\partial \mathbf{w}_1^T}\right]\right) - \frac{\mathbf{w}_1^T}{\|\mathbf{w}_1^T\|} \tag{23.21}$$

It suffices to consider the special case where only the two last samples are taken into account, so that the the delay $D = 2$. First, second-order correlations are removed from the data $\mathbf{R}$ by whitening. This can be done easily in terms of standard principal component analysis as explained in Chapter 6. After whitening, the subsequent separating matrix will be orthogonal, and thus the second term in Eq. (23.16) disappears, yielding the cost function

$$\mathcal{J}(\mathbf{y}) \sim \sum \log \det \mathbf{C}_k \tag{23.22}$$

with the 2×2 autocorrelation matrices given by

$$\mathbf{C}_k = \begin{bmatrix} 1 & \text{E}[y_k(m)y_k(m-1)] \\ \text{E}[y_k(m)y_k(m-1)] & 1 \end{bmatrix} \tag{23.23}$$

In this case, the separating vectors $\mathbf{w}_i^T$ can be found by maximizing sequentially $\text{E}[y_i(m)y_i(m-1) + y_i(m-1)y_i(m)]$, which is the first-order correlation coefficient of y_i. It follows that the function to be maximized becomes

$$J(\mathbf{w}) = \mathbf{w}^T\text{E}[\mathbf{r}_m\mathbf{r}_{m-1}^T + \mathbf{r}_{m-1}\mathbf{r}_m^T]\mathbf{w} \tag{23.24}$$

So the separating vector $\mathbf{w}_1^T$ corresponding to the most important path is given by the principal eigenvector of the matrix in Eq. (23.24). We have used a symmetric expression for the correlation coefficients in order to avoid asymmetry when the

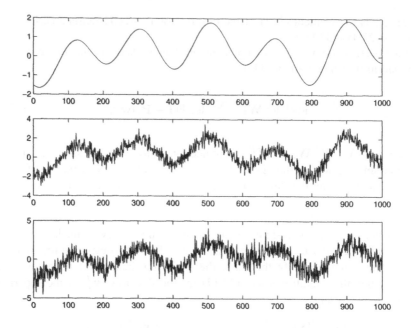

Fig. 23.4 The original fading process (top), its estimate given by our method (middle), and estimate given by the blind MMSE method (bottom). The signal-to-noise ratio was 10 dB.

observed data set is finite. This usually improves the estimation accuracy. Finally, we separate the desired channel coefficients by computing the quantities

$$a_{11} = \mathbf{w}_1^T \tilde{\mathbf{r}} \tag{23.25}$$

where $\tilde{\mathbf{r}}$ denotes whitened data vector $\mathbf{r}$. This is done for all the N data vectors contained in (23.8).

23.3.3 Comparisons and discussion

We have compared the method described and derived above to a well-performing standard method used in multiuser detection, namely the minimum mean-square error estimator [452, 287]. In the MMSE method, the desired signal is estimated (up to a scaling) from the formula

$$a_{MMSE} = \mathbf{g_1}^T \mathbf{U}_s \mathbf{\Lambda}_s^{-1} \mathbf{U}_s^T \mathbf{r} \tag{23.26}$$

where $\mathbf{\Lambda}_s$ and $\mathbf{U}_s$ are the matrices containing (in the same order) the eigenvalues and the respective eigenvectors of the data correlation matrix $\mathbf{R}\mathbf{R}^T/N$. The vector $\mathbf{g}_1$ is a column of the matrix $\mathbf{G}$ defined in (23.10) corresponding to the desired user's bit $b_{1,m}$, that is, either $\underline{\mathbf{g}}_{11}$ or $\overline{\mathbf{g}}_{11}$. The quantity a_{MMSE} is again computed for all the N data vectors $\mathbf{r}_1, \dots, \mathbf{r}_N$. If the pilot signal consists of ones, then a_{MMSE} provides estimates of the channel coefficients.

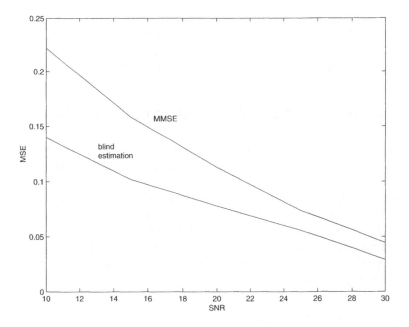

Fig. 23.5   The mean-square errors of the MMSE method and our method as the function of the signal-to-noise ratio. The number of users was $K = 6$.

The algorithms were tested in a simulation using length $C = 31$ quasiorthogonal gold codes (see [378]). The number of users was $K = 6$, and the number of transmission paths was $L = 3$. The powers of the channel paths were -5, -5, and 0 dB respectively for every user, and the signal-to-noise ratio (SNR) varied from 30 dB to 10 dB with respect to the main path. Only the real part of the data was used. The observation interval was $N = 1000$ symbols long.

We compared our algorithm with the blind MMSE method, where the pilot signal corresponding to the first user consisted of ones. The fading coefficients corresponding to the strongest path were estimated using both methods. Fig. 23.4 shows the original fading process and the estimated ones, giving an idea of the achieved accuracy. The figure shows that our method provides somewhat more accurate estimates than the MMSE method, though the estimated fading process is noisy. Fig. 23.5 presents numerical values of the average mean-square error (MSE) as a function of SNR. The complexity minimization based method performs clearly better than the MMSE method especially at lower signal-to-noise ratios. The convergence of the gradient approach took place in this case in 10–15 iterations for the learning parameters $\mu = 1$ and $\alpha = 0.5$.

In its current form, the proposed method needs training symbols for providing a temporally correlated structure for the desired user's signal. If the channel varies rapidly during the training phase, the method is not able to estimate the channel as the data modulation is on. This is because the temporal correlatedness of the desired signal is lost. A future research topic is to overcome this problem. Finally, we point out that instead of the simplified complexity minimization approach applied here, one could have as well tried other blind separation methods based on the time structure of the sources. Such methods are discussed in Chapter 18.

23.4   BLIND SEPARATION OF CONVOLVED CDMA MIXTURES *

Now consider estimation of the desired user's symbol process using a blind source separation method developed for convolutive mixtures. Such methods have been discussed earlier in Chapter 19. The model consists of a linear transformation of both the independent variables (the transmitted symbols) and their delayed version, where the delay is one time unit. For separating mixtures of delayed and convolved independent sources, we use an extension of the natural gradient method based on the information maximization principle [79, 13, 268, 426]. Experiments show that the proposed method has quite competitive detection capabilities compared with conventional symbol estimation methods.

23.4.1   Feedback architecture

The vector model (23.5) can be written

$$
\mathbf{r}_m = \sum_{k=1}^{K} \left[b_{k,m-1} \sum_{l=1}^{L} a_l \underline{\mathbf{g}}_{kl} \right] + \sum_{k=1}^{K} \left[b_{km} \sum_{l=1}^{L} a_l \overline{\mathbf{g}}_{kl} \right] + \mathbf{n}_m
\tag{23.27}
$$

This model differs slightly from the fading channel model used in the previous section in that the channel is now assumed to stay constant during the block of N symbols. Hence, the fading coefficients a_l depend only on the path l but not on the symbol index m. This type of channel is called *block fading*. As in the previous section, we use only the real part of the complex-valued data. This allows application of ICA and BSS methods developed for real-valued data to CDMA.

The model (23.27) can be further expressed in the matrix-vector form [99]

$$
\mathbf{r}_m = \mathbf{G}_0 \mathbf{b}_m + \mathbf{G}_1 \mathbf{b}_{m-1} + \mathbf{n}_m
\tag{23.28}
$$

where $\mathbf{G}_0$ and $\mathbf{G}_1$ are $C \times K$ mixing matrices corresponding to the original and the one time unit delayed symbols. The column vectors of $\mathbf{G}_0$ and $\mathbf{G}_1$ are given by the early and the late parts of the coding vectors multiplied by the fading coefficients, respectively:

$$
\mathbf{G}_0 = \left[\sum_{l=1}^{L} a_l \overline{\mathbf{g}}_{1l}, \cdots, \sum_{l=1}^{L} a_l \overline{\mathbf{g}}_{Kl} \right]
\tag{23.29}
$$

$$
\mathbf{G}_1 = \left[\sum_{l=1}^{L} a_l \underline{\mathbf{g}}_{1l}, \cdots, \sum_{l=1}^{L} a_l \underline{\mathbf{g}}_{Kl} \right]
\tag{23.30}
$$

The symbol vector $\mathbf{b}_m$ contains the binary symbols (information bits) of the K users at time index m:

$$\mathbf{b}_m = [b_{1m}, b_{2m}, \dots, b_{Km}]^T \tag{23.31}$$

The vector $\mathbf{b}_{m-1}$ is defined quite similarly.

Eq. (23.28) shows that our CDMA signal model represents a linear mixture of delayed and convolved sources in the special case where the maximum time delay is one unit. Assuming that all the mixing matrices (users' codes) and symbol sequences are unknown makes the separation problem blind. One method for solving this convolutive BSS problem is to consider a feedback architecture. Assuming that the users are independent of each other, we can apply to the convolutive BSS problem the principle of entropy maximization discussed earlier in Section 9.3. The weights of the network can be optimized using the natural gradient algorithm extended for convolutive mixtures in [13, 79, 268, 426].

The data vectors $\mathbf{r}_m$ are preprocessed by simultaneously whitening them and reducing their dimension to K (the number of users). Using PCA whitening (Chapter 6), the whitened data matrix becomes

$$\tilde{\mathbf{R}} = \mathbf{\Lambda}_s^{-\frac{1}{2}} \mathbf{U}_s^T \mathbf{R} \tag{23.32}$$

We can now write the whitened version of Eq. (23.28)

$$\mathbf{v}_m = \mathbf{H}_0 \mathbf{b}_m + \mathbf{H}_1 \mathbf{b}_{m-1} \tag{23.33}$$

where $\mathbf{v}_m$ is the whitened input vector, and $\mathbf{H}_0$ and $\mathbf{H}_1$ are whitened square $K \times K$ matrices corresponding to the rectangular matrices $\mathbf{G}_0$ and $\mathbf{G}_1$. From (23.33), the symbol vector $\mathbf{b}_m$ can be expressed in terms of the whitened data vector $\mathbf{v}_m$ and the previously estimated symbol vector $\mathbf{b}_{m-1}$ as

$$\mathbf{b}_m = \mathbf{H}_0^{-1}(\mathbf{v}_m - \mathbf{H}_1 \mathbf{b}_{m-1}) \tag{23.34}$$

The arising network architecture is depicted in Fig. 23.6.

23.4.2 Semiblind separation method

Based on this feedback architecture we propose the following algorithm for blind symbol detection in a CDMA system.

1. Initialize randomly the matrices $\mathbf{H}_0$ and $\mathbf{H}_1$.

2. Compute updates for the matrices $\mathbf{H}_0$ and $\mathbf{H}_1$ from the formulas [79, 268, 426]

$$\Delta \mathbf{H}_0 = -\mathbf{H}_0(\mathbf{I} + \mathbf{q}_m \mathbf{b}_m^T) \tag{23.35}$$

$$\Delta \mathbf{H}_1 = -(\mathbf{I} + \mathbf{H}_1)\mathbf{q}_m \mathbf{b}_{m-1}^T \tag{23.36}$$

Here $\mathbf{I}$ is a $K \times K$ unit matrix, and $\mathbf{q}_m = f(\mathbf{b}_m)$ is a nonlinearly transformed symbol vector $\mathbf{b}_m$. The nonlinear function f is typically a sigmoidal or cubic nonlinearity, and it is applied componentwise to the elements of $\mathbf{b}_m$.

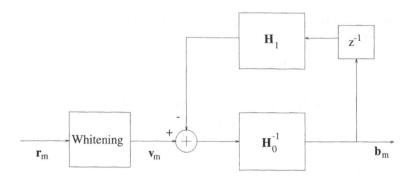

Fig. 23.6 A feedback network for a convolutive CDMA signal model

3. Compute new estimates for the matrices $\mathbf{H}_0$ and $\mathbf{H}_1$ using the rule

$$\mathbf{H}_i \leftarrow \mathbf{H}_i + \mu\Delta\mathbf{H}_i, \quad i = 1, 2 \tag{23.37}$$

where μ is a small learning parameter.

4. Compute new estimate of the symbol vector $\mathbf{b}_m$ from Eq. (23.34).

5. If the matrices $\mathbf{H}_0$ and $\mathbf{H}_1$ have not converged, return back to step 2.

6. Apply the sign nonlinearity to each component of the final estimate of the symbol vector $\mathbf{b}_m$. This quantizes the estimated symbols to the bits $+1$ or -1.

7. Identify the desired user's symbol sequence which best fits the training sequence.

If some prior information on the desired user's transmission delays is available, it can be utilized in the initialization step 1. The update rules in step 2 have been adapted for our special case of only one unit delay from the more general convolutive mixture algorithms described in [268, 426]. Because of the feedback, the choice of the learning parameter μ in step 3 is essential for the convergence. We have used a constant μ, but more sophisticated iteration dependent choices would probably make the convergence faster. In step 5, convergence is verified in terms of the mean-square-error matrix norm. Since the transmission system is binary differential, step 6 provides the most probable value for the estimated symbol. The detector estimates the symbols of all users up to a permutation. Therefore a pilot training sequence is needed in step 7 to identify the desired user. Hence the method presented above is an example of a semiblind separation approach.

23.4.3 Simulations and discussion

The method introduced in the previous section has been compared in [99] with two standard methods used in multiuser detection, namely the matched filter (MF) and the minimum mean-square error (MMSE) estimator [378, 444, 382, 287]. The matched filter estimator is simply

$$b_{MF} = \text{sign}(\mathbf{g_1}^T \mathbf{r}) \tag{23.38}$$

which is again computed for all the data vectors $\mathbf{r}_1, \dots, \mathbf{r}_N$. Here $\mathbf{g_1}$ is a column of the matrix $\mathbf{G}_0$ defined in (23.29) that corresponds to the desired user's bit $b_{1,m}$. Similarly, the (linear) MMSE symbol estimator is given by

$$b_{MMSE} = \text{sign}(\mathbf{g_1}^T \mathbf{U}_s \mathbf{\Lambda}_s^{-1} \mathbf{U}_s^T \mathbf{r}) \tag{23.39}$$

The matrices $\mathbf{\Lambda}_s$, $\mathbf{U}_s$ were defined earlier below Eq. (23.26).

It is noteworthy that the formula (23.39) is otherwise the same as (23.26), but now it provides bit estimates b_{MMSE} instead of fading channel coefficient estimates a_{MMSE}. The reason for this is that in the previous section the quantities to be estimated were products of bits and fading channel coefficients, or elements of the vector $\mathbf{f}_m$ defined in (23.11), and the bits were all ones in the pilot training sequence. On the other hand, in this section the vector $\mathbf{g_1}$ also contains the fading coefficients a_l (which stay constant during the observation interval) because of the definition of the matrix $\mathbf{G}_0$ in (23.29).

The algorithms were tested using quasiorthogonal gold codes [378] of length $C = 31$. The number of users was either $K = 4$ or $K = 8$, and the number of transmission paths was $L = 3$. The powers of the channel paths were, respectively, $-5, -5$ and 0 dB for every user, and the signal-to-noise ratio (SNR) varied from 30 dB to 0 dB with respect to the main path. Only the real part of the data was used. The observation interval was $N = 500$. A pilot training sequence of length $P = 20$ was compared with the preambles of the separated sources for identifying the desired user. A constant learning parameter $\mu = 0.05$ was used. Convergence took about 20 iterations in the above environment.

The experimental results shown in Fig. 23.7 give the bit-error-rates (BERs) for the compared three methods at different SNRs in the more difficult case of $K = 8$ users. The results are qualitatively similar for $K = 4$ users [99], and are therefore not shown here. Figure 23.7 indicates that the proposed semiblind convolutive separation method yields clearly better detection results than the widely used matched filter or linear minimum mean-square error estimators. The basic reason for this improved performance is that standard methods such as the MF and MMSE estimators do not exploit the independence of the received signals. The MMSE estimator makes the detected sources uncorrelated [287]. Even this much weaker assumption improves the performance clearly compared with the simple matched filter. In the studied scenario the independence of the received signals is a reasonable assumption, and its power becomes apparent by inspecting the results in Fig. 23.7.

We have in this section considered a batch method based on a feedback architecture for symbol detection, but an adaptive version could have been used instead. The batch method has the advantage that two observation vectors are used for estimating the current symbol vector, which improves the estimation. There is also no need for synchronization as is the case with the MMSE and MF methods. This is because dif-

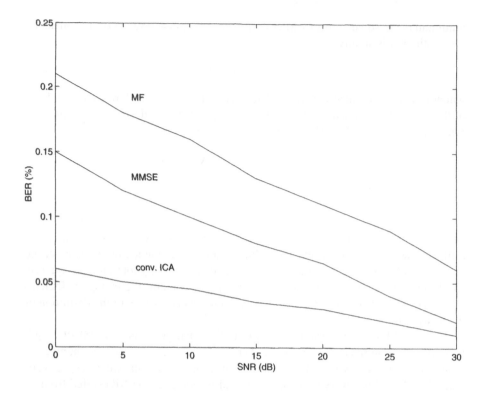

Fig. 23.7 The bit-error-rate (BER) for the convolutive mixture ICA, minimum mean-square error (MMSE), and matched filter (MF) methods. The number of users was $K = 8$.

ferent users' path delays are implicitly estimated simultaneously in the basis vectors of the mixing matrices.

23.5 IMPROVING MULTIUSER DETECTION USING COMPLEX ICA *

A general drawback of ICA methods in CDMA applications is that it is difficult or even impossible to utilize well the available prior information on the problem. However, quite generally in any problem it is highly desirable to apply prior information whenever available, because it usually improves estimation accuracy and overall performance if taken into account properly. In this section, a feasible solution to this issue is presented by using ICA as an additional processing element attached to existing standard receiver structures.

In this section, two types of receiver structures, RAKE-ICA and MMSE-ICA [382], are studied in a block fading CDMA downlink environment. Numerical results indicate that the performance of RAKE and subspace MMSE detectors can be greatly improved in terms of ICA postprocessing. This is mainly due to the facts that ICA efficiently utilizes the independence of the original signals, and that ICA does not explicitly depend on erroneous timing or channel estimation. On the other hand, the RAKE and subspace MMSE estimators can apply prior information on the CDMA problem. Since these estimators are complex-valued, an ICA method fitted to complex data must be used. To this end, complex FastICA algorithm (see Section 20.3 and [47]) is used.

This section is based on the references [382, 383]. The interested reader can find more information on ICA assisted CDMA reception in them.

23.5.1 Data model

The continuous time signal model is otherwise the same as in (23.3), but the fading coefficients a_{lm} in (23.3) are now replaced by the complex coefficients a_l. Thus each path l has its own coefficient a_l which is assumed to remain constant during the data block of N symbols b_{km}, $m = 1, 2, \ldots, N$. Another difference is that the processing window is now two symbols long while its length in the previous sections is one symbol. Hence the samples are collected into $2C$-dimensional vectors

$$\mathbf{r}_m = (r[mC], r[mC + 1], \ldots, r[(m + 2)C - 1])^T \qquad (23.40)$$

instead of the C-dimensional data vectors (23.4).

Since sampling is asynchronous with respect to the symbols, the vector sample $\mathbf{r}_m$ in (23.40) usually contains information on *three* successive symbols $b_{k,m-1}, b_{km}$, and $b_{k,m+1}$. The two symbols long data window has the advantage that it always contains one complete symbol. Similarly as in (23.27), the vectors (23.40) can be expressed in the well-known form

$$\mathbf{r}_m = \sum_{k=1}^{K} \left[b_{k,m-1} \sum_{l=1}^{L} a_l \underline{\mathbf{g}}_{kl} + b_{km} \sum_{l=1}^{L} a_l \mathbf{g}_{kl} + b_{k,m+1} \sum_{l=1}^{L} a_l \overline{\mathbf{g}}_{kl} \right] + \mathbf{n}_m \qquad (23.41)$$

The "early" and "late" code vectors $\underline{\mathbf{g}}_{kl}$ and $\overline{\mathbf{g}}_{kl}$ are defined quite similarly as in Eqs. (23.6) and (23.7). Now they just contain more zeros so that these vectors become $2C$-dimensional. However, the late code vector $\overline{\mathbf{g}}_{kl}$ is now associated with the symbol $b_{k,m+1}$, and for the middle symbol b_{km} the "middle" code vectors

$$\mathbf{g}_{kl} = [\mathbf{0}_{d_l}^T, s_k[1], \ldots, s_k[C], \mathbf{0}_{C-d_l}^T]^T \qquad (23.42)$$

are defined, where $\mathbf{0}_{d_l}^T$ is a row vector having d_l zeros as its elements.

Again, the data vectors (23.41) can be represented more compactly as

$$\mathbf{r}_m = \mathbf{G}\mathbf{b}_m + \mathbf{n}_m \qquad (23.43)$$

which has the form of a noisy linear ICA model. The $2C \times 3K$ dimensional code matrix $\mathbf{G}$ corresponds to the mixing matrix $\mathbf{A}$. It is assumed to have full rank, and contains the code vectors and path strengths:

$$\mathbf{G} = \left[\sum_{l=1}^{L} a_l \underline{\mathbf{g}}_{1l}, \sum_{l=1}^{L} a_l \mathbf{g}_{1l}, \sum_{l=1}^{L} a_l \overline{\mathbf{g}}_{1l}, \cdots \right.$$

$$\left. , \sum_{l=1}^{L} a_l \underline{\mathbf{g}}_{Kl}, \sum_{l=1}^{L} a_l \mathbf{g}_{Kl}, \sum_{l=1}^{L} a_l \overline{\mathbf{g}}_{Kl} \right] \qquad (23.44)$$

The $3K$-dimensional symbol vector

$$\mathbf{b}_m = [b_{1,m-1}, b_{1m}, b_{1,m+1}, \cdots, b_{K,m-1}, b_{Km}, b_{K,m+1}]^T$$

$$(23.45)$$

contains the symbols, and corresponds to the vector **s** of independent (or roughly independent) sources. Note that both the code matrix **G** and the symbol vector $\mathbf{b}_m$ consists of subsequent triplets corresponding to the early, middle, and late parts.

23.5.2   ICA based receivers

In the following, we consider in more detail two ways of initializing the ICA iteration. For a more thorough discussion, see [382]. The starting point for the receiver development is to look at the noiseless whitened data[1]

$$\mathbf{z}_m = \mathbf{V}\mathbf{b}_m = \mathbf{\Lambda}_s^{-\frac{1}{2}} \mathbf{U}_s^H \mathbf{G}\mathbf{b}_m \qquad (23.46)$$

where $\mathbf{\Lambda}_s$ and $\mathbf{U}_s$ are matrices containing (in the same order) the $3K$ principal eigenvalues and -vectors of the data autocorrelation matrix $E\{\mathbf{r}_m \mathbf{r}_m^H\}$, respectively. It is easy to see that for the data model (23.43), the whitening matrix **V** becomes orthonormal: $\mathbf{V}\mathbf{V}^H = \mathbf{I}$, because the symbols are uncorrelated: $E\{\mathbf{b}_m \mathbf{b}_m^H\} = \mathbf{I}$, and the whitened data vectors satisfy the condition $\mathbf{I} = E\{\mathbf{z}_m \mathbf{z}_m^H\}$.

It suffices to estimate one column of the whitening matrix **V**, say its second column $\mathbf{v}_2$. This is because we can then estimate the symbols b_{1m} of the desired user (user $k = 1$) by applying the vector $\mathbf{v}_2$ as follows:

$$\mathbf{v}_2^H \mathbf{z}_m = \mathbf{v}_2^H \mathbf{V}\mathbf{b}_m = [0100\cdots0]\mathbf{b}_m = b_{1m} \qquad (23.47)$$

From the definitions of **V** and **G** we see that

$$\mathbf{v}_2 = \mathbf{\Lambda}_s^{-\frac{1}{2}} \mathbf{U}_s^H \sum_{l=1}^{L} a_{1l} \mathbf{g}_{1l} \qquad (23.48)$$

which is exactly the subspace MMSE detector [451] for dispersive channels.

---

[1]Because the data is now complex-valued, the transpose T must be replaced by the Hermitian operator H, which equals transposition and complex conjugation.

Equation (23.48) can be applied to separating the desired symbol b_{1m}, but it uses only second-order statistics. In addition, the subspace parameters as well as the path gains and delays are always subject to estimation errors, degrading the performance of (23.48) in separation. But we can improve the separation capability of the estimator (23.48) by using ICA as a postprocessing tool. This is possible, since the independence of the original sources is not utilized in deriving (23.48). Moreover, it is meaningful to apply ICA by using the subspace MMSE estimator (23.48) as the starting point, because it already identifies the desired user. This identification is not possible by using ICA alone. The proposed DS-CDMA receiver structure, which we call MMSE-ICA, consists of a subspace MMSE detector refined by ICA iterations. The complex FastICA algorithm discussed earlier in Section 20.3 and in [47] is a natural choice for the ICA postprocessing method. It can deal with complex-valued data, and extracts one independent component at a time, which suffices in this application.

Alternatively, known or estimated symbols can be used for initializing the ICA iteration. This follows from the uncorrelatedness of the symbols, since then we have $E\{z_m b_{1m}\} = v_2$, leading again to the subspace MMSE detector. Because training symbols are not necessarily implemented in all the DS-CDMA systems, it is preferable to first use the traditional RAKE receiver [378, 382] or multipath correlator for detecting symbols. The RAKE estimator is nothing but a simple extension of the matched filter for several paths. Alternatively, one can initially detect the symbols by using the MMSE method for symbol outputs. The symbol estimates obtained initially in terms of the RAKE detector or MMSE symbol estimator are then refined using the complex FastICA algorithm. These structures are henceforth called as RAKE-ICA and MMSEbit-ICA detectors, respectively. Global convergence of the complex FastICA algorithm has been proved in [382]. The sign indeterminacy in the sources estimated by any ICA method is removed by a comparator element, which chooses the sign according to the RAKE receiver or subspace MMSE detector, respectively.

The proposed receiver structures are summarizes below. In step 1, an initial estimate of the kth whitened code vector v_k is computed using one of the three standard detection methods mentioned earlier. Steps 2-5 give the procedure for improving this initial estimate using the complex FastICA algorithm [47].

ICA-based blind interference suppression schemes   [382] Let k be the index of the desired user, and z_m the whitened data vector corresponding to the symbol vector b_m. The constant γ is 2 for complex-valued symbols, and $\gamma = 3$ for real-valued symbols (in the latter case, the data itself is complex, but the symbols are real). An estimate is denoted by the hat $\hat{}$. Then the iterative algorithms for blind interference suppression are as follows.

1. Initialize $\mathbf{w}(0) = \mathbf{v}_k / \|\mathbf{v}_k\|$, where

   (a) MMSE-ICA: $\mathbf{v}_k = \hat{\mathbf{\Lambda}}_s^{-1/2} \hat{\mathbf{U}}_s^H \sum_{l=1}^{\hat{L}} \hat{a}_l \hat{\mathbf{c}}_{kl}$.

   (b) RAKE-ICA: $\mathbf{v}_k = E\{z_m \hat{b}_{km}^{RAKE}\}$.

   (c) MMSEbit-ICA: $\mathbf{v}_k = E\{z_m \hat{b}_{km}^{MMSE}\}$.

Let $t = 1$.

2. Compute one iteration of the complex FastICA algorithm [47]:

$$\mathbf{w}(t) = \mathrm{E}\{\mathbf{z}_m(\mathbf{w}(t-1)^H\mathbf{z}_m)^*|\mathbf{w}(t-1)^H\mathbf{z}_m|^2\} - \gamma\mathbf{w}(t-1)$$
(23.49)

3. Divide $\mathbf{w}(t)$ by its norm.

4. If $|\mathbf{w}(t)^H\mathbf{w}(t-1)|$ is not close enough to 1, set $t = t + 1$, and go back to step 2.

5. Output the vector $\mathbf{w} = \epsilon\mathbf{w}(t)$, where $\epsilon = \mathrm{sign}(Re[\mathbf{w}(0)^H\mathbf{w}(t)])$.

23.5.3   Simulation results

The algorithms were tested using simulated DS-CDMA downlink data with a block fading channel.

In the first experiment gold codes of the length $C = 31$ were used. The length of the block was $M = 500$ binary phase shift keying (BPSK) symbols. The channel was fixed during the block period. The number of users was $K = 20$, and the multiple access interference (MAI) per user was 5 dB. Hence, the total interference power was 17.8 dB. The number of paths was $L = 3$. The path gains were gaussian distributed with a zero mean, and the path delays were chosen randomly from the interval $\{0, 1, \ldots, (C - 1)/2\}$. The delays and the path gains were assumed to be known. The signal-to-noise ratio (in the chip matched filter output) varied with respect to the desired user from 5 dB to 35 dB, and 10000 independent trials were made. A constant $\gamma = 3$ was used in the ICA iteration.

Figure 23.8 shows the achieved bit-error-rates (BERs) for the methods as a function of the SNR. The performance of RAKE is quite modest due to the near–far situation. Consequently, RAKE-ICA is able to improve the performance of the RAKE method only marginally. Subspace MMSE detector suffers from interference floor at higher SNRs. One reason for this is inaccurate estimation of the signal subspace. Even though MMSE-ICA uses the same estimate for the signal subspace, it is able to exploit the statistical independence of the source signals, and seems to follow quite closely the equal length optimal MMSE receiver, denoted as MMSE bound.

Figure 23.9 shows the corresponding block-error-rates (BLER) for these methods. A block is correctly estimated if *all* the symbols in the block are estimated correctly. For speech and data services that do not require real-time processing, raw BLER of 10^{-1} is sufficient. For real-time data services, raw BLER of the order of 10^{-2} is required. Figure 23.9 shows that RAKE-ICA actually improves the performance of the RAKE method quite remarkably with respect to BLER, even though the overall BER has not been improved that much.

More numerical experiments, including simulation results for the case of purely complex data ($\gamma = 2$), can be found in [382, 383]. They clearly indicate that the estimates given by the RAKE and subspace MMSE detectors can be greatly improved by using ICA as a postprocessing tool.

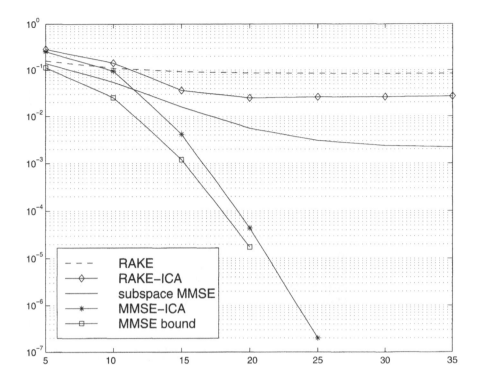

Fig. 23.8 Bit-error-rate as a function of SNR $(5, \ldots, 35 \text{ dB})$. The system includes $K = 20$ users with the average MAI of 5 dB per interfering user. BPSK data is used.

23.6 CONCLUDING REMARKS AND REFERENCES

In this chapter, we have applied several quite different extensions of basic ICA (or BSS) techniques to short-code CDMA data. It can be concluded that ICA often provides significant performance gains in CDMA applications. Basically, this results from the fact that standard CDMA detection and estimation methods do not exploit the powerful but realistic independence assumption. At best, they utilize the much weaker uncorrelatedness condition for the received source signals.

CDMA techniques are currently studied extensively in telecommunications, because they will be used in a form or another in future high performance mobile communications systems. A specific feature of all telecommunications applications of ICA is that they are almost always semiblind problems. The receiver has more or less prior information on the communication system, typically at least the spreading code of the desired user is known. This prior information should be combined in a suitable way with blind ICA techniques for achieving optimal results. Another important design feature is that practical algorithms should not be computationally too demanding, making it possible to realize them in real time.

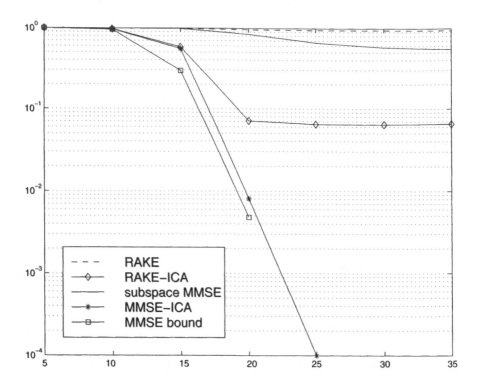

Fig. 23.9   Block-error-rate as a function of SNR $(5, \ldots, 35\,\mathrm{dB})$. The system includes $K = 20$ users with the average MAI of 5 dB per interfering user. BPSK data is used.

ICA methods also have been applied to CDMA data, for example, in references [223, 384, 100]. Other applications of blind source separation techniques to various communications problems can be found for example in [77, 111, 130, 435]. Related blind identification or blind equalization techniques are discussed in many papers; see for example [41, 73, 91, 122, 146, 143, 144, 158, 184, 224, 265, 276, 287, 351, 352, 361, 425, 428, 431, 439, 440] and the references therein. Blind identification techniques used in communications typically exploit second-order temporal statistics (or suitable explicit higher-order statistics) instead of ICA. The interested reader can find many more references on blind communications techniques in recent conference proceedings and journals dealing with telecommunications and statistical signal processing.

24

Other Applications

In this chapter, we consider some further applications of independent component analysis (ICA), including analysis of financial time series and audio signal separation.

24.1 FINANCIAL APPLICATIONS

24.1.1 Finding hidden factors in financial data

It is tempting to try ICA on financial data. There are many situations in which parallel financial time series are available, such as currency exchange rates or daily returns of stocks, that may have some common underlying factors. ICA might reveal some driving mechanisms that otherwise remain hidden.

In a study of a stock portfolio [22], it was found that ICA is a complementary tool to principal component analysis (PCA), allowing the underlying structure of the data to be more readily observed. If one could find the maximally independent mixtures of the original stocks, i.e., portfolios, this might help in minimizing the risk in the investment strategy.

In [245], we applied ICA on a different problem: the cashflow of several stores belonging to the same retail chain, trying to find the fundamental factors common to all stores that affect the cashflow. Thus, the effect of the factors specific to any particular store, i.e., the effect of the managerial actions taken at the individual store and in its local environment, could be analyzed.

In this case, the mixtures in the ICA model are parallel financial time series $x_i(t)$, with i indexing the individual time series, $i = 1, \ldots, m$ and t denoting discrete time.

We assume the instantaneous ICA model

$$x_i(t) = \sum_j a_{ij} s_j(t) \qquad (24.1)$$

for each time series $x_i(t)$. Thus the effect of each time-varying underlying factor or independent component $s_j(t)$ on the measured time series is approximately linear.

The assumption of having some underlying independent components in this specific application may not be unrealistic. For example, factors like seasonal variations due to holidays and annual variations, and factors having a sudden effect on the purchasing power of the customers, like price changes of various commodities, can be expected to have an effect on all the retail stores, and such factors can be assumed to be roughly independent of each other. Yet, depending on the policy and skills of the individual manager, e.g., advertising efforts, the effect of the factors on the cash flow of specific retail outlets are slightly different. By ICA, it is possible to isolate both the underlying factors and the effect weights, thus also making it possible to group the stores on the basis of their managerial policies using only the cash flow time series data.

The data consisted of the weekly cash flow in 40 stores that belong to the same retail chain, covering a time span of 140 weeks. Some examples of the original data $x_i(t)$ are shown in Fig. 24.1. The weeks of a year are shown on the horizontal axis, starting from the first week in January. Thus for example the heightened Christmas sales are visible in each time series before and during week 51 in both of the full years shown.

The data were first prewhitened using PCA. The original 40-dimensional signal vectors were projected to the subspace spanned by four principal components, and the variances were normalized to 1. Thus the dimension of the signal space was strongly decreased from 40. A problem in this kind of real world application is that there is no prior knowledge on the number of independent components. Sometimes the eigenvalue spectrum of the data covariance matrix can be used, as shown in Chapter 6, but in this case the eigenvalues decreased rather smoothly without indicating any clear signal subspace dimension. Then the only way is to try different dimensions. If the independent components that are found using different dimensions for the whitened data are the same or very similar, we can trust that they are not just artifacts produced by the compression, but truly indicate some underlying factors in the data.

Using the FastICA algorithm, four independent components (ICs) $s_j(t)$, $j = 1, ..., 4$ were estimated. As depicted in Fig. 24.2, the FastICA algorithm has found several clearly different fundamental factors hidden in the original data.

The factors have different interpretations. The topmost factor follows the sudden changes that are caused by holidays etc.; the most prominent example is Christmas time. The factor in the bottom row, on the other hand, reflects the slower seasonal variation, with the effect of the summer holidays clearly visible. The factor in the third row could represent a still slower variation, something resembling a trend. The last factor, in the second row, is different from the others; it might be that this factor follows mostly the relative competitive position of the retail chain with respect to its competitors, but other interpretations are also possible.

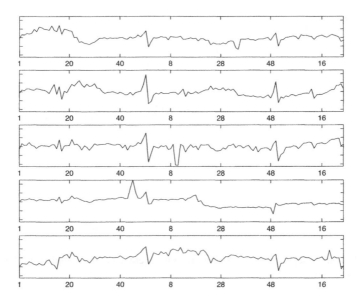

Fig. 24.1 Five samples of the 40 original cashflow time series (mean removed, normalized to unit standard deviation). Horizontal axis: time in weeks over 140 weeks. (Adapted from [245].)

If five ICs are estimated instead of four, then three of the found components stay virtually the same, while the fourth one separates into two new components. Using the found mixing coefficients a_{ij}, it is also possible to analyze the original time series and cluster them in groups. More details on the experiments and their interpretation can be found in [245].

24.1.2 Time series prediction by ICA

As noted in Chapter 18, the ICA transformation tends to produce component signals, $s_j(t)$, that can be compressed with fewer bits than the original signals, $x_i(t)$. They are thus more structured and regular. This gives motivation to try to predict the signals $x_i(t)$ by first going to the ICA space, doing the prediction there, and then transforming back to the original time series, as suggested by [362]. The prediction can be done separately and with a different method for each component, depending on its time structure. Hence, some interaction from the user may be needed in the overall prediction procedure. Another possibility would be to formulate the ICA contrast function in the first place so that it includes the prediction errors — some work along these lines has been reported by [437].

In [289], we suggested the following basic procedure:

1. After subtracting the mean of each time series and prewhitening (after which each time series has zero mean and unit variance), the independent components

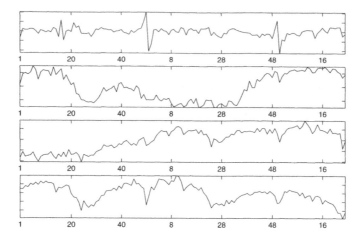

Fig. 24.2   Four independent components or fundamental factors found from the cashflow data. (Adapted from [245].)

$s_j(t)$, and the mixing matrix, **A**, are estimated using the FastICA algorithm. The number of ICs can be variable.

2. For each component $s_j(t)$, a suitable nonlinear filtering is applied to reduce the effects of noise — smoothing for components that contain very low frequencies (trend, slow cyclical variations), and high-pass filtering for components containing high frequencies and/or sudden shocks. The nonlinear smoothing is done by applying smoothing functions f_j on the source signals $s_j(t)$,

$$s_j^s(t) = f_j[s_j(t + r), \ldots, s_j(t), \ldots, s_j(t - k)]. \tag{24.2}$$

3. Each smoothed independent component is predicted separately, for instance using some method of autoregressive (AR) modeling [455]. The prediction is done for a number of steps into the future. This is done by applying prediction functions, g_j, on the smoothed source signals, $s_j^s(t)$:

$$s_j^p(t + 1) = g_j[s_j^s(t), s_j^s(t - 1), \ldots, s_j^s(t - q)] \tag{24.3}$$

The next time steps are predicted by gliding the window of length q over the measured and predicted values of the smoothed signal.

4. The predictions for each independent component are combined by weighing them with the mixing coefficients, a_{ij}, thus obtaining the predictions, $x_i^p(t)$, for the original time series, $x_i(t)$:

$$\mathbf{x}^p(t + 1) = \mathbf{A}\mathbf{s}^p(t + 1) \tag{24.4}$$

and similarly for $t + 2, t + 3, \ldots$.

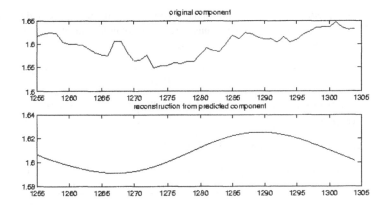

Fig. 24.3 Prediction of real-world financial data: the upper figure represents the actual future outcome of one of the original mixtures and the lower one the forecast obtained using ICA prediction for an interval of 50 values.

To test the method, we applied our algorithm on a set of 10 foreign exchange rate time series. Again, we suppose that there are some independent factors that affect the time evolution of such time series. Economic indicators, interest rates, and psychological factors can be the underlying factors of exchange rates, as they are closely tied to the evolution of the currencies. Even without prediction, some of the ICs may be useful in analyzing the impact of different external phenomena on the foreign exchange rates [22].

The results were promising, as the ICA prediction performed better than direct prediction. Figure 24.3 shows an example of prediction using our method. The upper figure represents one of the original time series (mixtures) and the lower one the forecast obtained using ICA prediction for a future interval of 50 time steps. The algorithm seemed to predict very well especially the turning points. In Table 24.1 there is a comparison of errors obtained by applying classic AR prediction to the original time series directly, and our method outlined above. The right-most column shows the magnitude of the errors when no smoothing is applied to the currencies.

While ICA and AR prediction are linear techniques, the smoothing was nonlinear. Using nonlinear smoothing, optimized for each independent component time series separately, the prediction of the ICs is more accurately performed and the results also are different from the direct prediction of the original time series. The noise in the time series is strongly reduced, allowing a better prediction of the underlying factors. The model is flexible and allows various smoothing tolerances and different orders in the classic AR prediction method for each independent component.

In reality, especially in real world time series analysis, the data are distorted by delays, noise, and nonlinearities. Some of these could be handled by extensions of the basic ICA algorithms, as reported in Part III of this book.

Table 24.1   The prediction errors (in units of 0.001) obtained with our method and the classic AR method. Ten currency time series were considered and five independent components were used. The amount of smoothing in classic AR prediction was varied.

		Errors					
Smoothing in AR prediction	2	0.5	0.1	0.08	0.06	0.05	0
ICA prediction	2.3	2.3	2.3	2.3	2.3	2.3	2.3
AR prediction	9.7	9.1	4.7	3.9	3.4	3.1	4.2

24.2   AUDIO SEPARATION

One of the original motivations for ICA research was the cocktail-party problem, as reviewed in the beginning of Chapter 7. The idea is that there are n sound sources recorded by a number of microphones, and we want to separate just one of the sources. In fact, often there is just one interesting signal, for example, a person speaking to the microphone, and all the other sources can be considered as noise; in this case, we have a problem of noise canceling. A typical example of a situation where we want to separate noise (or interference) from a speech signal is a person talking to a mobile phone in a noisy car.

If there is just one microphone, one can attempt to cancel the noise by ordinary noise canceling methods: linear filtering, or perhaps more sophisticated techniques like wavelet and sparse code shrinkage (Section 15.6). Such noise canceling can be rather unsatisfactory, however. It works only if the noise has spectral characteristics that are clearly different from those of the speech signal. One might wish to remove the noise more effectively by collecting more data using several microphones. Since in real-life situations the positions of the microphones with respect to the sources can be rather arbitrary, the mixing process is not known, and it has to be estimated blindly. In this case, we find the ICA model, and the problem is one of blind source separation.

Blind separation of audio signals is, however, much more difficult than one might expect. This is because the basic ICA model is a very crude approximation of the real mixing process. In fact, here we encounter almost all the complications that we have discussed in Part III:

- The mixing is not instantaneous. Audio signals propagate rather slowly, and thus they arrive in the microphones at different times. Moreover, there are echos, especially if the recording is made in a room. Thus the problem is more adequately modeled by a convolutive version of the ICA model (Chapter 19). The situation is thus much more complicated than with the separation of magnetoencephalographic (MEG) signals, which propagate fast, or with feature extraction, where no time delays are possible even in theory. In fact, even the basic convolutive ICA model may not be enough because the time delays may be fractional and may not be adequately modeled as integer multiples of the time interval between two samples.

- Typically, the recordings are made with two microphones only. However, the number of source signals is probably much larger than 2 in most cases, since the noise sources may not form just one well-defined source. Thus we have the problem of overcomplete bases (Chapter 16).

- The nonstationarity of the mixing is another important problem. The mixing matrix may change rather quickly, due to changes in the constellation of the speaker and the microphones. For example, one of these may be moving with respect to the other, or the speaker may simply turn his head. This implies that the mixing matrix must be reestimated quickly in a limited time frame, which also means a limited number of data. Adaptive estimation methods may alleviate this problem somewhat, but this is still a serious problem due to the convolutive nature of the mixing. In the convolutive mixing, the number of parameters can be very large: For example, the convolution may be modeled by filters of the length of 1000 time points, which effectively multiplies the number of parameters in the model by 1000. Since the number of data points should grow with the number of parameters to obtain satisfactory estimates, it may be next to impossible to estimate the model with the small number of data points that one has time to collect before the mixing matrix has changed too much.

- Noise may be considerable. There may be strong sensor noise, which means that we should use the noisy ICA model (Chapter 15). The noise complicates the estimation of the ICA model quite considerably, even in the basic case where noise is assumed gaussian. On the other hand, the effect of overcomplete bases could be modeled as noise as well. This noise may not be very gaussian, however, making the problem even more difficult.

Due to these complications, it may be that the prior information, independence and nongaussianity of the source signals, are not enough. To estimate the convolutive ICA model with a large number of parameters, and a rapidly changing mixing matrix, may require more information on the signals and the matrix. First, one may need to combine the assumption of nongaussianity with the different time-structure assumptions in Chapter 18. Speech signals have autocorrelations and nonstationarities, so this information could be used [267, 216]. Second, one may need to use some information on the mixing. For example, sparse priors (Section 20.1.3) could be used.

It is also possible that real-life speech separation requires sophisticated modeling of speech signals. Speech signals are highly structured, autocorrelations and nonstationarity being just the very simplest aspects of their time structure. Such approaches were proposed in [54, 15].

Because of these complications, audio separation is a largely unsolved problem. For a recent review on the subject, see [429]. One of the main theoretical problems, estimation of the convolutive ICA model, was described in Chapter 19.

24.3 FURTHER APPLICATIONS

Among further applications, let us mention

- Text document analysis [219, 229, 251]

- Radiocommunications [110, 77]

- Rotating machine monitoring [475]

- Seismic monitoring [161]

- Reflection canceling [127]

- Nuclear magnetic resonance spectroscopy [321]

- Selective transmission, which is a dual problem of blind source separation. A set of independent source signals are adaptively premixed prior to a nondispersive physical mixing process so that each source can be independently monitored in the far field [117].

Further applications can be found in the proceedings of the ICA'99 and ICA2000 workshops [70, 348].

References

1. K. Abed-Meraim and P. Loubaton. A subspace algorithm for certain blind identification problems. *IEEE Trans. on Information Theory*, 43(2):499–511, 1997.

2. L. Almeida. Linear and nonlinear ICA based on mutual information. In *Proc. IEEE 2000 Adaptive Systems for Signal Processing, Communications, and Control Symposium (AS-SPCC)*, pages 117–122, Lake Louise, Canada, October 2000.

3. S.-I. Amari. Neural learning in structured parameter spaces—natural Riemannian gradient. In *Advances in Neural Information Processing Systems 9*, pages 127–133. MIT Press, 1997.

4. S.-I. Amari. Natural gradient works efficiently in learning. *Neural Computation*, 10(2):251–276, 1998.

5. S.-I. Amari. Natural gradient learning for over- and under-complete bases in ICA. *Neural Computation*, 11(8):1875–1883, 1999.

6. S.-I. Amari. Estimating functions of independent component analysis for temporally correlated signals. *Neural Computation*, 12(9):2083–2107, 2000.

7. S.-I. Amari. Superefficiency in blind source separation. *IEEE Trans. on Signal Processing*, 47(4):936–944, April 1999.

8. S.-I. Amari and J.-F. Cardoso. Blind source separation—semiparametric statistical approach. *IEEE Trans. on Signal Processing*, 45(11):2692–2700, 1997.

9. S.-I. Amari, T.-P. Chen, and A. Cichocki. Stability analysis of adaptive blind source separation. *Neural Networks*, 10(8):1345–1351, 1997.

10. S.-I. Amari and A. Cichocki. Adaptive blind signal processing—neural network approaches. *Proceedings of the IEEE*, 86(10):2026–2048, 1998.

11. S.-I. Amari, A. Cichocki, and H. H. Yang. Blind signal separation and extraction: Neural and information-theoretic approaches. In S. Haykin, editor, *Unsupervised Adaptive Filtering*, volume 1, pages 63–138. Wiley, 2000.

12. S.-I. Amari, A. Cichocki, and H.H. Yang. A new learning algorithm for blind source separation. In *Advances in Neural Information Processing Systems 8*, pages 757–763. MIT Press, 1996.

13. S.-I. Amari, S. C. Douglas, A. Cichocki, and H. H. Yang. Novel on-line adaptive learning algorithms for blind deconvolution using the natural gradient approach. In *Proc. IEEE 11th IFAC Symposium on System Identification, SYSID-97*, pages 1057–1062, Kitakyushu, Japan, 1997.

14. T. W. Anderson. *An Introduction to Multivariate Statistical Analysis*. Wiley, 1958.

15. J. Anemüller and B. Kollmeier. Amplitude modulation decorrelation for convolutive blind source separation. In *Proc. Int. Workshop on Independent Component Analysis and Blind Signal Separation (ICA2000)*, pages 215–220, Helsinki, Finland, 2000.

16. B. Ans, J. Hérault, and C. Jutten. Adaptive neural architectures: detection of primitives. In *Proc. of COGNITIVA'85*, pages 593–597, Paris, France, 1985.

17. A. Antoniadis and G. Oppenheim, editors. *Wavelets in Statistics*. Springer, 1995.

18. J.J. Atick. Entropy minimization: A design principle for sensory perception? *Int. Journal of Neural Systems*, 3:81–90, 1992. Suppl. 1992.

19. H. Attias. Independent factor analysis. *Neural Computation*, 11(4):803–851, 1999.

20. H. Attias. Learning a hierarchical belief network of independent factor analyzers. In *Advances in Neural Information Processing Systems*, volume 11, pages 361–367. MIT Press, 1999.

21. A. D. Back and A. C. Tsoi. Blind deconvolution of signals using a complex recurrent network. In J. Vlontzos, J. Hwang, and E. Wilson, editors, *Neural Networks for Signal Processing*, volume 4, pages 565–574, 1994.

22. A. D. Back and A. S. Weigend. A first application of independent component analysis to extracting structure from stock returns. *Int. J. on Neural Systems*, 8(4):473–484, 1997.

23. P. F. Baldi and K. Hornik. Learning in linear neural networks: A survey. *IEEE Trans. on Neural Networks*, 6(4):837–858, 1995.

24. Y. Bar-Ness. Bootstrapping adaptive interference cancellers: Some practical limitations. In *The Globecom Conf.*, pages 1251–1255, Miami, 1982. Paper F3.7.

25. D. Barber and C. Bishop. Ensemble learning in Bayesian neural networks. In M. Jordan, M. Kearns, and S. Solla, editors, *Neural Networks and Machine Learning*, pages 215–237. Springer, 1998.

26. H. B. Barlow. Possible principles underlying the transformations of sensory messages. In W. A. Rosenblith, editor, *Sensory Communication*, pages 217–234. MIT Press, 1961.

27. H. B. Barlow. Single units and sensation: A neuron doctrine for perceptual psychology? *Perception*, 1:371–394, 1972.

28. H. B. Barlow. Unsupervised learning. *Neural Computation*, 1:295–311, 1989.

29. H. B. Barlow. What is the computational goal of the neocortex? In C. Koch and J.L. Davis, editors, *Large-Scale Neuronal Theories of the Brain*. MIT Press, Cambridge, MA, 1994.

30. H. B. Barlow, T.P. Kaushal, and G.J. Mitchison. Finding minimum entropy codes. *Neural Computation*, 1:412–423, 1989.

31. A. Barros, A. Mansour, and N. Ohnishi. Removing artifacts from electrocardiographic signals using independent component analysis. *Neurocomputing*, 22:173–186, 1998.

32. A. K. Barros, R. Vigário, V. Jousmäki, and N. Ohnishi. Extraction of event-related signals from multi-channel bioelectrical measurements. *IEEE Trans. on Biomedical Engineering*, 47(5):583–588, 2000.

33. J. Basak and S.-I. Amari. Blind separation of uniformly distributed signals: A general approach. *IEEE Trans. on Neural Networks*, 10(5):1173–1185, 1999.

34. A. J. Bell. Information theory, independent component analysis, and applications. In S. Haykin, editor, *Unsupervised Adaptive Filtering, Vol. I*, pages 237–264. Wiley, 2000.

35. A. J. Bell and T. J. Sejnowski. A non-linear information maximization algorithm that performs blind separation. In *Advances in Neural Information Processing Systems 7*, pages 467–474. The MIT Press, Cambridge, MA, 1995.

36. A.J. Bell and T.J. Sejnowski. An information-maximization approach to blind separation and blind deconvolution. *Neural Computation*, 7:1129–1159, 1995.

37. A.J. Bell and T.J. Sejnowski. Learning higher-order structure of a natural sound. *Network*, 7:261–266, 1996.

38. A.J. Bell and T.J. Sejnowski. The 'independent components' of natural scenes are edge filters. *Vision Research*, 37:3327–3338, 1997.

39. S. Bellini. Bussgang techniques for blind deconvolution and equalization. In S. Haykin, editor, *Blind Deconvolution*, pages 8–59. Prentice Hall, 1994.

40. A. Belouchrani and M. Amin. Blind source separation based on time-frequency signal representations. *IEEE Trans. on Signal Processing*, 46(11):2888–2897, 1998.

41. A. Belouchrani and M. Amin. Jammer mitigation in spread spectrum communications using blind sources separation. *Signal Processing*, 80(4):723–729, 2000.

42. A. Belouchrani and J.-F. Cardoso. Maximum likelihood source separation by the expectation-maximization technique: deterministic and stochastic implementation. In *Proc. Int. Symp. on Nonlinear Theory and its Applications (NOLTA'95)*, pages 49–53, Las Vegas, Nevada, 1995.

43. A. Belouchrani, K. Abed Meraim, J.-F. Cardoso, and E. Moulines. A blind source separation technique based on second order statistics. *IEEE Trans. on Signal Processing*, 45(2):434–444, 1997.

44. S. Bensley and B. Aazhang. Subspace-based channel estimation for code division multiple access communication systems. *IEEE Trans. on Communications*, 44(8):1009–1020, 1996.

45. P. Berg and M. Scherg. A multiple source approach to the correction of eye artifacts. *Electroencephalography and Clinical Neurophysiology*, 90:229–241, 1994.

46. D. Bertsekas. *Nonlinear Programming*. Athenas Scientific, Belmont, MA, 1995.

47. E. Bingham and A. Hyvärinen. A fast fixed-point algorithm for independent component analysis of complex-valued signals. *Int. J. of Neural Systems*, 10(1):1–8, 2000.

48. C. M. Bishop. *Neural Networks for Pattern Recognition*. Clarendon Press, 1995.

49. C. M. Bishop, M. Svensen, and C. K. I. Williams. GTM: The generative topographic mapping. *Neural Computation*, 10:215–234, 1998.

50. B. B. Biswal and J. L. Ulmer. Blind source separation of multiple signal sources of fMRI data sets using independent component analysis. *J. of Computer Assisted Tomography*, 23(2):265–271, 1999.

51. S. Blanco, H. Garcia, R. Quian Quiroga, L. Romanelli, and O. A. Rosso. Stationarity of the EEG series. *IEEE Engineering in Medicine and Biology Magazine*, pages 395–399, 1995.

52. H. Bourlard and Y. Kamp. Auto-association by multilayer perceptrons and singular value decomposition. *Biological Cybernetics*, 59:291–294, 1988.

53. G. Box and G. Tiao. *Bayesian Inference in Statistical Analysis*. Addison-Wesley, 1973.

54. M.S. Brandstein. On the use of explicit speech modeling in microphone array applications. In *Proc. IEEE Int. Conf. on Acoustics, Speech and Signal Processing (ICASSP'98)*, pages 3613–3616, Seattle, Washington, 1998.

55. H. Broman, U. Lindgren, H. Sahlin, and P. Stoica. Source separation: A TITO system identification approach. *Signal Processing*, 73:169–183, 1999.

56. C. H. M. Brunia, J. Möcks, and M. Van den Berg-Lennsen. Correcting ocular artifacts—a comparison of several methods. *J. of Psychophysiology*, 3:1–50, 1989.

57. G. Burel. Blind separation of sources: a nonlinear neural algorithm. *Neural Networks*, 5(6):937–947, 1992.

58. X.-R. Cao and R.-W. Liu. General approach to blind source separation. *IEEE Trans. on Signal Processing*, 44(3):562–571, 1996.

59. V. Capdevielle, C. Serviere, and J.Lacoume. Blind separation of wide-band sources in the frequency domain. In *Proc. IEEE Int. Conf. on Acoustics, Speech and Signal Processing (ICASSP'95)*, volume 3, pages 2080–2083, Detroit, Michigan, 1995.

60. J.-F. Cardoso. Blind identification of independent signals. In *Proc. Workshop on Higher-Order Specral Analysis*, Vail, Colorado, 1989.

61. J.-F. Cardoso. Source separation using higher order moments. In *Proc. IEEE Int. Conf. on Acoustics, Speech and Signal Processing (ICASSP'89)*, pages 2109–2112, Glasgow, UK, 1989.

62. J.-F. Cardoso. Eigen-structure of the fourth-order cumulant tensor with application to the blind source separation problem. In *Proc. IEEE Int. Conf. on Acoustics, Speech and Signal Processing (ICASSP'90)*, pages 2655–2658, Albuquerque, New Mexico, 1990.

63. J.-F. Cardoso. Iterative techniques for blind source separation using only fourth-order cumulants. In *Proc. EUSIPCO*, pages 739–742, Brussels, Belgium, 1992.

64. J.-F. Cardoso. Infomax and maximum likelihood for source separation. *IEEE Letters on Signal Processing*, 4:112–114, 1997.

65. J.-F. Cardoso. Blind signal separation: statistical principles. *Proceedings of the IEEE*, 9(10):2009–2025, 1998.

66. J.-F. Cardoso. Multidimensional independent component analysis. In *Proc. IEEE Int. Conf. on Acoustics, Speech and Signal Processing (ICASSP'98)*, Seattle, WA, 1998.

67. J.-F. Cardoso. On the stability of some source separation algorithms. In *Proc. Workshop on Neural Networks for Signal Processing (NNSP'98)*, pages 13–22, Cambridge, UK, 1998.

68. J.-F. Cardoso. High-order contrasts for independent component analysis. *Neural Computation*, 11(1):157–192, 1999.

69. J. F. Cardoso. Entropic contrasts for source separation: Geometry and stability. In S. Haykin, editor, *Unsupervised Adaptive Filtering*, volume 1, pages 139–189. Wiley, 2000.

70. J.-F. Cardoso, C. Jutten, and P. Loubaton, editors. *Proc. of the 1st Int. Workshop on Independent Component Analysis and Signal Separation, Aussois, France, January 11-15, 1999*. 1999.

71. J.-F. Cardoso and B. Hvam Laheld. Equivariant adaptive source separation. *IEEE Trans. on Signal Processing*, 44(12):3017–3030, 1996.

72. J.-F. Cardoso and A. Souloumiac. Blind beamforming for non Gaussian signals. *IEE Proceedings-F*, 140(6):362–370, 1993.

73. L. Castedo, C. Escudero, and A. Dapena. A blind signal separation method for multiuser communications. *IEEE Trans. on Signal Processing*, 45(5):1343–1348, 1997.

74. N. Charkani and Y. Deville. Self-adaptive separation of convolutively mixed signals with a recursive structure, part i: Stability analysis and optimization of the asymptotic behaviour. *Signal Processing*, 73:225–254, 1999.

75. T. Chen, Y. Hua, and W. Yan. Global convergence of Oja's subspace algorithm for principal component extraction. *IEEE Trans. on Neural Networks*, 9(1):58–67, 1998.

76. P. Chevalier. Optimal separation of independent narrow-band sources: Concept and performance. *Signal Processing*, 73:27–47, 1999.

77. P. Chevalier, V. Capdevielle, and P. Comon. Performance of HO blind source separation methods: Experimental results on ionopheric HF links. In *Proc. Int. Workshop on Independent Component Analysis and Signal Separation (ICA'99)*, pages 443–448, Aussois, France, 1999.

78. S. Choi and A. Cichocki. Blind signal deconvolution by spatio-temporal decorrelation and demixing. In *Proc. IEEE Workshop on Neural Networks for Signal Processing (NNSP·97)*, pages 426–435, Amelia Island, Florida, September 1997.

79. A. Cichocki, S.-I. Amari, and J. Cao. Blind separation of delayed and convolved sources with self-adaptive learning rate. In *Proc. Int. Symp. on Nonlinear Theory and Applications (NOLTA'96)*, pages 229–232, Kochi, Japan, 1996.

80. A. Cichocki, S. C. Douglas, and S.-I. Amari. Robust techniques for independent component analysis with noisy data. *Neurocomputing*, 22:113–129, 1998.

81. A. Cichocki, J. Karhunen, W. Kasprzak, and R. Vigário. Neural networks for blind separation with unknown number of sources. *Neurocomputing*, 24:55–93, 1999.

82. A. Cichocki and L. Moszczynski. A new learning algorithm for blind separation of sources. *Electronics Letters*, 28(21):1986–1987, 1992.

83. A. Cichocki and R. Unbehauen. *Neural Networks for Signal Processing and Optimization*. Wiley, 1994.

84. A. Cichocki and R. Unbehauen. Robust neural networks with on-line learning for blind identification and blind separation of sources. *IEEE Trans. on Circuits and Systems*, 43(11):894–906, 1996.

85. A. Cichocki, R. Unbehauen, and E. Rummert. Robust learning algorithm for blind separation of signals. *Electronics Letters*, 30(17):1386–1387, 1994.

86. A. Cichocki, L. Zhang, S. Choi, and S.-I. Amari. Nonlinear dynamic independent component analysis using state-space and neural network models. In *Proc. Int. Workshop on Independent Component Analysis and Signal Separation (ICA'99)*, pages 99–104, Aussois, France, 1999.

87. R. R. Coifman and D. L. Donoho. Translation-invariant de-noising. Technical report, Department of Statistics, Stanford University, Stanford, California, 1995.

88. P. Comon. Separation of stochastic processes. In *Proc. Workshop on Higher-Order Specral Analysis*, pages 174 – 179, Vail, Colorado, 1989.

89. P. Comon. Independent component analysis—a new concept? *Signal Processing*, 36:287–314, 1994.

90. P. Comon. Contrasts for multichannel blind deconvolution. *Signal Processing Letters*, 3(7):209–211, 1996.

91. P. Comon and P. Chevalier. Blind source separation: Models, concepts, algorithms and performance. In S. Haykin, editor, *Unsupervised Adaptive Filtering, Vol. I*, pages 191–235. Wiley, 2000.

92. P. Comon and G. Golub. Tracking a few extreme singular values and vectors in signal processing. *Proceedings of the IEEE*, 78:1327–1343, 1990.

93. P. Comon, C. Jutten, and J. Hérault. Blind separation of sources, Part II: Problems statement. *Signal Processsing*, 24:11–20, 1991.

94. P. Comon and B. Mourrain. Decomposition of quantics in sums of powers of linear forms. *Signal Processing*, 53(2):93–107, 1996.

95. D. Cook, A. Buja, and J. Cabrera. Projection pursuit indexes based on orthonormal function expansions. *J. of Computational and Graphical Statistics*, 2(3):225–250, 1993.

96. G.W. Cottrell, P.W. Munro, and D. Zipser. Learning internal representations from gray-scale images: An example of extensional programming. In *Proc. Ninth Annual Conference of the Cognitive Science Society*, pages 462–473, 1987.

97. T. M. Cover and J. A. Thomas. *Elements of Information Theory*. Wiley, 1991.

98. R. Cristescu, J. Joutsensalo, J. Karhunen, and E. Oja. A complexity minimization approach for estimating fading channels in CDMA communications. In *Proc. Int. Workshop on Independent Component Analysis and Blind Signal Separation (ICA2000)*, pages 527–532, Helsinki, Finland, June 2000.

99. R. Cristescu, T. Ristaniemi, J. Joutsensalo, and J. Karhunen. Blind separation of convolved mixtures for CDMA systems. In *Proc. Tenth European Signal Processing Conference (EUSIPCO2000)*, pages 619–622, Tampere, Finland, 2000.

100. R. Cristescu, T. Ristaniemi, J. Joutsensalo, and J. Karhunen. Delay estimation in CDMA communications using a Fast ICA algorithm. In *Proc. Int. Workshop on Independent Component Analysis and Blind Signal Separation (ICA2000)*, pages 105–110, Helsinki, Finland, 2000.

101. S. Cruces and L. Castedo. Stability analysis of adaptive algorithms for blind source separation of convolutive mixtures. *Signal Processing*, 78(3):265–276, 1999.

102. I. Daubechies. *Ten Lectures on Wavelets*. Society for Industrial and Applied Math., Philadelphia, 1992.

103. J. G. Daugman. Complete disrete 2-D Gabor transforms by neural networks for image analysis and compression. *IEEE Trans. on Acoustics Speech and Signal Processing*, 36(7):1169–1179, 1988.

104. G. Deco and W. Brauer. Nonlinear higher-order statistical decorrelation by volume-conserving neural architectures. *Neural Networks*, 8:525–535, 1995.

105. G. Deco and D. Obradovic. *An Information-Theoretic Approach to Neural Computing*. Springer Verlag, 1996.

106. S. Degerine and R. Malki. Second-order blind separation of sources based on canonical partial innovations. *IEEE Trans. on Signal Processing*, 48(3):629–641, 2000.

107. N. Delfosse and P. Loubaton. Adaptive blind separation of independent sources: a deflation approach. *Signal Processing*, 45:59–83, 1995.

108. N. Delfosse and P. Loubaton. Adaptive blind separation of convolutive mixtures. In *Proc. IEEE Int. Conf. on Acoustics, Speech and Signal Processing (ICASSP'97)*, pages 2940–2943, 1996.

109. P. Devijver and J. Kittler. *Pattern Recognition: A Statistical Approach*. Prentice Hall, 1982.

110. Y. Deville and L. Andry. Application of blind source separation techniques to multi-tag contactless identification systems. In *Proc. Int. Symp. on Nonlinear Theory and its Applications (NOLTA'95)*, pages 73–78, Las Vegas, Nevada, 1995.

111. Y. Deville, J. Damour, and N. Charkani. Improved multi-tag radio-frequency identification systems based on new source separation neural networks. In *Proc. Int. Workshop on Independent Component Analysis and Blind Source Separation (ICA'99)*, pages 449–454, Aussois, France, 1999.

112. K. I. Diamantaras and S. Y. Kung. *Principal Component Neural Networks: Theory and Applications*. Wiley, 1996.

113. K. I. Diamantaras, A. P. Petropulu, and B. Chen. Blind two-input-two-output FIR channel identification based on second-order statistics. *IEEE Trans. on Signal Processing*, 48(2):534–542, 2000.

114. D. L. Donoho. On minimum entropy deconvolution. In *Applied Time Series Analysis II*, pages 565–608. Academic Press, 1981.

115. D. L. Donoho. Nature vs. math: Interpreting independent component analysis in light of recent work in harmonic analysis. In *Proc. Int. Workshop on Independent Component Analysis and Blind Signal Separation (ICA2000)*, pages 459–470, Helsinki, Finland, 2000.

116. D. L. Donoho, I. M. Johnstone, G. Kerkyacharian, and D. Picard. Wavelet shrinkage: asymptopia? *Journal of the Royal Statistical Society, Ser. B*, 57:301–337, 1995.

117. S. C. Douglas. Equivariant adaptive selective transmission. *IEEE Trans. on Signal Processing*, 47(5):1223–1231, May 1999.

118. S. C. Douglas and S.-I. Amari. Natural-gradient adaptation. In S. Haykin, editor, *Unsupervised Adaptive Filtering*, volume 1, pages 13–61. Wiley, 2000.

119. S. C. Douglas, A. Cichocki, , and S.-I. Amari. A bias removal technique for blind source separation with noisy measurements. *Electronics Letters*, 34:1379–1380, 1998.

120. S. C. Douglas and A. Cichocki. Neural networks for blind decorrelation of signals. *IEEE Trans. on Signal Processing*, 45(11):2829–2842, 1997.

121. S. C. Douglas, A. Cichocki, and S.-I. Amari. Multichannel blind separation and deconvolution of sources with arbitrary distributions. In *Proc. IEEE Workshop on Neural Networks for Signal Processing (NNSP'97)*, pages 436–445, Amelia Island, Florida, 1997.

122. S. C. Douglas and S. Haykin. Relationships between blind deconvolution and blind source separation. In S. Haykin, editor, *Unsupervised Adaptive Filtering*, volume 2, pages 113–145. Wiley, 2000.

123. F. Ehlers and H. Schuster. Blind separation of convolutive mixtures and an application in automatic speech recognition in a noisy environment. *IEEE Trans. on Signal Processing*, 45(10):2608–2612, 1997.

124. B. Emile and P. Comon. Estimation of time delays between unknown colored signals. *Signal Processing*, 69(1):93–100, 1998.

125. R. M. Everson and S. Roberts. Independent component analysis: a flexible nonlinearity and decorrelating manifold approach. *Neural Computation*, 11(8):1957–1983, 1999.

126. R. M. Everson and S. J. Roberts. Particle filters for non-stationary ICA. In M. Girolami, editor, *Advances in Independent Component Analysis*, pages 23–41. Springer-Verlag, 2000.

127. H. Farid and E. H. Adelson. Separating reflections from images by use of independent component analysis. *J. of the Optical Society of America*, 16(9):2136–2145, 1999.

128. H. G. Feichtinger and T. Strohmer, editors. *Gabor Analysis and Algorithms*. Birkhauser, 1997.

129. W. Feller. *Probability Theory and Its Applications*. Wiley, 3rd edition, 1968.

130. M. Feng and K.-D. Kammayer. Application of source separation algorithms for mobile communications environment. In *Proc. Int. Workshop on Independent Component Analysis and Blind Source Separation (ICA'99)*, pages 431–436, Aussois, France, January 1999.

131. D.J. Field. What is the goal of sensory coding? *Neural Computation*, 6:559–601, 1994.

132. S. Fiori. Blind separation of circularly distributed sources by neural extended APEX algorithm. *Neurocomputing*, 34:239–252, 2000.

133. S. Fiori. Blind signal processing by the adaptive activation function neurons. *Neural Networks*, 13(6):597–611, 2000.

134. J. Fisher and J. Principe. Entropy manipulation of arbitrary nonlinear mappings. In J. Principe et al., editor, *Neural Networks for Signal Processing VII*, pages 14–23. IEEE Press, New York, 1997.

135. R. Fletcher. *Practical Methods of Optimization*. Wiley, 2nd edition, 1987.

136. P. Földiák. Adaptive network for optimal linear feature extraction. In *Proc. Int. Joint Conf. on Neural Networks*, pages 401 – 406, Washington DC, 1989.

137. J. H. Friedman and J. W. Tukey. A projection pursuit algorithm for exploratory data analysis. *IEEE Trans. of Computers*, c-23(9):881–890, 1974.

138. J.H. Friedman. Exploratory projection pursuit. *J. of the American Statistical Association*, 82(397):249–266, 1987.

139. C. Fyfe and R. Baddeley. Non-linear data structure extraction using simple Hebbian networks. *Biological Cybernetics*, 72:533–541, 1995.

140. M. Gaeta and J.-L. Lacoume. Source separation without prior knowledge: the maximum likelihood solution. In *Proc. EUSIPCO'90*, pages 621–624, 1990.

141. W. Gardner. *Introduction to Random Processes with Applications to Signals and Systems*. Macmillan, 1986.

142. A. Gelman, J. Carlin, H. Stern, and D. Rubin. *Bayesian Data Analysis*. Chapman & Hall/CRC Press, Boca Raton, Florida, 1995.

143. G. Giannakis, Y. Hua, P. Stoica, and L. Tong. *Signal Processing Advances in Communications, vol. 1: Trends in Channel Estimation and Equalization*. Prentice Hall, 2001.

144. G. Giannakis, Y. Hua, P. Stoica, and L. Tong. *Signal Processing Advances in Communications, vol. 2: Trends in Single- and Multi-User Systems*. Prentice Hall, 2001.

145. G. Giannakis, Y. Inouye, and J. M. Mendel. Cumulant-based identification of multi-channel moving-average processes. *IEEE Trans. on Automat. Contr.*, 34:783–787, July 1989.

146. G. Giannakis and C. Tepedelenlioglu. Basis expansion models and diversity techniques for blind identification and equalization of time-varying channels. *Proceedings of the IEEE*, 86(10):1969–1986, 1998.

147. X. Giannakopoulos, J. Karhunen, and E. Oja. Experimental comparison of neural algorithms for independent component analysis and blind separation. *Int. J. of Neural Systems*, 9(2):651–656, 1999.

148. M. Girolami. An alternative perspective on adaptive independent component analysis algorithms. *Neural Computation*, 10(8):2103–2114, 1998.

149. M. Girolami. *Self-Organising Neural Networks - Independent Component Analysis and Blind Source Separation*. Springer-Verlag, 1999.

150. M. Girolami, editor. *Advances in Independent Component Analysis*. Springer-Verlag, 2000.

151. M. Girolami and C. Fyfe. An extended exploratory projection pursuit network with linear and nonlinear anti-hebbian connections applied to the cocktail party problem. *Neural Networks*, 10:1607–1618, 1997.

152. D. N. Godard. Self-recovering equalization and carrier tracking in two-dimensional data communication systems. *IEEE Trans. on Communications*, 28:1867–1875, 1980.

153. G. Golub and C. van Loan. *Matrix Computations*. The Johns Hopkins University Press, 3rd edition, 1996.

154. R. Gonzales and P. Wintz. *Digital Image Processing*. Addison-Wesley, 1987.

155. A. Gorokhov and P. Loubaton. Subspace-based techniques for blind separation of convolutive mixtures with temporally correlated sources. *IEEE Trans. Circuits and Systems I*, 44(9):813–820, 1997.

156. A. Gorokhov and P. Loubaton. Blind identification of MIMO-FIR systems: A generalized linear prediction approach. *Signal Processing*, 73:105–124, 1999.

157. R. Gray and L. Davisson. *Random Processes: A Mathematical Approach for Engineers*. Prentice Hall, 1986.

158. O. Grellier and P. Comon. Blind separation of discrete sources. *IEEE Signal Processing Letters*, 5(8):212–214, 1998.

159. S. I. Grossman. *Elementary Linear Algebra*. Wadsworth, 1984.

160. P. Hall. Polynomial projection pursuit. *The Annals of Statistics*, 17:589–605, 1989.

161. F.M. Ham and N.A. Faour. Infrasound signal separation using independent component analysis. In *Proc. 21st Seismic Reseach Symposium: Technologies for Monitoring the Comprehensive Nuclear-Test-Ban Treaty*, Las Vegas, Nevada, 1999.

162. M. Hämäläinen, R. Hari, R. Ilmoniemi, J. Knuutila, and O. V. Lounasmaa. Magnetoencephalography—theory, instrumentation, and applications to noninvasive studies of the working human brain. *Reviews of Modern Physics*, 65(2):413–497, 1993.

163. F.R. Hampel, E.M. Ronchetti, P.J. Rousseuw, and W.A. Stahel. *Robust Statistics*. Wiley, 1986.

164. L. K. Hansen. Blind separation of noisy image mixtures. In M. Girolami, editor, *Advances in Independent Component Analysis*, pages 161–181. Springer-Verlag, 2000.

165. R. Hari. Magnetoencephalography as a tool of clinical neurophysiology. In E. Niedermeyer and F. Lopes da Silva, editors, *Electroencephalography. Basic principles, clinical applications, and related fields*, pages 1035–1061. Baltimore: Williams & Wilkins, 1993.

166. H. H. Harman. *Modern Factor Analysis*. University of Chicago Press, 2nd edition, 1967.

167. T. Hastie and W. Stuetzle. Principal curves. *Journal of the American Statistical Association*, 84:502–516, 1989.

168. M. Hayes. *Statistical Digital Signal Processing and Modeling*. Wiley, 1996.

169. S. Haykin. *Modern Filters*. Macmillan, 1989.

170. S. Haykin, editor. *Blind Deconvolution*. Prentice Hall, 1994.

171. S. Haykin. *Adaptive Filter Theory*. Prentice Hall, 3rd edition, 1996.

172. S. Haykin. *Neural Networks - A Comprehensive Foundation*. Prentice Hall, 2nd edition, 1998.

173. S. Haykin, editor. *Unsupervised Adaptive Filtering, Vol. 1: Blind Source Separation*. Wiley, 2000.

174. S. Haykin, editor. *Unsupervised Adaptive Filtering, Vol. 2: Blind Deconvolution*. Wiley, 2000.

175. Z. He, L. Yang, J. Liu, Z. Lu, C. He, and Y. Shi. Blind source separation using clustering-based multivariate density estimation algorithm. *IEEE Trans. on Signal Processing*, 48(2):575–579, 2000.

176. R. Hecht-Nielsen. Replicator neural networks for universal optimal source coding. *Science*, 269:1860–1863, 1995.

177. R. Hecht-Nielsen. Data manifolds, natural coordinates, replicator neural networks, and optimal source coding. In *Proc. Int. Conf. on Neural Information Processing*, pages 41–45, Hong Kong, 1996.

178. J. Hérault and B. Ans. Circuits neuronaux à synapses modifiables: décodage de messages composites par apprentissage non supervisé. *C.-R. de l'Académie des Sciences*, 299(III-13):525–528, 1984.

179. J. Hérault, C. Jutten, and B. Ans. Détection de grandeurs primitives dans un message composite par une architecture de calcul neuromimétique en apprentissage non supervisé. In *Actes du Xème colloque GRETSI*, pages 1017–1022, Nice, France, 1985.

180. G. Hinton and D. van Camp. Keeping neural networks simple by minimizing the description length of the weights. In *Proc. of the 6th Annual ACM Conf. on Computational Learning Theory*, pages 5–13, Santa Cruz, CA, 1993.

181. S. Hochreiter and J. Schmidhuber. Feature extraction through LOCOCODE. *Neural Computation*, 11(3):679–714, 1999.

182. H. Holma and A. Toskala, editors. *WCDMA for UMTS*. Wiley, 2000.

183. L. Holmström, P. Koistinen, J. Laaksonen, and E. Oja. Comparison of neural and statistical classifiers - taxonomy and two case studies. *IEEE Trans. on Neural Networks*, 8:5–17, 1997.

184. M. Honig and V. Poor. Adaptive interference suppression. In *Wireless Communications: Signal Processing Perspectives*, pages 64–128. Prentice Hall, 1998.

185. H. Hotelling. Analysis of a complex of statistical variables into principal components. *J. of Educational Psychology*, 24:417 – 441, 1933.

186. P. O. Hoyer and A. Hyvärinen. Independent component analysis applied to feature extraction from colour and stereo images. *Network: Computation in Neural Systems*, 11(3):191–210, 2000.

187. P. O. Hoyer and A. Hyvärinen. Modelling chromatic and binocular properties of V1 topography. In *Proc. Int. Conf. on Neural Information Processing (ICONIP'00)*, Taejon, Korea, 2000.

188. P.J. Huber. *Robust Statistics*. Wiley, 1981.

189. P.J. Huber. Projection pursuit. *The Annals of Statistics*, 13(2):435–475, 1985.

190. M. Huotilainen, R. J. Ilmoniemi, H. Tiitinen, J. Lavaikainen, K. Alho, M. Kajola, and R. Näätänen. The projection method in removing eye-blink artefacts from multichannel MEG measurements. In C. Baumgartner et al., editor, *Biomagnetism: Fundamental Research and Clinical Applications (Proc. Int. Conf. on Biomagnetism)*, pages 363–367. Elsevier, 1995.

191. J. Hurri, A. Hyvärinen, and E. Oja. Wavelets and natural image statistics. In *Proc. Scandinavian Conf. on Image Analysis '97*, Lappeenranta, Finland, 1997.

192. A. Hyvärinen. A family of fixed-point algorithms for independent component analysis. In *Proc. IEEE Int. Conf. on Acoustics, Speech and Signal Processing (ICASSP'97)*, pages 3917–3920, Munich, Germany, 1997.

193. A. Hyvärinen. One-unit contrast functions for independent component analysis: A statistical analysis. In *Neural Networks for Signal Processing VII (Proc. IEEE Workshop on Neural Networks for Signal Processing)*, pages 388–397, Amelia Island, Florida, 1997.

194. A. Hyvärinen. Independent component analysis for time-dependent stochastic processes. In *Proc. Int. Conf. on Artificial Neural Networks (ICANN'98)*, pages 135–140, Skövde, Sweden, 1998.

195. A. Hyvärinen. Independent component analysis in the presence of gaussian noise by maximizing joint likelihood. *Neurocomputing*, 22:49–67, 1998.

196. A. Hyvärinen. New approximations of differential entropy for independent component analysis and projection pursuit. In *Advances in Neural Information Processing Systems*, volume 10, pages 273–279. MIT Press, 1998.

197. A. Hyvärinen. Fast and robust fixed-point algorithms for independent component analysis. *IEEE Trans. on Neural Networks*, 10(3):626–634, 1999.

198. A. Hyvärinen. Fast independent component analysis with noisy data using gaussian moments. In *Proc. Int. Symp. on Circuits and Systems*, pages V57–V61, Orlando, Florida, 1999.

199. A. Hyvärinen. Gaussian moments for noisy independent component analysis. *IEEE Signal Processing Letters*, 6(6):145–147, 1999.

200. A. Hyvärinen. Sparse code shrinkage: Denoising of nongaussian data by maximum likelihood estimation. *Neural Computation*, 11(7):1739–1768, 1999.

201. A. Hyvärinen. Survey on independent component analysis. *Neural Computing Surveys*, 2:94–128, 1999.

202. A. Hyvärinen. Complexity pursuit: Separating interesting components from time-series. *Neural Computation*, 13, 2001.

203. A. Hyvärinen, R. Cristescu, and E. Oja. A fast algorithm for estimating overcomplete ICA bases for image windows. In *Proc. Int. Joint Conf. on Neural Networks*, pages 894–899, Washington, D.C., 1999.

204. A. Hyvärinen and P. O. Hoyer. Emergence of phase and shift invariant features by decomposition of natural images into independent feature subspaces. *Neural Computation*, 12(7):1705–1720, 2000.

205. A. Hyvärinen and P. O. Hoyer. Emergence of topography and complex cell properties from natural images using extensions of ICA. In *Advances in Neural Information Processing Systems*, volume 12, pages 827–833. MIT Press, 2000.

206. A. Hyvärinen, P. O. Hoyer, and M. Inki. Topographic independent component analysis. *Neural Computation*, 13, 2001.

207. A. Hyvärinen, P. O. Hoyer, and E. Oja. Image denoising by sparse code shrinkage. In S. Haykin and B. Kosko, editors, *Intelligent Signal Processing*. IEEE Press, 2001.

208. A. Hyvärinen and M. Inki. Estimating overcomplete independent component bases from image windows. 2001. submitted manuscript.

209. A. Hyvärinen and R. Karthikesh. Sparse priors on the mixing matrix in independent component analysis. In *Proc. Int. Workshop on Independent Component Analysis and Blind Signal Separation (ICA2000)*, pages 477–452, Helsinki, Finland, 2000.

210. A. Hyvärinen and E. Oja. A fast fixed-point algorithm for independent component analysis. *Neural Computation*, 9(7):1483–1492, 1997.

211. A. Hyvärinen and E. Oja. Independent component analysis by general nonlinear Hebbian-like learning rules. *Signal Processing*, 64(3):301–313, 1998.

212. A. Hyvärinen and E. Oja. Independent component analysis: Algorithms and applications. *Neural Networks*, 13(4-5):411–430, 2000.

213. A. Hyvärinen and P. Pajunen. Nonlinear independent component analysis: Existence and uniqueness results. *Neural Networks*, 12(3):429–439, 1999.

214. A. Hyvärinen, J. Särelä, and R. Vigário. Spikes and bumps: Artefacts generated by independent component analysis with insufficient sample size. In *Proc. Int. Workshop on Independent Component Analysis and Signal Separation (ICA'99)*, pages 425–429, Aussois, France, 1999.

215. S. Ikeda. ICA on noisy data: A factor analysis approach. In M. Girolami, editor, *Advances in Independent Component Analysis*, pages 201–215. Springer-Verlag, 2000.

216. S. Ikeda and N. Murata. A method of ICA in time-frequency domain. In *Proc. Int. Workshop on Independent Component Analysis and Signal Separation (ICA'99)*, pages 365–370, Aussois, France, 1999.

217. Y. Inouye. Criteria for blind deconvolution of multichannel linear time-invariant systems. *IEEE Trans. on Signal Processing*, 46(12):3432–3436, 1998.

218. Y. Inouye and K. Hirano. Cumulant-based blind identification of linear multi-input-multi-output systems driven by colored inputs. *IEEE Trans. on Signal Processing*, 45(6):1543–1552, 1997.

219. C. L. Isbell and P. Viola. Restructuring sparse high-dimensional data for effective retreval. In *Advances in Neural Information Processing Systems*, volume 11. MIT Press, 1999.

220. N. Japkowitz, S. J. Hanson, and M. A. Gluck. Nonlinear autoassociation is not equivalent to PCA. *Neural Computation*, 12(3):531–545, 2000.

221. B. W. Jervis, M. Coelho, and G. Morgan. Effect on EEG responses of removing ocular artifacts by proportional EOG subtraction. *Medical and Biological Engineering and Computing*, 27:484–490, 1989.

222. M.C. Jones and R. Sibson. What is projection pursuit ? *J. of the Royal Statistical Society, Ser. A*, 150:1–36, 1987.

223. J. Joutsensalo and T. Ristaniemi. Learning algorithms for blind multiuser detection in CDMA downlink. In *Proc. IEEE 9th International Symposium on Personal, Indoor and Mobile Radio Communications (PIMRC '98)*, pages 267–270, Boston, Massachusetts, 1998.

224. C. Johnson Jr., P. Schniter, I. Fijalkow, and L. Tong et al. The core of FSE-CMA behavior theory. In S. Haykin, editor, *Unsupervised Adaptive Filtering*, volume 2, pages 13–112. Wiley, New York, 2000.

225. T. P. Jung, C. Humphries, T.-W. Lee, S. Makeig, M. J. McKeown, V. Iragui, and T. Sejnowski. Extended ICA removes artifacts from electroencephalographic recordings. In *Advances in Neural Information Processing Systems*, volume 10. MIT Press, 1998.

226. C. Jutten. *Calcul neuromimétique et traitement du signal, analyse en composantes indépendantes*. PhD thesis, INPG, Univ. Grenoble, France, 1987. (in French).

227. C. Jutten. Source separation: from dusk till dawn. In *Proc. 2nd Int. Workshop on Independent Component Analysis and Blind Source Separation (ICA'2000)*, pages 15–26, Helsinki, Finland, 2000.

228. C. Jutten and J. Hérault. Blind separation of sources, part I: An adaptive algorithm based on neuromimetic architecture. *Signal Processing*, 24:1–10, 1991.

229. A. Kaban and M. Girolami. Clustering of text documents by skewness maximization. In *Proc. Int. Workshop on Independent Component Analysis and Blind Signal Separation (ICA2000)*, pages 435–440, Helsinki, Finland, 2000.

230. E. R. Kandel, J. H. Schwartz, and T. M. Jessel, editors. *The Principles of Neural Science*. Prentice Hall, 3rd edition, 1991.

231. J. Karhunen, A. Cichocki, W. Kasprzak, and P Pajunen. On neural blind separation with noise suppression and redundancy reduction. *Int. Journal of Neural Systems*, 8(2):219–237, 1997.

232. J. Karhunen and J. Joutsensalo. Representation and separation of signals using nonlinear PCA type learning. *Neural Networks*, 7(1):113–127, 1994.

233. J. Karhunen and J. Joutsensalo. Generalizations of principal component analysis, optimization problems, and neural networks. *Neural Networks*, 8(4):549–562, 1995.

234. J. Karhunen, S. Malaroiou, and M. Ilmoniemi. Local linear independent component analysis based on clustering. *Int. J. of Neural Systems*, 10(6), 2000.

235. J. Karhunen, E. Oja, L. Wang, R. Vigário, and J. Joutsensalo. A class of neural networks for independent component analysis. *IEEE Trans. on Neural Networks*, 8(3):486–504, 1997.

236. J. Karhunen, P. Pajunen, and E. Oja. The nonlinear PCA criterion in blind source separation: Relations with other approaches. *Neurocomputing*, 22:5–20, 1998.

237. K. Karhunen. Zur Spektraltheorie stochastischer Prozesse. *Ann. Acad. Sci. Fennicae*, 34, 1946.

238. E. Kaukoranta, M. Hämäläinen, J. Sarvas, and R. Hari. Mixed and sensory nerve stimulations activate different cytoarchitectonic areas in the human primary somatosensory cortex SI. *Experimental Brain Research*, 63(1):60–66, 1986.

239. M. Kawamoto, K. Matsuoka, and M. Ohnishi. A method for blind separation for convolved non-stationary signals. *Neurocomputing*, 22:157–171, 1998.

240. M. Kawamoto, K. Matsuoka, and M. Oya. Blind separation of sources using temporal correlation of the observed signals. *IEICE Trans. Fundamentals*, E80-A(4):695–704, 1997.

241. S. Kay. *Modern Spectral Estimation: Theory and Application*. Prentice Hall, 1988.

242. S. Kay. *Fundamentals of Statistical Signal Processing - Estimation Theory*. Prentice Hall, 1993.

243. M. Kendall. *Multivariate Analysis*. Griffin, 1975.

244. M. Kendall and A. Stuart. *The Advanced Theory of Statistics, Vols. 1–3*. Macmillan, 1976–1979.

245. K. Kiviluoto and E. Oja. Independent component analysis for parallel financial time series. In *Proc. Int. Conf. on Neural Information Processing (ICONIP'98)*, volume 2, pages 895–898, Tokyo, Japan, 1998.

246. H. Knuth. A bayesian approach to source separation. In *Proc. Int. Workshop on Independent Component Analysis and Signal Separation (ICA'99)*, pages 283–288, Aussois, France, 1999.

247. T. Kohonen. *Self-Organizing Maps*. Springer, 1995.

248. T. Kohonen. Emergence of invariant-feature detectors in the adaptive-subspace self-organizing map. *Biological Cybernetics*, 75:281–291, 1996.

249. V. Koivunen, M. Enescu, and E. Oja. Adaptive algorithm for blind separation from noisy time-varying mixtures. *Neural Computation*, 13, 2001.

250. V. Koivunen and E. Oja. Predictor-corrector structure for real-time blind separation from noisy mixtures. In *Proc. Int. Workshop on Independent Component Analysis and Signal Separation (ICA'99)*, pages 479–484, Aussois, France, 1999.

251. T. Kolenda, L. K. Hansen, and S. Sigurdsson. Independent components in text. In M. Girolami, editor, *Advances in Independent Component Analysis*, pages 235–256. Springer-Verlag, 2000.

252. M. A. Kramer. Nonlinear principal component analysis using autoassociative neural networks. *AIChE Journal*, 37(2):233–243, 1991.

253. H.J. Kushner and D.S. Clark. *Stochastic approximation methods for constrained and unconstrained systems*. Springer-Verlag, 1978.

254. J.-L. Lacoume and P. Ruiz. Sources identification: a solution based on cumulants. In *Proc. IEEE ASSP Workshop*, Minneapolis, Minnesota, 1988.

255. B. Laheld and J.-F. Cardoso. Adaptive source separation with uniform performance. In *Proc. EUSIPCO*, pages 183–186, Edinburgh, 1994.

256. R. H. Lambert. *Multichannel Blind Deconvolution: FIR Matrix Algebra and Separation of Multipath Mixtures*. PhD thesis, Univ. of Southern California, 1996.

257. R. H. Lambert and C. L. Nikias. Blind deconvolution of multipath mixtures. In S. Haykin, editor, *Unsupervised Adaptive Filtering, Vol. I*, pages 377–436. Wiley, 2000.

258. H. Lappalainen. Ensemble learning for independent component analysis. In *Proc. Int. Workshop on Independent Component Analysis and Signal Separation (ICA'99)*, pages 7–12, Aussois, France, 1999.

259. H. Lappalainen and A. Honkela. Bayesian nonlinear independent component analysis by multi-layer perceptrons. In M. Girolami, editor, *Advances in Independent Component Analysis*, pages 93–121. Springer-Verlag, 2000.

260. H. Lappalainen and J. W. Miskin. Ensemble learning. In M. Girolami, editor, *Advances in Independent Component Analysis*, pages 75–92. Springer-Verlag, 2000.

261. L. De Lathauwer. *Signal Processing by Multilinear Algebra*. PhD thesis, Faculty of Engineering, K. U. Leuven, Leuven, Belgium, 1997.

262. L. De Lathauwer, P. Comon, B. De Moor, and J. Vandewalle. Higher-order power method, application in independent component analysis. In *Proc. Int. Symp. on Nonlinear Theory and its Applications (NOLTA'95)*, pages 10–14, Las Vegas, Nevada, 1995.

263. L. De Lathauwer, B. De Moor, and J. Vandewalle. A technique for higher-order-only blind source separation. In *Proc. Int. Conf. on Neural Information Processing (ICONIP'96)*, Hong Kong, 1996.

264. M. LeBlanc and R. Tibshirani. Adaptive principal surfaces. *J. of the Amer. Stat. Assoc.*, 89(425):53 – 64, 1994.

265. C.-C. Lee and J.-H. Lee. An efficient method for blind digital signal separation of array data. *Signal Processing*, 77:229–234, 1999.

266. T. S. Lee. Image representation using 2D Gabor wavelets. *IEEE Trans. on Pattern Analysis and Machine Intelligence*, 18(10):959–971, 1996.

267. T.-W. Lee. *Independent Component Analysis - Theory and Applications*. Kluwer, 1998.

268. T.-W. Lee, A. J. Bell, and R. Lambert. Blind separation of delayed and convolved sources. In *Advances in Neural Information Processing Systems*, volume 9, pages 758–764. MIT Press, 1997.

269. T.-W. Lee, M. Girolami, A.J. Bell, and T.J. Sejnowski. A unifying information-theoretic framework for independent component analysis. *Computers and Mathematics with Applications*, 31(11):1–12, 2000.

270. T.-W. Lee, M. Girolami, and T. J. Sejnowski. Independent component analysis using an extended infomax algorithm for mixed sub-gaussian and super-gaussian sources. *Neural Computation*, 11(2):417–441, 1999.

271. T.-W. Lee, B.U. Koehler, and R. Orglmeister. Blind source separation of nonlinear mixing models. In *Neural Networks for Signal Processing VII*, pages 406–415. IEEE Press, 1997.

272. T.-W. Lee, M.S. Lewicki, M. Girolami, and T.J. Sejnowski. Blind source separation of more sources than mixtures using overcomplete representations. *IEEE Signal Processing Letters*, 4(5), 1999.

273. T.-W. Lee, M.S. Lewicki, and T.J. Sejnowski. ICA mixture models for unsupervised classification of non-gaussian sources and automatic context switching in blind signal separation. *IEEE Trans. Pattern Recognition and Machine Intelligence*, 22(10):1–12, 2000.

274. M. Lewicki and B. Olshausen. A probabilistic framework for the adaptation and comparison of image codes. *J. Opt. Soc. Am. A: Optics, Image Science, and Vision*, 16(7):1587–1601, 1998.

275. M. Lewicki and T.J. Sejnowski. Learning overcomplete representations. *Neural Computation*, 12:337–365, 2000.

276. T. Li and N. Sidiropoulos. Blind digital signal separation using successive interference cancellation iterative least squares. *IEEE Trans. on Signal Processing*, 48(11):3146–3152, 2000.

277. J. K. Lin. Factorizing multivariate function classes. In *Advances in Neural Information Processing Systems*, volume 10, pages 563–569. The MIT Press, 1998.

278. J. K. Lin. Factorizing probability density functions: Generalizing ICA. In *Proc. First Int. Workshop on Independent Component Analysis and Signal Separation (ICA'99)*, pages 313–318, Aussois, France, 1999.

279. J. K. Lin, D. G. Grier, and J. D. Cowan. Faithful representation of separable distributions. *Neural Computation*, 9(6):1305–1320, 1997.

280. U. Lindgren and H. Broman. Source separation using a criterion based on second-order statistics. *IEEE Trans. on Signal Processing*, 46(7):1837–1850, 1998.

281. U. Lindgren, T. Wigren, and H. Broman. On local convergence of a class of blind separation algorithms. *IEEE Trans. on Signal Processing*, 43:3054–3058, 1995.

282. R. Linsker. Self-organization in a perceptual network. *Computer*, 21:105–117, 1988.

283. M. Loève. Fonctions aléatoires du second ordre. In P. Lévy, editor, *Processus stochastiques et mouvement Brownien*, page 299. Gauthier - Villars, Paris, 1948.

284. D. Luenberger. *Optimization by Vector Space Methods*. Wiley, 1969.

285. J. Luo, B. Hu, X.-T. Ling, and R.-W. Liu. Principal independent component analysis. *IEEE Trans. on Neural Networks*, 10(4):912–917, 1999.

286. O. Macchi and E. Moreau. Adaptive unsupervised separation of discrete sources. *Signal Processing*, 73:49–66, 1999.

287. U. Madhow. Blind adaptive interference suppression for direct-sequence CDMA. *Proceedings of the IEEE*, 86(10):2049–2069, 1998.

288. S. Makeig, T.-P. Jung, A. J. Bell, D. Ghahramani, and T. Sejnowski. Blind separation of auditory event-related brain responses into independent components. *Proc. National Academy of Sciences (USA)*, 94:10979–10984, 1997.

289. S. Malaroiu, K. Kiviluoto, and E. Oja. Time series prediction with independent component analysis. In *Proc. Int. Conf. on Advanced Investment Technology*, Gold Coast, Australia, 2000.

290. S. G. Mallat. A theory for multiresolution signal decomposition: The wavelet representation. *IEEE Trans. on Pattern Analysis and Machine Intelligence*, 11:674–693, 1989.

291. Z. Malouche and O. Macchi. Adaptive unsupervised extraction of one component of a linear mixture with a single neuron. *IEEE Trans. on Neural Networks*, 9(1):123–138, 1998.

292. A. Mansour, C. Jutten, and P. Loubaton. Adaptive subspace algorithm for blind separation of independent sources in convolutive mixture. *IEEE Trans. on Signal Processing*, 48(2):583–586, 2000.

293. K. Mardia, J. Kent, and J. Bibby. *Multivariate Analysis*. Academic Press, 1979.

294. L. Marple. *Digital Spectral Analysis with Applications*. Prentice Hall, 1987.

295. G. Marques and L. Almeida. Separation of nonlinear mixtures using pattern repulsion. In *Proc. Int. Workshop on Independent Component Analysis and Signal Separation (ICA'99)*, pages 277–282, Aussois, France, 1999.

296. K. Matsuoka, M. Ohya, and M. Kawamoto. A neural net for blind separation of nonstationary signals. *Neural Networks*, 8(3):411–419, 1995.

297. M. McKeown, S. Makeig, S. Brown, T.-P. Jung, S. Kindermann, A.J. Bell, V. Iragui, and T. Sejnowski. Blind separation of functional magnetic resonance imaging (fMRI) data. *Human Brain Mapping*, 6(5-6):368–372, 1998.

298. G. McLachlan and T. Krishnan. *The EM Algorithm and Extensions*. Wiley-Interscience, 1997.

299. J. Mendel. *Lessons in Estimation Theory for Signal Processing, Communications, and Control*. Prentice Hall, 1995.

300. Y. Miao and Y. Hua. Fast subspace tracking and neural network learning by a novel information criterion. *IEEE Trans. on Signal Processing*, 46:1967–1979, 1998.

301. J. Miskin and D. J. C. MacKay. Ensemble learning for blind image separation and deconvolution. In M. Girolami, editor, *Advances in Independent Component Analysis*, pages 123–141. Springer-Verlag, 2000.

302. S. Mitra. *Digital Signal Processing: A Computer-Based Approach*. McGraw-Hill, 1998.

303. L. Molgedey and H. G. Schuster. Separation of a mixture of independent signals using time delayed correlations. *Physical Review Letters*, 72:3634–3636, 1994.

304. T. Moon. The expectation-maximization algorithm. *IEEE Signal Processing Magazine*, 13(6):47–60, 1996.

305. E. Moreau and O. Macchi. Complex self-adaptive algorithms for source separation based on higher order contrasts. In *Proc. EUSIPCO'94*, pages 1157–1160, Edinburgh, Scotland, 1994.

306. E. Moreau and O. Macchi. High order contrasts for self-adaptive source separation. *Int. J. of Adaptive Control and Signal Processing*, 10(1):19–46, 1996.

307. E. Moreau and J. C. Pesquet. Generalized contrasts for multichannel blind deconvolution of linear systems. *IEEE Signal Processing Letters*, 4:182–183, 1997.

308. D. Morrison. *Multivariate Statistical Methods*. McGraw-Hill, 1967.

309. J. Mosher, P. Lewis, and R. Leahy. Multidipole modelling and localization from spatio-temporal MEG data. *IEEE Trans. Biomedical Engineering*, 39:541–557, 1992.

310. E. Moulines, J.-F. Cardoso, and E. Gassiat. Maximum likelihood for blind separation and deconvolution of noisy signals using mixture models. In *Proc. IEEE Int. Conf. on Acoustics, Speech and Signal Processing (ICASSP'97)*, pages 3617–3620, Munich, Germany, 1997.

311. E. Moulines, J.-F. Cardoso, and S. Mayrargue. Subspace methods for blind identification of multichannel FIR filters. *IEEE Trans. on Signal Processing*, 43:516–525, 1995.

312. K.-R. Müller, P. Philips, and A. Ziehe. $JADE_{TD}$: Combining higher-order statistics and temporal information for blind source separation (with noise). In *Proc. Int. Workshop on Independent Component Analysis and Signal Separation (ICA'99)*, pages 87–92, Aussois, France, 1999.

313. J.-P. Nadal, E. Korutcheva, and F. Aires. Blind source processing in the presence of weak sources. *Neural Networks*, 13(6):589–596, 2000.

314. J.-P. Nadal and N. Parga. Non-linear neurons in the low noise limit: a factorial code maximizes information transfer. *Network*, 5:565–581, 1994.

315. A. Nandi, editor. *Blind Estimation Using Higher-Order Statistics*. Kluwer, 1999.

316. G. Nason. Three-dimensional projection pursuit. *Applied Statistics*, 44:411–430, 1995.

317. E. Niedermeyer and F. Lopes da Silva, editors. *Electroencephalography. Basic Principles, Clinical Applications, and Related fields*. Williams & Wilkins, 1993.

318. C. Nikias and J. Mendel. Signal processing with higher-order spectra. *IEEE Signal Processing Magazine*, pages 10–37, July 1993.

319. C. Nikias and A. Petropulu. *Higher-Order Spectral Analysis - A Nonlinear Signal Processing Framework*. Prentice Hall, 1993.

320. B. Noble and J. Daniel. *Applied Linear Algebra*. Prentice Hall, 3rd edition, 1988.

321. D. Nuzillard and J.-M. Nuzillard. Application of blind source separation to 1-d and 2-d nuclear magnetic resonance spectroscopy. *IEEE Signal Processing Letters*, 5(8):209–211, 1998.

322. D. Obradovic and G. Deco. Information maximization and independent component analysis: Is there a difference? *Neural Computation*, 10(8):2085–2101, 1998.

323. E. Oja. A simplified neuron model as a principal component analyzer. *J. of Mathematical Biology*, 15:267–273, 1982.

324. E. Oja. *Subspace Methods of Pattern Recognition*. Research Studies Press, England, and Wiley, USA, 1983.

325. E. Oja. Data compression, feature extraction, and autoassociation in feedforward neural networks. In *Proc. Int. Conf. on Artificial Neural Networks (ICANN'91)*, pages 737–745, Espoo, Finland, 1991.

326. E. Oja. Principal components, minor components, and linear neural networks. *Neural Networks*, 5:927–935, 1992.

327. E. Oja. The nonlinear PCA learning rule in independent component analysis. *Neurocomputing*, 17(1):25–46, 1997.

328. E. Oja. From neural learning to independent components. *Neurocomputing*, 22:187–199, 1998.

329. E. Oja. Nonlinear PCA criterion and maximum likelihood in independent component analysis. In *Proc. Int. Workshop on Independent Component Analysis and Signal Separation (ICA'99)*, pages 143–148, Aussois, France, 1999.

330. E. Oja and J. Karhunen. On stochastic approximation of the eigenvectors and eigenvalues of the expectation of a random matrix. *Journal of Math. Analysis and Applications*, 106:69–84, 1985.

331. E. Oja, J. Karhunen, and A. Hyvärinen. From neural principal components to neural independent components. In *Proc. Int. Conf. on Artificial Neural Networks (ICANN'97)*, Lausanne, Switzerland, 1997.

332. E. Oja, H. Ogawa, and J. Wangviwattana. Learning in nonlinear constrained Hebbian networks. In *Proc. Int. Conf. on Artificial Neural Networks (ICANN'91)*, pages 385–390, Espoo, Finland, 1991.

333. E. Oja, H. Ogawa, and J. Wangviwattana. Principal component analysis by homogeneous neural networks, part I: the weighted subspace criterion. *IEICE Trans. on Information and Systems*, E75-D(3):366–375, 1992.

334. T. Ojanperä and R. Prasad. *Wideband CDMA for Third Generation Systems*. Artech House, 1998.

335. B. A. Olshausen and D. J. Field. Emergence of simple-cell receptive field properties by learning a sparse code for natural images. *Nature*, 381:607–609, 1996.

336. B. A. Olshausen and D. J. Field. Natural image statistics and efficient coding. *Network*, 7(2):333–340, 1996.

337. B. A. Olshausen and D. J. Field. Sparse coding with an overcomplete basis set: A strategy employed by V1? *Vision Research*, 37:3311–3325, 1997.

338. B. A. Olshausen and K. J. Millman. Learning sparse codes with a mixture-of-gaussians prior. In *Advances in Neural Information Processing Systems*, volume 12, pages 841–847. MIT Press, 2000.

339. A. Oppenheim and R. Schafer. *Discrete-Time Signal Processing*. Prentice Hall, 1989.

340. S. Ouyang, Z. Bao, and G.-S. Liao. Robust recursive least squares learning algorithm for principal component analysis. *IEEE Trans. on Neural Networks*, 11(1):215–221, 2000.

341. P. Pajunen. Blind separation of binary sources with less sensors than sources. In *Proc. Int. Conf. on Neural Networks*, Houston, Texas, 1997.

342. P. Pajunen. Blind source separation using algorithmic information theory. *Neurocomputing*, 22:35–48, 1998.

343. P. Pajunen. *Extensions of Linear Independent Component Analysis: Neural and Information-theoretic Methods*. PhD thesis, Helsinki University of Technology, 1998.

344. P. Pajunen. Blind source separation of natural signals based on approximate complexity minimization. In *Proc. Int. Workshop on Independent Component Analysis and Signal Separation (ICA'99)*, pages 267–270, Aussois, France, 1999.

345. P. Pajunen, A. Hyvärinen, and J. Karhunen. Nonlinear blind source separation by self-organizing maps. In *Proc. Int. Conf. on Neural Information Processing*, pages 1207–1210, Hong Kong, 1996.

346. P. Pajunen and J. Karhunen. A maximum likelihood approach to nonlinear blind source separation. In *Proceedings of the 1997 Int. Conf. on Artificial Neural Networks (ICANN'97)*, pages 541–546, Lausanne, Switzerland, 1997.

347. P. Pajunen and J. Karhunen. Least-squares methods for blind source separation based on nonlinear PCA. *Int. J. of Neural Systems*, 8(5-6):601–612, 1998.

348. P. Pajunen and J. Karhunen, editors. *Proc. of the 2nd Int. Workshop on Independent Component Analysis and Blind Signal Separation, Helsinki, Finland, June 19-22, 2000*. Otamedia, 2000.

349. F. Palmieri and A. Budillon. Multi-class independent component analysis (mucica). In M. Girolami, editor, *Advances in Independent Component Analysis*, pages 145–160. Springer-Verlag, 2000.

350. F. Palmieri and J. Zhu. Self-association and Hebbian learning in linear neural networks. *IEEE Trans. on Neural Networks*, 6(5):1165–1184, 1995.

351. C. Papadias. Blind separation of independent sources based on multiuser kurtosis optimization criteria. In S. Haykin, editor, *Unsupervised Adaptive Filtering*, volume 2, pages 147–179. Wiley, 2000.

352. H. Papadopoulos. Equalization of multiuser channels. In *Wireless Communications: Signal Processing Perspectives*, pages 129–178. Prentice Hall, 1998.

353. A. Papoulis. *Probability, Random Variables, and Stochastic Processes*. McGraw-Hill, 3rd edition, 1991.

354. N. Parga and J.-P. Nadal. Blind source separation with time-dependent mixtures. *Signal Processing*, 80(10):2187–2194, 2000.

355. L. Parra. Symplectic nonlinear component analysis. In *Advances in Neural Information Processing Systems*, volume 8, pages 437–443. MIT Press, Cambridge, Massachusetts, 1996.

356. L. Parra. Convolutive BBS for acoustic multipath environments. In S. Roberts and R. Everson, editors, *ICA: Principles and Practice*. Cambridge University Press, 2000. in press.

357. L. Parra, G. Deco, and S. Miesbach. Redundancy reduction with information-preserving nonlinear maps. *Network*, 6:61–72, 1995.

358. L. Parra, G. Deco, and S. Miesbach. Statistical independence and novelty detection with information-preserving nonlinear maps. *Neural Computation*, 8:260–269, 1996.

359. L. Parra and C. Spence. Convolutive blind source separation based on multiple decorrelation. In *Proc. IEEE Workshop on Neural Networks for Signal Processing (NNSP'97)*, Cambridge, UK, 1998.

360. L. Parra, C.D. Spence, P. Sajda, A. Ziehe, and K.-R. Müller. Unmixing hyperspectral data. In *Advances in Neural Information Processing Systems 12*, pages 942–948. MIT Press, 2000.

361. A. Paulraj, C. Papadias, V. Reddy, and A.-J. van der Veen. Blind space-time signal processing. In *Wireless Communications: Signal Processing Perspectives*, pages 179–210. Prentice Hall, 1998.

362. K. Pawelzik, K.-R. Müller, and J. Kohlmorgen. Prediction of mixtures. In *Proc. Int. Conf. on Artificial Neural Networks (ICANN'96)*, pages 127 – 132. Springer, 1996.

363. B. A. Pearlmutter and L. C. Parra. Maximum likelihood blind source separation: A context-sensitive generalization of ICA. In *Advances in Neural Information Processing Systems*, volume 9, pages 613–619, 1997.

364. K. Pearson. On lines and planes of closest fit to systems of points in space. *Philosophical Magazine*, 2:559–572, 1901.

365. H. Peng, Z. Chi, and W. Siu. A semi-parametric hybrid neural model for nonlinear blind signal separation. *Int. J. of Neural Systems*, 10(2):79–94, 2000.

366. W. D. Penny, R. M. Everson, and S. J. Roberts. Hidden Markov independent component analysis. In M. Girolami, editor, *Advances in Independent Component Analysis*, pages 3–22. Springer-Verlag, 2000.

367. K. Petersen, L. Hansen, T. Kolenda, E. Rostrup, and S. Strother. On the independent component of functional neuroimages. In *Proc. Int. Workshop on Independent Component Analysis and Blind Signal Separation (ICA2000)*, pages 251–256, Helsinki, Finland, 2000.

368. D.-T. Pham. Blind separation of instantaneous mixture sources via an independent component analysis. *IEEE Trans. on Signal Processing*, 44(11):2768–2779, 1996.

369. D.-T. Pham. Blind separation of instantaneous mixture of sources based on order statistics. *IEEE Trans. on Signal Processing*, 48(2):363–375, 2000.

370. D.-T. Pham and J.-F. Cardoso. Blind separation of instantaneous mixtures of non-stationary sources. In *Proc. Int. Workshop on Independent Component Analysis and Blind Signal Separation (ICA2000)*, pages 187–193, Helsinki, Finland, 2000.

371. D.-T. Pham and P. Garrat. Blind separation of mixture of independent sources through a quasi-maximum likelihood approach. *IEEE Trans. on Signal Processing*, 45(7):1712–1725, 1997.

372. D.-T. Pham, P. Garrat, and C. Jutten. Separation of a mixture of independent sources through a maximum likelihood approach. In *Proc. EUSIPCO*, pages 771–774, 1992.

373. D. Pollen and S. Ronner. Visual cortical neurons as localized spatial frequency filters. *IEEE Trans. on Systems, Man, and Cybernetics*, 13:907–916, 1983.

374. J. Porrill, J. W. Stone, J. Berwick, J. Mayhew, and P. Coffey. Analysis of optical imaging data using weak models and ICA. In M. Girolami, editor, *Advances in Independent Component Analysis*, pages 217–233. Springer-Verlag, 2000.

375. K. Prank, J. Börger, A. von zur Mühlen, G. Brabant, and C. Schöfl. Independent component analysis of intracellular calcium spike data. In *Advances in Neural Information Processing Systems*, volume 11, pages 931–937. MIT Press, 1999.

376. J. Principe, N. Euliano, and C. Lefebvre. *Neural and Adaptive Systems - Fundamentals Through Simulations*. Wiley, 2000.

377. J. Principe, D. Xu, and J. W. Fisher III. Information-theoretic learning. In S. Haykin, editor, *Unsupervised Adaptive Filtering, Vol. I*, pages 265–319. Wiley, 2000.

378. J. Proakis. *Digital Communications*. McGraw-Hill, 3rd edition, 1995.

379. C. G. Puntonet, A. Prieto, C. Jutten, M. Rodriguez-Alvarez, and J. Ortega. Separation of sources: A geometry-based procedure for reconstruction of n-valued signals. *Signal Processing*, 46:267–284, 1995.

380. J. Rissanen. Modeling by shortest data description. *Automatica*, 14:465–471, 1978.

381. J. Rissanen. A universal prior for integers and estimation by minimum description length. *Annals of Statistics*, 11(2):416–431, 1983.

382. T. Ristaniemi. *Synchronization and Blind Signal Processing in CDMA Systems*. PhD thesis, University of Jyväskylä, Jyväskylä, Finland, 2000.

383. T. Ristaniemi and J. Joutsensalo. Advanced ICA-based receivers for DS-CDMA systems. In *Proc. IEEE Int. Conf. on Personal, Indoor, and Mobile Radio Communications (PIMRC'00)*, London, UK, 2000.

384. T. Ristaniemi and J. Joutsensalo. On the performance of blind source separation in CDMA downlink. In *Proc. Int. Workshop on Independent Component Analysis and Blind Source Separation (ICA'99)*, pages 437–441, Aussois, France, January 1999.

385. S. Roberts. Independent component analysis: Source assessment & separation, a Bayesian approach. *IEE Proceedings - Vision, Image & Signal Processing*, 145:149–154, 1998.

386. M. Rosenblatt. *Stationary Sequences and Random Fields*. Birkhauser, 1985.

387. S. Roweis. EM algorithms for PCA and SPCA. In M. I. Jordan, M. J. Kearns, and S. A. Solla, editors, *Advances in Neural Information Processing Systems*, volume 10, pages 626 – 632. MIT Press, 1998.

388. J. Rubner and P. Tavan. A self-organizing network for principal component analysis. *Europhysics Letters*, 10(7):693 – 698, 1989.

389. H. Sahlin and H. Broman. Separation of real-world signals. *Signal Processing*, 64(1):103–113, 1998.

390. H. Sahlin and H. Broman. MIMO signal separation for FIR channels: A criterion and performance analysis. *IEEE Trans. on Signal Processing*, 48(3):642–649, 2000.

391. T.D. Sanger. Optimal unsupervised learning in a single-layered linear feedforward network. *Neural Networks*, 2:459–473, 1989.

392. Y. Sato. A method for self-recovering equalization for multilevel amplitude-modulation system. *IEEE Trans. on Communications*, 23:679–682, 1975.

393. L. Scharf. *Statistical Signal Processing: Detection, Estimation, and Time Series Analysis*. Addison-Wesley, 1991.

394. M. Scherg and D. von Cramon. Two bilateral sources of the late AEP as identified by a spatio-temporal dipole model. *Electroencephalography and Clinical Neurophysiology*, 62:32 – 44, 1985.

395. M. Schervish. *Theory of Statistics*. Springer, 1995.

396. I. Schiessl, M. Stetter, J.W.W. Mayhew, N. McLoughlin, J.S.Lund, and K. Obermayer. Blind signal separation from optical imaging recordings with extended spatial decorrelation. *IEEE Trans. on Biomedical Engineering*, 47(5):573–577, 2000.

397. C. Serviere and V. Capdevielle. Blind adaptive separation of wide-band sources. In *Proc. IEEE Int. Conf. on Acoustics, Speech and Signal Processing (ICASSP'96)*, Atlanta, Georgia, 1996.

398. O. Shalvi and E. Weinstein. New criteria for blind deconvolution of nonminimum phase systems (channels). *IEEE Trans. on Information Theory*, 36(2):312–321, 1990.

399. O. Shalvi and E. Weinstein. Super-exponential methods for blind deconvolution. *IEEE Trans. on Information Theory*, 39(2):504:519, 1993.

400. S. Shamsunder and G. B. Giannakis. Multichannel blind signal separation and reconstruction. *IEEE Trans. on Speech and Aurdio Processing*, 5(6):515–528, 1997.

401. C. Simon, P. Loubaton, C. Vignat, C. Jutten, and G. d'Urso. Separation of a class of convolutive mixtures: A contrast function approach. In *Proc. Int. Conf. on Acoustics, Speech, and Signal Processing (ICASSP'99)*, Phoenix, AZ, 1999.

402. C. Simon, C. Vignat, P. Loubaton, C. Jutten, and G. d'Urso. On the convolutive mixture source separation by the decorrelation approach. In *Proc. Int. Conf. on Acoustics, Speech, and Signal Processing (ICASSP'98)*, pages 2109–2112, Seattle, WA, 1998.

403. E. P. Simoncelli and E. H. Adelson. Noise removal via bayesian wavelet coring. In *Proc. Third IEEE Int. Conf. on Image Processing*, pages 379–382, Lausanne, Switzerland, 1996.

404. E. P. Simoncelli and O. Schwartz. Modeling surround suppression in V1 neurons with a statistically-derived normalization model. In *Advances in Neural Information Processing Systems 11*, pages 153–159. MIT Press, 1999.

405. P. Smaragdis. Blind separation of convolved mixtures in the frequency domain. *Neurocomputing*, 22:21–34, 1998.

406. V. Soon, L. Tong, Y. Huang, and R. Liu. A wideband blind identification approach to speech acquisition using a microphone array. In *Proc. Int. Conf. ASSP-92*, volume 1, pages 293–296, San Francisco, California, March 23–26 1992.

407. H. Sorenson. *Parameter Estimation - Principles and Problems*. Marcel Dekker, 1980.

408. E. Sorouchyari. Blind separation of sources, Part III: Stability analysis. *Signal Processing*, 24:21–29, 1991.

409. C. Spearman. General intelligence, objectively determined and measured. *American J. of Psychology*, 15:201–293, 1904.

410. R. Steele. *Mobile Radio Communications*. Pentech Press, London, 1992.

411. P. Stoica and R. Moses. *Introduction to Spectral Analysis*. Prentice Hall, 1997.

412. J. V. Stone, J. Porrill, C. Buchel, and K. Friston. Spatial, temporal, and spatiotemporal independent component analysis of fMRI data. In R.G. Aykroyd K.V. Mardia and I.L. Dryden, editors, *Proceedings of the 18th Leeds Statistical Research Workshop on Spatial-Temporal Modelling and its Applications*, pages 23–28. Leeds University Press, 1999.

413. E. Ström, S. Parkvall, S. Miller, and B. Ottersten. Propagation delay estimation in asynchronous direct-sequence code division multiple access systems. *IEEE Trans. Communications*, 44:84–93, January 1996.

414. J. Sun. Some practical aspects of exploratory projection pursuit. *SIAM J. of Sci. Comput.*, 14:68–80, 1993.

415. K. Suzuki, T. Kiryu, and T. Nakada. An efficient method for independent component-cross correlation-sequential epoch analysis of functional magnetic resonance imaging. In *Proc. Int. Workshop on Independent Component Analysis and Blind Signal Separation (ICA2000)*, pages 309–315, Espoo, Finland, 2000.

416. A. Swami, G. B. Giannakis, and S. Shamsunder. Multichannel ARMA processes. *IEEE Trans. on Signal Processing*, 42:898–914, 1994.

417. A. Taleb and C. Jutten. Batch algorithm for source separation in post-nonlinear mixtures. In *Proc. First Int. Workshop on Independent Component Analysis and Signal Separation (ICA'99)*, pages 155–160, Aussois, France, 1999.

418. A. Taleb and C. Jutten. Source separation in post-nonlinear mixtures. *IEEE Trans. on Signal Processing*, 47(10):2807–2820, 1999.

419. C. Therrien. *Discrete Random Signals and Statistical Signal Processing*. Prentice Hall, 1992.

420. H.-L. Nguyen Thi and C. Jutten. Blind source separation for convolutive mixtures. *Signal Processing*, 45:209–229, 1995.

421. M. E. Tipping and C. M. Bishop. Mixtures of probabilistic principal component analyzers. *Neural Computation*, 11:443–482, 1999.

422. L. Tong, Y. Inouye, and R. Liu. A finite-step global convergence algorithm for the parameter estimation of multichannel MA processes. *IEEE Trans. on Signal Processing*, 40:2547–2558, 1992.

423. L. Tong, Y. Inouye, and R. Liu. Waveform preserving blind estimation of multiple independent sources. *IEEE Trans. on Signal Processing*, 41:2461–2470, 1993.

424. L. Tong, R.-W. Liu, V.C. Soon, and Y.-F. Huang. Indeterminacy and identifiability of blind identification. *IEEE Trans. on Circuits and Systems*, 38:499–509, 1991.

425. L. Tong and S. Perreau. Multichannel blind identification: From subspace to maximum likelihood methods. *Proceedings of the IEEE*, 86(10):1951–1968, 1998.

426. K. Torkkola. Blind separation of convolved sources based on information maximization. In *Proc. IEEE Workshop on Neural Networks and Signal Processing (NNSP'96)*, pages 423–432, Kyoto, Japan, 1996.

427. K. Torkkola. Blind separation of delayed sources based on information maximization. In *Proc. IEEE Int. Conf. on Acoustics, Speech and Signal Processing (ICASSP'96)*, pages 3509–3512, Atlanta, Georgia, 1996.

428. K. Torkkola. Blind separation of radio signals in fading channels. In *Advances in Neural Information Processing Systems*, volume 10, pages 756–762. MIT Press, 1998.

429. K. Torkkola. Blind separation for audio signals – are we there yet? In *Proc. Int. Workshop on Independent Component Analysis and Signal Separation (ICA'99)*, pages 239–244, Aussois, France, 1999.

430. K. Torkkola. Blind separation of delayed and convolved sources. In S. Haykin, editor, *Unsupervised Adaptive Filtering, Vol. I*, pages 321–375. Wiley, 2000.

431. M. Torlak, L. Hansen, and G. Xu. A geometric approach to blind source separation for digital wireless applications. *Signal Processing*, 73:153–167, 1999.

432. J. K. Tugnait. Identification and deconvolution of multichannel nongaussian processes using higher-order statistics and inverse filter criteria. *IEEE Trans. on Signal Processing*, 45:658–672, 1997.

433. J. K. Tugnait. Adaptive blind separation of convolutive mixtures of independent linear signals. *Signal Processing*, 73:139–152, 1999.

434. J. K. Tugnait. On blind separation of convolutive mixtures of independent linear signals in unknown additive noise. *IEEE Trans. on Signal Processing*, 46(11):3117–3123, November 1998.

435. M. Valkama, M. Renfors, and V. Koivunen. BSS based I/Q imbalance compensation in communication receivers in the presence of symbol timing errors. In *Proc. Int. Workshop on Independent Component Analysis and Blind Signal Separation (ICA2000)*, pages 393–398, Espoo, Finland, 2000.

436. H. Valpola. Nonlinear independent component analysis using ensemble learning: Theory. In *Proc. Int. Workshop on Independent Component Analysis and Blind Signal Separation (ICA2000)*, pages 251–256, Helsinki, Finland, 2000.

437. H. Valpola. Unsupervised learning of nonlinear dynamic state-space models. Technical Report A59, Lab of Computer and Information Science, Helsinki University of Technology, Finland, 2000.

438. H. Valpola, X. Giannakopoulos, A. Honkela, and J. Karhunen. Nonlinear independent component analysis using ensemble learning: Experiments and discussion. In *Proc. Int. Workshop on Independent Component Analysis and Blind Signal Separation (ICA2000)*, pages 351–356, Helsinki, Finland, 2000.

439. A.-J. van der Veen. Algebraic methods for deterministic blind beamforming. *Proceedings of the IEEE*, 86(10):1987–2008, 1998.

440. A.-J. van der Veen. Blind separation of BPSK sources with residual carriers. *Signal Processing*, 73:67–79, 1999.

441. A.-J. van der Veen. Algebraic constant modulus algorithms. In G. Giannakis, Y. Hua, P. Stoica, and L. Tong, editors, *Signal Processing Advances in Wireless and Mobile Communications, Vol. 2: Trends in Single-User and Multi-User Systems*, pages 89–130. Prentice Hall, 2001.

442. J. H. van Hateren and D. L. Ruderman. Independent component analysis of natural image sequences yields spatiotemporal filters similar to simple cells in primary visual cortex. *Proc. Royal Society, Ser. B*, 265:2315–2320, 1998.

443. J. H. van Hateren and A. van der Schaaf. Independent component filters of natural images compared with simple cells in primary visual cortex. *Proc. Royal Society, Ser. B*, 265:359–366, 1998.

444. S. Verdu. *Multiuser Detection*. Cambridge University Press, 1998.

445. R. Vigário. Extraction of ocular artifacts from EEG using independent component analysis. *Electroenceph. Clin. Neurophysiol.*, 103(3):395–404, 1997.

446. R. Vigário, V. Jousmäki, M. Hämäläinen, R. Hari, and E. Oja. Independent component analysis for identification of artifacts in magnetoencephalographic recordings. In *Advances in Neural Information Processing Systems*, volume 10, pages 229–235. MIT Press, 1998.

447. R. Vigário, J. Särelä, V. Jousmäki, M. Hämäläinen, and E. Oja. Independent component approach to the analysis of EEG and MEG recordings. *IEEE Trans. Biomedical Engineering*, 47(5):589–593, 2000.

448. R. Vigário, J. Särelä, V. Jousmäki, and E. Oja. Independent component analysis in decomposition of auditory and somatosensory evoked fields. In *Proc. Int. Workshop on Independent Component Analysis and Signal Separation (ICA'99)*, pages 167–172, Aussois, France, 1999.

449. R. Vigário, J. Särelä, and E. Oja. Independent component analysis in wave decomposition of auditory evoked fields. In *Proc. Int. Conf. on Artificial Neural Networks (ICANN'98)*, pages 287–292, Skövde, Sweden, 1998.

450. L.-Y. Wang and J. Karhunen. A unified neural bigradient algorithm for robust PCA and MCA. *Int. J. of Neural Systems*, 7(1):53–67, 1996.

451. X. Wang and H. Poor. Blind equalization and multiuser detection in dispersive CDMA channels. *IEEE Trans. on Communications*, 46(1):91–103, 1998.

452. X. Wang and H. Poor. Blind multiuser detection: A subspace approach. *IEEE Trans. on Information Theory*, 44(2):667–690, 1998.

453. M. Wax and T. Kailath. Detection of signals by information-theoretic criteria. *IEEE Trans. on Acoustics, Speech and Signal Processing*, 33:387–392, 1985.

454. A. Webb. *Statistical Pattern Recognition*. Arnold, 1999.

455. A. S. Weigend and N.A. Gershenfeld. Time series prediction. In *Proc. of NATO Advanced Research Workshop on Comparative Time Series Analysis*, Santa Fe, New Mexico, 1992.

456. E. Weinstein, M. Feder, and A. V. Oppenheim. Multi-channel signal separation by decorrelation. *IEEE Trans. on Signal Processing*, 1:405–413, 1993.

457. R. A. Wiggins. Minimum entropy deconvolution. *Geoexploration*, 16:12–35, 1978.

458. R. Williams. Feature discovery through error-correcting learning. Technical report, University of California at San Diego, Institute of Cognitive Science, 1985.

459. J. O. Wisbeck, A. K. Barros, and R. G. Ojeda. Application of ICA in the separation of breathing artifacts in ECG signals. In *Proc. Int. Conf. on Neural Information Processing (ICONIP'98)*, pages 211–214, Kitakyushu, Japan, 1998.

460. G. Wubbeler, A. Ziehe, B. Mackert, K. Müller, L. Trahms, and G. Curio. Independent component analysis of noninvasively recorded cortical magnetic dc-field in humans. *IEEE Trans. Biomedical Engineering*, 47(5):594–599, 2000.

461. L. Xu. Least mean square error reconstruction principle for self-organizing neural nets. *Neural Networks*, 6:627–648, 1993.

462. L. Xu. Bayesian Kullback Ying-Yang dependence reduction theory. *Neurocomputing*, 22:81–111, 1998.

463. L. Xu. Temporal BYY learning for state space approach, hidden markov model, and blind source separation. *IEEE Trans. on Signal Processing*, 48(7):2132–2144, 2000.

464. L. Xu, C. Cheung, and S.-I. Amari. Learned parameter mixture based ICA algorithm. *Neurocomputing*, 22:69–80, 1998.

465. W.-Y. Yan, U. Helmke, and J. B. Moore. Global analysis of Oja's flow for neural networks. *IEEE Trans. on Neural Networks*, 5(5):674 – 683, 1994.

466. B. Yang. Projection approximation subspace tracking. *IEEE Trans. on Signal Processing*, 43(1):95–107, 1995.

467. B. Yang. Asymptotic convergence analysis of the projection approximation subspace tracking algorithm. *Signal Processing*, 50:123–136, 1996.

468. H. H. Yang and S.-I. Amari. Adaptive on-line learning algorithms for blind separation: Maximum entropy and minimum mutual information. *Neural Computation*, 9(7):1457–1482, 1997.

469. H. H. Yang, S.-I. Amari, and A. Cichocki. Information-theoretic approach to blind separation of sources in non-linear mixture. *Signal Processing*, 64(3):291–300, 1998.

470. D. Yellin and E. Weinstein. Criteria for multichannel signal separation. *IEEE Trans. on Signal Processing*, 42:2158–2167, 1994.

471. D. Yellin and E. Weinstein. Multichannel signal separation: Methods and analysis. *IEEE Trans. on Signal Processing*, 44:106–118, 1996.

472. A. Yeredor. Blind separation of gaussian sources via second-order statistics with asymptotically optimal weighting. *IEEE Signal Processing Letters*, 7(7):197–200, 2000.

473. A. Yeredor. Blind source separation via the second characteristic function. *Signal Processing*, 80:897–902, 2000.

474. K. Yeung and S. Yau. A cumulant-based super-exponential algorithm for blind deconvolution of multi-input multi-output systems. *Signal Processing*, 67(2):141–162, 1998.

475. A. Ypma and P. Pajunen. Rotating machine vibration analysis with second-order independent component analysis. In *Proc. Int. Workshop on Independent Component Analysis and Signal Separation (ICA'99)*, pages 37–42, Aussois, France, 1999.

476. T. Yu, A. Stoschek, and D. Donoho. Translation- and direction- invariant denoising of 2-D and 3-D images: Experience and algorithms. In *Proceedings of the SPIE, Wavelet Applications in Signal and Image Processing IV*, pages 608–619, 1996.

477. S. Zacks. *Parametric Statistical Inference*. Pergamon, 1981.

478. C. Zetzsche and G. Krieger. Nonlinear neurons and high-order statistics: New approaches to human vision and electronic image processing. In B. Rogowitz and T.V. Pappas, editors, *Human Vision and Electronic Imaging IV (Proc. SPIE vol. 3644)*, pages 2–33. SPIE, 1999.

479. L. Zhang and A. Cichocki. Blind separation of filtered sources using state-space approach. In *Advances in Neural Information Processing Systems*, volume 11, pages 648–654. MIT Press, 1999.

480. Q. Zhang and Y.-W. Leung. A class of learning algorithms for principal component analysis and minor component analysis. *IEEE Trans. on Neural Networks*, 11(1):200–204, 2000.

481. A. Ziehe and K.-R. Müller. TDSEP—an efficient algorithm for blind separation using time structure. In *Proc. Int. Conf. on Artificial Neural Networks (ICANN'98)*, pages 675–680, Skövde, Sweden, 1998.

482. A. Ziehe, K.-R. Müller, G. Nolte, B.-M. Mackert, and G. Curio. Artifact reduction in magnetoneurography based on time-delayed second-order correlations. *IEEE Trans. Biomedical Engineering*, 47(1):75–87, 2000.

483. A. Ziehe, G. Nolte, G. Curio, and K.-R. Müller. OFI: Optimal filtering algorithms for source separation. In *Proc. Int. Workshop on Independent Component Analysis and Blind Signal Separation (ICA2000)*, pages 127–132, Helsinki, Finland, 2000.

Index

Printed and bound by CPI Group (UK) Ltd, Croydon, CR0 4YY

27/10/2024

14580332-0004